eMaintenance

eMaintenance

Essential Electronic Tools for Efficiency

Diego Galar and Uday Kumar

ELSEVIER

ACADEMIC PRESS
An imprint of Elsevier

Academic Press is an imprint of Elsevier
125 London Wall, London EC2Y 5AS, United Kingdom
525 B Street, Suite 1800, San Diego, CA 92101-4495, United States
50 Hampshire Street, 5th Floor, Cambridge, MA 02139, United States
The Boulevard, Langford Lane, Kidlington, Oxford OX5 1GB, United Kingdom

Library of Congress Cataloging-in-Publication Data
A catalog record for this book is available from the Library of Congress

British Library Cataloguing-in-Publication Data
A catalogue record for this book is available from the British Library

ISBN: 978-0-12-811153-6

Publisher: Matthew Deans
Acquisition Editor: Brian Guerin
Editorial Project Manager: Edward Payne
Production Project Manager: Mohana Natarajan
Designer: Mark Rogers

Typeset by TNQ Books and Journals

Contents

Chapter 1

Sensors and Data Acquisition

1.1 SENSORS IN MAINTENANCE AND THE NEED TO INTEGRATE INFORMATION

The diagnostic and prognostic needs a broad range of organizations may be met in a cost-effective manner and with minimal risk by leveraging three critical areas. The first is a framework that can efficiently integrate reusable algorithms. The second is communications, in particular, wireless technologies. The third is an infrastructure for the system-level integration of distributed intelligent system elements. System-level integration provides the foundation for robust systems capable of generating actionable diagnostics and prognostic plans, hardware platforms, sensor modules, database information, and intelligent devices in an automated system.

The integration of these three factors, together with advanced prognostic algorithms, is the basis for a new machinery maintenance paradigm. This paradigm employs targeted, task-specific sensors and algorithms integrated in a framework that is readily expanded and adapted to changing operational needs. Open industry-standard systems and interfaces are fundamental to this framework. Such systems can readily be integrated with a plant's equipment and coupled with various IT, operational planning, finance, and control systems.

Various components and functional capabilities of diagnostic and prognostic systems may be organized as a framework of system components or building blocks. Such a framework provides a scheme for identifying and organizing essential information about a system and a structure for defining standard interfaces between system elements and their functional requirements.

The requirements for capturing machinery health information, interpreting the data, and acting on the results of the analysis may be arranged in a hierarchy (Table 1.1). This hierarchy ranges from simply capturing data from a particular process or machine at the most basic level to interpreting sampled data to detecting a fault that has occurred and predicting what fault or faults will occur. Higher levels of the hierarchy provide a range of capabilities to automatically react to novel faults or to react before an anticipated failure. Higher levels, beginning with diagnosis and prognosis, typically require the

eMaintenance. http://dx.doi.org/10.1016/B978-0-12-811153-6.00001-4

TABLE 1.1 Hierarchy of Intelligent Machines

1. Data acquisition
2. Monitor
3. Detect
4. Diagnose
5. Prognosis
6. Prognostics and control
7. System-level prognosis and control
8. Dynamic optimization/multiobjective control
9. Adaptive/Reconfigurable *Order of increasing complexity/cost/economic benefit*

Adapted from Discenzo, F.M., Loparo, K.A., Chung, D., Twarowski, A., 2001a. Intelligent Sensor Nodes Enable A New Generation of Machinery Diagnostics and Prognostics. Virginia Beach, Virginia, April 2–5, 2001.

integration of multiple data sources, as well as knowledge of the process equipment and operating state or context (Discenzo et al., 2001a).

Throughout the working world, there is a move to intelligent devices and distributed intelligence. Intelligent components may be found at the structural level (e.g., smart materials), at the sensor level, or at the device level (e.g., embedded intelligence). There are no international standards for sensors, but this should happen in the near future. IEEE 1451 establishes a standard for a transducer (sensor/actuator) interface. It seeks to move data acquisition, distributed sensing, and control to an open system by establishing a framework and data elements for "smart" transducers. Included in this standard is a specification to facilitate sensor identification, calibration, documentation, sensor replacement, and network integration (O'Mara, 2000). Research in self-validating sensors at the University of Oxford over the last 12 years has been directed at defining intelligent sensors, which dynamically sense their own condition and provide information on the quality or validity of the sensor value returned (Henry, 2000). This is an important emerging area, with a standard proposed at the draft level by British Standards Institution for data quality metrics (Wood, 2000). Information on data quality becomes critical as we move to higher levels of the hierarchy of intelligent machines: automatic control, decision-making, and autonomous machines (Discenzo et al., 2001a).

Data critical to establishing the health of equipment may be captured and stored in an application-specific manner to accommodate essential memory or timing constraints. Preferably, machinery data should be organized in an open, industry standard format, such as that defined by Manufacturers Information Management Open Systems Alliance (MIMOSA). The data format and definitions are the result of many years of work by an international team of

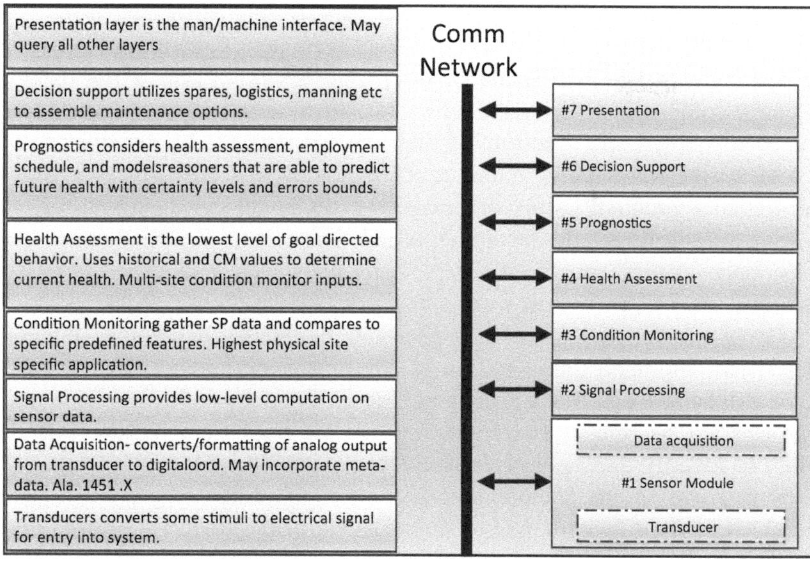

FIGURE 1.1 Open system architecture for condition-based maintenance. *Adapted from Adapted from Discenzo, F.M., Loparo, K.A., Chung, D., Twarowski, A., 2001a. Intelligent Sensor Nodes Enable A New Generation of Machinery Diagnostics and Prognostics. Virginia Beach, Virginia, April 2–5, 2001.*

MIMOSA sponsors and members. This standard is open and accessible to the public (Discenzo et al., 2001a).

More recently, a group of industry and academic partners teamed together to develop an Open Systems Architecture for Condition-Based Maintenance (OSA-CBM). This program is part of the Dual Use Science & Technology (DU&ST) program with joint industry–government funding (BAA 98–023). This effort, building on the work of MIMOSA, has resulted in an operational framework for machinery diagnostics in an open-system, layered framework, as shown in Fig. 1.1 (Discenzo et al., 2001b).

The integration of intelligent sensors and self-validating sensors in an open operational framework as specified by MIMOSA–OSA-CBM promotes the development and deployment of distributed intelligent sensors across a broad range of applications (Discenzo et al., 2001a).

1.1.1 Sensors Put Intelligence Into Maintenance

It seems like everything we buy today is intelligent. Thermostats adjust themselves based on weather conditions. Cars alert us to upcoming maintenance needs. Alarm systems can be armed or disarmed from a mobile phone. Wristbands track fitness progress.

Thanks to the Internet of Things, the same technologies that have revolutionized the way we do everyday things can now enhance the way we manage maintenance in the plant environment. We do not need a new plant or the latest equipment to derive benefit.

Broadly stated, the spectrum of maintenance ranges from passive to predictive. At the passive end, maintenance is approached primarily as a fix—a reaction to a problem, issue, or breakdown. Potential disadvantages of this reactive approach include downtime, production delays, inefficiencies, added costs, poor quality, increased scrap, warranty issues, and product recalls, making it easy to see why it's not a best practice.

At the predictive end of the spectrum, maintenance employs ongoing analysis to determine the condition of equipment and predict when maintenance will be needed. The more compelling selling points of this sophisticated approach are its role in preventing unexpected equipment failures and its ability to identify precisely when maintenance should be executed.

Somewhere in the middle is condition monitoring. A practice that is quickly becoming fundamental to maintenance management, it utilizes periodic assessments to analyze the health of equipment. When executed consistently, it enables an understanding of a more periodic state of the equipment—a heartbeat of its condition, if you will, to inform maintenance planning.

Condition monitoring identifies trends that may indicate a change in the way equipment is performing. Such changes often serve as early warnings of impending issues. Recognizing them can be the key to triggering action that prevents failure and avoids its consequences.

Not so long ago, the only option for monitoring the condition of equipment was to do it manually. Anyone who has spent any time in a plant knows the process can be arduous, especially for operations with significant equipment inventory. Visually reviewing gauges, entering data into spreadsheets, or going into hard-to-reach or dangerous areas is time-consuming and vulnerable to human error. Worse yet, it frequently is not proactive enough to prevent a fault or failure.

Sensors are remaking this manual process. When they are deployed to monitor such elements as equipment pressure, temperature, or humidity, they can pinpoint conditions that may be indicative of a specific problem (Calvo and Hannifin Parker Corporation et al., 2016).

1.1.1.1 Sensing Maintenance Needs

Sensors are a cost-effective way to take a snapshot of what is happening with a piece of equipment—down to the second, if that is what an organization needs. The technology is fast and efficient, enabling the monitoring of more process points and more machines. And it's suitable for use in a variety of applications. In fluid power, for example, sensors can be applied for condition monitoring of injection molding units, metal forming and fabrication equipment, conveyor systems, dispensing systems, robotic assembly, and hydraulic power units, to name a few.

To continue the example, fluid power sensors can assess systems that use oil, water, or gas either to transfer liquid or gas from one point to another or to generate power. Pressure, temperature, and humidity are critical in these systems, especially when it comes to quality. Too much or too little pressure can wreak havoc on production. If temperatures do not remain consistent—in the tools, lines, or atmosphere—the quality may be adversely affected. Sensors can provide enough data to enable the identification of trends that show the condition of a piece of equipment is changing. This insight can trigger corrective action, before something catastrophic occurs. Sensor-driven condition monitoring is especially useful in hard-to-reach or inaccessible areas where on-board gauges may be hard to reach or view. It's also valuable in hazardous environments because it allows the user to keep a safe distance and eliminates the need for a special safety-related clearance or process.

Because replacing manual monitoring with sensor technology enables the optimization of assets and processes, productivity arguably reaps the biggest reward. The first productivity gains come with the elimination of the need for staff to perform manual reads of equipment conditions, netting a significant improvement in labor efficiency. The second gains, likely even more significant, come with the prevention of downtime.

According to the US Department of Energy, minimizing downtime can drive up asset availability by as much as 30% and increase production by as much as 25%. By identifying issues before they escalate, organizations can also decrease maintenance costs and avoid potentially dangerous situations (Calvo and Hannifin Parker Corporation et al., 2016).

1.1.1.2 Going Wireless

When organizations learn how sensors can be applied for maintenance management and condition monitoring, they cannot deny the benefits. But the add-on, and often high, investment in the installation and use of traditional wired sensors can make the cost of implementation prohibitive.

Wireless technology is a viable alternative. Wireless options make sensor solutions surprisingly affordable. As consumer technologies push wireless into the mainstream, the cost of materials for wireless sensor acquisition continues to drop, especially components such as Bluetooth radios. In addition, because wireless does not require strenuous installation or specialized skills, implementation is more affordable.

The accessibility and ease of implementation makes wireless sensors ideal for placement in many types of production machines to monitor a variety of operating parameters. In other words, reduced cost is not the only benefit of going wireless. The sensing technology is easier to install than its wired counterparts, and it eliminates the risk of physical connection faults. The addition of access points is simplified, and the commissioning routine does not require laborious programming and data-collection devices, such as a programmable logic controller. Once installed, a wireless sensor allows instant and easy access to data for a variety of stakeholders via user-friendly software and apps (Calvo and Hannifin Parker Corporation et al., 2016).

1.1.1.3 Enhancing Equipment, Both New and Old

Some of today's newest equipment comes with on-board sensor technology. Older equipment simply does not have the same bells and whistles—but this equipment still accounts for the vast majority of the assets in plants around the world.

Fortunately, we do not need a new asset to implement sensor-based condition monitoring. Sensor technology is so flexible and adaptable that it can be implemented for maintenance management at any point throughout the lifespan of equipment. Even for new technology, organizations can consider aftermarket application of additional test nodes to increase their monitoring potential.

Wireless condition monitoring technology can readily supplement an existing wired system without being directly integrated into it. More importantly, cost-effective wireless sensors provide users with the flexibility to add more nodes, i.e., those making the most sense to their specific applications, to the process points. This capability was not available to industrial organizations as little as 3 years ago.

The collection of data through sensor technology is just the beginning. The data mean nothing if they are not utilized. Accordingly, the other critical element of a sensor-based condition monitoring system is the software application that houses and provides access to the data. These applications make information available on multiple devices. They can also be configured to generate dashboards that transform data into insights that drive maintenance management decisions. Additional support for decision-making can be provided by features such as alarms or email alerts that notify users of certain conditions in real time and allow them to act (Calvo and Hannifin Parker Corporation et al., 2016).

1.1.1.4 Taking Intelligence to the Next Level

Effective, disciplined condition monitoring lays the groundwork for—and is the precursor to—predictive maintenance. In fact, the data generated through condition monitoring are mission-critical for a predictive approach. For organizations that aspire to move along the maintenance management spectrum, thus coming closer to predictive maintenance, a sensor-driven condition monitoring program can be part of a deliberate roadmap.

In the short term, the options for how condition monitoring data can be used are becoming more robust. For example, organizations may be gathering maintenance data from a variety of systems. It is also likely those systems are not interconnected and their dashboards are exclusive. But incremental improvements in technology now allow us to marry data sets. An organization can overlay what is happening with its hydraulic equipment, for instance, with weather data or information on the utilization of energy. This enables the optimization of various factors that can affect production, such as ambient temperature conditions of the production floor that impact production equipment or energy-saving discounts offered by utility suppliers when energy utilization during peak times can be minimized and validated.

It's possible to make maintenance management as intelligent as some of the consumer technology that is so commonplace today. Sensor technology is the powerhouse that can make it happen (Calvo and Hannifin Parker Corporation, 2016).

1.1.2 Basic Sensor Technology

A sensor is a device that converts a physical phenomenon into an electrical signal. As such, sensors represent part of the interface between the physical world and the world of electrical devices, such as computers. The other part of this interface is represented by actuators, which convert electrical signals into physical phenomena.

In recent years, the electronics industry has developed an enormous capability for information processing. The most significant example is the personal computer. In addition, the availability of inexpensive microprocessors is having a tremendous impact on the design of embedded computing products, ranging from automobiles to microwave ovens to toys. Products that use microprocessors for control of functionality are now widely available. In automobiles, such capability is necessary to achieve compliance with pollution restrictions. In other cases, it simply offers an inexpensive performance advantage.

All of these microprocessors need electrical input voltages to receive instructions and information, so the use of sensors in a wide variety of products has grown concomitantly with the availability of inexpensive microprocessors. In addition, since the output of the sensor is an electrical signal, sensors tend to be characterized in the same way as electronic devices. The data sheets for many sensors are formatted just like electronic product data sheets.

However, there are many formats, and there is no international standard for sensor specifications. The system designer will encounter a variety of interpretations of sensor performance parameters, and it can be confusing. It is important to realize that this confusion is not due to an inability to explain the meaning of the terms; rather, different parts of the sensor community have grown comfortable using these terms differently (Wilson, 2005).

1.1.2.1 Sensor Data Sheets

It is important to understand the function of the data sheet to deal with this variability. The data sheet is primarily a marketing document. It is typically designed to highlight the positive attributes of a particular sensor and emphasize some of its potential uses; it may neglect to comment on some of its negative characteristics. In many cases, the sensor has been designed to meet a particular performance specification for a specific customer, so the data sheet will concentrate on the performance parameters of greatest interest to this customer. In this case, the vendor and customer might have grown accustomed to unusual definitions for certain sensor performance parameters. Potential new users must recognize this situation and interpret things accordingly. Odd definitions may be encountered here and there, and most sensor data sheets are missing some pieces of information that are of interest to particular applications (Wilson, 2005).

1.1.2.2 Sensor Performance Characteristics

The following are some of the more important sensor characteristics:

1.1.2.2.1 Transfer Function

The transfer function shows the functional relationship between physical input signal and electrical output signal. Usually, this relationship is represented as a graph showing the relationship between the input and output signal, and the details of this relationship may constitute a complete description of the sensor characteristics. For expensive sensors's that are individually calibrated, this might take the form of a certified calibration curve.

1.1.2.2.2 Sensitivity

The sensitivity is defined in terms of the relationship between physical input signal and electrical output signal. It is generally the ratio between a small change in electrical signal to a small change in physical signal. As such, it may be expressed as the derivative of the transfer function with respect to the physical signal. Typical units are volts/kelvin, millivolts/kilopascal, etc. A thermometer will have "high sensitivity" if a small temperature change results in a large voltage change.

1.1.2.2.3 Span or Dynamic Range

The range of input physical signals that may be converted to electrical signals by the sensor is the dynamic range or span. Signals outside this range are expected to cause unacceptably large inaccuracy. This span or dynamic range is usually specified by the sensor supplier as the range over which other performance characteristics described in the data sheets are expected to apply. Typical units are kelvin, pascal, newtons, etc.

1.1.2.2.4 Accuracy or Uncertainty

Uncertainty is generally defined as the largest expected error between actual and ideal output signals. Typical units are kelvin. Sometimes this is quoted as a fraction of the full-scale output or a fraction of the reading. For example, a thermometer might be guaranteed accurate to within 5% of full scale output (FSO). "Accuracy" is generally considered by metrologists to be a qualitative term, while "uncertainty" is quantitative. For example, one sensor might have better accuracy than another if its uncertainty is 1% and the other's is 3%.

1.1.2.2.5 Hysteresis

Some sensors do not return to the same output value when the input stimulus is cycled up or down. The width of the expected error in terms of the measured quantity is defined as the hysteresis. Typical units are kelvin or percent of FSO.

1.1.2.2.6 Nonlinearity (Often Called Linearity)

Nonlinearity refers to maximum deviation of linear transfer function over a specified dynamic range. There are several measures of this error. The most common compares the actual transfer function with the "best straight line," which lies midway between the two parallel lines that encompass the entire transfer function over the specified dynamic range of the device. This comparative method is popular because it makes most sensors look good. Other reference lines may be used, so the user should be careful to compare using the same reference.

1.1.2.2.7 Noise

All sensors produce some output noise in addition to the output signal. In some cases, the noise of the sensor is less than the noise of the next element in the electronics, or less than the fluctuations in the physical signal, in which case, it is not important. In many other cases, the noise of the sensor limits the performance of the sensor. Noise is generally distributed across the frequency spectrum. Many common noise sources produce a white noise distribution; this means that the spectral noise density is the same at all frequencies. Johnson noise in a resistor is a good example of such a noise distribution. For white noise, the spectral noise density is characterized in volts/Root (Hz). A distribution of this nature adds noise to a measurement with amplitude proportional to the square root of the measurement bandwidth. Since there is an inverse relationship between the bandwidth and measurement time, the noise decreases with the square root of the measurement time.

1.1.2.2.8 Resolution

The resolution of a sensor is defined as the minimum detectable signal fluctuation. Since fluctuations are temporal phenomena, there is some relationship between the timescale of the fluctuation and the minimum detectable amplitude. Therefore, the definition of resolution must include some information about the nature of the measurement being carried out. Many sensors are limited by noise with a white spectral distribution. In these cases, the resolution may be specified in units of physical signal/root (Hz). The actual resolution for a particular measurement may be obtained by multiplying this quantity by the square root of the measurement bandwidth. Sensor data sheets generally quote resolution in units of signal/root (Hz), or they give a minimum detectable signal for a specific measurement. If the shape of the noise distribution is also specified, it is possible to generalize these results to any measurement.

1.1.2.2.9 Bandwidth

All sensors have finite response times to an instantaneous change in physical signal. In addition, many sensors have decay times representing the time taken after a step change in physical signal for the sensor output to decay to its original value. The reciprocal of these times corresponds to the upper and lower cut-off frequencies, respectively. The bandwidth of a sensor is the frequency range between these two frequencies (Wilson, 2005).

1.1.3 Role of Sensors and Objectives of Sensing

The role of sensor systems in mechanical manufacturing generally comprises sensing, transformation/conversion, signal processing, and decision-making, as shown in Fig. 1.1. The output of the sensor system is either given to the operator via a human–machine interface or directly utilized to control the machine. Objectives, requirements, demands, boundary conditions, signal processing, communication techniques, and the human–machine interface of the sensor system are described in this section (Tönshoff and Inasaki, 2001; Fig. 1.2).

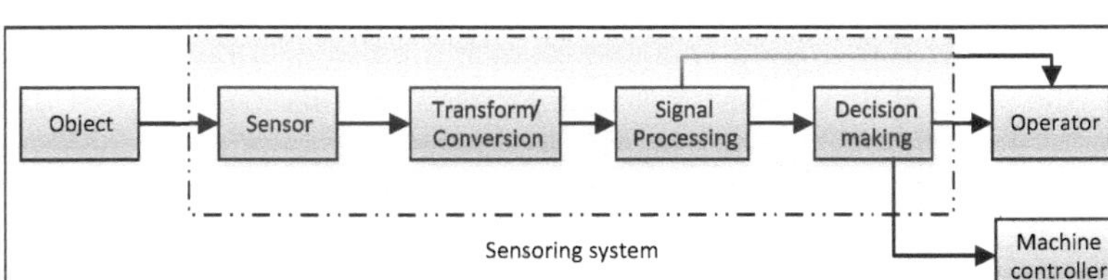

FIGURE 1.2 Basic composition of sensor system for mechanical manufacturing. *Adapted from Tönshoff, H.K., Inasaki, I., 2001. Sensors Applications Volume 1-Sensors in Manufacturing Sensors in Manufacturing. Copyright © 2001 Wiley-VCH Verlag GmbH. ISBN: 3-527-29558-5 (Hardcover); 3-527-60002-7 (Electronic).*

The objectives of general mechanical manufacturing and the items to be sensed and monitored are summarized in Table 1.1, together with the specific purposes of sensing and monitoring. Some items can be directly sensed with a particular sensor and, at the same time, can be used to estimate other properties. For instance, the cutting force is sensed with a tool dynamometer to monitor the cutting state, but the information can be used to estimate the wear of the cutting tool (Tönshoff and Inasaki, 2001).

Almost all kinds of machining processes require sensing and monitoring to maintain reliability and to avoid abnormal states. Table 1.2 gives a summary of the answers to a questionnaire given to machine tool users on the machining processes that require monitoring (Technical Committee on Integrated Manufacturing Systems et al., 1980). The responses show that monitoring is considered imperative, especially when weak tools are used, such as in tapping, drilling, and end milling (Tönshoff and Inasaki, 2001; Table 1.3).

TABLE 1.2 Objects, Items, and Purposes of Sensing

Object of Sensing and Monitoring	Items to be Sensed	Purpose of Sensing and Monitoring
Work	State of work clamping Geometrical and dimensional accuracy Surface roughness Surface quality	Maintain high quality Avoid damage and loss of work
Machining process	Force (torque, thrust) Heat generation Temperature Vibration Noise and sound State of chip	Maintain normal machining process Predict and avoid abnormal state
Tool	Tool edge position Wear Damage including chipping, breakage, and others	Manage tool changing time, including dressing Avoid damage or deterioration of work
Machine tool and auxiliary facility	Malfunction Vibration Deformation (elastic, thermal)	Maintain normal condition of machine tool and assure high accuracy
Environment	Ambient temperature change External vibration Condition of cutting fluid	Minimize environmental effect

Adapted from Tönshoff, H.K., Inasaki, I., 2001. Sensors Applications Volume 1-Sensors in Manufacturing Sensors in Manufacturing. Copyright © 2001 Wiley-VCH Verlag GmbH. ISBN: 3-527-29558-5 (Hardcover); 3-527-60002-7 (Electronic).

TABLE 1.3 Machining Processes Requiring Sensing

Kind of Machining	Number of Answers	Percentage
Tapping	67	19.8
Drilling	66	19.2
End milling	55	16.8
Internal turning	51	15.1
External turning	30	8.9
Face milling	25	7.4
Parting	17	5.0
Thread cutting	13	3.9
Others[a]	15	4.4
Total	338	100

[a]*Grinding, reaming, deep hole boring, etc.*
Adapted from Tönshoff, H.K., Inasaki, I., 2001. Sensors Applications Volume 1-Sensors in Manufacturing Sensors in Manufacturing. Copyright © 2001 Wiley-VCH Verlag GmbH. ISBN: 3-527-29558-5 (Hardcover); 3-527-60002-7 (Electronic).

1.1.4 Distributed Intelligent Sensors

The development of intelligent sensors and open systems is ongoing, building on developments in software architectures, sensor technologies, and networks moving toward open, public standards. Despite the rapid pace of development, the deployment of intelligent devices in manufacturing and commercial operations remains limited because of the high cost of installation and the cost and complexity of the processing and analysis of the massive amount of real-time data received from the hundreds or thousands of sensors. Typically, data are received from remote sensors, not information (e.g., actionable information or health information).

Developments in wired networks and wireless networks are promising to reduce or even eliminate the wiring costs of distributed sensors. Bit-oriented networks such as DeviceNet are effective for local-scale integration of plant sensors. For higher bandwidth and extended networking, TCP/IP is more attractive, particularly when the higher cost of network access and transport can be justified by many remote intelligent nodes (Discenzo et al., 2001a).

Wiring costs are often a significant component in the installation of manufacturing and commercial systems. These costs may be eliminated with the use of wireless communications technology, albeit at the expense of needed radio links. There is a variety of radio links; a link is selected based on the bandwidth need, reliability, distance, interoperability, and network architecture. In the United States, various programs funded by the Department of Energy, Department of Commerce, and the Defense Advance Research

Projects Agency (DARPA) support the development of wireless technology for smart devices (Manges, 2000).

Recently, there has been significant interest in low-cost wireless networks. Much of this interest is driven by the enormous commercial potential for wireless consumer products. Driven by this potential, the development of a Bluetooth communications standard may provide low cost, 2.4 GHz wireless data links for local sensor networks and data acquisition (Browne, 2000).

Wireless sensor networks (WSNs) promise to enable numerous applications for sensing and control, including machinery monitoring for intelligent maintenance management. Rockwell Science Center has created a development environment to test wireless sensing applications (Agre et al., 1999). Its Wireless Integrated Networked Sensing (WINS) platform includes the following:

- Bidirectional RF communications hardware and sophisticated networking protocols,
- Processing and memory with a multitasking, real-time operating system for sensor data acquisition, signal conditioning, and algorithmic processing of the data, and
- Support for multiple sensor inputs, including wide bandwidth accelerometers for vibration monitoring.

The integration of these technologies allows the easy installation of remote access sensors that can be configured in a variety of ways. The individual units, or nodes on the network, receive sensor data from attached sensors, process the sensor data, and send back messages to the end-user, informing him/her about the condition of the equipment or process that is being monitored. Implicit in this architecture is a degree of distributed processing that can range from individual WINS nodes processing the data from their sensors and interpreting those data to higher level algorithms that perform system-wide diagnostics using preprocessed sensor data from other WINS nodes (Clare et al., 1999).

The current WINS communications network operates in the 900 MHz RF band regulated for unlicensed use in the United States for transmission powers under 1 W (spread spectrum) or 1 MW (single frequency). Limited transmit power is required for operation at these frequencies, as the range of each unit is nominally 100 m. In addition, there may be other path losses from absorbers or reflectors between the WINS nodes and the end-user gateway node, and these may be dynamic. Hence, there is a clear need for a robust network system.

Rockwell Science Center has launched a commercial product, Highly Deployable Remote Access (HiDRA), based on the WINS technology (see Fig. 1.3). Included in this product is a network protocol that supports broadcast and unicast multihop communication links. The multihop protocols allow messages to be routed from nodes that are out of the RF communications range of the end-user gateway node through intermediate nodes that are within the RF range of each other. The system is self-configurable, so that the user

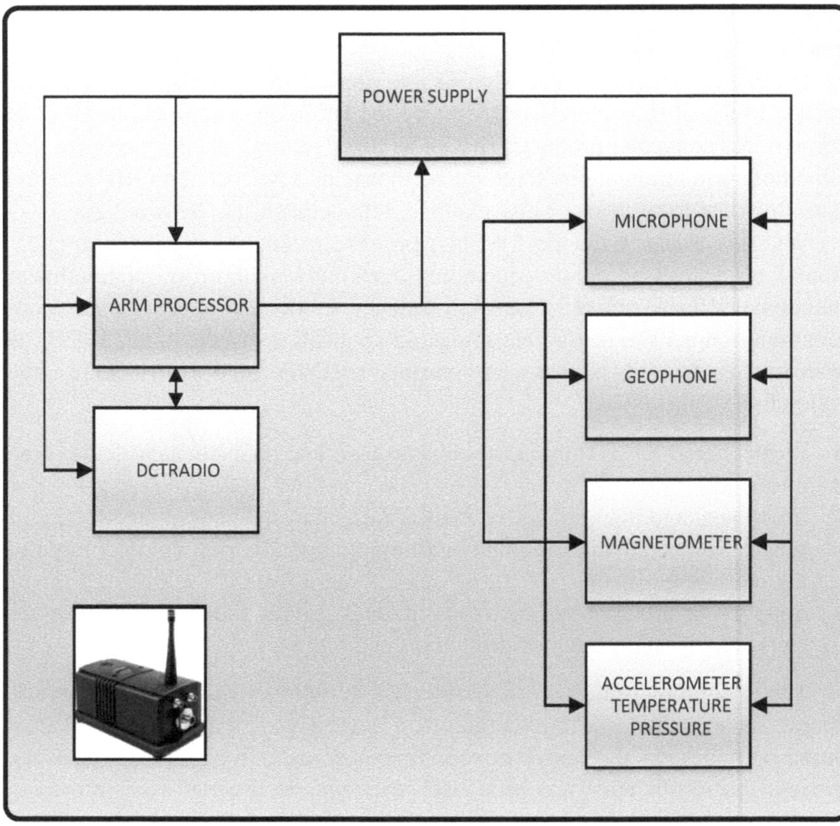

FIGURE 1.3 Rockwell Science Center HiDRA node. *Adapted from Discenzo, F.M., Loparo, K.A., Chung, D., Twarowski, A., 2001a. Intelligent Sensor Nodes Enable A New Generation of Machinery Diagnostics and Prognostics. Virginia Beach, Virginia, April 2–5, 2001.*

does not have to spend time setting up the network; it is done automatically at start-up. It is also dynamic, so if conditions change in the RF environment of the HiDRA node, the routing table for relaying a message through several nodes to arrive at the desired destination will also change.

The WINS system has been deployed for over a year monitoring bearing health in the heating, ventilation, and air conditioning facility serving the two plants at Rockwell Science Center in 1000 Oaks. Ten WINS nodes are mounted on 50–75 hp motors, housed in the cooling fluid pump room; these drive the pumps supplying the cooling fluids for the buildings at Rockwell Science Center. Accelerometers are mounted to the motor casing over the bearing locations of the motors; they measure the vibration at these locations. A temperature sensor monitors the temperature of the motor case close to the position of the accelerometer. Each WINS node does a spectral analysis of the vibration signals it receives; it then computes bearing health status indicators that it transmits through an RF link to a base-station WINS node connected to an internet server (Discenzo et al., 2001a).

1.1.5 Infrastructure for Intelligent Systems

Distributed intelligent sensor nodes capable of processing data from multiple sensors provide unique opportunities for data fusion and cooperative processing. Our objective is to put as much information as possible into each node and leverage the capabilities for processing complex algorithms in parallel and in collaboration with other sensor nodes. This permits us to establish accurate, dynamic, and robust models for diagnostics and prognostics. The following section outlines recent developments in model-based and nonlinear analysis methods. These new diagnostic/prognostic and modeling tools are considered foundational, as they provide a basis for future self-organizing sensor networks and dynamically reconfigurable systems.

There are a variety of approaches to the development of fault detection, diagnosis, and prognosis algorithms; each has unique advantages and limitations when implemented as a distributed processing node (Discenzo et al., 2001a).

Model-based approaches combine physical modeling of the system with experimental data to determine a mathematical relationship between the occurrence of a fault and the characteristics of a measured quantity in the system. Extended model-based techniques may employ a family of models relating each of the individual faults and their severity. The residuals from each of the model-based observers are combined and then integrated with other information available from causal modeling, signal processing, expert systems, etc. to arrive at a decision on the current operating status of the system. A diagram of the fault detection and diagnosis system implemented for rotating machinery is shown in Fig. 1.4. The core of this technology is an array of nonlinear filter/model-based observer blocks that combine the output from a suite of observers. For details of this structure and operation, refer to Loparo and Adams (1998) and Loparo et al. (2000).

An important aspect of the system shown in Fig. 1.4 is the ability to integrate information from a variety of different sources, such as multiple sensor nodes, for comprehensive fault detection and fault decisions. This system, designed for a rolling element bearing, uses novel algorithms for the detection and diagnosis of faults. These algorithms are particularly well-suited for implementation at distributed sensor nodes (Discenzo et al., 2001a).

One study of bearing fault detection used model-based techniques with a fault detection filter. To isolate the specific fault (e.g., ball, inner, and outer race defects), it developed a time-frequency analysis method. Researchers tested the method using experimental data collected from an induction motor system (Lou et al., 2000a). Another novel approach integrates sliding mode observers and fault detection filters for the detection and isolation of faults in rolling element bearings (Loparo et al., 2000).

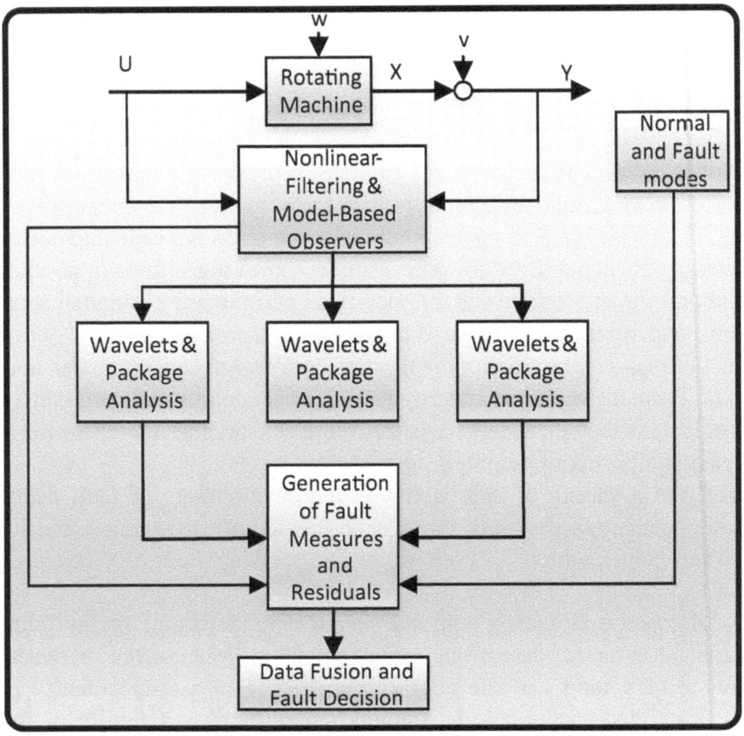

FIGURE 1.4 Fault detection and diagnosis system for rotating machinery. *Adapted from Discenzo, F.M., Loparo, K.A., Chung, D., Twarowski, A., 2001a. Intelligent Sensor Nodes Enable A New Generation of Machinery Diagnostics and Prognostics. Virginia Beach, Virginia, April 2–5, 2001.*

A new method has been developed to detect and diagnose defects in ball bearings using the wavelet transform (Lou et al., 2000b). The signature produced by damage on the DB2 wavelet is used for the wavelet decomposition of the preprocessed vibration signals. A set of feature vectors is then defined based on the wavelet decompositions.

Finally, it is possible to detect and diagnose mechanical faults in rolling element bearings using vibration data and knowledge of the bearing defect frequencies. For a particular bearing geometry, inner raceway, outer raceway, and rolling element faults generate vibration spectra with unique frequency components. The bearing defect frequencies are linear functions of the rotating speed of the shaft. Outer race and inner race frequencies are also linear functions of the number of balls in the bearing. The operating speed changes with load and is often unknown and/or unmeasurable. In addition, even if the type of bearing in the machine is known, the number of balls in the bearing may be unknown. Thus, estimating the running speed and the number of balls in the bearing is required for failure detection and diagnosis methods that rely on knowledge of the defect frequencies of the bearing.

In a study, we developed and implemented separate algorithms for estimating the rotational speed and the number of balls in a bearing using vibration data (Discenzo et al., 2001a). More specifically, we used spectral information obtained from the fast Fourier transform of the vibration data to obtain the estimates and used this information to calculate the bearing defect frequencies.

Our estimation algorithms were tested using experimental data consisting of vibration signals gathered from a transducer mounted on the drive-end bearing of an induction motor. The induction motor was operated under four different load conditions (four different running speeds), and three types of single point defects (inner race, outer race and ball) were introduced into the drive-end bearing. The test results proved the algorithms were very reliable; when they were integrated with an envelope detection algorithm, fault detection and diagnosis were also reliable (Ocak et al., 2000). These core capabilities are well suited for implementation in a distributed agent-based architecture (Discenzo et al., 2001a).

Agent technology expands the notion of distributed computing, it may change the problem of distributing the computation load into a situation where the solution method is both localized (autonomous) and collaborative and adaptive (cooperative/goal oriented). Within the collection of software agents, each agent will have a local goal or agenda. In this sense, each module is autonomous. In addition, there is an overarching goal or system objective that each individual module must accommodate and the collective efforts of the multiple agents must strive to meet. The collection of such agents is termed a holonic system, based on the Greek *holos*, meaning total or whole (Koestler, 1989; Brussel and Hendrik, 1994).

The multiagent approach provides an extremely powerful framework for integrating partial solutions into more complex and more sophisticated diagnostic, decision-making, and control structures (Balasubramanian et al., 1999; Maturana et al., 2000). This concept is particularly useful for machinery diagnostics and prognostics. For example, remote intelligent sensor nodes may efficiently monitor critical system components such as bearings. If an excessive vibration level or temperature level is observed, information on the degraded operation and reduced component lifetime may be relayed to a central information system. In addition, specific data on abnormal operation may be exchanged with neighboring smart sensor nodes. This will permit the nodes to collaborate, exchanging data and analysis to jointly establish a more accurate, complete hypothesis of the root cause of a fault, such as a bent shaft. This will also maintain a more robust and accurate system model essential for accurately predicting the future operating state of the machinery and estimated time until failure. Wireless technologies will permit the wide-scale deployment of smart sensor nodes across many system components. This, in turn, will lead to more accurate system models and prognostic estimates at a much lower cost. It will also enable very flexible and easily reconfigured monitoring and diagnostic systems.

The availability of complete and accurate process information and superior failure prediction accuracy directly addresses some key concerns expressed by a broad range of major manufacturers. Simply stated, they need the following:

1. To know overall process health, and
2. To have accurate prognostic information.

The integration of this new and accurate information into existing control and information, plant monitoring, scheduling, and maintenance systems will result in unprecedented levels of plant performance and derive superior economic value from installed equipment (Discenzo et al., 2001a).

1.2 SENSOR FUSION

Sensor fusion is the process of using information from several different sensors to estimate the state of a dynamic system. The resulting estimate is, in some senses, better than it would be if the sensors were used individually. Here better means more accurate, more reliable, more available, and safer. Furthermore, the resulting estimate may be obtained by using data from different types of sensors. Fig. 1.5 illustrates the basic concept of the sensor fusion framework. Traditionally, systems have been stand-alone, with one or several sensors transmitting information to a single application. With a sensor fusion approach, it might be possible to remove one sensor and still perform the same tasks or to add new applications without the need to add new sensors (Lundquist, 2011).

The main components of a sensor fusion framework are shown in the middle box of Fig. 1.5. The framework receives measurements from several sensors, fuses them, and produces one state estimate, which can be used by several applications. Sensor fusion may reduce cost, system complexity, and the number of components involved, while increasing accuracy and confidence of sensing (Lundquist, 2011).

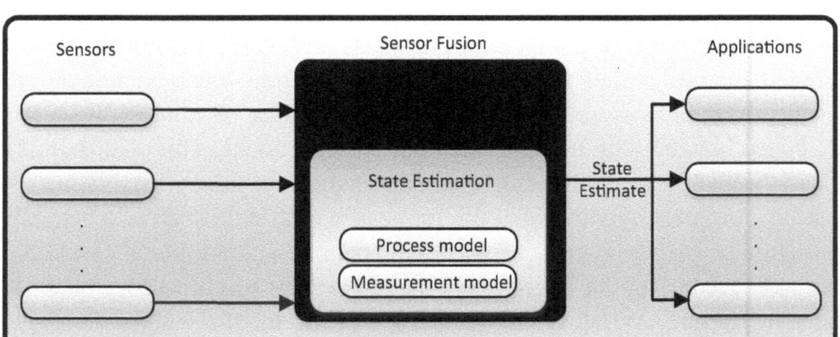

FIGURE 1.5 Sensor fusion framework. *Adapted from Lundquist, C., 2011. Sensor Fusion for Automotive Applications. Department of Electrical Engineering Linköping University, SE—581 83 Linköping, Sweden. Linköping Studies in Science and Technology. Dissertations. No. 1409.*

1.2.1 Principles of Sensor Fusion

There is some confusion in the terminology used for fusion systems. The terms "sensor fusion," "data fusion," "information fusion," "multisensor data fusion," and "multisensor integration" have been widely used in the technical literature to refer to a variety of techniques, technologies, systems, and applications that use data derived from multiple information sources. Fusion applications range from real-time sensor fusion for the navigation of mobile robots to the off-line fusion of human or technical strategic intelligence data (Rothman and Denton, 1991).

Several attempts have been made to define and categorize fusion terms and techniques. Wald et al. propose using "data fusion" as the general term for fusion (Wald, 1998). However, while the concept of data fusion is easy to understand, its exact meaning varies from one scientist to another. Wald uses "data fusion" for a formal framework that comprises means and tools for the alliance of data originating from different sources. It aims to obtain information of superior quality, but the exact definition of superior quality depends on the application. The term "data fusion" is used in this sense by the Geoscience and Remote Sensing Society [U.S. Department of Defense (DoD), 1991] and by many authors researching motion tracking, remote sensing, and mobile robots. Unfortunately, the term has different meanings in other work (Smestad, 2001). In some fusion models, "data fusion" is used to denote fusion of raw data (Dasarathy, 1997).

Classic books on fusion include *Multisensor Data Fusion* (Waltz and Llinas, 1990) and *Mathematical Techniques in Multisensor Data Fusion* (Hall, 1992). Both propose an extended term, "multisensor data fusion," defined as the technology on how to combine data from multiple (and possible diverse) sensors to make inferences about a physical event, activity, or situation. However, both books also use "data fusion" to mean "multisensor data fusion" (Hall, 1992).

To avoid confusion, Dasarathy decided to use "information fusion" to designate the fusion of any kind of data (Dasarathy, 2001). "Information fusion" had not been used extensively before and, thus, was not associated with any single aspect of the fusion domain. The fact that "information fusion" is also applicable in the context of data mining and database integration is not necessarily a negative, as the meaning is unaltered: information fusion is an all-encompassing term covering all aspects of the fusion field (except nuclear fusion or fusion in the music world) (Elmenreich, 2002).

The following definition of information fusion appears on the homepage of the International Society of Information Fusion (http://isif.org/):

> *Information Fusion encompasses theory, techniques and tools conceived and employed for exploiting the synergy in the information acquired from multiple sources (sensor, databases, information gathered by human, etc.) such that the resulting decision or action is in some sense better (qualitatively or quantitatively, in terms of accuracy, robustness, etc.) than would be possible if any of these sources were used individually without such synergy exploitation.*

In contrast, sensor fusion is the combining of sensory data or data derived from sensory data such that the resulting information is in some sense better than would be possible if these sources were used individually (Elmenreich, 2002).

The data sources for fusion do not need to be identical sensors. McKee distinguishes between direct fusion, indirect fusion, and fusion of the outputs (McKee, 1993). Direct fusion means the fusion of sensor data from a set of heterogeneous or homogeneous sensors, soft sensors, and history values of sensor data, while indirect fusion uses information sources like a priori knowledge about the environment and human input. Therefore, sensor fusion describes direct fusion systems, while information fusion also encompasses indirect fusion processes.

Since "data fusion" is a standard term in the scientific community for earth image data processing, we should avoid the stand-alone term "data fusion" when we mean "low-level data fusion". Unless "data fusion" is meant as it is proposed by the earth science community, a prefix such as "low-level" or "raw" is adequate.

The sensor fusion definition above does not require that inputs be produced by multiple sensors; it only means that sensor data or data derived from sensor data must be combined. For example, the definition encompasses sensor fusion systems with a single sensor that takes multiple measurements at different instants, which are then combined (Elmenreich, 2002).

Another frequently used term is multisensor integration—the synergistic use of sensor data for the accomplishment of a task by a system. Sensor fusion is different from multisensor integration in the sense that it includes the actual combination of sensory information in one representational format (Singhal, 1997; Luo and Kay, 1989). The difference between sensor fusion and multisensor integration is outlined in Fig. 1.6. The circles S_1, S_2, and S_3 depict physical sensors that provide an interface to the process environment. Block diagram 1(A) shows that the sensor data are converted by a sensor fusion block into a respective representation of the variables of the process environment (Elmenreich, 2002).

These data are then used by a control application. In contrast, Fig. 1.1 (B) shows multisensor integration, where the different sensor data are directly processed by the control application (Elmenreich, 2002).

1.2.2 Motivation for Sensor Fusion

Systems using sensor fusion methods have a number of benefits over single-sensor systems. A single-sensor measurement generally suffers from the following problems (Elmenreich, 2002):

Sensor deprivation: The breakdown of a sensor element causes a loss of perception of the desired object.

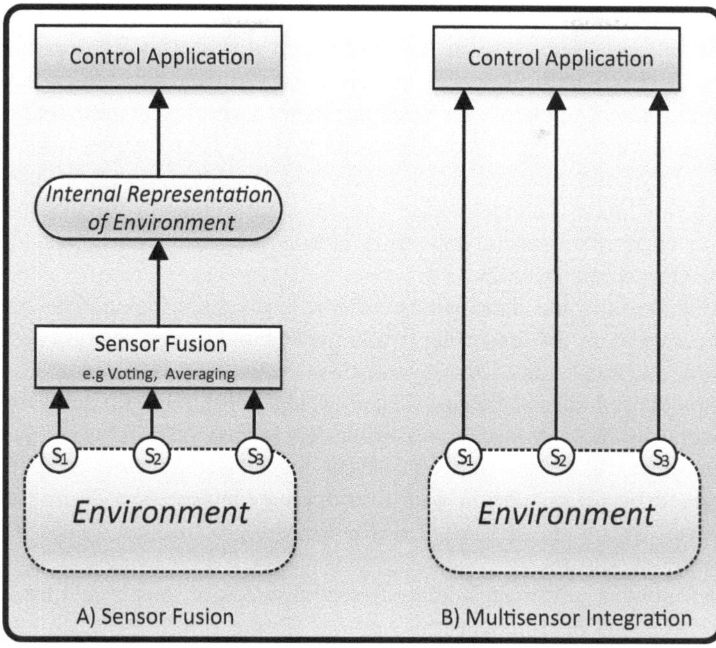

FIGURE 1.6 Block diagram of sensor fusion and multisensor integration. *Adapted from Elmenreich, W., November 19, 2002. An Introduction to Sensor Fusion. Institut für Technische Informatik Vienna University of Technology, Austria wil@vmars.tuwien.ac.at. Research Report 47/ 2001.*

Limited spatial coverage: An individual sensor usually covers a restricted region. For example, a reading from a boiler thermometer only provides an estimation of the temperature near the thermometer and may fail to correctly render the average water temperature in the boiler.

Limited temporal coverage: Some sensors need a particular set-up time to perform and to transmit a measurement, thus limiting the maximum frequency of measurements.

Imprecision: Measurements from individual sensors are limited to the precision of the employed sensing element.

Uncertainty: Uncertainty, in contrast to imprecision, depends on the object being observed rather than the observing device. Uncertainty arises when features are missing (e.g., occlusions), when the sensor cannot measure all relevant attributes of the percept or when the observation is ambiguous (Murphy, 1996). A single-sensor system is unable to reduce uncertainty in its perception because of its limited view of the object (Foote and Huebner, 1995).

As an example, consider a distance sensor mounted at the rear of a car to assist backing into a parking space. The sensor can provide information about objects in front of the sensor, not beside it; thus, the spatial coverage is limited. In addition, the sensor has an update time of 1 s; this is limited temporal coverage and is significant for a human driver. Finally, the sensor does not provide unlimited precision; for example, its measurements could be 2 cm off the actual distance to the object. Uncertainty arises, for example, if the object behind the rear of the car is a small motorcycle; the driver cannot be sure if the sensor beam hits it and delivers a correct measurement with the specified precision or if it misses it, delivering a value suggesting a much different distance (Elmenreich, 2002).

One solution to the listed problems is to use sensor fusion. The standard way to compensate for sensor deprivation is to build a fault-tolerant unit of at least three identical units with a voter (Von Neumann, 1956) or at least two units showing fail-silent behavior (Kopetz et al., 1990). Fail-silent means that a component produces either correct results or, in case of failure, no results at all. In a sensor fusion system, robust behavior against sensor deprivation can be achieved by using sensors with overlapping views of the desired object. This works with a set of sensors of the same type and also with a suite of heterogeneous sensors.

The following advantages derive from the fusion of sensor data from a set of heterogeneous or homogeneous sensors (Bosse et al., 1996; Grossmann, 1998):

Robustness and reliability: Multiple sensor suites have an inherent redundancy, which enables the system to provide information even in case of partial failure.

Extended spatial and temporal coverage: One sensor can look where others cannot; therefore, it can perform a measurement while others cannot.

Increased confidence: A measurement of one sensor is confirmed by the measurements of other sensors covering the same domain.

Reduced ambiguity and uncertainty: Joint information reduces the set of ambiguous interpretations of the measured value.

Robustness against interference: By increasing the dimensionality of the measurement space (e.g., measuring the desired quantity with optical sensors and ultrasonic sensors), the system becomes less vulnerable against interference.

Improved resolution: When multiple independent measurements of the same property are fused, the resolution of the resulting value is better than a single sensor's measurement.

Rao et al. compared the performance of sensor measurements obtained from an appropriate fusing process to the measurements of a single sensor (Rao, 1998). According to them, an optimal fusing process can be designed if the distribution function describing measurement errors of one particular sensor is precisely known. This optimal fusing process performs at least as well as the best single sensor (Elmenreich, 2002).

A further advantage of sensor fusion is the possibility of reducing system complexity. In a traditionally designed system, the sensor measurements are fed into the application, which, in turn, has to cope with a big number of imprecise, ambiguous, and incomplete data streams. In a system where sensor data are preprocessed by fusion methods, the input to the controlling application can be standardized independently of the employed sensor types, thus facilitating application implementation and providing the possibility of modifications in the sensor system with respect to the number and type of employed sensors without modifications of the application software (Elmenreich and Pitzek, 2001).

1.2.3 Limitations of Sensor Fusion

We now have the ability to fuse multisensory data into reliable and feature-rich recognition, but sensor fusion is not an omnipowerful. Fowler gave harsh criticism in 1979:

> *One of the grabbiest concepts around is synergism. Conceptual application of synergism is spread throughout military systems but is most prevalent in the 'multisensor' concept. This is a great idea provided the input data are [of] a good quality. Massaging a lot of crummy data doesn't produce good data; it just requires a lot of extra equipment and may even reduce the quality of the output by introducing time delays and/or unwarranted confidence. [...] It takes more than correlation and fusion to turn sows' ears into silk purses.*
>
> Fowler (1979).

Since this was published, many people have tried to prove the opposite. Nahin and Pokoski (1980) present theoretical proof that the addition of sensors improves performance, using majority vote and maximum likelihood theory for decision fusion. Elmenreich et al. define performance as the probability of making the right decision but without considering the effect on processing power and communication Elmenreich and Peti (2002).

In contrast, Tenney and Sandell (1981) consider communication bandwidth for distributed fusion architectures. They show that a distributed system is suboptimal in its communication effort in comparison to a completely centralized processing scheme.

An essential criterion for the possible benefit of sensor fusion is a comprehensive set of performance measures. Theil, Kester, and Boss present measures of performance for the fields of detection, tracking, and classification. They suggest measuring the quality of the output data and the reaction time (Theil et al., 2000).

Dasarathy (2001) investigates the benefits of increasing the number of inputs to a sensor fusion process. Although the analysis is limited to the augmentation of two-sensor systems by an extra sensor, the work shows that increasing the number of sensors may lead to a performance gain or loss depending on the sensor fusion algorithm.

It can be concluded from the knowledge on sensor fusion performance that in spite of the great potential of sensor fusion, some slight skepticism about "perfect" or "optimal" fusion methods is appropriate (Elmenreich, 2002).

1.2.4 Types of Sensor Fusion

The following paragraphs categorize sensor fusion applications:

1.2.4.1 C^3I Versus Embedded Real-Time Applications

There is an important dichotomy in research on sensor fusion for command, control, communications, and intelligence (C^3I)–oriented applications and sensor fusion targeted at real-time embedded systems. The C^3I-oriented research focuses on intermediate and high-level sensor fusion issues, while on-board applications concentrate on low-level fusion. Table 1.4 compares some central aspects of C^3I and embedded fusion applications (Rothman and Denton, 1991).

1.2.4.2 Three-Level Categorization

Fusion processes are often categorized in a three-level model distinguishing low, intermediate, and high level fusion:

Low-level fusion or raw data fusion (see Section 2 on the double meaning of "data fusion") combines several sources of raw data to produce new data that are expected to be more informative than the inputs.

TABLE 1.4 Comparison of C^3I and Embedded fusion Applications

	Onboard Fusion	C^3I Fusion
Timescale	Milliseconds	Seconds...minutes
Data type	Sensor data	Also linguistic or symbolic data
Man–machine interaction	Optional	Frequently
Database size	Small to moderate	Large to very large
Level of abstraction	Low	High

Adapted from Elmenreich, W., November 19, 2002. An Introduction to Sensor Fusion. Institut für Technische Informatik Vienna University of Technology, Austria wil@vmars.tuwien.ac.at. Research Report 47/2001.

Intermediate-level fusion or feature-level fusion combines various features, such as edges, corners, lines, textures, or positions, into a feature map that may then be used for segmentation and detection.

High-level fusion, also called decision fusion, combines decisions from several experts. Methods of decision fusion include voting, fuzzy-logic, and statistical methods (Elmenreich, 2002).

1.2.4.3 Categorization Based on Input/Output

Dasarathy has proposed a refined categorization based on the three-level model (Dasarathy, 1997). In the Dasarathy model, fusion processes can be categorized based on the abstraction level of the processes' input and output data (Elmenreich, 2002).

The motivation for the Dasarathy model is the existence of fusion paradigms where the input and output of the fusion process belong to different levels. Examples include feature selection and extraction, as the processed data come from the raw data level and the results belong to the feature level. For example, pattern recognition and pattern processing operate between the feature and decision levels. These ambiguous fusion paradigms are sometimes assigned according to the level of their input data and sometimes according to the level of their output data.

To avoid these categorization problems, Dasarathy extends the three-level view to five fusion categories defined by their input/output characteristics. Fig. 1.7 depicts the relation between the three-level categorization and the Dasarathy model.

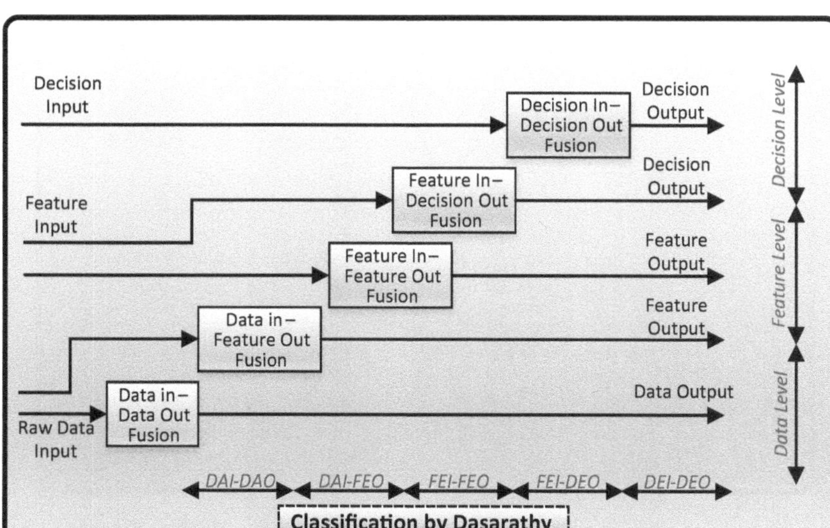

FIGURE 1.7 Alternative fusion characterizations based on input/output. *Adapted from Dasarathy, B.V., January 1997. Sensor fusion potential exploitation-innovative architectures and illustrative applications. Proceedings of the IEEE, 85, 24–38.*

1.2.4.4 Categorization Based on Sensor Configuration

Sensor fusion networks can also be categorized according to the type of sensor configuration. Durrant-Whyte (1988) distinguishes the following three types of sensor configuration:

Complementary: A sensor configuration is called complementary if the sensors do not directly depend on each other but can be combined to give a more complete image of the phenomenon under observation. This resolves the incompleteness of sensor data. An example of a complementary configuration is the employment of multiple cameras, each observing disjunctive parts of a room (Hoover and Olsen, 2000). Generally, fusing complementary data is easy, as the data from independent sensors can be appended to each other (Brooks and Iyengar, 1998).

Sensors S_2 and S_3 in Fig. 1.8 represent a complementary configuration, as each sensor observes a different part of the environment space (Elmenreich, 2002).

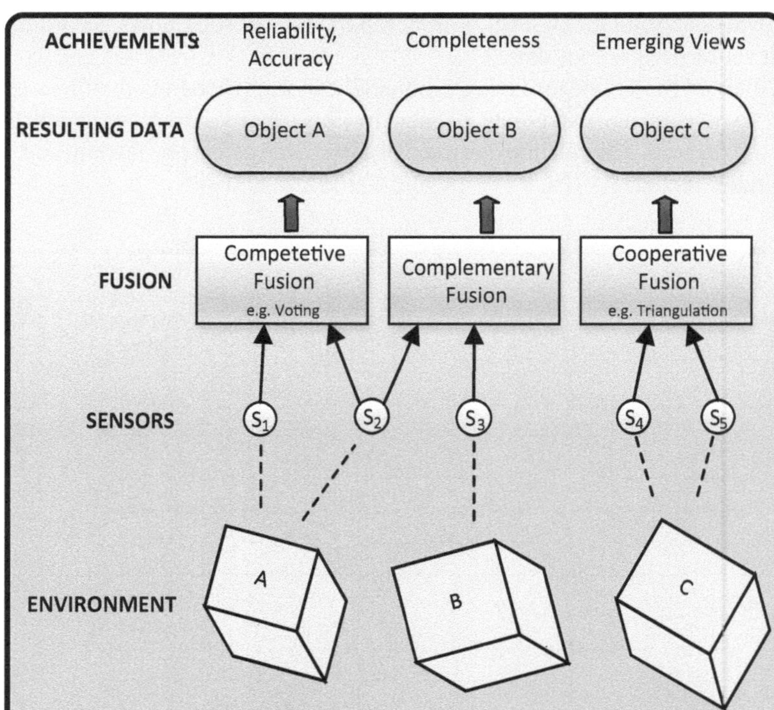

FIGURE 1.8 Competitive, complementary, and cooperative fusion. *Adapted from Elmenreich, W., November 19, 2002. An Introduction to Sensor Fusion. Institut für Technische Informatik Vienna University of Technology, Austria wil@vmars.tuwien.ac.at. Research Report 47/2001.*

Competitive: Sensors are configured competitively if each sensor delivers independent measurements of the same property. Visser and Groen (1999) distinguish two possible competitive configurations—the fusion of data from different sensors or the fusion of measurements from a single sensor taken at different instants. Competitive sensor configuration is also called a redundant configuration (Luo and Kay, 1989).

A special case of competitive sensor fusion is fault tolerance. Fault tolerance requires an exact specification of the service and the failure modes of the system. If a fault is covered by the fault hypothesis, the system still has to provide its specified service. Examples of fault-tolerant configurations are N-modular redundancy (Nelson, 1990) and other schemes where a certain number of faulty components are tolerated (Pease et al., 1980; Marzullo, 1990).

In contrast to fault tolerance, competitive configurations can provide robustness to a system. Robust systems provide a degraded level of service in the presence of faults. While this degradation is weaker than the achievement of fault tolerance, the respective algorithms perform better in terms of resource needs and work well with heterogeneous data sources (Bakken et al., 2001). An example of an architecture that supports heterogeneous competitive sensors can be found in Parhami (1991) and Elmenreich and Peti (2002). Both use confidence tags to indicate the dependability of an observation.

Sensors S_1 and S_2 in Fig. 1.8 represent a competitive configuration, where both sensors redundantly observe the same property of an object in the environment space (Elmenreich, 2002).

Cooperative: A cooperative sensor network uses the information provided by two independent sensors to derive information that would not be available from the single sensors. An example of a cooperative sensor configuration is stereoscopic vision—by combining two-dimensional images from two cameras at slightly different viewpoints, a three-dimensional image of the observed scene is derived. According to Brooks and Iyengar, cooperative sensor fusion is the most difficult to design because the resulting data are sensitive to inaccuracies in all individual participating sensors (Brooks and Iyengar, 1998). Thus, in contrast to competitive fusion, cooperative sensor fusion generally decreases accuracy and reliability (Elmenreich, 2002).

Sensors S_4 and S_5 in Fig. 1.8 represent a cooperative configuration. Both sensors observe the same object, but the measurements are used to form an emerging view of object C that could not have been derived from the measurements of S_4 or S_5 alone.

These three categories of sensor configuration are not mutually exclusive. Many applications implement aspects of more than one of the three types. An example of such hybrid architecture is the application of multiple cameras to monitor a given area. In regions covered by two or more cameras, the sensor configuration can be competitive or cooperative. For regions observed by only one camera, the sensor configuration is complementary (Elmenreich, 2002).

1.2.5 Architectures for Sensor Fusion

Because sensor fusion models depend on the application, there is still no generally accepted model of sensor fusion (Bedworth and O'Brien, 1999). According to Kam, Zhu, and Kalata, it is unlikely that one technique or one architecture will provide a uniformly superior solution (Kam et al., 1997). In this section, we look at some well-known architectures (Elmenreich, 2002).

1.2.5.1 Joint Directors of Laboratories Fusion Architecture

A frequently referred to fusion model originates from the US JDL (Joint Directors of Laboratories). It was proposed in 1985 under the guidance of the US Department of Defense. The JDL model proposed by Waltz and Llinas (1990) comprises five levels of data processing and a database; all are interconnected. The five levels are not meant to be processed in a strict order and can be executed concurrently. Fig. 1.9 depicts the top level of the JDL data fusion process model.

The elements of the JDL model include the following:

Sources: Information comes from a variety of data sources, including sensors, a priori information, databases, and human input.

Source preprocessing (*Level 0*): The task is to reduce the processing load of the fusion processes by prescreening and allocating data to appropriate processes (Bedworth and O'Brien, 1999).

Object refinement (*Level 1*): This level performs data alignment (transformation of data to a consistent reference frame and units) and association (using correlation methods); it also tracks actual and future positions of objects and makes identifications using classification methods.

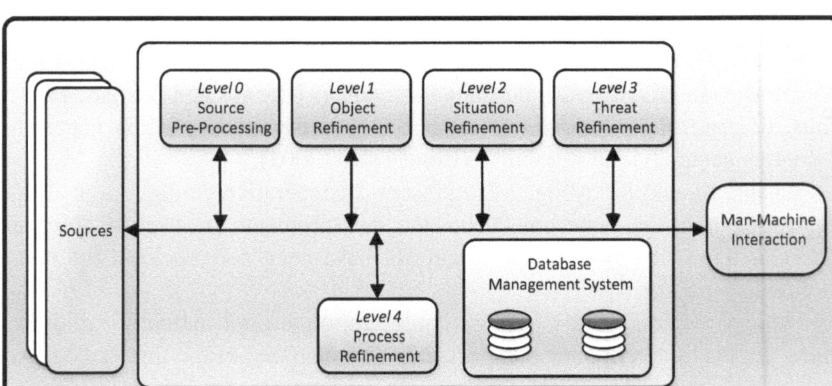

FIGURE 1.9 Joint Directors of Laboratories fusion model. *Adapted from Llinas, J., Hall, D.L., May–June 1998. An introduction to multi-sensor data fusion. Proceedings of the 1998 IEEE International Symposium on Circuits and Systems, 6, 537–540.*

Situation refinement (*Level 2*): Situation refinement attempts to find a contextual description of the relationship between objects and observed events.
Threat refinement (*Level 3*): Based on a priori knowledge and predictions about the future situation, this processing level tries to draw inferences about vulnerabilities and opportunities for operation.
Process refinement (*Level 4*): Process refinement is a metaprocess that monitors system performance (e.g., real-time constraints) and reallocates sensors and sources to achieve particular mission goals.
Database management system: The task of the database management system is to monitor, evaluate, add, update, and provide information for fusion processes.
Man–machine interaction: This provides an interface for human input and the communication of fusion results to operators and users.

The JDL model is very popular for fusion systems. Despite its origin in the military, it can be used in commercial applications. The JDL model can also categorize processes related to a fusion system.

However, the model has the following drawbacks:

- It is a data-centered or information-centered model, making it difficult to extend or reuse applications built with this model.
- The model is very abstract, making it difficult to properly interpret its parts or to apply it appropriately to specific problems.
- The model is helpful for common understanding but does not guide a developer in identifying the methods that should be used (Llinas and Hall, 1998); thus, the model does not help in developing an architecture for a real system.

The basic JDL model has been improved and extended for various applications. Waltz and colleagues show that the model does not address multi-image fusion problems; they present an extension that includes the fusion of image data (Waltz, 1995). Steinberg, Bowman, and White propose revisions and expansions of the JDL model involving broadening the functional model, relating the taxonomy to fields beyond the original military focus, and integrating a data fusion tree architecture model for system description, design, and development (Steinberg et al., 1999).

1.2.5.2 Waterfall Fusion Process Model

The waterfall model proposed by Markin et al. (1997) emphasizes the processing functions on the lower levels. Fig. 1.10 depicts the processing stages of the waterfall model. The stages relate to levels 0, 1, 2, and 3 of the JDL model as follows: sensing and signal processing correspond to source preprocessing (level 0); feature extraction and pattern processing match object refinement (level 1); situation assessment is similar to situation refinement (level 2); decision-making corresponds to threat refinement (level 3) (Elmenreich, 2002).

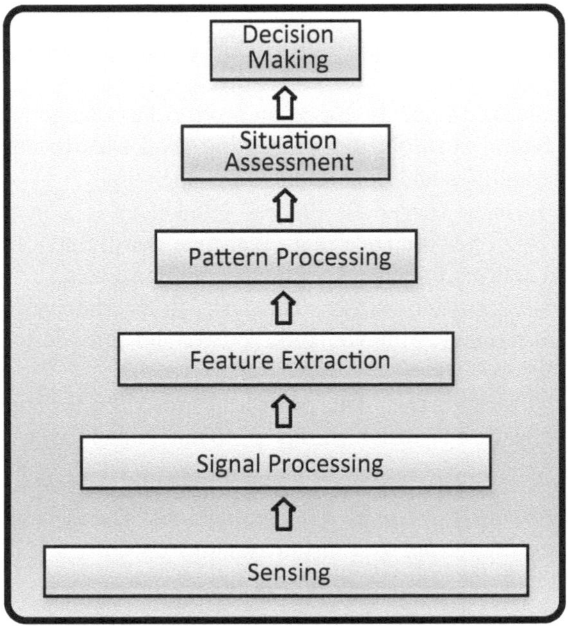

FIGURE 1.10 Waterfall fusion process model. *Adapted from Markin, M., Harris, C., Bernhardt, M., Austin, J., Bedworth, M., Greenway, P., Johnston, R., Little, A., Lowe, D., 1997. Technology Foresight on Data Fusion and Data Processing. Publication, The Royal Aeronautical Society.*

Because it is so similar to the JDL model, the waterfall model has the same drawbacks. Although it is more exact in analyzing the fusion process than other model, it's major limitation is the omission of any feedback data flow. The waterfall model has been used in the defense data fusion community in the United Kingdom but has not been adopted elsewhere to any significant extent (Bedworth and O'Brien, 1999).

1.2.5.3 Boyd Model

Boyd has proposed a cycle consisting of four stages (Boyd, 1987). The Boyd control cycle or observe, orientate, decide, and act (OODA) loop (depicted in Fig. 1.11) represents the classic decision-support mechanism in military information operations. Because decision-support systems for situational awareness are tightly coupled with fusion systems (Bass, 2000), the Boyd loop has also been used for sensor fusion.

When Bedworth and O'Brien compared the stages of the Boyd loop to the JDL model (Bedworth and O'Brien, 1999), they found the following:

Observe: This stage is broadly comparable to source preprocessing in the JDL model.

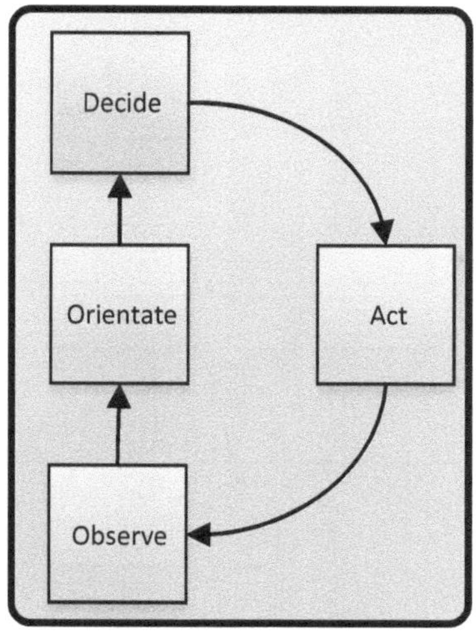

FIGURE 1.11 Boyd (or observe, orientate, decide, and act) loop (Boyd, 1987).

Orientate: This stage correspond to functions of levels 1, 2, and 3 of the JDL model.

Decide: This stage is comparable to level 4 of the JDL model (process refinement).

Act: Since the JDL model does not close the loop by taking the actuating part of the interaction into account, this stage has no direct counterpart in the JDL model.

The Boyd model represents the stages of a closed control system and gives an overview of the overall task of a system, but it lacks an appropriate structure for identifying and separating different sensor fusion tasks (Elmenreich, 2002).

1.2.5.4 LAAS Architecture

The Laboratoire d'Analyse et d'Architecture des Systemes (LAAS) architecture was developed as an integrated architecture to use in the design and implementation of mobile robots for real-time and code reuse. Because mobile robot systems often employ sensor fusion methods, it is worth taking a quick look at the elements of the LAAS architecture (depicted in Fig. 1.12) (Alami et al., 1998).

The architecture consists of the following levels:

Logical robot level: The task of the logical robot level is to establish a hardware-independent interface between the physical sensors and actuators and the functional level.

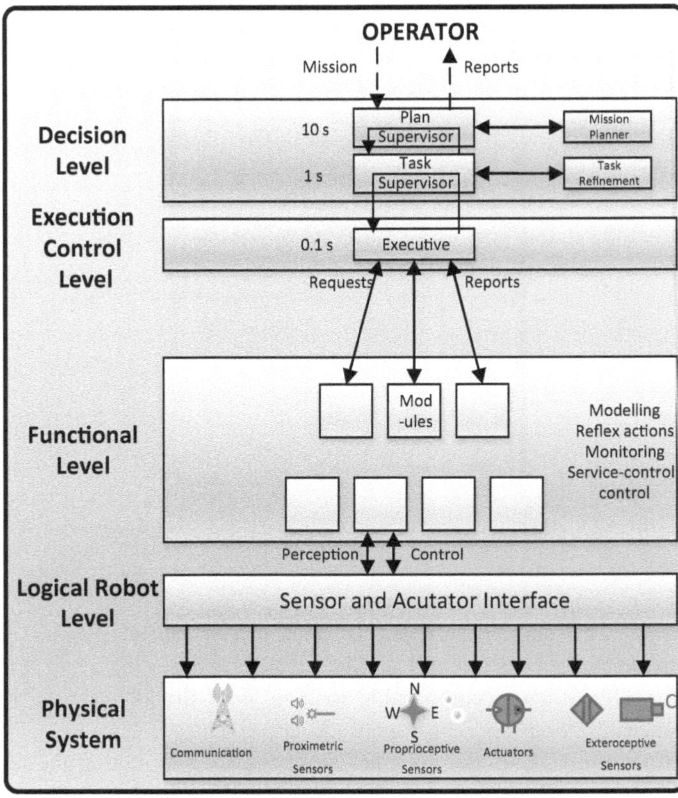

FIGURE 1.12 Laboratoire d'Analyse et d'Architecture des Systemes architecture. *Adapted from Alami, R., Chatila, R., Fleury, S., Ghallab, M., Ingrand, F., April 1998. An architecture for autonomy. International Journal of Robotics Research, 17 (4), 315–337.*

Functional level: The functional level includes all the basic built-in robot action and perception capabilities. The processing functions, such as image processing, obstacle avoidance, and control loops, are encapsulated into separate controllable communicating modules.

Execution control level: The execution control level controls and co-ordinates the execution of the functions provided by the modules according to the task requirements.

Decision level: The decision level includes the capabilities of producing the task plan and supervising its execution while, at the same time, remaining reactive to other events from the execution control level. Depending on the application, the decision level can be composed of several layers that provide different representation abstractions and have different temporal properties.

The LAAS architecture maps low-level and intermediate-level sensor fusion to modules at the functional level. High-level sensor fusion is represented in the decision level. The timing requirements are different at the decision level and the functional level (refer to Section 1.2.4 on the difference between C^3I and real-time applications). While the architecture provides a good way to partition large systems into modules, it does not provide an appropriate real-time communication and representation of the fused data at the levels above the functional level. In contrast to the JDL model, the LAAS architecture guides a designer in implementing reusable modules as part of a real-time application (Elmenreich, 2002).

1.2.5.5 Omnibus Model

The omnibus model was introduced in 1999 by Bedworth and O'Brien. Fig. 1.13 depicts its architecture. Unlike the JDL model, the omnibus model defines the ordering of processes and makes the cyclic nature explicit. It uses a general terminology that does not assume applications are defense-oriented. The model's cyclic structure is comparable to the Boyd loop, but it provides a much more finely grained structuring of the processing levels. The model is intended to be used multiple times in the same application recursively at two

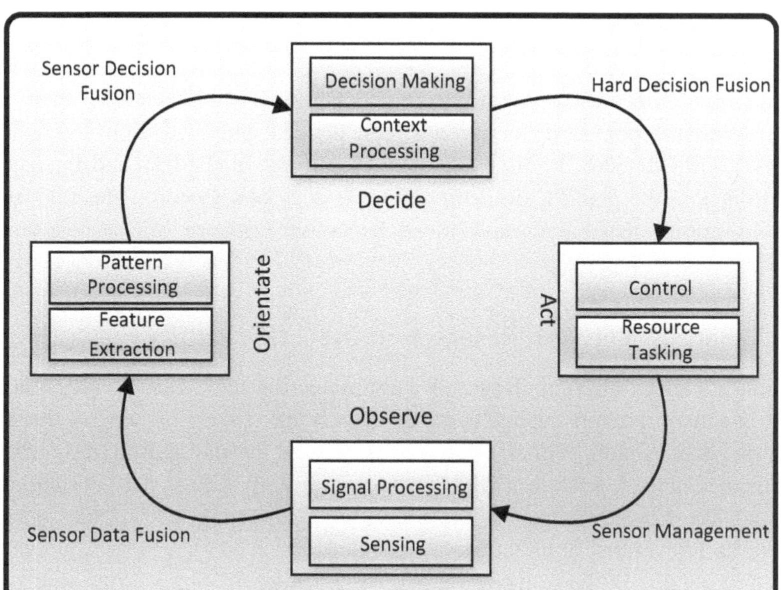

FIGURE 1.13 Omnibus model. *Adapted from Bedworth, M.D., O'Brien, J., July 1999. The omnibus model: a new architecture for data fusion? In: Proceedings of the 2nd International Conference on Information Fusion (FUSION 99), Helsinki, Finnland.*

different levels of abstraction. First, the model is used to characterize and structure the overall system. Second, these structures are used to model the single subtasks of the system.

Although the hierarchical separation of the sensor fusion tasks is very sophisticated in the omnibus model, the model does not support a horizontal partitioning into tasks that reflect distributed sensing and data processing. Thus, it does not support decomposition into modules that can be separately implemented, separately tested, and reused for different applications (Elmenreich, 2002).

1.3 SENSOR NETWORKS: A DISTRIBUTED APPROACH IN LARGE ASSETS

The development of sensor networks requires technologies from three research areas: sensing, communication, and computing (including hardware, software, and algorithms). Thus, combined and separate advancements in each of these areas have driven research in sensor networks. An example of an early sensor network is the radar networks used in air traffic control. The national power grid is another example; with its many sensors, it can be viewed as one large sensor network. Both systems require specialized computers and communication capabilities and were created long before the term "sensor networks" came into vogue (Chong and Kumar, 2003).

Sensor networks are a promising approach in a variety of applications, such as monitoring safety and security of buildings and spaces, measuring traffic flows, and tracking environmental pollutants. The ongoing miniaturization of computing devices armed with wireless technologies influences our everyday lives. With the popularity of laptops, cell phones, personal digital assistants (PDAs), GPS devices, Radio Frequency Identification (RFID), and intelligent electronics in the post-PC era, computing devices have become cheaper, more mobile, more distributed, and more pervasive (Bokare and Ralegaonkar, 2012).

1.3.1 Sensor Network Research in the 21st Century

Recent advances in computing and communication have caused a significant shift in sensor network research, paradoxically bringing it closer to achieving the original vision. Small and inexpensive sensors based on microelectromechanical system (MEMS) technology, wireless networking, and inexpensive low-power processors allow the deployment of wireless *ad hoc* networks for various applications.

Once again, DARPA has emerged as a leader in innovation, starting a research program on sensor networks to leverage the latest technological advances. The recently concluded DARPA Sensor Information Technology

(SensIT) program had two key research and development thrusts. First, it developed new networking techniques. Today's networking techniques, developed for voice and data and relying on a fixed infrastructure do not suffice for battlefield use. In the battlefield context, these sensor devices or nodes should be ready for rapid deployment, in an *ad hoc* fashion and in highly dynamic environments. Second, it looked at networked information processing, i.e., how to extract useful, reliable, and timely information from a deployed sensor network. This implies leveraging the distributed computing environment created by these sensors to process signals and information in the network; it also implies the ability to dynamically and interactively query and task the sensor network.

SensIT suggests new possibilities for sensors. Current systems such as the Tactical Automated Security System for perimeter security are dedicated rather than programmable (Corella, 2003). They use technologies based on transmit-only nodes and a long-range detection paradigm. SensIT networks have new capabilities. They are interactive and programmable with dynamic tasking and querying. In addition, a multitasking feature in the system allows multiple simultaneous users. Finally, since detection ranges are much shorter in a sensor system, the software and algorithms can exploit the proximity of devices to threats to dramatically improve the accuracy of detection and tracking. The software and the overall system design support low latency, energy-efficient operation, built-in autonomy and survivability, and low probability of detection of operation. As a result, a network of SensIT nodes can support the detection, identification, and tracking of threats, as well as targeting and communication, both within and outside the network, such as in an overhead asset (Chong and Kumar, 2003).

1.3.2 Technology Trends

Current sensor networks can exploit technologies not available 20 years ago and perform functions that were not even dreamed of at that time. Sensors, processors, and communication devices are getting smaller and cheaper. Commercial companies such as Ember, Crossbow, and Sensoria are building and deploying small sensor nodes and systems; in the near future, our daily lives will be enhanced by a network of small, embedded sensor nodes. Commercial off-the-shelf PDAs using Palm or Pocket PC operating systems already offer significant computing power in a small package and are widely available. These can easily be "ruggedized" to become processing nodes in a sensor network. Some even have built-in sensing capabilities, such as cameras. These powerful processors can be hooked to MEMS devices and machines with extensive databases and communication platforms to usher in a new era of technologically sophisticated sensor nets (Chong and Kumar, 2003).

Wireless networks based on IEEE 802.11 standards can now provide bandwidths approaching those of wired networks. Noting the low expense and high capabilities of sensor networks, the IEEE has set IEEE 802.15 standard for personal area networks (PANs), with "personal networks" defined as having a radius of 5–10 m. Networks of short-range sensors are the ideal technology to be employed in PANs. The IEEE encouragement of the development of technologies and algorithms for such short ranges ensures the continued development of low-cost sensor nets (Heile, 2001). Furthermore, increases in chip capacity and processor production capabilities have reduced the energy per bit requirement for both computing and communication. Sensing, computing, and communications can now be performed on a single chip, further reducing the cost and allowing deployment in ever larger numbers (Chong and Kumar, 2003).

Looking into the future, we predict advances in MEMS technology will produce sensors that are even more capable and versatile. For example, Dust Inc., Berkeley, California, a company that sprang from the late 1990s Smart Dust research project at the University of California, Berkeley, is building MEMS sensors that can sense and communicate and yet are tiny enough to fit inside a cubic millimeter. A Smart Dust optical mote uses MEMS to aim submillimeter-sized mirrors for communications. Smart Dust sensors can be deployed using a 3 f10 mm "wavelet" shaped like a maple tree seed and dropped to float to the ground (Kahn et al., 1999). A wireless network of these ubiquitous, low-cost, disposable microsensors can provide close-range sensing capabilities in many novel applications.

Table 1.5 compares three generations of sensor nodes; Fig. 1.14 shows their sizes (Chong and Kumar, 2003).

1.3.3 Wireless Sensor Network

The emergence of WSNs is the latest manifestation of Moore's Law, i.e., the trend toward the miniaturization and ubiquity of computing devices. Typically, a wireless sensor node (or simply sensor node) includes sensing, computing, communication, actuation, and power components.

A WSN consists of spatially distributed autonomous sensors to monitor physical or environmental conditions, such as temperature, sound, vibration, pressure, motion, or pollutants and to cooperatively pass their data through the network to a main location. The more modern networks are bidirectional, thus enabling control of sensor activity. Such networks are used in many industrial and consumer applications, including industrial process monitoring and control, machine health monitoring, and so on (Fig. 1.15).

Typically, a sensor node is a tiny device that includes three basic components: a sensing subsystem for data acquisition from the physical surrounding environment, a processing subsystem for local data processing and storage, and a wireless communication subsystem for data transmission. In addition,

TABLE 1.5 Three Generations of Sensor Nodes

	Yesterday (1980s–90s)	Today (2000–03)	Tomorrow (2010)
Manufacturer	Custom contractors, e.g., for TRSS	Commercial: Crossbow Technology Inc., Sensoria Corp., Ember Corp.	Dust Inc., and others to be formed
Size	Large shoe box and up	Pack of cards to small shoe box	Dust particle
Weight	Kilograms	Grams	Negligible
Node architecture	Separate sensing, processing, and communication	Integrated sensing, processing, and communication	Integrated sensing, processing, and communication
Topology	Point-to-point, star	Client server, peer to peer	Peer to peer
Power supply lifetime	Large batteries; hours, days, and longer	AA batteries; days to weeks	Solar; months to years
Deployment	Vehicle-placed or air-drop single sensors	Hand-emplaced	Embedded, "sprinkled" left behind

Adapted from Chong, C-Y., Kumar, S.P., August 2003. Sensor networks: evolution, opportunities, and challenges. Proceedings of the IEEE, vol. 91, 8.

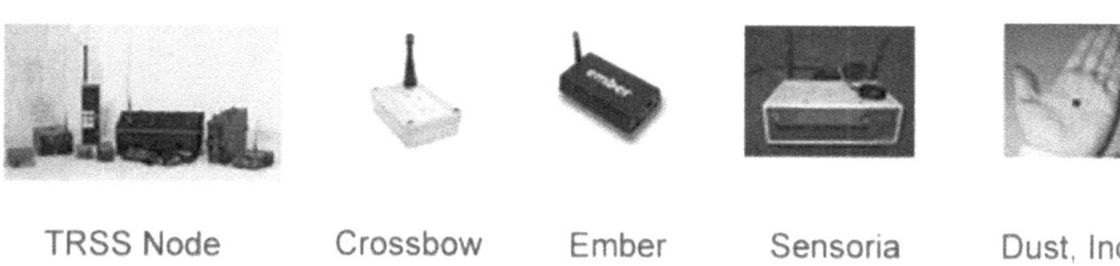

FIGURE 1.14 Three generations of sensor nodes. *Adapted from Chong, C-Y., Kumar, S.P., August 2003. Sensor networks: evolution, opportunities, and challenges. Proceedings of the IEEE, vol. 91, 8.*

a power source supplies the energy needed by the device to perform the programmed task. This power source often consists of a battery with a limited energy budget. There are many different kinds of sensors: pressure, accelerometer, camera, thermal, microphone, etc. They monitor different conditions at different locations, noting such things as temperature, humidity, vehicular movement, lightning condition, pressure, soil makeup, noise levels, the presence or absence of certain kinds of objects, mechanical stress levels on attached objects, and current characteristics such as speed, direction, and size

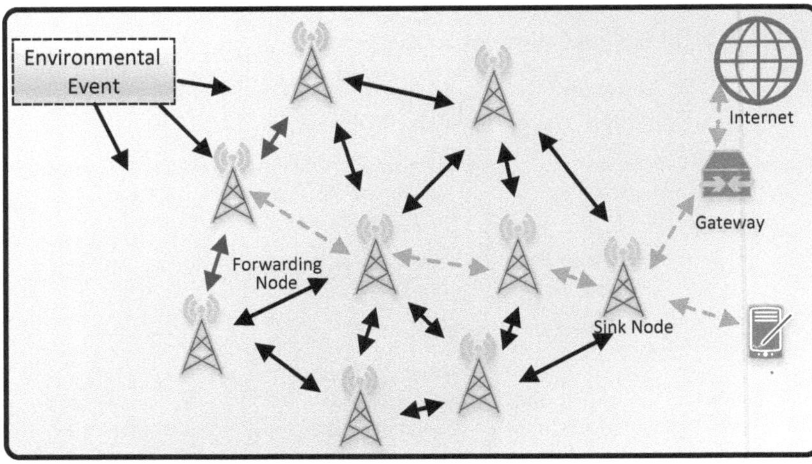

FIGURE 1.15 Wireless sensor network. *Adapted from Bokare, M., Ralegaonkar, A., December–January 2012. Wireless sensor network: a promising approach for distributed sensing tasks. Excel Journal of Engineering Technology and Management Science I 1, ISSN 2249–9032.*

of an object. Normally a sensor node combines the abilities to compute, communicate, and sense (Bokare and Ralegaonkar, 2012).

1.3.3.1 Sensor Node Architecture

A sensor node typically consists of five main parts. First, one or more sensors gather data from the environment. Second, the central unit, a microprocessor, manages the tasks. Third, a transceiver (included in the communication module in Fig. 1.16) communicates with the environment, and fourth, a memory is used to store temporary data or data generated during processing. Finally, a battery supplies all parts with energy (see Fig. 1.16). To assure a sufficiently long network lifetime, energy efficiency in all parts of the network is crucial. Because of this, data processing tasks are often spread over the network: i.e., nodes cooperate in transmitting data to the sinks. Although most sensors have a traditional battery, there is some early stage research on the production of sensors without batteries, using technologies similar to those used for passive RFID chips without batteries (Bokare and Ralegaonkar, 2012).

The ongoing development of sensor nodes is influenced by the following (Bokare and Ralegaonkar, 2012):

- Increasing complexity of microchips,
- High performance, wireless networking technologies,
- Combining digital signal processing (DSP) and sensor data acquisition,
- Ongoing development of MEMS, and
- Availability of high-performance development tools.

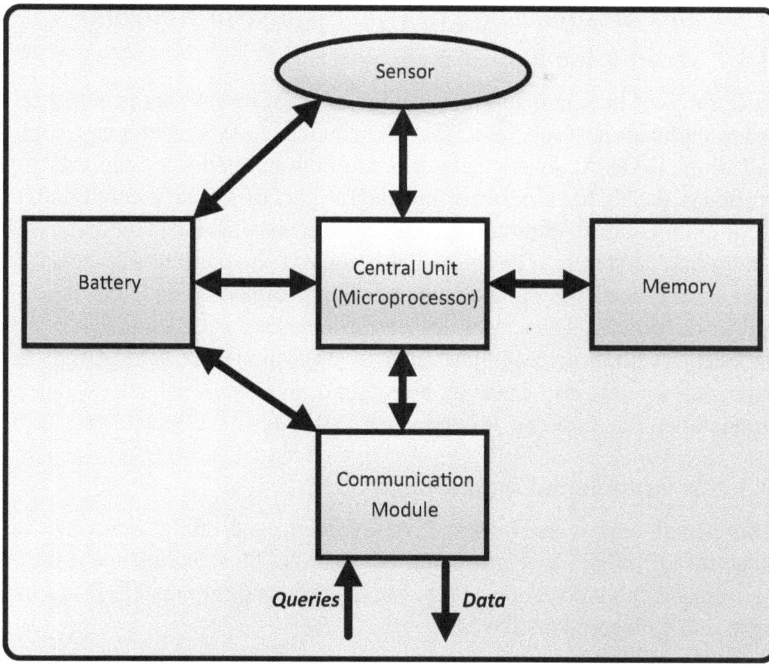

FIGURE 1.16 Sensor node architecture. *Adapted from Bokare, M., Ralegaonkar, A., December–January 2012. Wireless sensor network: a promising approach for distributed sensing tasks. Excel Journal of Engineering Technology and Management Science I 1, ISSN 2249–9032.*

1.3.3.2 Characteristics of Wireless Sensor Networks

The main characteristics of a WSN include the following (Bokare and Ralegaonkar, 2012):

- Power consumption constraints for nodes using batteries or energy harvesting
  - Ability to cope with node failures
  - Mobility of nodes
  - Dynamic network topology
  - Communication failures
  - Heterogeneity of nodes
  - Scalability to large scale of deployment
  - Ability to withstand harsh environmental conditions
  - Ease of use
- Unattended operation
  - Power consumption

1.3.3.3 Fields of Application of Wireless Sensor Networks

1.3.3.3.1 Security and Surveillance

Because most of the elemental knowledge of sensor networks is based on the defense application, especially the Distributed Sensor Networks and the SensIT from DARPA, sensor networks are now applied very successfully in the military. WSNs have become an integral part of military command, control, communications, computing, intelligence, surveillance, reconnaissance, and targeting systems. The battlefield context is particularly specialized because of the need for rapid deployment, self-organization, fault tolerance, and network security. The sensor devices or nodes are useful for such things as battlefield surveillance; reconnaissance of opposing forces; targeting; battle damage assessment; and nuclear, biological, and chemical attack detection reconnaissance (Bokare and Ralegaonkar, 2012; Fig. 1.17).

1.3.3.3.2 Environmental Monitoring

Environmental sensor networks have evolved, and there are now many applications of WSNs to earth science research. This includes sensing volcanoes, oceans, glaciers, forests, etc. Some other major areas are listed below (Bokare and Ralegaonkar, 2012).

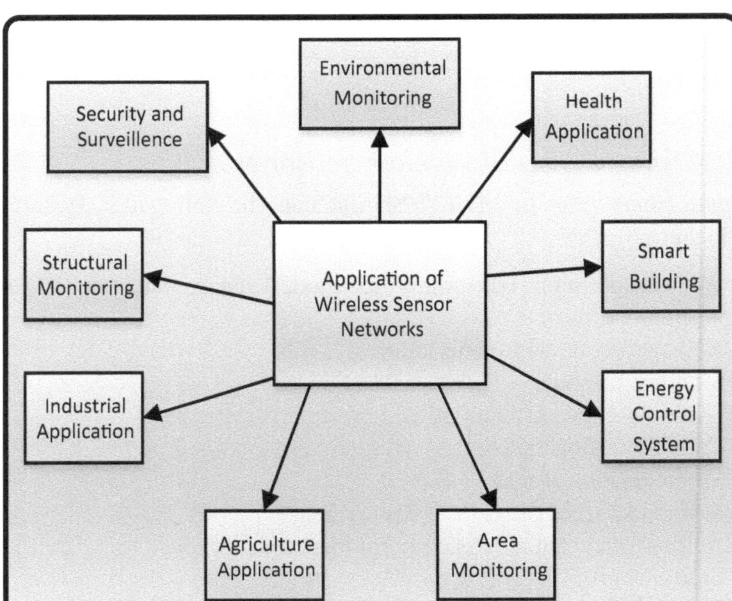

FIGURE 1.17 Fields of applications of wireless sensor network. *Adapted from Bokare, M., Ralegaonkar, A., December–January 2012. Wireless sensor network: a promising approach for distributed sensing tasks. Excel Journal of Engineering Technology and Management Science I 1, ISSN 2249–9032.*

- Air Pollution Monitoring:
  WSNs have been deployed in several cities to monitor the concentration of dangerous gases. These can take advantage of *ad hoc* wireless links, making them more mobile for testing readings in different areas.
- Forest Fire Detection:
  A network of sensor nodes can be installed in a forest to detect fires. The nodes can be equipped with sensors to measure temperature, humidity, and gases produced by fires in the trees or vegetation. Early detection is crucial; with WSNs, the fire fighters know when a fire starts and how/where it is spreading.
- Landslide Detection:
  A landslide detection system makes use of a WSN to detect the slight movements of soil and changes in various parameters that may occur before or during a landslide. With the data, it may be possible to predict the occurrence of a landslide long before it actually happens.

1.3.3.4 Health Applications

Sensor networks are widely used in health care. In some modern hospitals, sensor networks are constructed to monitor patients' physiological data, to control the drug administration, and to track and monitor patients and doctors. Some even use RFIDs. In long-term nursing homes, cameras, pressure sensors, orientation sensors, and sensors are widely used, for example, to detect falls or level of consciousness or to monitor vital signs and dietary/exercise programs. These applications reduce personnel cost and expedite the reaction time in emergencies. In the Smart Sensors and Integrated Microsystems project, retina prosthesis chips consisting of 100 microsensors are implanted in the human eye. This allows patients with no vision or limited vision to see at an acceptable level. Wireless communication is required for feedback, image identification, and validation.

Other health care applications include glucose level monitors, organ monitors, cancer detectors, and general health monitors. The idea of embedding wireless biomedical sensors inside the human body is promising, but there are many challenges to overcome before this is a reality: the system must be ultra-safe and reliable, it must require minimal maintenance, and it must harness energy from body heat. Researchers are hopeful that with more research, a better quality of life can be achieved and medical costs reduced (Bokare and Ralegaonkar, 2012).

1.3.3.5 Smart Buildings

Home applications will be part of our normal lives in the future. Imagine the following scenario. You get home after a day of work. At the front door, the sensor detects someone is opening the door. It tells the electric kettle to boil water and the air conditioner to turn on. Five sensors around the room measure the temperature to keep it constant. You sit down on the sofa. The light on the table automatically turns on because the pressure sensor under the cushion has detected your weight. The TV also comes on because a sensor has monitored that someone is sitting in front of it. The lights in your hall, in the bathroom, and on the balcony all have sensors and can be turned on or off automatically. Your windows have vibratory sensors connected to police to prevent break and enter theft (Bokare and Ralegaonkar, 2012). This scenario is not unlikely in the not-so-distant future. We already have most of the technology.

1.3.3.6 Energy Control Systems

Societal-scale sensor networks can greatly improve the efficiency of the energy-provision chain, which consists of three components: energy generation, energy distribution, and energy consumption infrastructure (Bokare and Ralegaonkar, 2012).

1.3.3.7 Area Monitoring

Area monitoring is a common application of WSNs. In area monitoring, the WSN is deployed over a region where some phenomenon is to be monitored. A civilian example is the geo-fencing of gas or oil pipelines (Bokare and Ralegaonkar, 2012).

1.3.3.8 Agriculture Applications

1.3.3.8.1 Agriculture

The use of WSNs is increasingly common in the agricultural industry; a wireless network frees the farmer of the need to maintain wiring in a difficult environment. Gravity-fed water systems can use pressure transmitters to monitor water tank levels, pumps can be controlled using wireless I/O devices, and water use can be measured and wirelessly transmitted back to a central control center for billing. Automated irrigation enables more efficient water use and reduces waste.

1.3.3.8.2 Greenhouse Monitoring

WSNs are used to control the temperature and humidity levels inside commercial greenhouses. When the temperature and humidity drop below specific levels, the greenhouse manager is notified via email or cell phone text message; alternatively, host systems can trigger misting systems, open vents, turn on fans, or control a wide variety of system responses (Bokare and Ralegaonkar, 2012).

1.3.3.9 Industrial Applications

WSNs have been developed for machinery CBM as they offer significant cost savings and enable new functionalities. In wired systems, the installation of enough sensors is often limited by the cost of wiring. Previously inaccessible locations, rotating machinery, hazardous or restricted areas, and mobile assets can now be reached with wireless sensors. There are many opportunities to use WSNs in the water/wastewater industries (Bokare and Ralegaonkar, 2012).

1.3.3.10 Structural Monitoring

Wireless sensors can be used to monitor the movement in buildings and infrastructure, such as bridges, flyovers, embankments, tunnels, etc., thus enabling engineers to monitor assets remotely without the need for costly site visits. An additional advantage is having access to daily data; traditionally these data were collected weekly or monthly in physical site visits, involving either road or rail closure in some cases. Wireless monitoring is also far more accurate than any visual inspection (Bokare and Ralegaonkar, 2012).

1.3.4 New Applications of Sensor Networks

As has been noted, research on sensor networks was originally confined to military applications. Military sensor networks range from large-scale acoustic surveillance systems for ocean surveillance to small networks of unattended ground sensors for ground target detection. However, the availability of low-cost sensors and communication networks has resulted in the development of many other applications outside the military, from infrastructure security to industrial sensing. The following are a few examples (Chong and Kumar et al., 2003).

1.3.4.1 Infrastructure Security

Sensor networks can be used for infrastructure security and counterterrorism applications. Critical buildings and facilities, such as power plants and communication centers, have to be protected from terrorist attack. Networks of video, acoustic, and other sensors can be deployed around these facilities to provide early detection of possible threats. Improved coverage and detection and a reduced false alarm rate can be achieved by fusing the data from multiple sensors. Even though fixed sensors connected by a fixed communication network protect most facilities, wireless ad hoc networks provide more flexibility and additional coverage when needed. Sensor networks can also be used to detect biological, chemical, and nuclear attacks. Examples of such networks can be found in Hills (2001); other uses of sensor networks appear in Chong and Kumar (2003).

1.3.4.2 Environment and Habitat Monitoring

Environment and habitat monitoring is a natural candidate for sensor networks, as the variables to be monitored, e.g., temperature, are usually distributed over a large region (Steere et al., 2000). The Center for Embedded

Network Sensing, Los Angeles, California, has a focus on environmental and habitat monitoring. It uses environmental sensors to study vegetation response to climatic trends and diseases; in addition, acoustic and imaging sensors identify, track, and measure the population of birds and other species. On a very large scale, the System for the Vigilance of the Amazon provides environmental monitoring, drug trafficking monitoring, and air traffic control for the Amazon Basin (Jensen, 2002). Sponsored by the government of Brazil, this large sensor network consists of different types of interconnected sensors, including radar, imagery, and environmental sensors. The imagery sensors are space-based, with radar located on aircraft and environmental sensors on the ground. The communication network connecting the sensors operates at different speeds. For example, high-speed networks connect sensors on satellites and aircraft, while low-speed networks connect the ground-based sensors (Chong and Kumar, 2003).

1.3.4.3 Industrial Sensing

Commercial industries have long been interested in sensing as a means of lowering cost and improving machine (and perhaps user) performance and maintainability. Monitoring machine "health" by determining vibration or wear and lubrication levels and the insertion of sensors into regions inaccessible to humans are two examples of industrial applications of sensors. Several years ago, the IEEE and the National Institute for Standards and Technology (NIST) launched the P1451 Smart Transducer Interface Standard to enable full plug-and-play of sensors and networks in industrial environments (Lee, 2001). Factories have continued to automate production and assembly lines with remote sensing nets, implementing sophisticated online quality control tests enabled by the sensors. Remote wireless sensors, in particular, can allow a factory to be instrumented to ensure and maintain compliance with federal safety and guidelines while keeping installation costs low (Chong and Kumar, 2003).

Spectral sensors are one example of sensing in an industrial environment. From simple optical devices such as obtrudes and pH probes to true spectral devices that can function as miniature spectrometers, optical sensors can replace existing instruments and measure material properties and composition. Optical sensing is also facilitated by miniaturization, as low-cost charge-coupled device, array devices, and microengineering enable smaller, smarter sensors. The goal of this and other industrial sensing is to permit multipoint or matrix sensing: inputs from hundreds or thousands of sensors fed into databases that can be queried in any number of ways to show real-time information on a large or small scale (Chong and Kumar, 2003).

1.3.4.4 Traffic Control

Sensor networks have been used for vehicle traffic monitoring and control for quite a while. Most traffic intersections have either overhead or buried sensors to detect vehicles and control traffic lights. Video cameras are frequently used to monitor road segments with heavy traffic, with the video sent to human operators at central locations. However, these sensors and the communication network that connects them are costly; thus, traffic monitoring is generally limited to a few critical points. Inexpensive wireless *ad hoc* networks will completely change traffic monitoring and control. Cheap sensors with embedded networking capability can be deployed at every road intersection to detect and count vehicle traffic and estimate its speed. The sensors will communicate with neighboring nodes to develop a "global traffic picture," which can be queried by human operators or automatic controllers to generate control signals (Chong and Kumar, 2003).

A more radical concept has sensors attached to each vehicle (Estrin et al., 1999). As the vehicles pass each other, they exchange summary information on the location of traffic jams and the speed and density of traffic, information that may be generated by ground sensors. These summaries propagate from vehicle to vehicle and can be used by drivers to avoid traffic jams and plan alternative routes (Chong and Kumar, 2003).

1.4 SMART SENSORS

Smart sensors permit the analog processing of signals recorded by sensors: digital representations of the analog signal, address, and data transfer through a bidirectional digital bus; and manipulation and computation of the sensor-derived data (Najafi, 1991).

The term "smart sensor" has been used by various researchers in a number of different contexts, ranging from sensors incorporating a few active devices to provide a more reliable interface to the sensor to enhance the quality of the sensed signal, to integrated sensors incorporating a sophisticated electronic circuit block, including both digital and analog circuitry that helps transform the sensor from a passive component into the "smart" periphery of a control/instrumentation system (Najafi, 1991). However, over the past few years, a broad consensus has been achieved, whereby a smart sensor is defined as one that is capable of the following (Brignell, 1989; Giachino, 1986):

1. Providing a digital output;
2. Communicating through a bidirectional digital bus;
3. Being accessed through a specific address; and
4. Executing commands and logical functions.

In the context of this book, a smart sensor is defined as one possessing these features. In addition, it is desirable for the smart sensor to perform such functions as compensation of secondary parameters (e.g., temperature), failure prevention and detection, self-testing, auto calibration, and various computationally intensive operations. The development of such sensors will greatly enhance the capabilities of many control and instrumentation systems with gaps in their interface with the external world.

Fig. 1.18 shows the elements of a typical electronic control system. Sensors provide analog information on the system being monitored, taking the information through signal conditioning circuits to a microprocessor-based controller. The processor interprets the information, makes appropriate decisions (perhaps in conjunction with higher level control), and implements those decisions via the actuators. Sensors currently represent the weakest link in the development of most next-generation instrumentation and control systems. Where sensors exist at all, they are frequently unreliable, rarely attain an accuracy of 8 bits and may cost more than the processor. This situation is especially true in biomedicine, where the problems are particularly difficult. Only in the past few years has this situation begun to change (Najafi, 1991).

In contrast to smart sensors, primitive sensors are devices or materials with some electrical property that changes with some physical phenomenon. An example of a primitive sensor is a cadmium sulfide photoresistor whose resistance changes with light intensity. With the right circuit and program, microcontroller light measurements are possible. Other examples of common primitive sensors are current/voltage output temperature sensors, microphone transducers, and the potentiometer, a rotational position sensor (Lindsay, 2003).

Inside every smart sensor is one or more primitive sensors and support circuitry. What makes a smart sensor "smart" is the additional, built-in

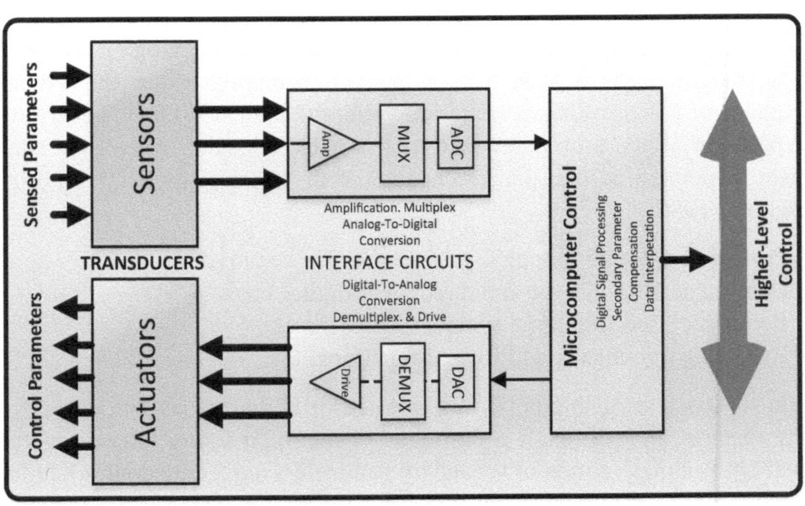

FIGURE 1.18 Typical electronic control system. *Adapted from Wise and Najafi, 1988.*

electronics. The electronics allow these sensors to do one or more of the following (Lindsay, 2003):

- Preprocess measured values into meaningful quantities.
  - Communicate measurements with digital signals and communication protocols.
  - Orchestrate the actions of primitive circuits and sensors to "take" measurements.
  - Make decisions and initiate action based on sensed conditions, independent of a microcontroller.
  - Remember calibration or configuration settings.

1.4.1 Structure of Smart Sensor

The basic architectural components of a smart sensor node are shown in Fig. 1.19. Changes in parameters are sensed by the sensing unit, and digital signals are generated by the signal conditioning circuitry from electrical signals. Analog to digital conversion is performed, and this input is given to processing units or application programs. Task processing is done by the memory unit; communication with base station or sensors or sinks in WSN is the job of the transceiver (Chaudhari and Dharavath, 2014).

The five main parts of a sensor node are the following:

- Central unit: the central unit is a microprocessor that manages the tasks.
- Battery: this is the source of energy.

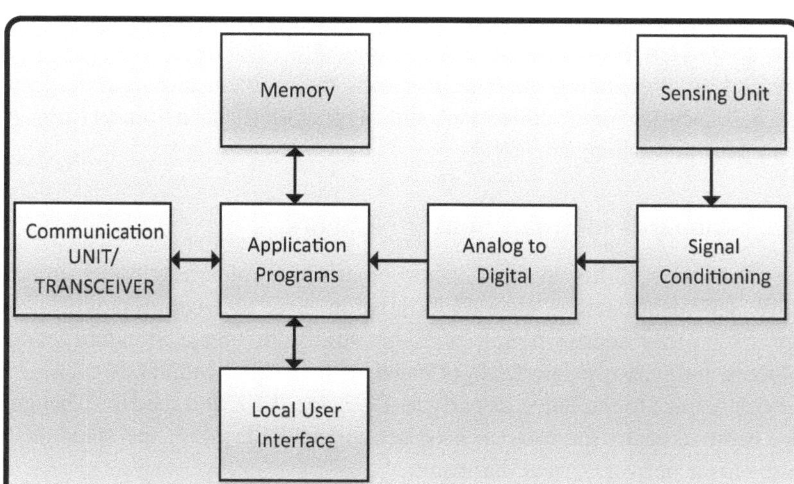

FIGURE 1.19 Basic architectural components of smart sensor node (Chaudhari and Dharavath, 2014).

- Transceiver: the transceiver interacts with the environment and collects data.
- Memory: this unit stores or processes data.
- Communication module: the module includes transceivers; it forwards queries and data to and from the central module.

Energy efficiency in all parts of a sensor network is crucial for long network lifetime. Nodes in the sensor network cooperate to spread the data processing task and send the processed information to sinks. RFID chips with no batteries are used to reduce the overhead costs of the power supply of all nodes.

Sensors are used to monitor parameters related to lighting conditions, noise levels, humidity, vehicle movement, soil makeup, mechanical stress levels, presence or absence of certain type of objects, and other properties.

Sensor node mechanisms can be thermal, visual, acoustic, radar, magnetic, or seismic. Self-identification and self-diagnosis is possible for a sensor node. The three mechanisms of sensor nodes are as follows:

- Working toward a predefined target by a line of sight (e.g., visual sensors).
- Working with space and time relationships to a target (e.g., seismic sensors).
- Working in a wavelike propagation with possible bending (e.g., acoustic sensors).

Unlike address-centric networks, sensor networks are data-centric. In such a network, queries are directed to a region containing a cluster of sensor nodes rather than specific sensor addresses. Data in a cluster (i.e., of a similar kind) are collected and aggregated locally. The aggregator node performs the functions of summarizing and analyzing the aggregated data in a cluster; this helps to reduce the communication bandwidth. The level of accuracy is increased by aggregating the data; this ends the problem of data redundancy. Fig. 1.20 shows data collection and aggregation in a sensor network (Chaudhari and Dharavath, 2014).

1.4.2 Standards of Smart Sensor Network

Other than their ability to sense, some other pluses of sensor nodes are reliability, self-identification, ease of installation, time and coordination knowledge with other sensor nodes, and the ability to create standard control protocols and network interfaces (Chaudhari and Dharavath, 2014).

With respect to the latter, as early as 1993, the IEEE and the NIST began to work on a standard for smart sensor networks. IEEE 1451, the standard for smart sensor networks, was the result.

The basic architecture of IEEE 1451 (Conway, 2003) is shown in Fig. 1.21. Major components are transducer electronic data sheet, transducer

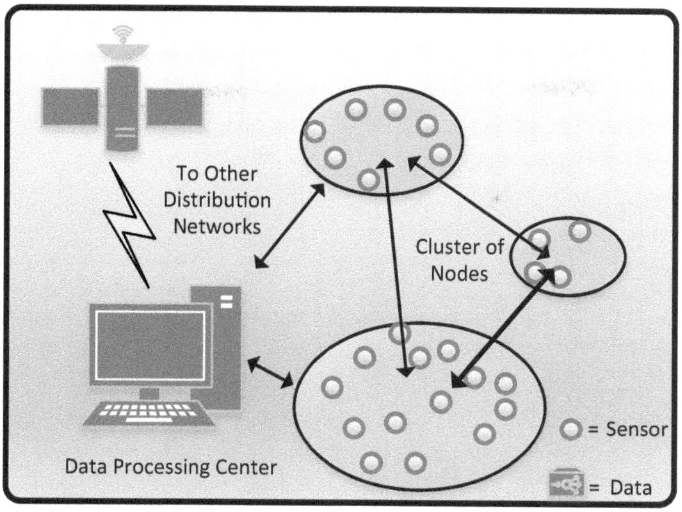

FIGURE 1.20 Data collection and aggregation in a wireless sensor network. *Adapted from Pathan, A-S.K., Hong, C.S., Lee, H-W., February 20–22, 2006. Smartening the environment using wireless sensor networks in a developing Country.*

intelligent interface, smart transducer interface module, and network capable application processor, as shown in Fig. 1.21. IEEE 1451 gives the formalized concept of a smart sensor: a smart sensor provides extra functions beyond those necessary for generating a correct representation of the sensed quantity (Frank, 2000). A general model of a smart sensor appears in Fig. 1.22. A smart sensor's objectives include integrating and maintaining the distributed sensor system, thus yielding cost-effectiveness; measuring smartly and intelligently; creating a common platform for computing, controlling, and communication toward a common goal; and interfacing various sensors of different types. A virtual sensor is the physical sensor/transducer plus the associated signal conditioning and DSP required to obtain reliable estimates of the required sensory information. The virtual sensor is a component of the smart sensor.

1.4.3 Importance and Adoption of Smart Sensor

The presence of a controller/processor in a smart sensor has led to corrections for undesirable sensor characteristics, including input offset and span variation, nonlinearity, and cross-sensitivity. As these are carried in software, no additional hardware is required, and calibration becomes an electronic process. Thus, it is possible to calibrate the batches of a sensor during production

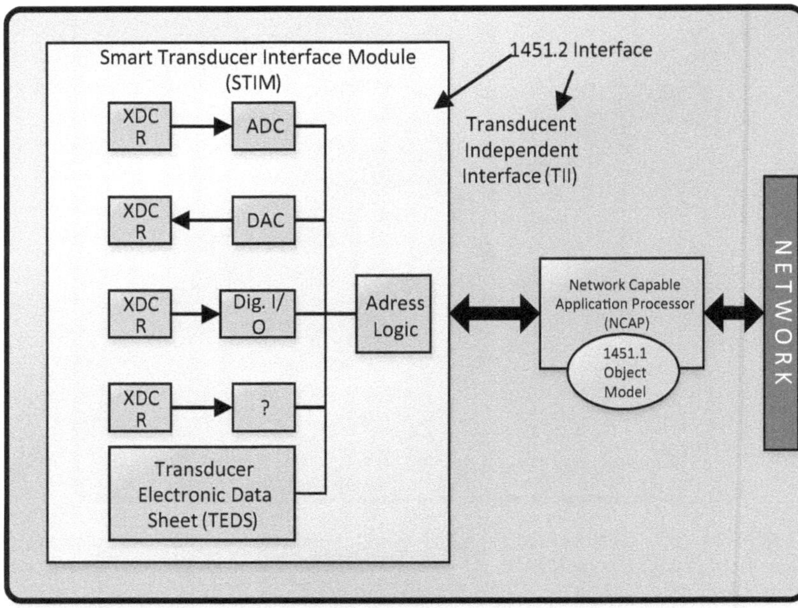

FIGURE 1.21 IEEE 1451 standard for smart sensor network. *ADC*, analog to digital converter; *DAC*, digital to analog converter; *XDCR*, cross datacenter replication. *Adapted from Lewis, F.L., 2005. Wireless Sensor Networks. Associate Director for Research Head, Advanced Controls, Sensors, and MEMS Group Automation and Robotics Research Institute, The University of Texas at Arlington.*

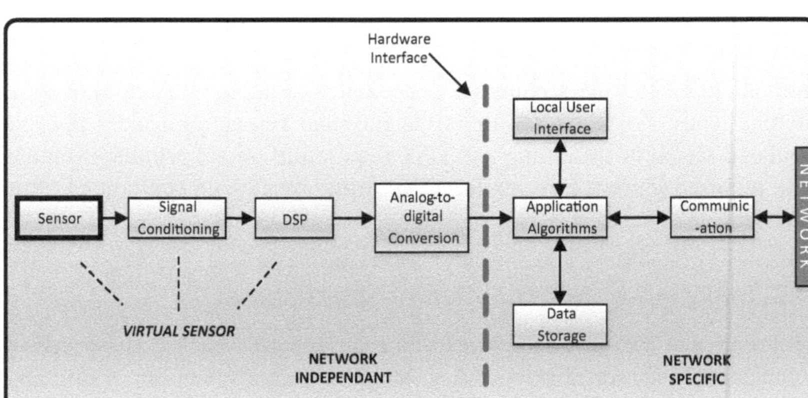

FIGURE 1.22 General model of a smart sensor. *DSP*, digital signal processing. *Adapted from Lewis, F.L., 2005. Wireless Sensor Networks. Associate Director for Research Head, Advanced Controls, Sensors, and MEMS Group Automation and Robotics Research Institute, The University of Texas at Arlington.*

without the need to remove the sensor from its current environment or test fixture (Smith and Bowen, 1995).

1.4.3.1 Cost Improvement

The main cost of the sensor is in its setup; this can be reduced by reducing the effort of setup and by removing repetitive testing (Dubey, 2002).

1.4.3.2 Reduced Cost of Bulk Cables and Connectors

The use of smart sensors has significantly reduced the cost of bulk cables and connectors needed to connect different blocks (i.e., electronic circuits) (Dubey, 2002).

1.4.3.3 Remote Diagnostics

Because the processor is in the package, it is possible to have digital communication via a standard bus and a built in self-test. This is very helpful in production tests of integrated circuits. The diagnostic can simply be a rules-based program running in the sensor (Smith and Bowen, 1995).

1.4.3.4 Enhancement of Application

Smart sensors enhance the following applications (Giachino, 1986):

- Self-calibration
- Computation
- Communication
- Multisensing

1.4.3.4.1 Self-Calibration

Self-calibration means adjusting some parameter of a sensor during fabrication, either gain or offset or both. Self-calibration adjusts the deviation of the output of a sensor from the desired value when the input is at minimum; alternatively, it can be an initial adjustment of gain. Because adjustments usually change with time, devices require recalibration. If it is difficult to recalibrate units once they are in service, the manufacturer overdesigns to ensure the device will operate within its specifications during its service life. This problem is solved by a smart sensor; its built-in microprocessor has the correction functions in its memory.

1.4.3.4.2 Computation

Computation allows us to obtain the average, variance, and standard deviation for a set of measurements. This is easy with a smart sensor. Its computational ability allows it to compensate for environmental changes such as temperature and to correct for changes in offset and gain.

1.4.3.4.3 Communication

Communication is the means of exchanging or conveying information, something easily accomplished by a smart sensor. This is very helpful, especially as the sensor can broadcast information about its own status and any measurement uncertainty.

1.4.3.4.4 Multisensing

Some smart sensors can measure more than one physical or chemical variable simultaneously. A single smart sensor can measure pressure, temperature, humidity, gas flow, infrared, chemical reactions, surface acoustic vapor, etc. (Giachino, 1986).

1.4.3.5 System Reliability

System reliability is significantly improved with the use of smart sensors, first, because of the reduction in system wiring and, second, because of the ability of the sensor to diagnose its own faults and their effect (Giachino, 1986).

1.4.3.6 Better Signal to Noise Ratio

The electrical output of most sensors is very weak, and if this is transmitted through long wires, noise may be coupled. This problem can be avoided by employing a smart sensor (Dubey, 2002).

1.4.3.7 Sensor Improvement

- Nonlinearity: Many sensors show some nonlinearity; by using on-chip feedback systems or look-up tables, we can improve linearity.
- Cross-sensitivity: Most sensors show an undesirable sensitivity to strain and temperature. Incorporating relevant sensing elements and circuits on the same chip can reduce the cross-sensitivity.
- Offsetting: Offset adjustment requires expensive trimming procedures, and even this tends to drift. This is reduced by using a sensitivity reduction method.
- Parameter drift and component values: These are functions of time and can be solved by automatic calibration (Middelhoer and Hoogerwerf, 1985).

1.4.4 General Architecture of Smart Sensor

It is easy to propose a general architecture of a smart sensor based on its definition and functions. The definition of smart sensor suggests it is similar to a data acquisition system, the only difference being the presence of a complete system on a single silicon chip. In addition, it has on-chip offset and temperature compensation. The general architecture of a smart sensor has the following important components:

- Sensing element/transduction element,
- Amplifier,
- Sample and hold,
- Analog multiplexer,
- Analog to digital converter (ADC),
- Offset and temperature compensation,
- Digital to analog converter,
- Memory,
- Serial communication, and
- Processor.

1.4.5 Description of Smart Sensor Architecture

The generalized architecture of a smart sensor is depicted in Fig. 1.23. In the figure, A1, A2...An and S/H1, S/H2...S/Hn are the amplifiers and sample and hold circuit corresponding to different sensing elements, respectively.

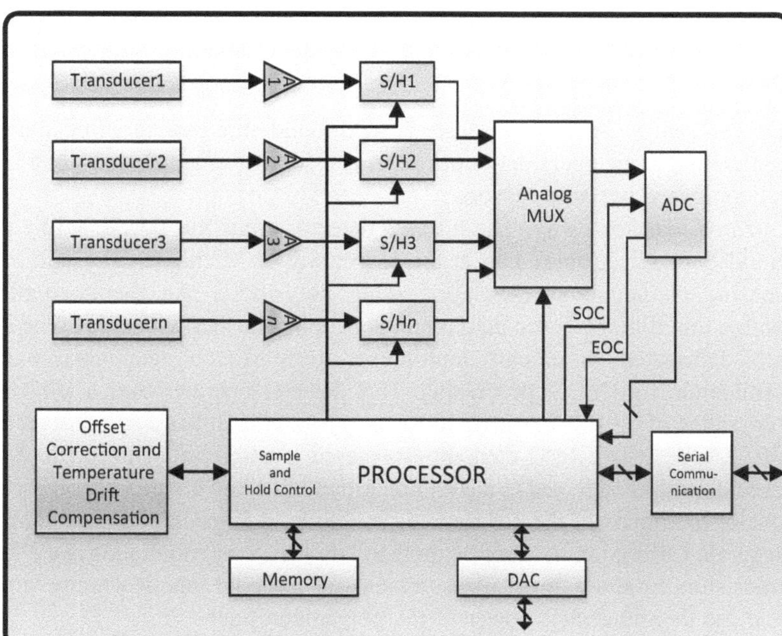

FIGURE 1.23 General architecture of smart sensor. *ADC*, analog to digital converter; *EOC*, end of conversion signal; *DAC*, digital to analog converter; *MUX*, Multiplexer; *SOC*, start of conversion signal; *S/H*, sample and hold. *Adapted from Dubey, D., 2002. Smart sensors. M.Tech.credit seminar report, electronic systems group, EE Dept, IIT Bombay, Submitted November 2002. 02307402.*

To get a digital form of an analog signal, the analog signal is periodically sampled (its instantaneous value is acquired by the circuit); that constant value is held and is converted into digital words. Any type of analog to digital converter (ADC) must contain or be preceded by a circuit that holds the voltage at the input to the ADC constant during the entire conversion.

Conversion times vary widely, from nanoseconds (for flash ADCs) to microseconds (successive approximation ADC) to hundreds of microseconds (for dual slope integrator ADCs). An ADC starts conversion when it receives a start of conversion signal from the processor; after conversion is complete, it gives an end of conversion signal to the processor. Outputs of all the sample and hold circuits are multiplexed together so that we can use a single ADC; this will reduce the cost of the chip. Offset compensation and correction require one ADC to measure a reference voltage and another for the zero. Dedicating two channels of the multiplexer and using only one ADC for whole system can avoid the addition of the second ADC. This is helpful in offset correction and zero compensation of gain due to temperature drifts of the acquisition chain.

In addition, the smart sensor includes an internal memory so it can store the data and program required (Dubey, 2002).

1.4.6 Varieties of Smart Sensors

Some of the smart sensors developed at various research institutes are the following:

Optical sensor: An optical sensor is used to measure exposure in cameras, optical angle encoders, and optical arrays. Similar examples are load-cell silicon-based pressure sensors.

Infrared detector array: The infrared detector array was developed by the Solid State Electronics Lab at the University of Michigan. The infrared-sensing element was developed using polysilicon–Au thermocouples, with a thin film dielectric diaphragm to support the thermocouples. The lab also fabricated an on-chip multiplexer using silicon gate metal oxide semiconductor (MOS) processing. This detector operates over a temperature range of 0–100°C with a 10-ms response time. It has a responsiveness of 12 V/W. It is a 16*2 element–staggered linear array, with one lead of each detector connected to a common ground line and another connected to one of the inputs of the 16*1 analog multiplexer. This chip also contains a separate calibration thermopile, polysilicon resistors, and diodes and MOS transistors to allow direct measurements of the cold junction temperature and the thermoelectric power of the polysilicon lines.

Accelerometer: The accelerometer fabricated at the IBM Research laboratory at San Jose, California, consists of a sensing element and electronics on silicon. The accelerometer itself is a metal-coated SiO_2 cantilever beam fabricated on a silicon chip where the capacitance between the beam and the substrate provides the output signal.

Integrated multisensor: An integrated multisensor chip developed at the Electronics Research Laboratory at the University of California contains MOS devices for signal conditioning with an on-chip sensor, a gas flow sensor, an infrared-sensing array, a chemical reaction sensor, a cantilever beam, an accelerometer, a surface acoustic wave vapor sensor, a tactile sensor array, and an infrared charge–coupled device imager. This chip was fabricated using conventional silicon planer processing, silicon micromachining, and thin deposition techniques (Giachino, 1986).

1.4.7 Smart Sensors for Condition Based Maintenance

CBM is a well-known concept; its advantages include its ability to generate prognostics for failure avoidance and its reduction of maintenance costs.

That being said, another breakthrough in CBM is expected with the development of information and communications technology and embedded systems. This technology has the potential to supply integrated chips implementing all the necessary circuitry to manage field data capture, data processing, local diagnosis, local feedback (where possible), and information transfer to the upper control levels. In the near future, smart sensors will exploit new technologies of microsensors (MEMS) and wireless communication with the computing power of a microprocessor.

The new technologies will support the decision-making process and reduce costs; by properly monitoring a plant, they will help to manage maintenance, production, and logistic issues. Service-oriented architecture (SOA) is presently being analyzed by many researchers; it promises to be an interesting solution for many issues related to controlling the plant. Some European-funded SOA projects and the issues related to monitoring plant condition are mentioned by Pala et al. (2014).

Smart sensors may be adapted to monitor activities; they can be used to support CBM, in particular, state detection (SD). A CBM program has the following modules (MIMOSA OSA-CBM, Bengtsson, 2003):

1. Data acquisition module: collects data and properly formats them to store/transmit information to upper levels.
2. Data manipulation module: cleans and preprocesses data; typical operations are normalization, smoothing, outlier removal, missing data imputation, etc.
3. SD module: monitors the machine state by checking if the machine parameters are compliant with target ranges; it can generate alarms and warnings when compliancy is not reached.
4. Health assessment (HA) module: receives data from the SD module or other HA modules and detects if the health state of the system or subsystem has degraded; it can also suggest possible fault causes.

5. Prognostic assessment (PA) module: works on the results of the previous modules; it is used to calculate the future health of an asset and the remaining useful life by taking into account the future usage profile.
6. Advisory generation (AG) module: receives data from HA and PA modules and generates suggestions on recommended action(s) related to the maintenance of how to run the asset under actual conditions.
7. Presentation module: presents the outputs from the SD, HA, PA, and AG modules to the user through a human–machine interface.

At a glance, the importance of smart sensors for maintenance rests on the possibility of building or configuring custom devices dedicated to equipment monitoring and diagnostics. Smart sensors are enhanced by the use of MEMS technology, as it allows many different types of microsensors to be included in a device or a single chip. In other words, the broad spectrum of applications and the power of a comprehensive data acquisition system are made available in the small and self-contained package of the smart sensor (Garetti et al., 2007). This allows data acquisition and initial data manipulation to be concentrated in the sensors, leaving SD to other levels of the architecture (Pala et al., 2014).

1.4.7.1 What is OSA-CBM?

The OSA-CBM specification is a standard architecture for moving information in a CBM system. A closer look reveals it can reduce costs, improve interoperability, increase competition, incorporate design changes, and ensure further cooperation in the realm of CBM.

Project managers implementing CBM systems must take on the task of integrating a wide variety of software and hardware components, as well as developing a framework for these components. OSA-CBM simplifies this process by specifying a standard architecture and framework. It describes the six functional blocks of CBM systems and the interfaces between those blocks. The architecture provides a means to integrate many disparate components and eases the process by specifying the inputs and outputs of the components. In short, it gives a standardized information delivery system for CBM. It describes the information that is moved around and shows how to move it. It also has built-in metadata to describe the processing.

OSA-CBM is defined using Unified Modeling Language and is designed as a "multitechnological implementation," meaning that it separates the information that can be exchanged in a CBM system from the technical interfaces system integrators use to communicate the information. Vendors and integrators can implement the architecture using the appropriate technology for their environment. For example, while aircraft health management vendors may elect to use a "real-time" implementation, a vendor developing a portable maintenance aid may elect to implement the standard using XML and web services (MIMOSA OSA-CBM).

1.4.7.2 Benefits of OSA-CBM

Cost: OSA-CBM can provide significant cost savings because system integrators and vendors will not have to spend time creating new or proprietary architectures. Savings will also come from not being committed to single vendors developing entire CBM systems. As the standard is broken into functional components, multiple vendors may compete to develop select blocks of functionality.

Specialization: When vendors are not constrained to providing an entire CBM system, they can concentrate on one or a few areas. The increase in specialization will allow better algorithms and technologies to be developed. Smaller companies that could not provide an entire CBM system can now specialize in one or more of the functional blocks.

Competition: OSA-CBM allows all vendors to use the same input and output interfaces. The separation of functionality from how the information is presented to other applications allows direct comparison of the developed functionalities. Competition now occurs at a functional level, not a system or total solution level.

Cooperation: Competition may increase, but so too can cooperation. The sectioning of CBM into separate independent blocks will allow vendors to work on separate modules. Because the architecture also defines the interfaces, each module will be able to communicate with the others seamlessly if developed using the same technologies (MIMOSA OSA-CBM).

1.5 ENERGY HARVESTING FOR SENSORS AND CONFIGURATION ISSUES

Energy harvesting, also known as "energy scavenging," "parasitic energy," or "microgenerators" in the literature, is a process performed by a conversion mechanism for generating electric power from available ambient energy sources. Incident light, thermal gradients, machine vibrations, and human body functions are well-known examples of ambient energy sources and receive attention from many researchers. Since energy harvesting systems offer a maintenance-free, long-lasting, green power supply for many portable, low-powered electronic devices, they are likely to become an essential part of power management systems (Erdogan, 2011).

1.5.1 Energy Harvesting Sources

Popular ambient energy sources available in our environment include the following:

- Mechanical (vibration, pressure...etc.)
- Thermal (energy from heat)

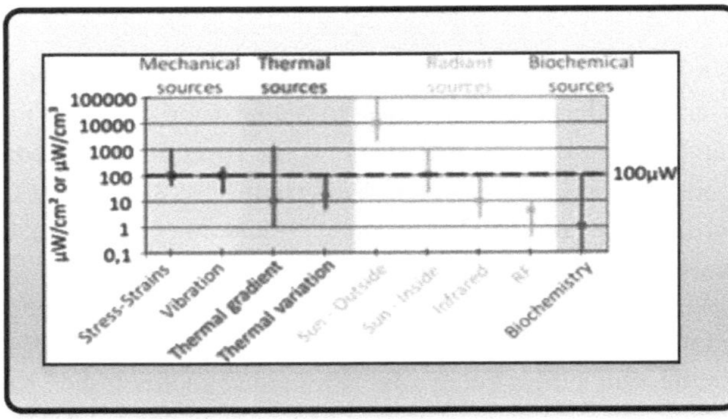

FIGURE 1.24 Ambient energy power source before conversion *Adapted from Boisseau, S., Despesse, G., Ahmed Seddik, B., 2012. Electrostatic Conversion for Vibration Energy Harvesting, Small-Scale Energy Harvesting, Intech.*

- Radiated (solar, infrared RF)
- Chemical
- Nuclear etc.

Each of these sources is characterized by different power density. Fig. 1.24 gives a breakdown of the energy density before conversion (Nwogu, 2015).

As might be expected, radiant sources perform better outdoors than indoors. The ambient energy must be selected based on the sensor's environment. A concise energy harvesting block diagram is shown in Fig. 1.25 (Nwogu, 2015).

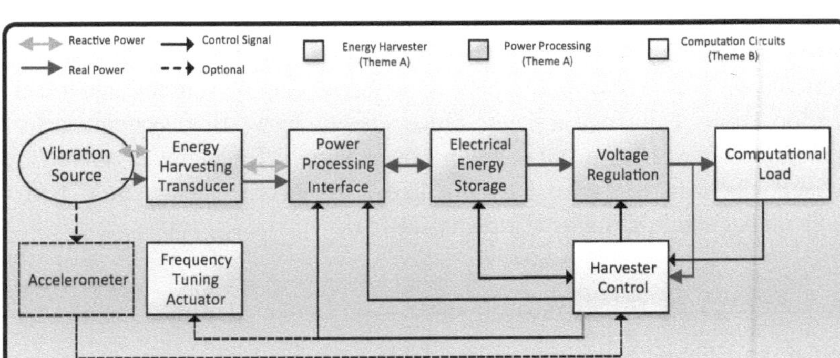

FIGURE 1.25 Energy harvesting system block diagram. *Adapted from Nwogu, K., 2015. Energy Harvesting and Storage: The Catalyst to the Power Constraint for Leveraging Internet of Things (IoT) on Trains. Theses, Dissertations, & Student Research in Computer Electronics & Engineering. Paper 33. University of Nebraska – Lincoln.*

1.5.2 Energy Harvesting for Microelectromechanical Systems

The growing interest in the field of energy harvesting systems is also due to developments in related technologies such as MEMS technology, WSN technology, very large scale integration (VLSI) design technology, and complementary metal oxide semiconductor (CMOS) fabrication technology. Sensors are emerging as a promising application of energy harvesting techniques, wherein these state-of-the-art technologies work together, as shown in Fig. 1.26 (Erdogan, 2011).

While advances in microfabrication techniques have allowed the development of various MEMS sensors that integrate intelligent electronic control systems with a mechanical system on the microscale, VLSI design and CMOS fabrication techniques have reduced power requirements of the sensors to the range of tens to hundreds of microwatts. Such low power dissipation opens up the possibility of powering sensors by scavenging ambient energy from the environment, eliminating the need for batteries and extending the lifetimes indefinitely (Williams and Yates, 1995).

In addition, WSNs consisting of large numbers of sensor nodes capable of wireless communication make it possible to locate sensor nodes in remote and sometimes unrecoverable locations. With the ongoing development of WSNs, long-lasting wireless power supplies have become a must in applications; it is now impractical to use and replace batteries, even long-lived types. The labor needed and other costs associated with changing hundreds of batteries make systems such more expensive and less efficient. Furthermore, because the size of sensor nodes is critical, nodes not bigger than a dime are required in many applications where batteries might be too bulky.

In Table 1.6, some battery types and most applicable energy conversion mechanisms are compared with respect to their long-term and short-term power densities. Power density, or the amount of average energy generated

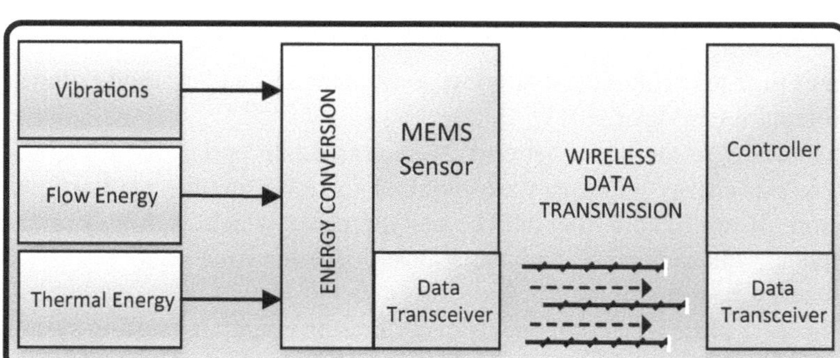

FIGURE 1.26 Wireless self-powered sensor nodes. *MEMS*, microelectromechanical Systems. *Adapted from Erdogan, G., 2011. Energy Harvesting for Micro-Electromechanical-Systems (MEMS). Conference Proceedings, Chicago.*

TABLE 1.6 Comparison of Energy Sources

	Power Density (μW/cm³) 1 year Lifetime	Power Density (μW/cm³) 10 year Lifetime
Solar (outdoors)	15,000—Direct sun, 150—Cloudy day	15,000—Direct sun, 150—Cloudy day
Solar (indoors)	6—Office desk	6—Office desk
Vibrations (piezoelectric conversion)	250	250
Vibrations (electrostatic conversion)	50	50
Acoustic noise	0.003 at 75 dB, 0.96 at 100 dB	0.003 at 75 dB, 0.96 at 100 dB
Temperature gradient	15 at 10 °C gradient	15 at 10 °C gradient
Shoe inserts	330	330
Batteries (nonrechargeable lithium)	45	3.5
Batteries (rechargeable lithium)	7	0
Hydrocarbon fuel (microheat engine)	333	33
Fuel cells (methanol)	280	28

Adapted from Williams, C.B., Yates, R.B., 1995. Analysis of a micro-electric generator for microsystems. Proceedings of the Transducers 95/Eurosensors IX 341–344.

per unit time and volume, is the most convenient and widely used criterion in the literature (Erdogan, 2011).

According to the table, batteries are a reasonable option for 1-year applications, but energy harvesters are required for long-lifetime applications. In addition, if we ignore the direct sunlight case, which is highly ambient dependent, "piezoelectric" and "shoe insert" mechanisms stand out as vibration energy conversion mechanisms.

Fig. 1.27 places the components of a vibration energy harvesting system in a flow chart. This flow chart can be generalized for all energy harvesting systems in which an energy source, a conversion device, a conditioning circuit, and an electric load are the main components of the general energy harvesting system (Erdogan, 2011).

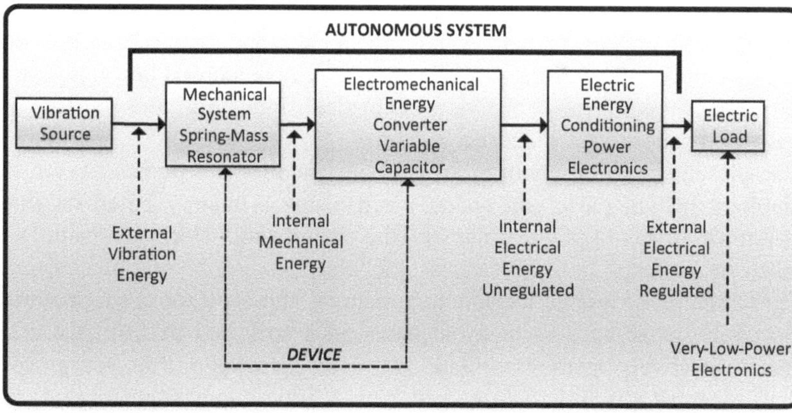

FIGURE 1.27 Vibration energy harvesting system components. *Adapted from Erdogan, G., 2011. Energy Harvesting for Micro-Electromechanical-Systems (MEMS). Conference Proceedings, Chicago.*

The general system basically aims to accomplish five consecutive tasks:

- Collect the maximum energy from the energy source.
- Convert the ambient energy into electric energy efficiently.
- Rectify and store the maximum amount of electric energy.
- Regulate the output voltage level depending on the application.
- Transmit the electric energy to the load when it is required.

Constructing mathematical models and manufacturing prototypes to estimate or ameliorate the efficiency and the performance of an energy conversion mechanism are major areas of research. Many energy harvesting mechanisms are at a very early stage of being prototyped; more efficient systems will be obtained in the future by optimizing the tasks listed above.

The availability of the ambient sources, the power densities of the converters, the duty cycles, and the power needs of the electric loads, however, are the primary limitations of energy harvesting systems. For example, solar cells can generate excellent power densities in direct sunlight, but they need to be optimized for conditions of dim light or no light at all. Thermoelectric energy converters need large energy gradients to generate substantial power. Power delivery and user comfort are critical when generating power by means of body functions such as breathing, blood pressure, walking, etc. Since there are abundant and continuous vibration sources, such as machinery, waves, and wind, vibration energy converters seem to be the most promising mechanism thus far (Erdogan, 2011).

Energy harvesting from a vibration source for low power electronic devices and sensors is an appealing idea, as there are various kinds of vibration sources around us, ranging from the wind and sea waves to human body motion and vibrating machinery. Vibration sources are usually preferable to incident light or thermal gradient energy sources requiring an appropriate operating time and running condition. Therefore, research programs have focused on "vibration to electric energy converters" for various medical, industrial, and military applications for more than a decade.

Despite the large variety of prototypes designed for this purpose so far, the technology behind these conversion mechanisms is mainly based on three well-known effects in physics, namely, the electrostatic, electromagnetic, and piezoelectric effects. In brief, electrostatic, electromagnetic, and piezoelectric designs require a variable capacitor, a magnet, and a piezoelectric material, respectively, inducing a voltage on plates, in a coil, and between the electrodes, respectively, as they oscillate. However, the design of an energy converter, especially in microscale, becomes a little more sophisticated and therefore attractive for the researchers when the system emerges as a vibration energy dissipation problem needing to be examined for various aspects to achieve a maximum power density and efficiency. While some generators have already been fabricated using MEMS techniques, others have been made on a mesoscale with the intention of later miniaturizing the devices using MEMS (Roundy et al., 2003).

1.5.3 Harvesting Methods

1.5.3.1 Solar Energy

Typically, a sensor node that uses energy from nature consists of several blocks, including an energy harvesting module, a storage unit such as a supercapacitor, a sensing element, a microcontroller, and a transmitter, as shown in Fig. 1.28 (Akbari, 2014).

Photovoltaic technology generates direct current from semiconductors when they are illuminated by photons. Most of today's solar cells are made of crystalline silicon (c-Si). Silicon is the most widely used material in the production of photovoltaic modules. Mono-crystal, multi-crystal, micro-crystal, and amorphous silicons are the most widely used in the production of photovoltaic cells (Jayarama Reddy, 2012).

Nanotechnology-based semiconductors, including nanowires, nanocrystals, nanorods, and nanodots, have been opening up new possibilities for scientists to design solar cells.

Solar energy can generate enough power for wireless sensors (Akbari, 2014).

Solar energy harvesting provides direct DC voltage; therefore, additional circuit rectifications are not required. This type of energy scavenging is free of emissions, as it does not produce contaminants or bypass products that are harmful to the environment (Ma and Bondade, 2011).

One of the limitations of solar energy is its dependency on solar radiation, which can be degraded in areas where enough sunlight is not available. Solar

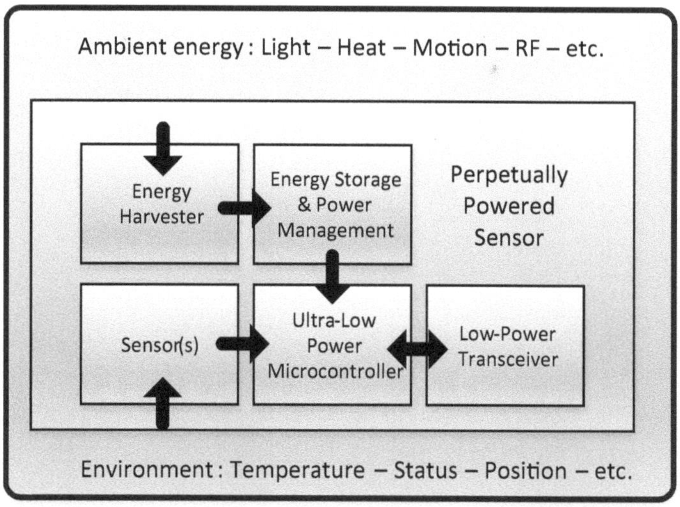

FIGURE 1.28 Wireless sensor network with an energy harvesting module. *Adapted from Raju, M., Grazier, M., 2010. Energy Harvesting. ULP Meets Energy Harvesting: A game-Changing Combination for Design Engineers. Texas Instrument. http://www.ti.com.cn/cn/lit/wp/slyy018a/slyy018a.pdf.*

energy is characterized by its variable nature. Another issue associated with solar cells, and a topic of ongoing research, is the cooling of solar cells and recovery of heat (Akbari, 2014).

1.5.3.2 Piezoelectricity

The word piezoelectricity links the Greek word *piezo* meaning pressure with electricity. The main advantage of piezoelectric materials, as shown in Fig. 1.29, is the high amount of voltage they can provide. Some materials having a piezoelectric effect are quartz, soft and hard lead zirconate titanate piezo ceramics (PZT-5H and PZT5A), barium titanate ($BaTiO_3$), and polyvinylidene fluoride (PVDF) (Beeby and White, 2010).

In the piezoelectric effect, the usable output voltage can be obtained directly from the material; there is no need to apply multistage postprocessing to generate the desired amount of voltage (Erturk and Inman, 2011). Piezoelectric materials require dynamic forces to retain the output voltage; a notable drawback of piezoelectric sensors is their inability to respond to static loads (Karwowski and Marras, 1998).

1.5.3.3 Radio Frequency Energy

Harvesting RF energy can permit the wireless charging of a sensor node (Smith, 2013).

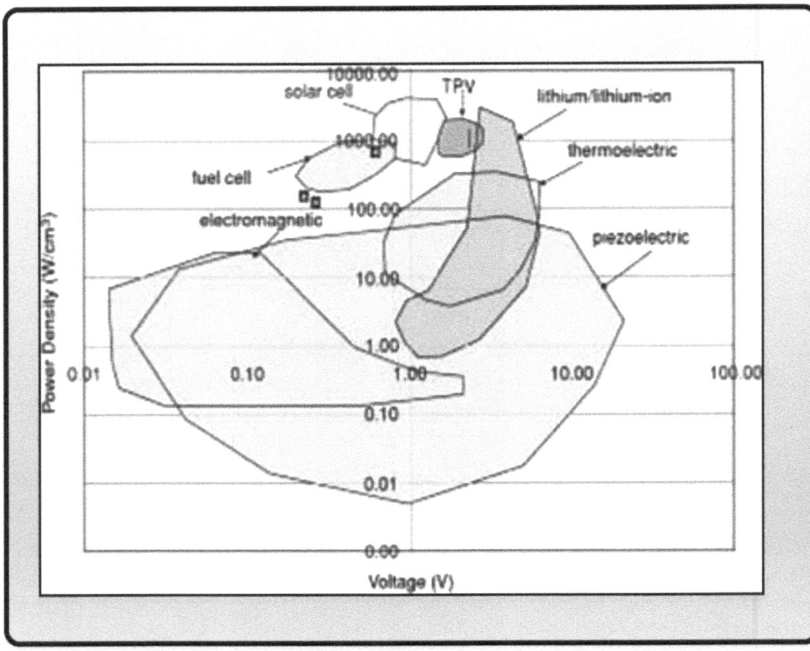

FIGURE 1.29 Voltage generated by piezoelectric material compared to other harvesting sources. *TPV*, novel thermophotovoltaic. *Adapted from Erturk, A., Inman, D.J., 2011. Piezoelectric Energy Harvesting. UK, John Wiley & Sons, 2011, 2–3.*

With a transmitter set, it is possible to harvest energy from radio waves. The advantage of this alternative is that the scavenging mechanism can be flexible, and it is possible to control the amount of transferred energy by making it continuous, i.e., on regular intervals; alternatively, the amount of radiated energy can be adjusted according to the requirements of the relevant application (Ostaffe and Tollefson, 2011).

Note that according to the Friis equation, the amount of power collected at the receiver side is not equal to the exact value sent by transmitter (Vias et al., 2012).

1.5.3.4 Thermal Energy

Thermal energy produces electricity when there is a temperature difference.

Most thermoelectric generators consist of an array of p- and n-type semiconductors in series. Heat from a hot source reaches the hot shoe and is conducted to the semiconductors. The hot shoe has a high thermal conductivity and a high electrical conductivity. Electrons are released from n-type semiconductors and establish a current through p-type semiconductors. Cold shoes transfer the current to the electric load (Kroos and Potter, 2014).

Commercially available thermoelectric generators require a temperature difference of 10–200 °C.

This type of energy is able to provide DC power continuously, but temperature differentials can be difficult to generate in enclosed environments (Akbari, 2014).

1.5.3.5 Wind Energy

Wind energy is a remarkable macroscale source of electricity. Countries such as Germany, Spain, and Denmark are now providing a significant amount of their required electricity by using wind energy.

It has possibilities for high-power electronics but is also relevant for miniaturized devices (Akbari, 2014).

Park and colleagues tested different wind turbine blade configurations (two, three, and six blade propellers) within a wind speed interval of 3.5–7 m/s. They found the propeller with six blades outputs the maximum power. The authors proposed adding a solar panel to the harvesting system in case the amount of energy provided by wind energy does not suffice (Park et al., 2012).

An experiment on the feasibility of using wind energy as a complementary system for wireless sensor systems tested two types of blades (Akbari, 2014). The measured wind speed was 3.7 and 4.5 m/s. In the first case, the length of each blade was 4.35 cm and in the second one, 8.85 cm. Fig. 1.30 shows power generated by the wind turbine as a function of load. The peak power for the

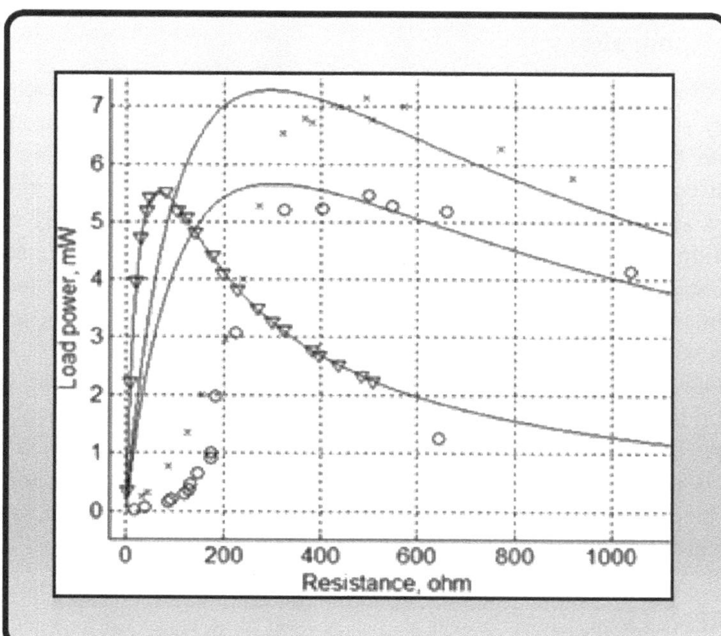

FIGURE 1.30 Power versus resistance for two types of propellers. *Adapted from Akbari, S., 2014. Energy harvesting for wireless sensor networks review. "MATI" – Russian State Technological University Orshanskaya 3, 121552, Moscow, Russia. Proceedings of the 2014 Federated Conference on Computer Science and Information Systems 987–992.*

propeller with smaller blades at the speed of 4.5 m/s is 7.2 mW. At 3.7 m/s, the peak value of power for the first type of blade is 5.5 mW and for the second type of blade, the maximum amount of power at the speed of 4.5 m/s is approximately 5.5 mW. As the rotor of the DC motor starts rotating, the AC current flows in a circuit consisting of motor windings and the load.

Self-induction takes place as a result. This effect can be seen in Fig. 1.30 for the first type of propeller when the load resistance is less than 500 Ω; in the meantime, the experimental points and the approximation do not converge. The experiment indicates that a microturbine based on a DC motor can be represented as a voltage source and an internal resistance according to the Thevenin theorem. Two factors causing less power were wind speed and motor. In Fig. 1.30, the larger propeller provides maximum output at smaller values of load (~60 Ω) and the smaller one reaches its power peak values at higher loads (~250 Ω). In other words, depending on the load value, the necessary blade can be chosen (Park et al., 2012).

This model is valid only in cases when the induced current is greater than the self-induction current. Wind is a free energy provided by nature, and there is no limit to its use. It produces no polluting emissions of any kind. In some environments, wind energy cannot be considered as a primary source of electricity; for example, if its behavior is intermittent, this causes instability in the power system (Wagner and Mathur, 2013).

1.5.4 Applications

The above alternative sources can be used, along with batteries, as complementary systems. If enough energy can be harvested from the environment, they can be employed as primary power sources. Pipeline monitoring in industrial complexes is a remarkable example of the possible deployment of wireless sensors (Yu and Guo, 2012). It is quite plausible to employ energy harvesting techniques as primary or secondary power supplies for the sensors in this scenario. In the event of battery replacement, the harvesting technologies increase the possibility of continuous operation of the system, and the process will not be stopped (Akbari, 2014).

Bridge health monitoring is another example discussed by Reyer. In the Jindo bridge project, 8 sensors out of 70 are equipped with solar panels. The charging process of the solar system is working well, except for one node, which is not receiving enough sunlight because of its location. In the next deployment, a more sensitive solar panel may be used or another type of energy harvesting may be considered (Cao and Liu, 2012).

REFERENCES

Agre, J.R., Clare, L.P., Pottie, G., Romanov, N., 1999. Development platform for distributed microsensor networks. In: Proceedings of SPIE's 13th Annual International Symposium on Aerospace/Defense Sensing, Simulation, and Controls Conference, Orlando, FL, April 1999.

Akbari, S., 2014. Energy harvesting for wireless sensor networks review. In: Proceedings of the 2014 Federated Conference on Computer Science and Information Systems, pp. 987–992. "MATI" – Russian State Technological University Orshanskaya 3, 121552, Moscow, Russia.

Alami, R., Chatila, R., Fleury, S., Ghallab, M., Ingrand, F., April 1998. An architecture for autonomy. International Journal of Robotics Research 17 (4), 315–337.

Bakken, D.E., Zhan, Z., Jones, C.C., Karr, D.A., 2001. Middleware support for voting and data fusion. In: Proceedings of the International Conference on Dependable Systems and Networks, pp. 453–462. Gothenburg, Sweden.

Balasubramanian, S., Maturana, F., Vasko, D., Lenner, J., October 5–8, 1999. A Hybrid Autonomous Control Architecture, ISA-tech/1999 Conference, Philadelphia, PA.

Bass, T., May 2000. 2000. Intrusion detection systems and multisensor data fusion: creating cyberspace situational awareness. Communications of the ACM 43 (4), 99–105.

Bedworth, M.D., O'Brien, J., July 1999. The omnibus model: a new architecture for data fusion?. In: Proceedings of the 2nd International Conference on Information Fusion (FUSION' 99), Helsinki, Finnland.

Beeby, S., White, N.M., 2010. Energy Harvesting for Autonomous Systems. Norwood, MA, pp. 99–100.

Bengtsson, M., August 27–29, 2003. Standardization issues in condition based maintenance. In: Condition Monitoring and Diagnostic Engineering Management – Proceedings of the 16th International Congress.

Boisseau, S., Despesse, G., Ahmed Seddik, B., 2012. Electrostatic Conversion for Vibration Energy Harvesting. Small-Scale Energy Harvesting, Intech.

Bokare, M., Ralegaonkar, A., December–January 2012. Wireless sensor network: a promising approach for distributed sensing tasks. Excel Journal of Engineering Technology and Management Science I (1), 2249–9032.

Bosse, E., Roy, J., Grenier, D., May 1996. Data fusion concepts applied to a suite of dissimilar sensors. Canadian Conference on Electrical and Computer Engineering (2), 692–695.

Boyd, J.R., May 1987. A Discourse on Winning and Losing. Unpublished Set of Briefing Slides. Air University Library, Maxwell AFB, AL, USA.

Brignell, J.E., 1989. Smart sensors: a comprehensive survey. Ed. W., Gopel, J., Hesse, J.N., Zemel, vol, 1. New York, VCH.

Brooks, R.R., Iyengar, S.S., 1998. Multi-Sensor Fusion: Fundamentals and Applications. Prentice Hall, New Jersey.

Browne, J., July 2000. Wireless sensors leverage bluetooth communications standard. Wireless Systems Design Vol. 5 (7). ISSN 1089-5566, Penton Media Inc.

Brussel, V., Hendrik, 1994, December 1, 1994. Holonic Manufacturing Systems, The Vision Matching Problem, First European Conference on Holonic Manufacturing Systems. Germany, Hannover.

Calvo, M., Hannifin Parker Corporation, 2016. Sensors Drive Intelligence into Maintenance, 08/04/2016. http://www.plantengineering.com/single-article/sensors-drive-intelligence-into-maintenance/c4d668a6bc99a1feee97f88cda6e67eb.html.

Cao, J., Liu, X., 2012. Structural Health Monitoring Using Wireless Sensor Networks in Mobile and Pervasive Computing in Construction. Chimay, J., Anumba, Xiangyu Wang, Ed. UK, John Wiley & Sons, 224, DOI: 10.1002/9781118422281.ch11.

Chaudhari, M., Dharavath, S., January 2014. Study of smart sensors and their applications. International Journal of Advanced Research in Computer and Communication Engineering 3 (1).

Chong, C.-Y., Kumar, S.P., August 2003. Sensor networks: evolution, opportunities, and challenges. Proceedings of the IEEE 91 (No. 8).

Clare, L.P., Pottie, G., Agre, J.R., 1999. Self-organizing distributed microsensor networks. In: Proceedings of SPIE's 13th Annual International Symposium on Aerospace/Defense Sensing, Simulation, and Controls Conference, Orlando, FL, April 1999.

Conway, H., 2003. University of Limerick. http://wwww.ul.ie/~pei.

Corella, J., 2003. Tactical Automated Security System (TASS): Air Force Expeditionary Security. Presented at the SPIE Conf. Unattended Ground Sensor Technologies and Applications, Orlando, FL.

Dasarathy, B.V., January 1997. 1997. Sensor fusion potential exploitation-innovative architectures and illustrative applications. Proceedings of the IEEE 85, 24–38.

Dasarathy, B.V., 2001. Information fusion – what, where, why, when, and how? Information Fusion 2 (2), 75–76.

Discenzo, F.M., Loparo, K.A., Chung, D., Twarowski, A., 2001a. Intelligent Sensor Nodes Enable a New Generation of Machinery Diagnostics and Prognostics. Virginia Beach, Virginia, April 2–5, 2001.

Discenzo, F.M., Nickerson, W., Mitchell, C.E., Keller, K.J., 2001b. Open systems architecture enables health management for next generation system monitoring and maintenance. Development Program White Paper from Open System Architecture for Condition-based Maintenance (OSA-CBM). Development Group.

Dubey, D., 2002. Smart sensors. M.Tech. Credit Seminar Report, Electronic Systems Group, EE Dept, IIT Bombay, Submitted November 2002. 02307402.

Durrant-Whyte, H.F., December 1988. Sensor models and multisensor integration. International Journal of Robotics Research 7 (6), 97–113.

Elmenreich, W., November 19, 2002. An Introduction to Sensor Fusion. Institut für Technische Informatik Vienna University of Technology, Austria. Research Report 47/2001. wil@vmars.tuwien.ac.at.

Elmenreich, W., Peti, P., May 2002. Achieving dependability in a time-triggered network by sensor fusion. In: Proceedings of the 6th IEEE International Conference on Intelligent Engineering Systems (INES), pp. 167–172. Opatija, Croatia.

Elmenreich, W., Pitzek, S., November–December 2001. Using sensor fusion in a time-triggered network. In: Proceedings of the 27th Annual Conference of the IEEE Industrial Electronics Society, 1, pp. 369–374. Denver, CO, USA.

Erdogan, G., 2011. Energy harvesting for micro-electromechanical-systems (MEMS). In: Conference Proceedings, Chicago.

Erturk, A., Inman, D.J., 2011. Piezoelectric Energy Harvesting. John Wiley & Sons, UK, pp. 2–3.

Estrin, D., Govindan, R., Heidemann, J., Kumar, S., 1999. Next century challenges: scalable coordination in sensor networks. In: Proc. Int. Conf. Mobile Computing and Networking (MOBICOM), 1999, pp. 263–270.

Foote, K.E., Huebner, D.J., 1995. Error, Accuracy, and Precision. Technical Report, the Geographer's Craft Project. Department of Geography. University of Texas at Austin.

Fowler, C.A., January 1979. Comments on the cost and performance of military systems. IEEE Transactions on Aerospace and Electronic Systems 15, 2–10.

Frank, R., 2000. Understanding Smart Sensors, second ed. Artech House, Inc., Norwood, MA.

Garetti, M., Macchi, M., Terzi, S., Fumagalli, L., 2007. Investigating the organizational business models of maintenance when adopting self-diagnosing and self-healing ICT systems in multi-site contexts. In: Proceedings of the IFAC Conference on Cost Effective Automation in Networked Product Development and Manufacturing IFAC-cea, vol. 7.

Giachino, J.M., 1986. Smart sensors. Sensor and Actuators 10, 239–248.

Grossmann, P., 1998. Multisensor data fusion. The GEC Journal of Technology 15, 27–37.

Hall, D.L., 1992. Mathematical Techniques in Multi-Sensor Data Fusion. Artech House, Norwood, Massachusetts.

Heile, B., 2001. Working Group for Wireless Personal Area Networks (WPAN). IEEE 802.15. http://grouper.ieee.org/groups/802/15/.

Henry, M., October 2000. Self-validating digital coriolis mass flow meter. Computing & Control Engineering Journal 219–227. ISSN 0956–3385.

Hills, R., July/August, 2001. Sensing for danger. Science Technology. http://www.llnl.gov/str/JulAug01/Hills.html.

Hoover, A., Olsen, B.D., April 2000. Sensor network perception for mobile robotics. In: Proceedings of the IEEE International Conference on Robotics and Automation, San Francisco, CA, pp. 342–347.

IEEE 1451, October 2001. A Standard Smart Transducer Interface. Sensors Expo, Philadelphia. http://ieee1451.nist.gov/Workshop_04Oct01/1451_overview.pdf.

Jayarama Reddy, P., 2012. Solar Power Generation: Technology, New Concepts & Policy. CRC Press/Balkema, Leiden, The Netherlands, pp. 2–3.

Jensen, D., June 2002. SIVAM: communication, navigation and surveillance for the Amazon. Avionics Magazine. http://www.aviationtoday.com/reports/avionics/previous/0602/0602sivam.htm.

Kahn, J.M., Katz, R.H., Pister, K.S.J., 1999. Mobile networking for smart dust. In: Proc. ACM/IEEE Int. Conf. Mobile Computing and Networking (Mobi Com), pp. 271–278.

Kam, M., Zhu, X., Kalata, P., January 1997. Sensor fusion for mobile robot navigation. Proceedings of the IEEE 85 (1), 108–119.

Karwowski, W., Marras, W.S., 1998. The Occupational Ergonomics Handbook, 1998. CRC Press, Boca Raton, FL, p. 570.

Koestler, A., 1989. The Ghost in the Machine, Arkana Books.

Kopetz, H., Kantz, H., Grunsteidl, G., Puschner, P., Reisinger, J., June 1990. 1990. Tolerating transient faults in MARS. In: Proceedings of the 20th. Symposium on Fault Tolerant Computing, Newcastle upon Tyne, UK.

Kroos, K., Potter, M., 2014. Thermodynamics for engineers. Stamford CT: Cengage Learning 2014, 472.

Lee, K., 2001. Wireless sensing and IEEE 1451. In: Presented at the Sensor Conf./Expo 2001, Chicago, IL.

Lewis, F.L., 2005. Wireless Sensor Networks. Associate Director for Research Head, Advanced Controls, Sensors, and MEMS Group Automation and Robotics Research Institute. The University of Texas at Arlington.

Lindsay, A., 2003. Smart Sensors & Applications. Parallax, Inc.

Llinas, J., Hall, D.L., May–June 1998. An introduction to multi-sensor data fusion. In: Proceedings of the 1998 IEEE International Symposium on Circuits and Systems, 6, pp. 537–540.

Loparo, K.A., Adams, M.L., April 1998. Development of machinery monitoring and diagnostic methods. In: Proceedings 52nd Meeting of the Society for Machinery Failure Prevention (Virgina Beach).

Loparo, K.A., Adams, M.L., Lin, W., Abdel-Magied, M.F., Afshari, N., October 2000. fault detection and diagnosis of rotating machinery. IEEE Transactions on Industrial Electronics 47 (No. 5).

Lou, X., Loparo, K.A., Discenzo, F.M., Yoo, J., Twarowski, A., August 2000a. A model-based technique for the detection of bearing faults. ICANOV.

Lou, X., Loparo, K.A., Discenzo, F.M., Yoo, J., Twarowski, A., August 2000b. A wavelet based technique for bearing diagnostics. ICANOV.

Lundquist, C., 2011. Sensor Fusion for Automotive Applications. Department of Electrical Engineering Linköping University, SE–581 83 Linköping, Sweden. Linköping Studies in Science and Technology. Dissertations. No. 1409.

Luo, R.C., Kay, M., September–October 1989. Multisensor integration and fusion in intelligent systems. IEEE Transactions on Systems, Man, and Cybernetics 19 (5), 901–930.

Ma, D., Bondade, R., 2011. Reconfigurable Switched-capacitor Power Converters: Principles and Designs for Self-Powered Microsystems. Springer, New York, p. 3. http://dx.doi.org/10.1007/978-1-4614-4187-8.

Manges, W., April 2000. Intelligent wireless sensors for industrial manufacturing. Sensors vol. 17 (4), 44–55.

Markin, M., Harris, C., Bernhardt, M., Austin, J., Bedworth, M., Greenway, P., Johnston, R., Little, A., Lowe, D., 1997. Technology Foresight on Data Fusion and Data Processing. Publication, The Royal Aeronautical Society.

Marzullo, K., November 1990. Tolerating failures of continuous-valued sensors. ACM Transactions on Computer Systems 8 (4), 284–304.

Maturana, F., Balasubramanian, S., Vasko, D., 2000. An Autonomous Cooperative System for High Speed Sorting Systems, Accepted for Publication And Presentation in the Holomas 2000 Workshop on Industrial Applications of Holonic and Multi-Agents Systems.

McKee, G.T., 1993. What can be fused? Multisensor fusion for computer vision. Nato Advanced Studies Institute Series F 99, 71–84.

Middelhoer, S., Hoogerwerf, A.C., 1985. Smart sensors when and where. Sensors and Actuators 8, 39–48.

Murphy, R.R., January 1996. 1996. Biological and cognitive foundations of intelligent sensor fusion. IEEE Transactions on Systems. Man and Cybernetics 26 (1), 42–51.

Nahin, P.J., Pokoski, J.L., May 1980. NCTR plus sensor fusion equals IFFN or can two plus two equal five? IEEE Transactions on Aerospace and Electronic Systems 16 (3), 320–337.

Najafi, K., 1991. Smart sensors. Center for integrated sensors and circuits, Department of electrical engineering and computer science, University of Michigan, Ann Arbor, MI 48109-2122, USA. Journal of Micromechanical and Microengineering 1, 86–102 (Printed in the UK).

Nelson, V.P., July 1990. Fault-tolerant computing: fundamental concepts. IEEE Computer 23 (7), 19–25.

Nwogu, K., 2015. Energy Harvesting and Storage: The Catalyst to the Power Constraint for Leveraging Internet of Things (IoT) on Trains. Theses, Dissertations, & Student Research in Computer Electronics & Engineering. Paper 33. University of Nebraska – Lincoln.

O'Mara, 2000, August 2000. Designing an IEEE 1451.2-compliant transducer. Sensors 17 (8), 46–55.

Ocak, H., Loparo, K.A., Discenzo, F.M., Yoo, J., Twarowski, A., August 2000. Estimating the running speed and bearing Defect frequencies of an induction motor from vibration data. ICANOV.

Ostaffe, H., Tollefson, J., December 14, 2011. Harvested RF Powers Remote Sensors. Contributed by Digi-Key Electronics. http://www.digikey.com/en-US/articles/techzone/2011/dec/harvested-rf-powers-remote-sensors.

Pala, S., Fumagalli, L., Garetti, M., Macchi, M., 2014. Smart sensors for condition based maintenance: a test case in the manufacturing industry. In: European conference of the prognostics and health management society. Department of Management, Economics and Industrial Engineering, Politecnico di Milano, P.zza Leonardo Da Vinci, 32, p. 20133 (Milano, Italy).

Parhami, B., 1991. A data-driven dependability assurance scheme with applications to data and design diversity. In: Avizienis, A., Laprie, J.C. (Eds.), Dependable Computing for Critical Applications, 4. Springer Verlag, Vienna, pp. 257–282.

Park, J.W., Jung Jo, H., Jo, H., Spencer, B.F., 2012. Feasibility study of micro-wind turbines for powering wireless sensors on a Cable-Stayed bridge. MDPI: Energies 5 (2012), 3450–3464. http://dx.doi.org/10.3390/en5093450.

Pathan, A.-S.K., Hong, C.S., Lee, H.-W., February 20–22, 2006. Smartening the Environment Using Wireless Sensor Networks in a Developing Country.

Pease, M., Shostak, R., Lamport, L., 1980. Reaching agreement in the presence of faults. Journal of the ACM 27 (2).

Raju, M., Grazier, M., 2010. Energy Harvesting. ULP Meets Energy Harvesting: A Game-Changing Combination for Design Engineers. Texas Instrument. http://www.ti.com.cn/cn/lit/wp/slyy018a/slyy018a.pdf.

Rao, N.S.V., July 1998. A fusion method that performs better than best sensor. In: Proceedings of the First International Conference on Multisource-multisensor Information Fusion, pp. 19–26.

Rothman, P.L., Denton, R.V., 1991. Fusion or confusion: knowledge or nonsense? SPIE Data Structures and Target Classification 1470, 2–12.

Roundy, S., Wright, P.K., Rabaey, J., July 1, 2003. A study of low level vibrations as a power source for wireless sensor nodes. Computer Communications 26 (11), 1131–1144.

Singhal, A., May 1997. Issues in Autonomous Mobile Robot Navigation. Survey Paper towards Partial Fulfillment of MS Degree Requirements, Computer Science Department. University of Rochester, Rochester, NY, 14627–0226.

Smestad, T., February 2001. Data Fusion – for Humans, Computers or Both? Translated article from Mikroskopet, Norwegian Defence Research Establishment.

Smith, J.R., 2013. Range Scaling of Wirelessly Powered Sensor Systems in Wirelessly Powered Sensor Networks and Computational RFID. Springer, New York, p. 3. http://dx.doi.org/10.1007/978-1-4419-6166-2.

Smith, G., Bowen, M., 1995. Considerations for the utilization of smart sensors. Sensors and Actuators A46–47, 521–524.

Steere, D., Baptista, A., McNamee, D., Pu, C., Walpole, J., 2000. Research challenges in environmental observation and forecasting systems. In: Proc. 6th Int. Conf. Mobile Computing and Networking (MOBICOMM), pp. 292–299.

Steinberg, A.N., Bowman, C.L., White, F.E., May 1999. Revisions to the JDL data fusion model. In: Proceedings of the 1999 IRIS Unclassified National Sensor and Data Fusion Conference (NSSDF).

Technical Committee on Integrated Manufacturing Systems, 1980. Questionnaire on Unmanned Operation and Cutting State Monitoring. JSPE Technical Committee on IMS, 1980 (in Japanese).

Tenney, R.R., Sandell Jr., N.R., July 1981. 1981. Detection with distributed sensors. IEEE Transactions on Aerospace and Electronic Systems 17 (4), 501–510.

Theil, A., Kester, L.J.H.M., Bossé, E., July 2000. On measures of performance to assess sensor fusion effectiveness. In: Proceedings of the 3rd International Conference on Information Fusion. France, Paris.

Tönshoff, H.K., Inasaki, I., 2001. Sensors Applications Volume 1-Sensors in Manufacturing Sensors in Manufacturing. Copyright © 2001 Wiley-vch Verlag GmbH. ISBNs: 3-527-29558-5 (Hardcover); 3-527-60002-7 (Electronic).

U. S. Department of Defense (DoD), 1991. Data Fusion Subpanel of the Joint Directors of Laboratories, Technical Panel for C3. Data Fusion Lexicon.

Vias, R., Nishimoto, H., Tentzeris, M., Kawahara, Y., Asami, T., June 2012. A Battery-less, Energy Harvesting Device for Long Range Scavenging of Wireless Power from Terrestrial TV Broadcasts. IEEE 2012 IMS Digest. Montreal, Canada. http://dx.doi.org/10.1109/MWSYM.2012.6259708.

Visser, A., Groen, F.C.A., August 1999. Organisation and Design of Autonomous Systems. Textbook, Faculty of Mathematics, Computer Science, Physics and Astronomy. University of Amsterdam, Kruislaan 403, NL-1098, SJ, Amsterdam.

Von Neumann, J., 1956. Probabilistic logics and the synthesis of reliable organisms from unreliable components. In: Shannon, C.E., McCarthy, J. (Eds.), Automata Studies. Princeton University Press, pp. 43–98.

Wagner, H.J., Mathur, J., 2013. Introduction to Wind Energy Systems: Basics, Technology and Operation. Springer, Berlin Heidelberg, p. 3. http://dx.doi.org/10.1007/978-3-642-32976-0.

Wald, L., 1998. A European proposal for terms of reference in data fusion. International Archives of Photogrammetry and Remote Sensing XXXII (7), 651–654.

Waltz, E., 1995. The principles and practice of image and spatial data fusion. In: Proceedings of the 8th National Data Fusion Conference, Dallas.

Waltz, E., Llinas, J., 1990. Multisensor Data Fusion. Artech House, Norwood, Massachusetts.

Williams, C.B., Yates, R.B., 1995. Analysis of a micro-electric generator for micro-systems. Proceedings of the Transducers 95/Eurosensors IX, 341–344.

Wilson, J.S., 2005. Sensor technology Handbook. In: in Chief (Ed.), Newnes Is an Imprint of Elsevier. Copyright © 2005, Elsevier Inc.

Wise, K.D., Najafi, K., 1988. VLSI sensors in medicine. In: Norman, G., Einspruch, Goid, R.D. (Eds.), VLSI. Academic, New York.

Wood, G., October 2000. UK activities in measurement validation and data quality. Computing & Control Engineering Journal 214–218. ISSN 0956–3385.

Yu, H., Guo, M., 2012. An effective oil and gas pipeline monitoring systems based on wireless sensor networks. In: International Conference on Information Security and Intelligence Control (ISIC), pp. 178–181.

FURTHER READING

Inc, P., 2006. Smart Sensors and Applications. Student Guide, VERSION 1.0. 3rd Printing, ISBN 1-928982-39-5.

MIMOSA OSA-CBM Specification, http://www.mimosa.org/mimosa-osa-cbm.

Reyer, M., Hurlebaus, S., Mander, J., Ozbulut, O.E., 2013. Design of a wireless sensor network for structural health monitoring of bridges. In: Subhas, C., Mukhopadhyay, Joe Air Jiang (Eds.), Wireless sensor networks and Ecological monitoring smart sensors, measurement and instrumentation in wireless sensor networks and Ecological monitoring, vol. 3. Springer, Berlin Heidelberg, pp. 195–216. http://dx.doi.org/10.1007/978-3-642-36365-8.

Chapter 2

Data Collection

2.1 DATA COLLECTION IN INDUSTRY

2.1.1 Data Needs for Industry Management

Industrial planners and managers need to understand the dynamics of the complete supply chain, i.e., operations, infrastructure, communities, and individuals involved, set policy and manage industrial assets. Data collection and analysis can provide information on how industry are likely to respond to different policies. Identify constraints on production, suggest the development of new factories or assess the effective of prices and cost changes.

Industrial management requires consideration of a variety of issues, all of which need to be addressed using information collected from technological, economic, and sociocultural sources. A factory is a complex system of interacting factors, including the state of certain resources, social and institutional constraints, economic conditions, and cultural beliefs. Integrated analysis, using a variety of data, is necessary for predictive assessments of future conditions and the outcomes of alternative management measures.

Over time, many different management issues will emerge in every industry. Many of these issues, can only be identified using information derived from a data collection program. For this reason, it is wise to develop a program that covers a wider range of variables than is required under current policy.

Industry is subject to increasing stress. For industry to develop, maintain, and improve, specific data are required.

The greatest threats to industrial assets are often loss of performance and degradation, making data collection on the effect of the working environment on the assets essential collection of data on production patterns is equally important. Economic and socio-cultural aspects must also be assessed. In all industries, it is people who utilize the assets and people whose behavior must be influenced to implement effective management measures (FAO, 1998).

2.1.2 Data Collection Strategy

Strategies for the design of data collection programs will vary between industries. Each will have its own characteristics, its own relative importance,

eMaintenance. http://dx.doi.org/10.1016/B978-0-12-811153-6.00002-6

and its own potential for the supply of data. In addition, some information is only obtainable from external sources, such as international market data.

Each industry will require its own strategy, including enumeration and sampling. Over time, some aspects of a data collection strategy may move from complete enumeration to sampling (or vice versa), particularly as knowledge is developed and requirements or resources change. Sampling strategies are often punctuated by complete enumeration from time to time to re-evaluate baseline data.

It is not feasible to construct a perfect strategy for any one subsector that will meet all its requirements for all time. Flexibility and the adoption of alternative approaches must be a key component of any strategy (FAO, 1998).

In general, however, any strategy will require the following steps:

- Evaluate existing data sets in relation to the objectives of the company, including accessibility of the data (i.e., computerized, on paper);
- Describe the operating characteristics of the sector or subsector (market, asset fleet, community, institutional environment), also known as the census or frame survey;
- Decide on the approach, i.e., complete enumeration or sampling, including cost–benefit and cost-effectiveness analysis and an evaluation of operational considerations (institutional, financial, and human resources);
- Design methods according to the approach adopted, including the form of stratification to be used in sampling;
- Implement a test phase to validate the method, including participation by other stakeholders;
- Establish a continuing feedback mechanism between data sources and data users to ensure that data types, quantity, quality, and origin are consistent with the requirements of the performance indicator in question.

2.1.2.1 Complete Enumeration and Sampling

2.1.2.1.1 Definitions

Data collection is the recording of one or more data variables (length, duration, etc.) from members of a population of "data units" (the population of assets). The two basic data collection approaches are possible:

- Complete enumeration, where all members of the whole population are measured;
- Sampling, where only some of the whole proportion is measured.

Manufacturing data usually collected by complete enumeration include asset registers and infrastructure information. Data sometimes collected by complete enumeration and sometimes by sample surveys include number of indicators, most of which are described in EN 15341. Data usually collected by sampling include (e.g., size–frequency data) (FAO, 1998).

2.1.2.1.2 Complete Enumeration or Sampling?

Both approaches collect data for a specified period, for example, a month or a year, to determine some statistic of interest. The applicability of either sampling or complete enumeration is determined by various criteria related to the type of data and to existing financial and human constraints.

A complete enumeration-based survey is often preferred for certain types of data, as it is expected to provide complete statistical coverage over space and time. However, a well-designed sample-based survey can often provide good estimates of important parameters at a fraction of the cost. Ultimately, the choice of approach will depend on the local circumstances.

Complete enumeration may be seen as desirable, but not attainable for operational reasons. In this case, an existing sampling program can be progressively expanded to provide more reliable and robust estimates, if human and logistics resources allow such expansion in a sustainable manner. Such progressive expansion is done in distinct phases.

Moving from one sampling scenario to another while retaining accuracy requires large increases in operational and logistical support and is not always feasible or desirable (FAO, 1998).

2.1.2.1.3 Stratification in Data Collection

Stratification reduces the error in sample estimates by systematically removing as much of the data variability as possible through the sampling design. This is achieved by dividing the sample population into groups or strata, where as much as possible of the variability in the population is represented in differences between the groups. They may be an obvious divide in many variables or clear logistical criteria supporting the choice of strata.

There are two major types of stratification in a data collection program:

- Subdivisions based on administrative, geographical, or temporal criteria may be imposed on the data collection program for reporting purposes and are not under the control of the survey designer. Conventionally, this type of subdivision is called a major stratum; examples of major strata include provinces of a country, months of the year, subdivisions based on specific research needs, etc. Major strata may be based on any combination of such criteria.
- Within a major stratum, there are usually subdivisions based on criteria chosen by the designer for the sole purpose of increasing the accuracy of the derived estimates. These subdivisions are chosen in such way as to partition the population into homogeneous subsets. They are conventionally called minor strata.

2.1.2.3.2 The Effect of Stratification

Stratification may be complicated by the need to reconcile two conflicting objectives:

- To select strata with the maximum degree of homogeneity;
- To minimize the number of strata (usually because of operational constraints).

However, by systematically varying the stratification in a pilot phase, the appropriate balance can often be found.

Once a stratification scheme has been decided on, the next problem is the allocation of sampling effort to the strata. The three basic rules (known as "Von Neumann allocation") for optimization are to allocate greater sampling effort to strata with

1. greater size,
2. greater variation, and
3. lower sampling cost.

The cost of a sample-based approach is mainly a function of its statistical coverage or sample size. When the sample size increases, the survey cost also increases, but the expected accuracy is higher. However, the increase in accuracy is not proportional to the sample size and, suffers from diminishing returns (FAO, 1998).

2.1.3 Data Collection Methods

Methods to collect or obtain data for statistical analysis can be qualitative or quantitative Qualitative ones are subjective and include, among others: questionnaires, interviews, direct observations. Quantitative ones are more fact-based and include, among others, work orders, sales reports, measured values (temperature, humidity etc.).

2.2 DATA CLEANING

Data cleaning also called "data cleansing" or "scrubbing" refers to detecting and removing errors and inconsistencies from data to improve their quality. Data quality problems appear even in single data collections, such as files and databases, e.g., due to misspellings during data entry, missing information, or other invalid data. When multiple data sources are integrated, e.g., in data warehouses, federated database systems or global web-based information systems, the need for data cleaning increases significantly because the sources often contain redundant data in different representations. To provide access to accurate and consistent data, it is necessary to consolidate different data representations and eliminate duplicate information (Rahm and Hai Do, 2000).

Data warehouses require and provide extensive support for data cleaning (Chaudhuri and Dayal, 1997; Jarke et al., 2000). They load and continuously refresh huge amounts of data from a variety of sources, so the probability of some of the sources containing "dirty data" is high. However, as data warehouses are used for decision making, the correctness of their data is vital to avoid erroneous conclusions. For instance, duplicated or missing information will produce incorrect or misleading statistics (garbage in, garbage out). Given the wide range of possible data inconsistencies and the sheer data volume, data cleaning is one of the biggest problems in data warehousing. During the

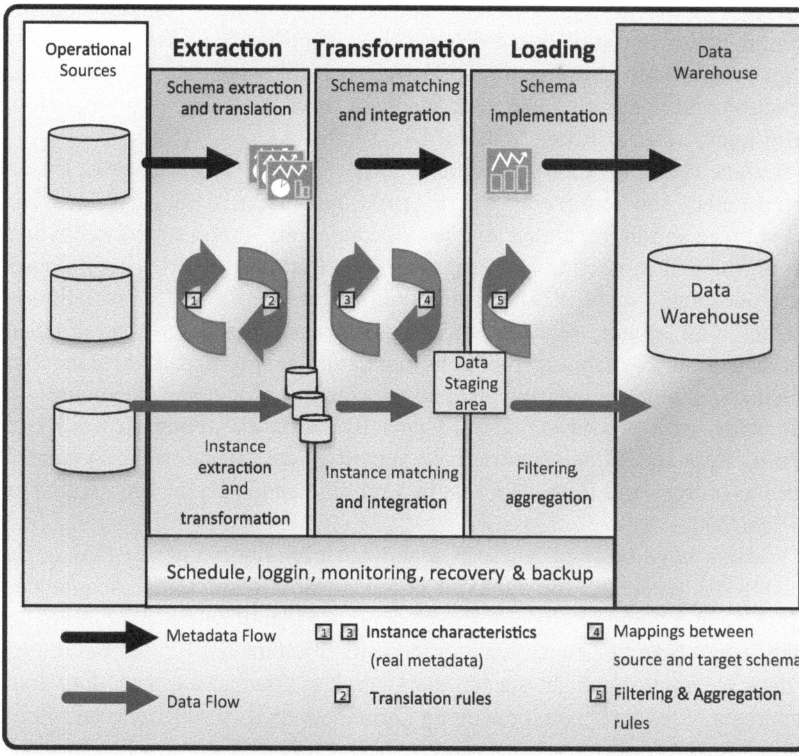

FIGURE 2.1 Steps in building a data warehouse: ETL process. *Adapted from Rahm, E., Hai Do, H., 2000. Data Cleaning: Problems and Current Approaches. University of Leipzig, Germany. Bulletin of the IEEE Computer Society Technical Committee on Data Engineering. Copyright 2000 IEEE.*

extraction–transformation–loading (ETL) process, illustrated in Fig. 2.1, further data transformations deal with translating, integrating, filtering, and aggregating data/schema to be stored in the warehouse. As indicated in the figure, data cleaning is typically performed in a separate data staging area before the transformed data are loaded into the warehouse. Numerous tools of varying functionality are available to support these tasks, but a significant portion of the cleaning and transformation work is done manually or by low-level programs that are difficult to write and maintain (Rahm and Hai Do, 2000).

The data transformation steps used by federated database systems and web-based information systems are similar to those used by data warehouses. Typically, there is a *wrapper* per data source for extraction and a *mediator* for integration (Roth and Schwarz, 1997; Wiederhold, 1992). These systems provide only limited support for data cleaning, as they focus on data transformations for schema translation and integration. Data are not pre-integrated, as they are for data warehouses, but need to be extracted from multiple sources, transformed, and combined during query runtime. The corresponding communication and processing delays can be significant, making it difficult to achieve acceptable response times. The effort needed for data cleaning during extraction and integration increases response times but is necessary to achieve useful query results (Rahm and Hai Do, 2000).

A data cleaning approach should satisfy several requirements. First, it should detect and remove all major errors and inconsistencies in individual data sources and in integrated sources. Second, it should be supported by tools to limit manual inspection and programming effort and be extensible to easily cover additional sources. Third, it should not be performed in isolation but together with schema-related data transformations based on comprehensive metadata. Fourth, mapping functions for data cleaning and other data transformations should be clearly specified and reusable for other data sources, as well as for query processing. Fifth, especially for data warehouses, a workflow infrastructure should be supported to execute all data transformation steps for multiple sources and large data sets in a reliable and efficient way (Rahm and Hai Do, 2000).

While a huge body of research deals with translation and integration, data cleaning has received little attention in the research community. A number of authors have focused on the problem of duplicate identification and elimination. Some research groups concentrate on general problems not limited but relevant to data cleaning, such as special data mining approaches, and data transformations based on schema matching (Abiteboul et al., 1999; Milo and Zohar, 1998). More recently, several research efforts have investigated a comprehensive and uniform treatment of data cleaning covering several transformation phases, specific operators, and implementation (Galhardas et al., 2000; Lee et al., 1999; Raman and Hellerstein, 1999).

2.2.1 Data Cleaning Problems

The major data quality problems to be solved by data cleaning and data transformation are closely related and, thus, should be treated in a uniform way. Data transformations are needed to support any changes in the structure, representation, or content of data (Rundensteiner, 1999). These transformations become necessary in many situations, e.g., when dealing with schema evolution, migrating a legacy system to a new information system, or integrating multiple data sources (Rahm and Hai Do, 2000).

Fig. 2.2 roughly distinguishes between single-source and multisource problems and between schema- and instance-related problems. Schema-level problems can be addressed at the schema level by an improved schema design (schema evolution), schema translation, or schema integration. Instance-level problems are errors and inconsistencies in the actual data content, which are not visible at the schema level. They are the primary focus of data cleaning. Fig. 2.2 also indicates some typical problems for the various

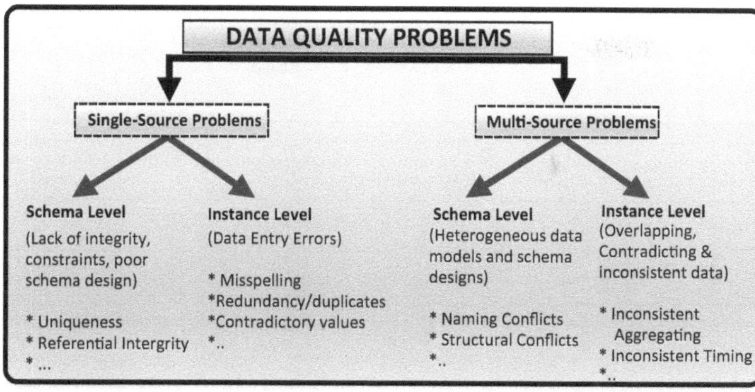

FIGURE 2.2 Classification of data quality problems in data sources. *Adapted from Borkar, V.R., Deshmukh, K., Sarawagi, S., 2000. Automatically extracting structure from free text addresses. IEEE Data Engineering Bulletin, 23.4, 27–32.*

cases. While not shown in the figure, single-source problems occur (with increased likelihood) in the multisource case, along with specific multisource problems (Rahm and Hai Do, 2000).

2.2.1.1 Single-Source Problems

The data quality of a source largely depends on the degree to which it is governed by schema and integrity constraints (ICs) controlling permissible data values. On the one hand, for sources without schema, such as files, there are few restrictions on what data can be entered and stored, giving rise to a high probability of errors and inconsistencies. Database systems, on the other hand, enforce restrictions of a specific data model (e.g., the relational approach requires simple attribute values, referential integrity, etc.), as well as application-specific ICs. Therefore, schema-related data quality problems occur because of a lack of appropriate model-specific or application-specific ICs, e.g., the data model has limitations, the schema are poorly designed, or only a few ICs are defined to limit the overhead for integrity control. Instance-specific problems relate to errors and inconsistencies that cannot be prevented at the schema level (e.g., misspellings) (Rahm and Hai Do, 2000).

Both schema- and instance-level problems can be differentiated with respect to problem scope, i.e., attribute (field), record, record type, and source. Examples of the various cases are shown in Tables 2.1 and 2.2. Note that uniqueness constraints specified at the schema level do not prevent duplicated instances, e.g., if information on the same real-world entity is entered twice with different attribute values (see example in Table 2.2).

As cleaning data sources is an expensive process, preventing dirty data from being entered is obviously an important way to reduce the cleaning problem. This requires an appropriate design of the database schema and

TABLE 2.1 Examples of Single-Source Problems at Schema Level (Violated Integrity Constraints)

Scope /Problem		Dirty Data	Reason /Remarks
Attribute	Illegal values	Bdate=31.14.82	Values outside of domain range
Record	Violated attribute dependencies	Age=31, b date=06.05.85	Age = (Current date – Birth Date) Should hold
Record type	Uniqueness violation	empX=(name="Jorge Villa", SSN="987654" empY=(name="Luis Sanchez", SSN= "987654"	Unique characteristics of SSN (Social Security Number) violated
Source	Referential integrity violation	emp=(name="Jorge Villa", deptno =139)	Referenced department (139) is not defined

Adapted from Borkar, V.R., Deshmukh, K., Sarawagi, S., 2000. Automatically extracting structure from free text addresses. IEEE Data Engineering Bulletin, 23.4, 27–32.

TABLE 2.2 Examples of Single-Source Problems at Instance Level

Scope/Problem		Dirty Data	Reason/Remarks
Attribute	Missing Values	phone=1234-567890	Unavailable values during data entry (dummy values or null)
	Misspellings	city="Nadrid"	Commonly typos, phonetic errors
	Abbreviations, Cryptic values	experience="X"; occupation="XY App."	
	Embedded values	nameX="J. Villa", nameY="Sanchez. L"	Commonly in a free-form field
	Misfielded values	city="Spain"	
Record	Violated attribute dependencies	city="Sevilla", zip=12345	City and zip should correspond
Record type	Word transpositions	nameX="J. Villa", nameY="Sanchez. L"	Usually in a free-form field
	Duplicated records	empX=(name="Jorge Villa",...); empY=(name="J. Villa",...)	Same employee represented twice due to some data entry errors
	Contradicting records	empX=(name="Jorge Villa", bdate=06.05.85); empY=(name="Jorge Villa", bdate=05.06.85)	The same real world entity is described by different values
Source	Wrong references	emp=(name="Jorge Villa", deptno=39)	Referenced department (39) is defined but wrong

Adapted from Borkar, V.R., Deshmukh, K., Sarawagi, S., 2000. Automatically extracting structure from free text addresses. IEEE Data Engineering Bulletin, 23.4, 27–32.

integrity constraints and the, data entry applications. In addition, the discovery of data cleaning rules during warehouse design can suggest improvements to the constraints enforced by existing schemas (Rahm and Hai Do, 2000).

2.2.1.2 Multi-source Problems

The problems present in single sources are aggravated when multiple sources need to be integrated. Each source may contain dirty data, and the data in the sources may be represented differently, overlap, or contradict because the sources are typically developed, deployed, and maintained independently to serve specific needs. This results in a large degree of heterogeneity with respect to data management systems, data models, schema designs, and the actual data (Rahm and Hai Do, 2000).

At the schema level, data model and schema design differences can be addressed by schema translation and schema integration, respectively. The main problems in schema designs are naming and structural conflicts (Batini et al., 1986; Kashyap and Sheth, 1996; Parent and Spaccapietra, 1998). Naming conflicts arise when the same name is used for different objects (homonyms) or different names are used for the same object (synonyms). Structural conflicts occur in many variations and are different representations of the same object in different sources, e.g., attribute versus table representation, different component structure, different data types, different ICs, etc. (Rahm and Hai Do, 2000).

Many conflicts appear only at the instance level (data conflicts). All problems from the single-source case can occur with different representations in different sources (e.g., duplicated records, contradicting records, etc.). Furthermore, even with the same attribute names and data types, there may be different value representations (e.g., for marital status) or different interpretations of the values (e.g., measurement units, such as dollar vs. euro) across sources. Finally, information in the sources may be provided at different aggregation levels (e.g., sales per product versus sales per product group) or refer to different points in time (e.g., current sales as of yesterday for source 1 versus as of last week for source 2).

A main problem in cleaning data from multiple sources is identifying overlapping data, especially matching records referring to the same real-world entity (e.g., customer). This problem is also called the object identity problem (Galhardas et al., 2000), duplicate elimination, or the merge/purge problem (Hernandez and Stolfo, 1998). Frequently, the information is only partially redundant, and the sources may complement each other by providing additional information about an entity. Thus, duplicate information should be purged and complementing information should be consolidated and merged to achieve a consistent view of real-world entities (Rahm and Hai Do, 2000).

The two sources in Fig. 2.3 are both in relational format but exhibit schema and data conflicts. At the schema level, we see name conflicts (synonyms customer/client, Cid/Cno, sex/gender) and structural conflicts (different representations for names and addresses). At the instance level, we note different gender representations ("0"/"1" versus "F"/"M") and presumably a duplicate record (Kristen Smith). The latter observation also reveals that while Cid/Cno is both source-specific identifiers, their contents are not comparable; different numbers (11/493) may refer to the same person, while different persons can have the same number (24). Solving these problems requires both schema five integration and data cleaning. The third table shows a possible solution. Note that the schema conflicts should be resolved first to allow data cleaning, especially the detection of duplicates based on a uniform representation of names and addresses and the matching of the gender/sex values (Rahm and Hai Do, 2000).

Customer (source 1)

CID	Name	Street	City	Sex
11	Maria Moreno	Calle Santiago 3	Zaragoza	0
24	Miguel Moreno	Calle Santiago 3	Zgz.	1

Client (source 2)

Cno	LastName	FirstName	Gender	Address	Phone/Fax
324	Moreno	Marita	F	Calle Santiago 3 de Zaragoza	411-124-4131/ 411-124-4156
24	Moreno	Manuel	M	Calle de la Santiago 3 de Madrid	111-222-3333

Customers (integrated target with cleaned data)

No.	LName	FName	Gender	Street	City	State	ZIP	Phone	Fax	CID	Cno
1	Moreno	Miguel	M	C.S	Zaragoza	-	11111			24	
2	Moreno	Maria	F	C.S	Zaragoza	-	11111	411-124-4131	411-124-4156	11	324
3	Moreno	Manuel	M	C.de.la.S	Madrid	-	22222	111-222-3333			24

FIGURE 2.3 Examples of multisource problems at schema and instance level. *Adapted from Borkar, V.R., Deshmukh, K., Sarawagi, S., 2000. Automatically extracting structure from free text addresses. IEEE Data Engineering Bulletin, 23.4, 27–32.*

2.2.2 Data Cleaning Approaches

In general, data cleaning involves several phases (Rahm and Hai Do, 2000):

- *Data analysis*: Detailed data analysis is required to detect which kinds of errors and inconsistencies are to be removed. In addition to a manual inspection of the data or data samples, analysis programs should be used to gain metadata on the data properties and detect data quality problems.
- *Definition of transformation workflow and mapping rules*: Depending on the number of data sources, their degree of heterogeneity and the "dirtiness" of the data, many data transformation and cleaning steps may have to be executed. A schema translation is sometimes used to map sources to a common data model; for data warehouses, a relational representation is typically used. Early data cleaning steps can correct single-source instance problems and prepare the data for integration. Later steps deal with schema/data integration and cleaning multisource instance problems, e.g., duplicates. For data warehousing, the control and data flow of transformation and cleaning steps should be specified within a workflow that defines the extraction, transformation, loading (ETL) process (see Fig. 2.1). As far as possible, the schema-related data transformations, as well as the cleaning steps, should be specified by a declarative query and mapping language to enable automatic generation of the transformation code. In addition, it should be possible to invoke user-written cleaning code and special purpose tools during a data transformation workflow. The transformation steps may request user feedback on data instances for which they have no built-in cleaning logic.
- *Verification*: The correctness and effectiveness of a transformation workflow and the transformation definitions should be tested and evaluated,

e.g., on a sample or copy of the source data, to improve the definitions if necessary. Multiple iterations of the analysis, design, and verification steps may be needed, e.g., since some errors only become apparent after applying some transformations.

- *Transformation*: Execution of the transformation steps either by running the ETL workflow for loading and refreshing a data warehouse or by answering queries on multiple sources.
- *Backflow of cleaned data*: After (single-source) errors are removed, the cleaned data should replace the dirty data in the original sources to give legacy applications the improved data and avoid redoing the cleaning in future data extractions. For data warehousing, the cleaned data are available from the data staging area (see Fig. 2.1).

The transformation process requires a large amount of metadata, such as schemas, instance-level data characteristics, transformation mappings, workflow definitions, etc. For consistency, flexibility, and ease of reuse, these metadata should be maintained in a database management system (DBMS) repository (Bernstein and Dayal, 1994). To support data quality, detailed information about the transformation process is recorded, both in the repository and in the transformed instances, notably, information about the completeness and freshness of source data and lineage information on the origin of transformed objects and the changes applied to them. For instance, in Fig. 2.3, the table Customers contains the attributes Cid and Cno, allowing us to trace the source records (Rahm and Hai Do, 2000).

2.2.2.1 Data Analysis

Metadata reflected in schemas are typically insufficient to assess the data quality of a source, especially if only a few ICs are enforced. Therefore, it is important to analyze the actual instances to obtain real (reengineered) metadata on data characteristics or unusual value patterns. These metadata help find data quality problems. They can also help identify attribute correspondences between source schemas (schema matching) based on which, automatic data transformations can be derived (Doan et al., 2000; Li and Clifton, 2000).

There are two related approaches to data analysis: data profiling and data mining. Data profiling focuses on the instance analysis of individual attributes. It derives information such as the data type, length, value range, discrete values and their frequency, variance, uniqueness, occurrence of null values, typical string pattern (e.g., for phone numbers), etc., providing an exact view of various quality aspects of the attribute. Table 2.3 shows examples of how these metadata can assist in the detection of data quality problems (Rahm and Hai Do, 2000).

Data mining reveals specific data patterns in large data sets, e.g., relationships between several attributes. This is the focus of descriptive data mining models, and it includes clustering, summarization, association

TABLE 2.3 Examples of the Use of Reengineered Metadata to Address Data Quality Problems

Problems	Metadata	Examples/Heuristics
Illegal values	Cardinality	E.g cardinality (gender) > 2 indicates problem
	Min/Max	
	Deviation, Variance	Deviation , variance of statistical values should not be higher than threshold
Misspellings	Attribute values	Sorting on values often brings misspelled values next to correct values
Missing values	Null values	Number/percentage of null values
	Attribute values + default values	
Varying value representations	Attribute values	Comparing attribute value set of a column of one table against that of a colum of another table
Duplicates	Cardinality + Uniqueness	Attribute = # rows should hold
	Attribute values	Sorting values by number of occurrences; more than 1 occurrence indicates duplicates

Adapted from Borkar, V.R., Deshmukh, K., Sarawagi, S., 2000. Automatically extracting structure from free text addresses. IEEE Data Engineering Bulletin, 23.4, 27–32.

discovery, and sequence discovery (Fayyad, 1998). As shown by Sapia et al. (1999), ICs among attributes, such as functional dependencies (FDs) or application-specific "business rules," can be derived and used to complete missing values, correct illegal values, and identify duplicate records across data sources. For example, an association rule with high confidence suggests there are data quality problems in instances violating this rule. To be precise, a confidence of 99% for rule "*total* = *quantity***unit price*" indicates that 1% of the records do not comply and may require closer examination (Rahm and Hai Do, 2000).

2.2.2.2 Defining Data Transformations

The data transformation process typically consists of multiple steps, with each step performing schema and instance-related transformations (mappings). To allow a data transformation and cleaning system to generate a transformation code and reduce the amount of self-programming, it is necessary to specify the required transformations in an appropriate language, e.g., supported by a graphical user interface. Various ETL tools (see Section 2.2.3) offer this functionality by supporting proprietary rule languages. A more general and flexible approach is the use of the standard query language (SQL) to perform data transformations and take advantage of application-specific language extensions, in particular, user-defined functions (UDFs) supported in SQL:99 (Haas et al., 1999; Hellerstein et al., 1999). UDFs can be implemented in SQL or a general purpose programming language with embedded SQL statements may be used. UDFs allow implementing a wide range of data transformations and support easy reuse for different transformation and query processing tasks. Furthermore, their execution by the DBMS can reduce data access cost and thus improve performance. Finally, UDFs are part of the SQL:99 standard and

```
CREATE VIEW: Customer2 (Lname, Fname, Gender, Street, State, ZIP, CID) AS

    SELECT: LastNameExtract (Name), FirstNameExtract (Name), Sex, Street,
            CityExtract (City), StateExtract (City), ZIPExtract (City), CID

    FROM: Customer
```

FIGURE 2.4 Example of transformation step definition. *Adapted from Borkar, V.R., Deshmukh, K., Sarawagi, S., 2000. Automatically extracting structure from free text addresses. IEEE Data Engineering Bulletin, 23.4, 27–32.*

should (eventually) be portable across many platforms and DBMSs (Rahm and Hai Do, 2000).

Fig. 2.4 shows a transformation step specified in SQL:99. The example refers to Fig. 2.3 and covers part of the necessary data transformation for the first source. The transformation defines a view on which further mappings can be performed. It performs a schema restructuring with additional attributes in the view obtained by splitting the name and address attributes of the source. The required data extractions are achieved by UDFs (shown in boldface). The UDF implementations can contain cleaning logic, e.g., to remove misspellings in city names or add missing zip codes (Rahm and Hai Do, 2000).

UDFs may still require a substantial implementation effort and might not support all necessary schema transformations. Simple and frequently needed functions, such as attribute splitting or merging, are not generically supported but need to be reimplemented in application-specific variations (see extract functions in Fig. 2.4). More complex schema restructurings (e.g., folding and unfolding of attributes) are not supported at all. To generically support schema-related transformations, language extensions such as SchemaSQL are required (Lakshmanan et al., 1996). Data cleaning at the instance level can also benefit from special language extensions such as a match operator supporting "approximate joins" (see below). System support for such powerful operators can greatly simplify the programming effort required for data transformations and improve performance. Some current research on data cleaning is investigating the usefulness and implementation of such query language extensions (Hellerstein et al., 1999; Raman and Hellerstein, 1999).

2.2.2.3 Conflict Resolution

A set of transformation steps must be specified and executed to resolve the various schema- and instance-level data quality problems of data sources. Several types of transformations can be performed on individual data sources to deal with single-source problems and to prepare for integration with other sources. In addition to a possible schema translation, preparatory steps typically include the following (Rahm and Hai Do, 2000):

- *Extraction of values from free-form attributes* (*attribute split*): Free-form attributes often capture multiple individual values that should be extracted to achieve a more precise representation and support further cleaning steps, such as instance matching and duplicate elimination. Typical examples are name and address fields (Table 2.2; Figs. 2.3 and 2.4). Required transformations in this step are the reordering of values within a field to deal with word transpositions and value extraction for attribute splitting.

- *Validation and correction*: This step examines each source instance for data entry errors and tries to correct them automatically as far as possible. Spell checking based on dictionary lookup is useful for identifying and correcting misspellings. Furthermore, dictionaries on geographic names and zip codes can help correct address data. Attribute dependencies (birthdate–age, total price–unit price/quantity, city–phone area code, etc.) can be used to detect problems and substitute missing values or correct wrong values.
- *Standardization*: To facilitate instance matching and integration, attribute values should be converted to a consistent and uniform format. For example, date and time entries should be put into a specific format; names and other string data should be converted to either upper or lower case, etc. Text data may be condensed and unified by performing stemming or removing prefixes, suffixes, and stop words. Finally, abbreviations and encoding schemes should be resolved consistently by consulting special synonym dictionaries or applying predefined conversion rules.

Dealing with multisource problems requires the restructuring of schemas to achieve schema integration, including steps such as splitting, merging, and folding and unfolding of attributes and tables. At the instance level, conflicting representations need to be resolved and overlapping data dealt with. The duplicate elimination task is typically performed after most other transformation and cleaning steps are complete, especially when single-source errors and conflicting representations are cleaned. It is performed either on two cleaned sources at a time or on a single already integrated data set. Duplicate elimination requires three steps: first, identifying (i.e., matching) similar records for the same real-world entity; second, merging similar records into one record containing all relevant attributes without redundancy; third, purging redundant records (Monge, 2000).

In the simplest case, there is an identifying attribute or attribute combination per record that can be used to match records, e.g., if different sources share the same primary key or if there are other common unique attributes. Instance matching between different sources is then achieved by a standard equijoin on the identifying attribute(s). In the case of a single data set, matches can be determined by sorting on the identifying attribute and checking if neighboring records match. Both cases lead to efficient implementation, even for large data sets. Unfortunately, without common key attributes or in the presence of dirty data, such straightforward approaches are often too restrictive. To determine most or all matches a "fuzzy matching" (approximate join) becomes necessary to find similar records based on a matching rule, e.g., specified declaratively or implemented by a user-defined function (Galhardas et al., 2000; Hellerstein et al., 1999).

2.2.3 Tool Support

A large variety of commercially available tools support data transformation and data cleaning tasks, especially for data warehousing. Some tools concentrate on a specific domain, such as cleaning name and address data, or a specific cleaning phase, such as data analysis or duplicate elimination. Because of their restricted domain, specialized tools typically perform very well but must be complemented by other tools to address the broad spectrum of transformation and cleaning problems. Other tools, e.g., ETL tools, provide comprehensive transformation and workflow capabilities to cover a large part of the data transformation and cleaning process. A general problem of ETL tools, however, is their limited interoperability; more specifically, proprietary application programming interfaces (APIs) and proprietary metadata formats make it difficult to combine the functionality of several tools (Do and Rahm, 2000).

2.2.3.1 Data Analysis and Reengineering Tools

Data analysis tools can be divided into data profiling and data mining tools. MIGRATIONARCHITECT (Evoke Software) is one of the few commercially available data profiling tools. For each attribute, it determines the following real metadata: data type, length, cardinality, discrete values and their percentage, minimum and maximum values, missing values, and uniqueness. MIGRATIONARCHITECT also assists in developing the target schema for data migration. Data mining tools, such as WIZRULE (WizSoft) and DATAMININGSUITE (InformationDiscovery), infer relationships among attributes and their values and compute a confidence rate indicating the number of qualifying rows. Notably, WIZRULE can reveal three kinds of rules: mathematical formula, if-then rules, and spelling-based rules indicating misspelled names, e.g., "value Edinburgh appears 52 times in field Customer; 2 case(s) contain similar value(s)." WIZRULE also automatically points to deviations from the set of discovered rules as suspected errors.

Data reengineering tools, e.g., INTEGRITY (Vality), use discovered patterns and rules to specify and perform cleaning transformations; put otherwise, they reengineer legacy data. In INTEGRITY, data instances undergo several analysis steps, including parsing, data typing, and pattern and frequency analysis. The result is a tabular representation of field contents, patterns and frequencies, based on which the pattern for standardizing data can be selected. To specify cleaning transformations, INTEGRITY provides a language, including a set of operators, for column transformations (e.g., move, split, delete) and row transformations (e.g., merge, split). Finally, INTEGRITY identifies and consolidates records using a statistical matching technique.

Automated weighting factors compute scores to rank matches based on which the user can select the real duplicates (Rahm and Hai Do, 2000).

2.2.3.2 Specialized Cleaning Tools

Specialized cleaning tools typically deal with a particular domain, mostly name and address data, or concentrate on the elimination of duplicates. The transformations are provided in advance in the form of a rule library or are supplied interactively by the user. Alternatively, data transformations can automatically be derived from schema matching tools, such as those described by Milo and Zohar (1998).

- *Special domain cleaning*: Names and addresses are recorded in many sources and typically have high cardinality. For example, finding customer matches is very important for customer relationship management. Many commercial tools, e.g., IDCENTRIC (FirstLogic), PUREINTEGRATE (Oracle), QUICKADDRESS (QASSystems), REUNION (PitneyBowes), and TRILLIUM (TrilliumSoftware), focus on cleaning this kind of data. They provide techniques such as extracting and transforming name and address information into individual standard elements, and validating street names, cities, and zip codes. In addition, they have a matching facility when data are cleaned , and they incorporate a huge library of prespecified rules dealing with the problems commonly found when processing such data. For example, TRILLIUM's extraction (parser) and matcher module contains over 200,000 business rules. The tools also provide facilities to customize or extend the rule library with user-defined rules for specific needs.
- *Duplicate elimination*: Sample tools for duplicate identification and elimination include DATACLEANSER (EDD), MERGE/PURGELIBRARY (Sagent/QMSoftware), MATCHIT (HelpITSystems), and MASTERMERGE (PitneyBowes). Data sources must already be cleaned for matching, but the tools support several approaches to matching attribute values; tools such as DATACLEANSER and MERGE/PURGE LIBRARY also allow user-specified matching rules to be integrated.

2.2.3.3 Extraction, Transformation, Loading Tools

Many commercial tools support the ETL process for data warehouses in a comprehensive way. They use a repository built on a data-based management system (DBMS) to manage all metadata about the data sources and to target schemas, mappings, script programs, etc., in a uniform way. Schemas and data are extracted from operational data sources via both native file and DBMS gateways, as well as standard interfaces such as open database connectivity (ODBC) and exploratory data analysis (EDA). Data transformations are defined with an easy-to-use graphical interface. To specify individual mapping steps, a proprietary rule language and a comprehensive library of predefined conversion functions are typically provided. The tools also support reusing existing transformation solutions, such as external C/C++ routines, by providing an interface to integrate them into the internal transformation library. Transformation processing is carried out either by an engine that interprets the specified transformations at runtime or by a compiled code. All engine-based tools (e.g., COPYMANAGER, DECISIONBASE, POWERMART, DATASTAGE, WAREHOUSEADMINISTRATOR), possess a scheduler and support workflows with complex execution dependencies among mapping jobs. A workflow may also invoke external tools, e.g., for specialized cleaning tasks such as name/address cleaning or duplicate elimination.

ETL tools typically have little built-in data cleaning capability but allow the user to specify cleaning functionality via a proprietary application program interface (API). There is usually no data analysis support to automatically detect data errors and inconsistencies. However, users can implement such logic with the metadata maintained and by determining content characteristics with the help of aggregation functions (sum, count, min, max, median, variance, deviation, etc.). The provided transformation library covers many data transformation and cleaning needs, such as data type conversions (e.g., date reformatting), string functions (e.g., split, merge, replace, substring search), arithmetic, scientific, and statistical functions, etc. The extraction of values from free-form attributes is not completely automatic; rather, the user must specify the delimiters separating subvalues.

The rule languages typically cover if-then and case constructs that help handling exceptions in data values, such as misspellings, abbreviations, missing or cryptic values, and values outside of range. These problems can also be addressed by using a table lookup construct and join functionality. Support for instance matching is typically restricted to the use of the join construct and some simple string matching functions, e.g., exact or wildcard matching and soundex, but user-defined field matching functions and functions for correlating field similarities can be programmed and added to the internal transformation library (Rahm and Hai Do, 2000).

2.2.4 Data Cleaning Overview

Data cleaning usually consists of two stages: error detection and error repairing.

2.2.4.1 Qualitative Error Detection

Given a dirty database instance, the first step is to detect anomalies or errors. Fig. 2.5 illustrates a taxonomy of qualitative error detection. Every error detection technique needs to address three main issues:

1. What to detect,
2. How to detect,
3. Where to detect.

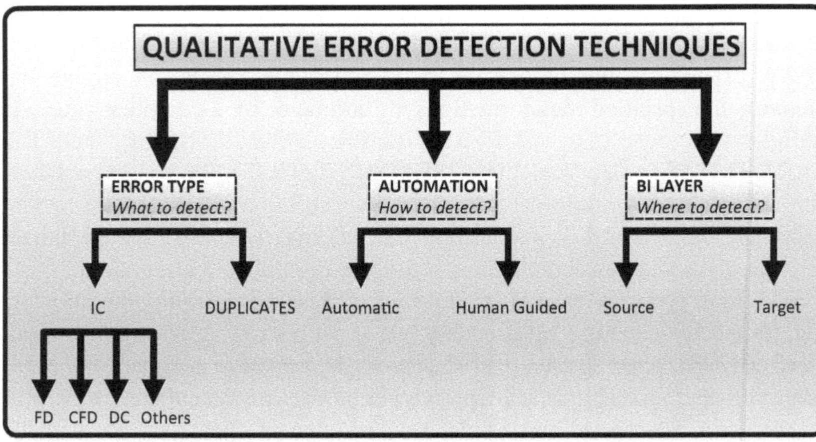

FIGURE 2.5 Classification of qualitative error detection techniques. *Adapted from Borkar, V.R., Deshmukh, K., Sarawagi, S., 2000. Automatically extracting structure from free text addresses. IEEE Data Engineering Bulletin, 23.4, 27–32.*

- *Error type* (*what to detect*): Qualitative error detection techniques can be classified according to which types of errors are captured, or in other words, the languages used to describe patterns or constraints of a data instance. A large body of work uses ICs to capture data quality rules to which the database should conform, including FDs (Bohannon et al., 2005) and denial constraints (Chu et al., 2013b). Duplicate records can be considered a violation of an IC (key constraint); the large body of work focusing on this problem separates it from other types of ICs (Chu et al., 2016).

As manually designing such ICs or patterns requires considerable domain expertise and is time-consuming, automatic discovery techniques are essential. These have been proposed for various ICs (Chu et al., 2013a). The IC discovery techniques can be divided into schema-driven and instance-driven approaches (Chu et al., 2016).

- *Automation* (*how to detect*): Approaches are classified according to whether and how humans are involved in the error detection process. Most techniques are fully automatic, for example, detecting violations of FDs (Bohannon et al., 2005), but some involve humans, for example, identifying duplicate records (Volkovs et al., 2014).
- *Business intelligence layer* (*where to detect*): Errors can happen in all stages of a business intelligence stack. For example, errors in the source database are often propagated through the data processing pipeline. While most error detection techniques detect errors in the original database, some errors can only be discovered much later in the data processing pipeline

(Chalamalla et al., 2014), where more semantics and business logic are available. For example, constraints on total budget can only be enforced after cost and expenses are aggregated.

Table 2.4 gives a sample of detection techniques, covering all categories of the proposed taxonomy.

2.2.4.2 Error Repairing

Once errors are detected, they obviously must be repaired. As in error detection, every technique needs to address three questions:

1. What to repair,
2. How to repair,
3. Where to repair.

Figure 2.6 depicts a taxonomy of data repairing techniques.

- *Repair target* (*what to repair*): Repairing algorithms requires making certain assumptions about the data and the quality rules, for example, (1) trusting the declared ICs and only updating data by removing errors (Chu et al., 2013b); (2) trusting the data completely and allowing the relaxation of the constraints (Golab et al., 2008), for example, to address schema evolution and obsolete business rules; and (3) exploring the possibility of changing both the data and the constraints (Beskales et al., 2010). Techniques that

TABLE 2.4 Sample of Qualitative Error Detection Techniques

	ERROR TYPE WHAT		AUTOMATION HOW		BI LAYER WHERE	
	IC	Data deduplication	Automatic	Human involved	Source	Target
FDs Value Modification (13)	✓		✓		✓	
Holistic Data Cleaning (22)	✓		✓		✓	
CrowdER (74)		✓		✓	✓	
Corleone (35)		✓		✓	✓	
Causality Analysis (58)	✓			✓		✓
Scorpion (79)	✓			✓		✓
DBRx (16)	✓		✓			✓

Adapted from Borkar, V.R., Deshmukh, K., Sarawagi, S., 2000. Automatically extracting structure from free text addresses. IEEE Data Engineering Bulletin, 23.4, 27–32.

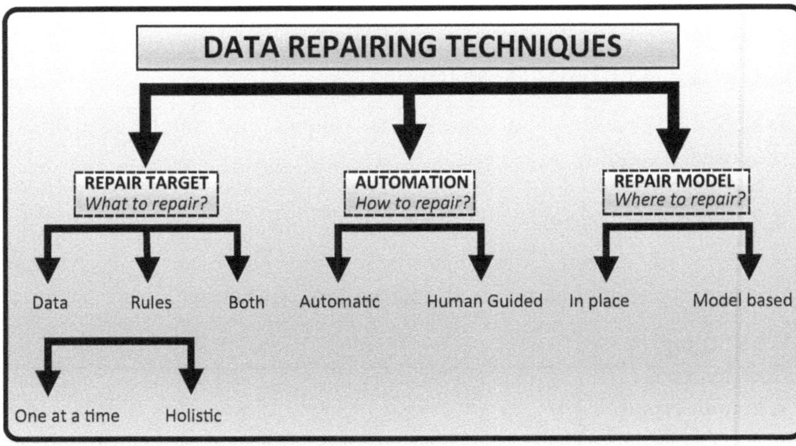

FIGURE 2.6 Classification of data repairing techniques. *Adapted from Borkar, V.R., Deshmukh, K., Sarawagi, S., 2000. Automatically extracting structure from free text addresses. IEEE Data Engineering Bulletin, 23.4, 27–32.*

trust the rules and change only the data can be further divided based on the driver of the repairing exercise, i.e., what types of errors are targeted. Most techniques repair one type of error only (one at a time), but certain emerging techniques consider the interactions among multiple types of errors and provide a holistic repair of the data (Chu et al., 2016).

- *Automation* (*how to repair*): Approaches can be classified based on the tools used in the repairing process. For example, repairing approaches can be classified according to whether and how humans are involved. Some techniques are fully automatic, for example, modifying the database, such that the distance between the original database I and the modified database I' 0 is minimized according to some cost function. Other techniques involve humans in the repairing process to verify the fixes, to suggest fixes, or to train machine learning models to carry out automatic repair decisions (Yakout et al., 2011).
- *Repair model* (*where to repair*): Approaches can be classified based on whether they change the database in situ or build a model to describe the possible repairs. Most proposed techniques repair the database in place, thus destroying the original database. For non in situ repairs, a model may be built to describe the various ways to repair the underlying database. Queries are answered by these repairing models using, for example, sampling from all possible repairs and other probabilistic query answering mechanisms (Beskales et al., 2010).

Table 2.5 shows a sample of data repairing techniques.

TABLE 2.5 Sample of Data Repairing Techniques

	REPAIR TARGET (What)				AUTOMATION (How)		REPAIR MODEL (Where)	
	Data–One at a time	Data- Holistic	Rules	Both	Automatic	Human involved	In place	Model based
FDs Value Modification (13)	✓				✓		✓	
FDs Hypergraph (46)	✓				✓		✓	
CFDs Value Modification	✓				✓		✓	
Holistic Data Cleaning (22)		✓			✓		✓	
LLUNATIC (33)		✓			✓		✓	
Record Matching & Data Repairing (31)		✓			✓		✓	
NADEEF (26)		✓			✓		✓	
Generate Optimal Tablaux (36)			✓		✓		✓	
Unified Repair (19)				✓	✓		✓	
Relative Trust (11)	✓			✓	✓		✓	
Continuous Data Cleaning (73)	✓			✓	✓		✓	
Potter's Wheel (64)	✓					✓	✓	
GDR (82)	✓					✓	✓	
KATARA (23)	✓					✓	✓	
DataTamer (69)	✓					✓	✓	
Editing Rules (30)	✓					✓	✓	
Sampling FDs Repairs (10)	✓							✓
Sampling Duplicates (12)	✓							✓

Adapted from Borkar, V.R., Deshmukh, K., Sarawagi, S., 2000. Automatically extracting structure from free text addresses. IEEE Data Engineering Bulletin, 23.4, 27–32.

2.2.5 Data Cleaning From a Statistical Perspective

As analytics become increasingly complex, it is important to understand the statistical implications of data cleaning. This section considers various techniques to improve accuracy or efficiency. It focuses on deduplication (duplication error), repairing missing and incorrect values (attribute error), and the removal of erroneous or irrelevant data (relevant error). It builds on the previously established taxonomy and employs data cleaning algorithms and systems based on statistics, as shown in Table 2.6 (Chu et al., 2016).

The categorization of capabilities of recently proposed data cleaning algorithms and systems (some may fit in multiple categories) can be summarized as the following: "use statistics" are those using a statistical model (e.g., a probabilistic graphical model) to identify and correct errors; "for statistics" are

TABLE 2.6 Capabilities of Recently Proposed Data Cleaning Algorithms and Systems

Type of Error	Operation	Uses Statistics	For Statistics
Attribute Error	Value imputation	(57, 72, 8, 81, 62, 23)	(75, 52, 43, 24)
	Data repairing	(1, 57, 8, 37)	(75, 52)
Duplicate Error	Entity Resolution	(39, 24, 81, 69)	(35, 37)
Relevance Error	Error/Outlier Removal	(27, 70, 28, 15, 34)	(52)

Adapted from Borkar, V.R., Deshmukh, K., Sarawagi, S., 2000. Automatically extracting structure from free text addresses. IEEE Data Engineering Bulletin, 23.4, 27–32.

explicitly designed to support data cleaning for aggregate analytics and advanced statistical analytics (Chu et al., 2016).

2.2.5.1 Data Cleaning With Statistics

There are several techniques to improve the efficiency or accuracy of data cleaning algorithms using statistical methods.

Active Learning in Crowdsourcing: Crowdsourcing is widely applied in industry for data cleaning (Marcus and Parameswaran, 2013). In academia, there is a growing consensus that crowds are difficult to scale; as a result, several recent works employ active learning to prioritize queries to the crowd (Gokhale et al., 2014; Haas et al., 2014; Mozafari et al., 2014). The basic idea is to formulate the human input to data cleaning as labels for a supervised learning technique (such as an support vector machine (SVM) or Random Forest). Active Learning is a class of algorithms that selects the most informative labels to acquire (Chu et al., 2016).

Other statistical methods: Other work uses statistical techniques to more accurately clean a dirty database. For example, the Eracer project shows how data cleaning on dirty relations can be posed as a two-step learning problem: first learning a graphical model to represent the relation and then creating a message-passing algorithm to resolve inconsistencies (Mayfield et al., 2010). The sensor network literature has several examples of statistical outlier detection and mitigation techniques. Finally, Yakout et al. (2013) employ machine learning (ML) to improve the reliability of data cleaning.

2.2.5.2 Data Cleaning for Statistical Analysis

While larger data sets can facilitate training more sophisticated ML models, systematic data errors, i.e., corruption affecting certain records disproportionately, can make model training unreliable. It has repeatedly been found that ML problems are highly sensitive to dirty data, even when using robust techniques (Krishnan et al., 2014, 2015a; Mahler et al., 2014), and the high dimensionality of these models leads to counterintuitive effects when trained after some types of data cleaning procedures. For example, in one fraud prediction example, simply merging inconsistent attributes before SVM model training improved the true positive detection probabilities from 62% to 91% (Krishnan et al., 2015a).

Aggregate queries: SampleClean notes that for aggregates such as *sum*, *count*, and *average*, there are diminishing returns for data cleaning, and it often suffices to clean small samples of data to estimate results with high accuracy. This problem has recently been extended to study aggregate queries on materialized views (Krishnan et al., 2015b).

Machine Learning: SampleClean was extended to study data cleaning that precedes machine learning model training in a system called ActiveClean (Krishnan et al., 2015a). ActiveClean employs selection techniques for the most valuable data and techniques to incrementally update ML models given newly clean data. One of the interesting findings is that progressive data cleaning and model training do not commute in an expected way. Suppose $k << N$ records are cleaned, but all the remaining dirty records are retained in the data set. Aggregates over mixtures of different populations of data can result in spurious relationships because of the well-known phenomenon Simpson's paradox (Simpson, 1951). Some of these problems are not apparent in one-dimensional analytics such as *sum*, *count*, and *average* but can lead to subtle biases in higher-dimensional statistical analysis (Chu et al., 2016).

Adaptive data analysis: The statistics community has studied some of these problems in a field called adaptive data analysis. Concepts such as multiple hypothesis testing and false discovery rate are highly relevant to the design of analysis tools (Dwork et al., 2015; Russo and Zou, 2015). This work considers the problem of false discoveries, where by analysts discover nonexistent trends in a data set simply because of statistical chance.

2.2.6 New Challenges

The emerging trends in data cleaning research include the following:

- *Scalability.* Scaling data cleaning techniques for the large and rapidly growing data sets of the Big Data era are an ongoing challenge. Current techniques include blocking for duplicate detection, sampling for data cleaning, and distributed data cleaning.
- *User engagement.* Although much research considers the use of humans to perform data deduplication, for example, through active learning, involving humans in other data cleaning tasks, such as repairing IC violations or taking user feedback to discover data quality rules, is yet to be explored.
- *Semistructured and unstructured data.* A significant portion of data is in semistructured formats, e.g., JSON, and unstructured formats, e.g., text documents. Data quality problems for semistructured and unstructured data remain largely unexplored (Chu et al., 2016).

- *New applications for streaming data.* There is renewed interest in data collected from vast collections of sensors and mobile devices. Gartner estimates there will be 26 billion devices on the Internet of Things by 2020 (Gartner, 2013). Five to ten years ago, the data management and quality challenges from distributed sensors were an important research topic. But most work on data cleaning in this domain relies on quantitative methods, such as outlier detection. One avenue for future work is to consider how qualitative data cleaning approaches will work on distributed streams of data (Chu et al., 2016).
- *Growing privacy and security concerns.* There are significant concerns about data privacy as more individual data are collected by enterprises. Data cleaning is, by nature, a task that requires examining and searching through raw data, and these data are restricted in some domains including finance and medicine. An important challenge is to reconcile the need for data provenance, access to unaggregated data, and privacy (Chu et al., 2016).

2.3 DATA SANITIZATION

Data sanitization is the process of making sensitive information in nonproduction databases safe for wider visibility (Net 2000 Ltd., 2010). Alternatively, it can be defined as the process of removing information from electronic media with the reasonable assurance that the information cannot be retrieved (Wargo, 2016).

2.3.1 Data Sanitization Techniques

Test and development teams need to work with databases that are structurally correct and functional copies of live environments. However, they do not necessarily need to be able to view security sensitive information. For test and development purposes, as long as the data look real, the actual record content is usually irrelevant. There are a variety of data sanitization techniques available—the pros and cons of some of the most useful are discussed below (Net 2000 Ltd., 2010).

2.3.1.1 Technique: NULL'ing Out

Simply deleting a column of data by replacing it with NULL values is an effective way of ensuring it is not inappropriately visible in test environments. Unfortunately, it is also one of the least desirable options from a test database standpoint. Test teams usually need to work on the data, or at least a realistic approximation of it. For example, it is very hard to write and test customer account maintenance forms if the customer's name, address, and contact details are all NULL values.

Verdict: The NULL'ing Out technique is useful in certain specific circumstances but rarely useful as the entire data sanitization strategy (Net 2000 Ltd., 2010).

2.3.1.2 Technique: Masking Data

Masking data means replacing certain fields with a mask character (such as an X). This disguises the data content, while preserving the same formatting on front end screens and reports. For example, a column of credit card numbers might look like this

4346 6454 0020 5379
4493 9238 7315 5787
4297 8296 7496 8724

After the masking operation, the information would appear as the following:

4346 XXXX XXXX 5379
4493 XXXX XXXX 5787
4297 XXXX XXXX 8724

The masking characters remove much of the sensitive content from the record while preserving the look and feel. Enough of the data must be masked to preserve security. It would not be hard to regenerate the original credit card number from a masking operation, such as 4297 8296 7496 87XX, as the numbers are generated with a specific and well-known checksum algorithm. Care must also be taken not to mask out potentially required information. A masking operation such as XXXX XXXX XXXX 5379 will strip the card issuer details from the credit card number. This may not be desirable.

Verdict: If the data are in a specific, invariable format, then masking is a powerful and fast data sanitization option. If numerous special cases must be dealt with, however, masking can be slow extremely complex to administer and can potentially leave some data items inappropriately masked (Net 2000 Ltd., 2010).

2.3.1.3 Technique: Substitution

Substitution consists of randomly replacing the contents of a column of data with information that looks similar but is completely unrelated to the real details. For example, the surnames in a customer database could be sanitized by replacing the real last names with surnames drawn from a random list.

Substitution is very effective in terms of preserving the look and feel of the existing data. The downside is that a large store of substitutable information must be maintained for each column to be substituted. For example, to sanitize surnames by substitution, a list of random last names must be available. Then to sanitize telephone numbers, a list of phone numbers must be available. The ability to generate known invalid data (phone numbers that will never work) can be a handy feature.

Substitution data can sometimes be very hard to find in large quantities. For example, if a million random street addresses are required, obtaining the substitution data can be a major exercise in itself.

Verdict: Substitution is quite powerful, works reasonably quickly, and preserves the look and feel of the data. But finding the required random data to substitute and developing the procedures to accomplish the substitution can be a major effort (Net 2000 Ltd., 2010).

2.3.1.4 Technique: Shuffling Records

Shuffling is similar to substitution except that the substitution data are derived from the column itself. Essentially, the data in a column are randomly moved between rows until there is no longer any reasonable correlation with the remaining information in the row.

There is a certain danger in the shuffling technique. It does not prevent people from asking questions such as "I wonder if so-and-so is on the supplier list?" In other words, the original data are still present, and meaningful questions can still be asked of it. Another consideration is the algorithm used to shuffle the data. If the shuffling method can be determined, the data can be easily "unshuffled." For example, if the shuffle algorithm simply runs down the table, swapping the column data between every group of two rows, it will not take much work by an interested party to revert things to their unshuffled state.

On the one hand, shuffling is rarely effective when used on small amounts of data. For example, if there are only five rows in a table, it will not be too difficult to figure out which of the shuffled data really belong to which row.

On the other hand, if a column of numeric data is shuffled, the sum and average of the column still work out to the same amount. This can sometimes be useful.

Verdict: Shuffle rules are best used on large tables and leave the look and feel of the data intact. They are fast and relatively simple to implement, as no new data need to be found, but care must be taken to use a sophisticated algorithm to randomize the shuffling of the rows (Net 2000 Ltd., 2010).

2.3.1.5 Technique: Number Variance

Number variance is useful on numeric data. Simply put, the algorithm involves modifying each number value in a column by some random percentage of its real value. This technique has the advantage of providing a reasonable disguise for the numeric data, while keeping the range and distribution of values in the column within viable limits. For example, a column of sales data might have a random variance of 10% placed on it. Some values will be higher, some lower but all will be not too far from their original range.

Verdict: The number variance technique is occasionally useful and can prevent attempts to correlate true records using known numeric data. This type of data sanitization must be used in conjunction with other options (Net 2000 Ltd., 2010).

2.3.1.6 Technique: Gibberish Generation

In general, when sanitizing data, it is essential to remove all imbedded references to the real data. For example, it is pointless to carefully remove real customer names and addresses, while leaving them intact in stored copies of correspondence in another table, especially if the original record can be determined via a simple join on a unique key.

Sanitizing "formless" nonspecific data, such as letters, memos, and notes, is one of the hardest techniques in data sanitization. Usually these types of fields are just substituted with a random quantity of equivalently sized gibberish or random words. If real-looking data are required, an elaborate substitution exercise must be undertaken or a few carefully hand built examples must be judiciously substituted to provide some representative samples.

Verdict: It is occasionally it is useful to be able to substitute quantities of random text. Gibberish generation is useful when needed but is not a widely applicable technique (Net 2000 Ltd., 2010).

2.3.1.7 Technique: Encryption/Decryption

This technique offers the option of leaving the data in place and visible to those with the appropriate key while remaining useless to anybody without the key. This seems to be a good option—yet, as with all techniques, it has its strengths and weaknesses.

The big plus is that the real data are available to anybody with the key; for example, administration personnel might be able to see personal details on their front end screens, but no one else would have this capability. This "optional" visibility is also this technique's biggest weakness. The encryption password only needs to escape once, and all the data are compromised. Even if we change the key and regenerate the test instances, stored or saved copies of the data are immediately available under the old password.

Encryption also destroys the formatting and the look and feel of the data. Encrypted data rarely look meaningful, in fact, they usually look like binary data. This sometimes leads to natural language steganography (NLS) character set issues when manipulating encrypted fields. Certain types of encryption impose constraints on the data format as well. For example, the Oracle Obfuscation toolkit requires that all data to be encrypted should have a length that is a multiple of eight characters. In effect, this means the fields must be extended with a suitable padding character, which must be stripped off at decryption time.

The strength of the encryption is another issue. Some encryption is more secure than others. According to the experts, most encryption systems can be broken—it is just a matter of time and effort. In other words, not very much will keep national security agencies from reading your files should they choose to do so. That being said, this may not be a big worry if the requirement is to simply protect proprietary business information.

Verdict: The security is dependent on the strength of the encryption used. It may not be suitable for high security requirements or where the encryption key cannot be secured. Encryption also destroys the look and feel of the sanitized data. The big plus is the selective access it offers (Net 2000 Ltd., 2010).

2.3.2 Data Sanitization Methods

A data sanitization method is the specific way a data destruction program or file shredder overwrites the data on a hard drive or other storage device. Data sanitization methods are often referred to as data erasure methods, data wipe methods, wipe algorithms, and data wipe standards. Most data destruction programs support multiple data sanitization methods.

Note: Other methods of destroying data not based on software overwriting are also called data sanitization methods, but most of the time, the term refers to software-based methods of erasing data (Fisher, 2016).

Some popular data sanitization methods used by data destruction programs are given in the following sections (Fisher, 2016).

2.3.2.1 Secure Erase

Secure Erase is the name given to a set of commands available from the firmware on parallel AT attachment (PATA)- and serial AT attachment (SATA)-based hard drives. Secure Erase commands are used as a data sanitization method to completely overwrite all data on a hard drive.

Once a hard drive has been erased using a program with Secure Erase firmware commands, no file recovery program, partition recovery program, or other data recovery method will be able to extract data from the drive (Fisher, 2016).

2.3.2.1.1 Secure Erase Wipe Method

The Secure Erase data sanitization method is implemented in the following way:

Pass 1: Writes a binary one or zero.

No verification of the overwrite is needed because the writing occurs from within the drive, so the drive's write fault detection prevents any misses. This makes Secure Erase very fast compared to other data sanitization methods and arguably more effective.

Some specific Secure Erase commands include SECURITY ERASE PREPARE and SECURITY ERASE UNIT (Fisher, 2016).

2.3.2.2 DoD 5220.22-M

DoD 5220.22-M is a software-based data sanitization method used in various file shredder and data destruction programs to overwrite existing information on a hard drive or other storage device.

Erasing a hard drive using the DoD 5220.22-M data sanitization method will prevent all software-based file recovery methods from lifting information from the drive and should also prevent most, if not all, hardware-based recovery methods.

The DoD 5220.22-M method is often incorrectly referenced as DoD 5220.2-M (.2-M instead of .22-M) (Fisher, 2016).

2.3.2.2.1 DoD 5220.22-M Wipe Method

The DoD 5220.22-M data sanitization method is usually implemented in the following way:

Pass 1: Writes a zero and verifies the write.
Pass 2: Writes a one and verifies the write.
Pass 3: Writes a random character and verifies the write.

There could be various iterations of DoD 5220.22-M, including DoD 5220.22-M (E), DoD 5220.22-M (ECE), or others. Each will probably use a character and its complement (as in 1 and 0) and varying frequencies of verifications.

While less common, another altered version of DoD 5220.22-M writes a 97 during the last pass instead of a random character (Fisher, 2016).

2.3.2.3 NCSC-TG-025

NCSC-TG-025 is a software-based data sanitization method used in some file shredder and data destruction programs to overwrite existing information on a hard drive or other storage device.

Erasing a hard drive using the NCSC-TG-025 data sanitization method will prevent all software-based file recovery methods from lifting information from the drive; it is also likely to prevent most hardware-based recovery methods from extracting information (Fisher, 2016).

2.3.2.3.1 NCSC-TG-025 Wipe Method

NCSC-TG-025 is similar to other data sanitization methods in that it passes over the data at least once to overwrite it with a zero, one, or random character. However, it is unlike Write Zero, as this method simply overwrites the information with a zero, nor is it similar to Random Data, which only uses random characters. Instead, NCSC-TG-025 combines zeros, ones, and random characters.

It is implemented in the following way:

Pass 1: Writes a zero and verifies the write.
Pass 2: Writes a one and verifies the write.
Pass 3: Writes a random character and verifies the write.

The NCSC-TG-025 data sanitization method is exactly the same as the DoD 5220.22-M method; therefore, variations in how it is implemented will be similar.

A program using this data wipe method will most likely verify that the data are successfully overwritten before moving on to the next pass. If the overwrite does not complete for some reason, the software will most likely redo that specific pass until it can verify that the data have been overwritten (Fisher, 2016).

2.3.2.4 AFSSI-5020

AFSSI-5020 is a software-based data sanitization method used in various file shredder and data destruction programs to overwrite existing information on a hard drive or other storage device. Erasing a hard drive using the AFSSI-5020 data sanitization method will prevent all software-based file recovery methods from lifting information from the drive and is also likely to prevent most hardware-based recovery methods from extracting information (Fisher, 2016).

2.3.2.4.1 AFSSI-5020 Wipe Method

The AFSSI-5020 data sanitization method is similar to the method in that it uses zeros, ones, and random characters, but different in the order and number of passes.

The AFSSI-5020 data wipe method is usually implemented in the following way:

Pass 1: Writes a zero.
Pass 2: Writes a one.
Pass 3: Writes a random character and verifies the write.

Iterations of the AFSSI-5020 data sanitization method may write a one for the first pass and a zero for the second (Fisher, 2016).

2.3.2.5 AR 380-19 Method

AR 380-19 is a software-based data sanitization method used in various file shredder and data destruction programs to overwrite existing information on a hard drive or other storage device.

Erasing a hard drive using the AR 380-19 data sanitization method will prevent all software-based file recovery methods from lifting information from the drive and is also likely to prevent most hardware-based recovery methods from extracting information (Fisher, 2016).

2.3.2.5.1 AR 380-19 Wipe Method

The AR 380-19 data sanitization method is usually implemented in the following way:

Pass 1: Writes a random character.
Pass 2: Writes a specified character (e.g., zero).
Pass 3: Writes the complement of the specified character (i.e., one) and verifies the write.

The AR 380-19 data sanitization method is sometimes used incorrectly by data destruction programs; it could implemented without a verification of the final pass or without a third pass at all (Fisher, 2016).

2.3.2.6 NAVSO P-5239-26

NAVSO P-5239-26 is a software-based data sanitization method used in various file shredder and data destruction programs to overwrite existing information on a hard drive or other storage device.

Erasing a hard drive using the NAVSO P-5239-26 data sanitization method will prevent all software-based file recovery methods from lifting information from the drive and is also likely to prevent most hardware-based recovery methods from extracting information (Fisher, 2016).

2.3.2.6.1 NAVSO P-5239-26 Wipe Method

The NAVSO P-5239-26 data sanitization method is usually implemented in the following way:

Pass 1: Writes a specified character (e.g., one).
Pass 2: Writes the complement of the specified character (e.g., zero).
Pass 3: Writes a random character and verifies the write.

The NAVSO P-5239-26 data sanitization method is the way most data destruction programs implement the standard. However, according to the actual specification, this is a less effective, "alternate method." The "preferred method" involves a more complicated overwriting pattern, discussed below (Fisher, 2016).

2.3.2.7 RCMP TSSIT OPS-II

RCMP TSSIT OPS-II is a software-based data sanitization method used in various file shredder and data destruction programs to overwrite existing information on a hard drive or other storage device (Fisher, 2016).

Erasing a hard drive using the RCMP TSSIT OPS-II data sanitization method will prevent all software-based file recovery methods from finding information on the drive and is also likely to prevent most hardware-based recovery methods from extracting information (Fisher, 2016).

2.3.2.7.1 RCMP TSSIT OPS-II Wipe Method

Some data sanitization methods will overwrite all the data with zeros, for example, Write Zero. Others may use ones as well, for example, Secure Erase, while still others use random characters, for example, the Random Data and Gutmann methods.

RCMP TSSIT OPS-II combines these methods and is usually implemented in the following way:

Pass 1: Writes a zero.
Pass 2: Writes one.

Pass 3: Writes a zero.
Pass 4: Writes one.
Pass 5: Writes a zero.
Pass 6: Writes one.
Pass 7: Writes a random character and verifies the write.

The RCMP TSSIT OPS-II data sanitization method is usually correctly used, but can be incorrectly implemented with random characters in place of some of the zero/one repeating passes in some programs.

Verifying the write in Pass 7 means the software using the data sanitization method will check that the storage device has actually been overwritten with random characters; the DoD 5220.22-M method does this after each of its passes. A verify pass will normally repeat itself if the verification check fails (Fisher, 2016).

2.3.2.8 CSEC ITSG-06

CSEC ITSG-06 is a software-based data sanitization method used in some file shredder and data destruction programs to overwrite existing information on a hard drive or other storage device.

Erasing a hard drive using the CSEC ITSG-06 data sanitization method will prevent all software-based file recovery methods from finding information on the drive and is also likely to prevent most hardware-based recovery methods from extracting information (Fisher, 2016).

2.3.2.8.1 CSEC ITSG-06 Wipe Method

The CSEC ITSG-06 data sanitization method is uses a combination of zeros and random characters, plus ones and is usually implemented in the following way:

Pass 1: Writes a one or zero.
Pass 2: Writes the complement of the previously written character (e.g., one if Pass 1 was zero).
Pass 3: Writes a random character and verifies the write.

CSEC ITSG-06 is identical to the NAVSO P-5239-26 data sanitization method. It is also similar to DoD 5220.22-M except that, as shown above, it does not verify the first two writes (Fisher, 2016).

2.3.2.9 HMG IS5

HMG IS5 (Infosec Standard 5) is a software-based data sanitization method used in some file shredder and data destruction programs to overwrite existing information on a hard drive or other storage device.

Erasing a hard drive using the HMG IS5 data sanitization method will prevent all software-based file recovery methods from finding information on the drive and is also likely to prevent most hardware-based recovery methods from extracting information.

This data wipe method comes in two similar versions: HMG IS5 Baseline and HMG IS5 Enhanced (Fisher, 2016).

2.3.2.9.1 HMG IS5 Wipe Method

Some data wipe methods only write a zero over the data, for example, Write Zero. Others such as Random Data only use random characters. However, HMG IS5 is a little different because it combines the two.

The HMG IS5 Baseline data sanitization method is usually implemented in the following way:

Pass 1: Writes a zero.
Pass 2: Writes a random character and verifies the write.

The HMG IS5 Enhanced wipe method normally works like this:

Pass 1: Writes a zero.
Pass 2: Writes a one.
Pass 3: Writes a random character and verifies the write.

HMG IS5 Enhanced is almost identical to DoD 5220.22-M except that the first two passes do not require verification. It's also very similar to CSEC ITSG-06, which writes either a one or zero for the first two passes and then finishes with a random character and verification.

When verification is required with an HMG IS5 pass, it means the program must verify that the data were overwritten. If the verification fails, the program will most likely redo that pass or give a notification that it did not complete it properly (Fisher, 2016).

2.3.2.10 ISM 6.2.92

ISM 6.2.92 is a software-based data sanitization method used in various file shredder and data destruction programs to overwrite existing information on a hard drive or other storage device.

Erasing a hard drive using the ISM 6.2.92 data sanitization method will prevent all software-based file recovery methods from lifting information from the drive and is also likely to prevent most hardware-based recovery methods from extracting information (Fisher, 2016).

2.3.2.10.1 ISM 6.2.92 Wipe Method

The ISM 6.2.92 data sanitization method is usually implemented in the following way:

Pass 1: Writes a random character and verifies the write.

If a drive is under 15 GB in size, the ISM 6.2.92 wipe method specifies that the drive being sanitized must be overwritten three times with a random character.

ISM 6.2.92 is a bit like the Random Data wipe method except Random Data normally does more than just one pass of random characters. Another difference is that ISM 6.2.92 does not require the pass to be verified.

If the pass is verified, this means the software implementing ISM 6.2.92 has made sure the data have been overwritten with random characters. If the pass does not complete properly, the software will prompt users to rerun it, or it might do so automatically (Fisher, 2016).

2.3.2.11 NZSIT 402

NZSIT 402 is a software-based data sanitization method used as a standard wipe method by the New Zealand government and any contractor or consultant providing services to the government.

Erasing a hard drive using the NZSIT 402 data sanitization method will prevent all software-based file recovery methods from lifting information from the hard drive and is also likely to prevent most hardware-based recovery methods from extracting information.

A number of file shredder and data destruction programs can be used to overwrite existing information on a hard drive or other storage device.

Note: This sanitization method is often written with a hyphen, as NZSIT 402 (Fisher, 2016).

2.3.2.11.1 NZSIT 402 Wipe Method

The NZSIT 402 data sanitization method is typically implemented in the following way:

Pass 1: Writes a random character and verifies the write.

In other words, such as the Random Data and Gutmann methods, NZSIT 402 writes a random character over every piece of information on the device, in contrast to other wipe methods, such as Write Zero, that only use zeros.

As part of the NZSIT 402 policy defined by the New Zealand government, the software must check to make sure everything has been overwritten, the "verify" part of the method.

Other data sanitization methods that are extremely similar to NZSIT 402 include ISM 6.2.92, HMG IS5, CSEC ITSG-06, NAVSO P-5239-26, and RCMP TSSIT OPS-II. Each writes a random character and then finishes by verifying the write.

A program using NZSIT 402 may make more than one pass over the drive, as in the Pfitzner method. This just means it will do the same thing one more time (or 10 more times, etc.). In additional passes, a random character is written over an already randomized piece of information.

If the software does not support multiple passes, the method can simply be run again, as many times as you like. This is true for NZSIT 402 and any other data sanitization method (Fisher, 2016).

2.3.2.12 VSITR

VSITR is a software-based data sanitization method used in some file shredder and data destruction programs to overwrite existing information on a hard drive or other storage device.

Erasing a hard drive using the VSITR data sanitization method will prevent all software-based file recovery methods from finding information on the drive and is also likely to prevent most hardware-based recovery methods from extracting information (Fisher, 2016).

2.3.2.12.1 VSITR Wipe Method

Data sanitization methods are supported by various applications but all use ones, zeros, random data, or a combination of all three. VSITR uses all three. Write Zero writes zeros over the data, and Random Data uses random characters, but VSITR combines both.

This is how the VSITR data sanitization method is usually implemented:

Pass 1: Writes a zero.
Pass 2: Writes a one.
Pass 3: Writes a zero.
Pass 4: Writes a one.
Pass 5: Writes a zero.
Pass 6: Writes a one.
Pass 7: Writes a random character.

VSITR's implementation is identical to the RCMP TSSIT OPS-II data sanitization method except VSITR lacks any verification (Fisher, 2016).

2.3.2.13 Gutmann Method

The Gutmann method was developed by Peter Gutmann in 1996 and is one of several software-based data sanitization methods used in some file shredder and data destruction programs to overwrite existing information on a hard drive or other storage device.

Using the Gutmann data sanitization method on a hard drive will prevent all software-based file recovery methods from finding information on the drive and is also likely to prevent most hardware-based recovery methods from extracting information (Fisher, 2016).

2.3.2.13.1 Gutmann Wipe Method

The Gutmann data sanitization method is often implemented in the following way:

Pass 1—35: Writes a random character.

The Gutmann method uses a random character for the first four and the last four passes, but then uses a complex pattern of overwriting from Pass 5 through Pass 31.

2.3.2.14 Schneier Method

The Schneier method is a software-based data sanitization method used in some file shredder and data destruction programs to overwrite existing information on a hard drive or other storage device.

Erasing a hard drive using the Schneier data sanitization method will prevent all software-based file recovery methods from finding information on the drive and is also likely to prevent most hardware-based recovery methods from extracting information.

2.3.2.14.1 Schneier Wipe Method

The Schneier method uses a combination of multiple passes of random characters as well as zeros and ones. It is normally implemented in the following way:

Pass 1: Writes a one.
Pass 2: Writes a zero.
Pass 3: Writes a stream of random characters.
Pass 4: Writes a stream of random characters.
Pass 5: Writes a stream of random characters.
Pass 6: Writes a stream of random characters.
Pass 7: Writes a stream of random characters.

Some programs may use the Schneier method with small variations. For example, some applications may support a verification after the first or last pass.

The verification confirms that the character, such as a one or random character, was written to the drive. If it was not, the software may send a notice or automatically restart and run through the passes again (Fisher, 2016).

2.3.2.15 Pfitzner Method

The Pfitzner method is a software-based data sanitization method created by Roy Pfitzner for erasing data from a hard drive or other storage device.

Using the Pfitzner data sanitization method will prevent all software-based file recovery methods from finding information on the drive and is also likely to prevent most hardware-based recovery methods from extracting information.

2.3.2.15.1 Pfitzner Wipe Method

While most software implements the Pfitzner method in the following way, some may modify it and use a smaller number of passes:

Pass 1—33: Writes a random character.

Tip: Random Data and Gutmann work in a very similar way to Pfitzner in that they both use only random characters to overwrite the data; the difference is how many passes are performed.

A "pass" is simply how many times the method is run. The Pfitzner method overwrites data with random characters, not once or twice but 33 different times.

In addition, most software will run the whole Pfitzner method more than once. If the method were run 50 times (definitely overkill), the software will have overwritten the drive not 33 times, but 1650 times (33 × 50)!

The software checks that the information was overwritten with random characters. If the verification process fails, the program will send a notification or automatically run the method again until it passes verification (Fisher, 2016).

2.3.2.16 Random Data Method

The Random Data method, sometimes called the random number method, is a software-based data sanitization method used in some file shredder and data destruction programs to overwrite existing information on a hard drive or other storage device.

Erasing a hard drive using the Random Data sanitization method will prevent all software-based file recovery methods from finding information on the drive and may also prevent most hardware-based recovery methods from extracting information (Fisher, 2016).

2.3.2.16.1 Random Data Wipe Method

The Random Data method, as the name suggests, only uses random characters.

It is implemented as follows:

Pass 1: Writes a random character.

Tip: A data sanitization method very similar to Random Data is NZSIT 402. It also writes random characters. However, it includes verification at the end of the pass.

Most data destruction tools with a Random Data method use it as a kind of do-it-yourself sanitization method, allowing the user to customize the number of passes. Therefore, this data wipe method may run as few as two passes or as many as 20 or 30, or more. It may also have the option of verification after each pass or after the final pass.

The program verifies that the data are overwritten with, random characters.

If verification fails, the program will either provide a prompt to rerun the task or automatically overwrite the data (Fisher, 2016).

2.3.2.17 Write Zero Method

Many file shredder and data destruction programs support the Write Zero software-based data sanitization method to overwrite existing data on a storage device such as a hard drive.

The Write Zero data sanitization method may not stop the most advanced hardware-based recovery methods from extracting at least some of the deleted data, but it is likely to prevent all software-based file recovery methods from lifting information from the drive.

Note: The Write Zero method is sometimes, and more accurately, referred to as the Single Overwrite method. It is also called zero-fill erase or zero fill (Fisher, 2016).

2.3.2.17.1 Write Zero Wipe Method

The Write Zero data sanitization method is, unsurprisingly, implemented in the following way:

Pass 1: Writes a zero.

Some implementations of the Write Zero method may include verification after the first pass, may write a character other than zero, or may write zeros over several passes, but these are not common (Fisher, 2016).

2.4 DATA COMPRESSION AND TRANSMISSION

Data transmission and storage cost money. The more information being dealt with, the more it costs. Even so, most digital data are not stored in the most compact form. Rather, they are stored in whatever way makes them easiest to use, such as ASCII text from word processors, binary code that can be executed on a computer, individual samples from a data acquisition system, etc. Typically, these easy-to-use encoding methods require data files about twice as large as are actually needed to represent the information.

Data compression is the general term for the various algorithms and programs developed to address this problem. A compression program is used to convert data from an easy-to-use format to one optimized for compactness. Likewise, an uncompression program returns the information to its original form. Techniques for data compression discussed in this section include three simple encoding techniques, run length, Huffman, and delta encoding, and two more elaborate procedures that have established themselves as industry standards, LZW, and JPEG (Smith, 2013).

2.4.1 Data Compression

Data compression is often referred to as coding, where coding is a very general term encompassing any special representation of data, which satisfies a given need. Information theory is defined as the study of efficient coding and its consequences, in the form of speed of transmission and probability of error (Ingels, 1971). Data compression may be viewed as a branch of information theory in which the primary objective is to minimize the amount of data to be transmitted.

A simple characterization of data compression is that it involves transforming a string of characters in some representation (such as ASCII) into a new string (of bits, for example), which contains the same information but whose length is as small as possible. Data compression has important applications in the areas of data transmission and data storage. Many data processing applications require storage of large volumes of data, and the number of such applications is constantly increasing as the use of computers extends to new disciplines. At the same time, the proliferation of computer communication networks is resulting in the massive transfer of data over communication links. Compressing data to be stored or transmitted reduces storage and/or communication costs. When the amount of data to be transmitted is reduced, the effect is that of increasing the capacity of the communication channel. Similarly, compressing a file to half of its original size is equivalent to doubling the capacity of the storage medium. It may then become feasible to store the data at a higher, thus faster, level of the storage hierarchy and reduce the load on the input/output channels of the computer system (Lelewer and Hirschberg, 1987).

2.4.1.1 *Fundamental Concept*

Data compression is an effective way of saving storage space and network bandwidth. Numerous compression schemes have been devised based on character encoding or on the detection of repetitive strings (Bell et al., 1989; Lelewer and Hirschberg, 1987), Many achieve data reduction rates of 2.3–2.5 bits per character for English text (Bell et al., 1989), i.e., compression factors of about 31/4. Since compression schemes are so successful for network bandwidth, the advantageous effects of data compression on I/O performance in database systems are obvious, i.e., its effects on disk space, bandwidth, and throughput. Arguably, however, the benefits of compression in database systems can be observed and exploited beyond I/O performance (Graefe and Shapiro, 1991).

This section provides a brief introduction to information theory, giving the definitions and assumptions necessary to a more comprehensive discussion and evaluation of data compression methods. The following string of characters is used to illustrate the concepts defined: *EXAMPLE* = *aa bbb cccc ddddd eeeeee fffffffgggggggg* (Lelewer and Hirschberg, 1987).

2.4.1.1.1 Definitions

A code is a mapping of source messages (words from the source alphabet alpha) into codewords (words of the code alphabet beta). The source messages are the basic units into which the string to be represented is partitioned. These basic units may be single symbols from the source alphabet, or they may be strings of symbols, such as string EXAMPLE, alpha = {a, b, c, d, e, f, g, space}.

For purposes of explanation, beta is taken to be {0, 1}. Codes can be categorized as block–block, block–variable, variable–block, or variable–variable;

SOURCE MESSAGE	CODEWORD
a	000
b	001
c	010
d	011
e	100
f	101
g	110
space	111

FIGURE 2.7 block–block code. *Adapted from Borkar, V.R., Deshmukh, K., Sarawagi, S., 2000. Automatically extracting structure from free text addresses. IEEE Data Engineering Bulletin, 23.4, 27–32.*

block–block indicates that the source messages and codewords are of fixed length, and variable–variable codes map variable–length source messages into variable–length codewords. A block–block code for EXAMPLE is shown in Fig. 2.7, and a variable–variable code is given in Fig. 2.8 If the string EXAMPLE were coded using Fig. 2.7 code, the length of the coded message would be 120; using Fig. 2.8, the length would be 30 (Lelewer and Hirschberg, 1987).

The oldest and most widely used codes, ASCII and EBCDIC, are examples of block–block codes, mapping an alphabet of 64 (or 256) single characters

SOURCE MESSAGE	CODEWORD
aa	0
bbb	1
cccc	10
ddddd	11
eeeeee	100
fffffff	101
gggggggg	110
space	111

FIGURE 2.8 variable–variable code. *Adapted from Borkar, V.R., Deshmukh, K., Sarawagi, S., 2000. Automatically extracting structure from free text addresses. IEEE Data Engineering Bulletin, 23.4, 27–32.*

onto 6-bit (or 8-bit) codewords. These are not discussed here, as they do not provide compression. We only consider the block—variable, variable—variable, and variable—block types.

When source messages of variable—length are allowed, the question of how a message *ensemble* (sequence of messages) is parsed into individual messages arises. Many of the algorithms described in what follows are *defined-word schemes*. That is, the set of source messages is determined prior to the invocation of the coding scheme. For example, in text file processing, each character may constitute a message, or messages may be defined to consist of alphanumeric and nonalphanumeric strings. In Pascal source code, each token may represent a message. All codes involving fixed-length source messages are, by default, defined-word codes. In *free-parse* methods, the coding algorithm itself parses the ensemble into variable—length sequences of symbols. Most of the known data compression methods are defined-word schemes; the free-parse model differs in a fundamental way from the classical coding paradigm.

A code is *distinct* if each codeword is distinguishable from every other one (i.e., the mapping from source messages to codewords is one-to-one). A distinct code is *uniquely decodable* if every codeword is identifiable when immersed in a sequence of codewords. Clearly, each of these features is desirable. The codes of Figs. 2.1 and 2.2 are both distinct, but the code of the former is not uniquely decodable. For example, the coded message 11 could be decoded as either *ddddd* or *bbbbbb*. A uniquely decodable code is a *prefix code* (or *prefix-free code*) if it has the prefix property, which requires that no codeword is a proper prefix of any other codeword. All uniquely decodable block—block and variable—block codes are prefix codes. The code with codewords {1, 100000, 00} is an example of a code that is uniquely decodable but does not have the prefix property. Prefix codes are *instantaneously decodable*; that is, they have the desirable property that the coded message can be parsed into codewords without the need for lookahead. Lookahead is required to decode a message encoded using the codeword set {1, 100000, 00}. For example, the first codeword of the message 1000000001 is 1, but this cannot be determined until the last (10th) symbol of the message is read (if the string of zeros had been of odd length, then the first codeword would have been 100000).

A *minimal* prefix code is a prefix code such that if x is a proper prefix of some codeword, then x *sigma* is either a codeword or a proper prefix of a codeword, for each letter *sigma* in *beta*. The set of codewords {00, 01, 10} is an example of a prefix code, which is not minimal. The fact that 1 is a proper prefix of the codeword 10 requires that 11 be either a codeword or a proper prefix of a codeword, and it is neither. Intuitively, the minimality constraint prevents the use of codewords that are longer than necessary. In the above example, the codeword 10 could be replaced by the codeword 1, yielding a minimal prefix code with shorter codewords.

The process of transforming a source ensemble into a coded message is *coding* or *encoding*. The encoded message may be referred to as an *encoding* of the source ensemble. The algorithm that constructs the mapping and uses it to transform the source ensemble is called the *encoder*. The *decoder* performs the inverse operation, restoring the coded message to its original form (Lelewer and Hirschberg, 1987).

2.4.1.2 Classification of Methods

In addition to being categorized with respect to message and codeword lengths, data compression methods are also classified as either static or dynamic. A static method is one in which the mapping from the set of messages to the set of codewords is fixed before transmission begins, so that a given message is represented by the same codeword every time it appears in the message ensemble.

The classic static defined-word scheme is Huffman coding (Huffman, 1952). In Huffman coding, the assignment of codewords to source messages is based on the probabilities with which the source messages appear in the message ensemble. Messages that appear more frequently are represented by short codewords; messages with smaller probabilities map to longer codewords. These probabilities are determined before transmission begins. A Huffman code for the ensemble EXAMPLE is given in Fig. 2.9. If EXAMPLE were coded using this Huffman mapping, the length of the coded message would be 117 (Lelewer and Hirschberg, 1987).

A code is *dynamic* if the mapping from the set of messages to the set of codewords changes over time. For example, dynamic Huffman coding involves computing an approximation to the probabilities of occurrence "on the fly," as the ensemble is being transmitted. The assignment of codewords to messages is based on the values of the relative frequencies of occurrence at each point in time. A message x may be represented by a short codeword early in the transmission because it occurs frequently at the beginning of the ensemble, even

SOURCE MESSAGE	PROBABILITY	CODEWORD
a	2/40	1001
b	3/40	1000
c	4/40	011
d	5/40	010
e	6/40	111
f	7/40	110
g	8/40	00
space	5/40	101

FIGURE 2.9 Huffman code for message EXAMPLE (code length = 117). *Adapted from Borkar, V.R., Deshmukh, K., Sarawagi, S., 2000. Automatically extracting structure from free text addresses. IEEE Data Engineering Bulletin, 23.4, 27–32.*

SOURCE MESSAGE	PROBABILITY	CODEWORD
a	2/6	10
b	3/6	0
space	1/6	11

FIGURE 2.10 Dynamic Huffman code table for prefix *aa bbb* of message EXAMPLE. *Adapted from Borkar, V.R., Deshmukh, K., Sarawagi, S., 2000. Automatically extracting structure from free text addresses. IEEE Data Engineering Bulletin, 23.4, 27–32.*

though its probability of occurrence over the total ensemble is low. Later, when the more probable messages begin to occur with higher frequency, the short codeword will be mapped to one of the higher probability messages, and x will be mapped to a longer codeword. Fig. 2.10 presents a dynamic Huffman code table corresponding to the prefix *aa bbb* of EXAMPLE. Although the frequency of *space over* the entire message is greater than that of b, at this point in time, b has higher frequency and is therefore mapped to the shorter codeword (Lelewer and Hirschberg, 1987).

Dynamic codes are also referred to in the literature as *adaptive*, in that they adapt to changes in ensemble characteristics over time. The fact that these codes adapt to changing characteristics is the source of their appeal. Some adaptive methods adapt to changing patterns in the source (Welch, 1984), while others exploit locality of reference (Bentley et al., 1986). Locality of reference is the tendency, common in a wide variety of text types, for a particular word to occur frequently for short periods of time then fall into disuse for long periods.

All of the adaptive methods are *one-pass* methods; only one scan of the ensemble is required. Static Huffman coding requires two passes: one pass to compute probabilities and determine the mapping, and a second pass for transmission. Thus, as long as the encoding and decoding times of an adaptive method are not substantially greater than those of a static method, the fact that an initial scan is not needed implies a speed improvement in the adaptive case. In addition, the mapping determined in the first pass of a static coding scheme must be transmitted by the encoder to the decoder. The mapping may preface each transmission (that is, each file sent), or a single mapping may be agreed on and used for multiple transmissions. In one-pass methods, the encoder defines and redefines the mapping dynamically, during transmission. The decoder must define and redefine the mapping in sympathy, in essence, "learning" the mapping as codewords are received.

An algorithm may be a *hybrid*, neither completely static nor completely dynamic. In a simple hybrid scheme, sender and receiver maintain identical *codebooks* containing k static codes. For each transmission, the sender must choose one of the k previously agreed-on codes and inform the receiver of his or her choice (by transmitting the "name" or number of the chosen code) (Lelewer and Hirschberg, 1987).

2.4.2 Data Compression Strategies

Table 2.7 shows two ways to categorize data compression algorithms. On (side a), the methods are classified as either lossless or lossy. A lossless technique means the restored data file is identical to the original. This is absolutely necessary for many types of data, for example, executable code, word processing files, tabulated numbers, etc. In such cases, we cannot afford to misplace even a single bit of this type of information. In comparison, data files that represent images and other acquired signals do not have to be kept in perfect condition for storage or transmission. All real-world measurements inherently contain a certain amount of noise. If the changes made to these signals resemble a small amount of additional noise, no harm is done. Compression techniques that allow this type of degradation are called lossy. This distinction is important because lossy techniques are much more effective at compression than lossless ones. The higher the compression ratio, the more noise added to the data (Smith, 2013).

Images transmitted over the World Wide Web are an excellent illustration of why data compression is important. Suppose we need to download a digitized color photograph over a computer's 33.6 kbps modem. If the image is not compressed (a TIFF file, for example), it will contain about 600 kbytes of data. If it has been compressed using a lossless technique (such as in the GIF format), it will be about one-half this size, or 300 kbytes. If lossy compression has been used (a JPEG file), it will be about 50 kbytes. The point is, the download times for these three equivalent files are 142, 71, and 12 s, respectively. That is a big difference! JPEG is the best choice for digitized photographs, while GIF is popular for drawn images, such as company logos with large areas of a single color.

In Table 2.7 (side b), the methods are classified according to a fixed or variable size of group taken from the original file and written to the compressed file. Most data compression programs operate by taking a group of

TABLE 2.7 Compression Classifications

LOSSLESS	LOSSY
Run-length	CS&Q
Huffman	JPEG
Delta	MPEG
LZW	

a) LOSSLESS or LOSSY

METHOD	GROUP SIZE: Input	Output
CS&Q	fixed	fixed
Huffman	fixed	variable
Arithmetic	variable	variable
Run-length, LZW	variable	fixed

b) Fixed or variable group size

Adapted from Borkar, V.R., Deshmukh, K., Sarawagi, S., 2000. Automatically extracting structure from free text addresses. IEEE Data Engineering Bulletin, 23.4, 27–32.

data from the original file, compressing it in some way, and then writing the compressed group to the output file. One of the techniques in this table is coarser sampling and/or quantization(CS&Q), coarser sampling and/or quantization. Suppose a digitized waveform, such as an audio signal that has been digitized to 12 bits is being compressed. Here we might read two adjacent samples from the original file (24 bits), discard one completely, discard the least significant 4 bits from the other sample, and then write the remaining 8 bits to the output file. With 24 bits in and 8 bits out, we have just implemented a 3:1 compression ratio using a lossy algorithm. While this is crude, it is very effective when used with a technique called transform compression.

The table also shows CS&Q to be a fixed-input fixed-output scheme. That is, a fixed number of bits are read from the input file and a smaller fixed number of bits are written to the output file. Other compression methods allow a variable number of bits to be read or written. Why are JPEG and MPEG not listed in this table? Simply stated, these are composite algorithms that combine many of the other techniques, making them too sophisticated for these simple categories (Smith, 2013).

2.4.3 A Data Compression Model

Before we can weigh the relative merits of the various data compression techniques, we must establish a framework for comparison. There are two dimensions along which each of the schemes discussed here may be measured: algorithm complexity and amount of compression. When data compression is used in a data transmission application, the goal is speed. Speed of transmission depends on the number of bits sent, the time required for the encoder to generate the coded message, and the time required for the decoder to recover the original ensemble. In a data storage application, although the degree of compression is the primary concern, the algorithm must be efficient for the scheme to be practical. A static scheme has three algorithms: the map construction algorithm, the encoding algorithm, and the decoding algorithm. A dynamic scheme has two: the encoding algorithm and the decoding algorithm (Lelewer and Hirschberg, 1987).

Three common measures of compression are redundancy (Shannon and Weaver, 1949), average message length (Huffman, 1952) and compression ratio (Rubin, 1976; Ruth and Kreutzer, 1972). These measures are defined below. Related to each of these measures are assumptions about the characteristics of the source. It is generally assumed in information theory that all statistical parameters of a message source are known with perfect accuracy (Gilbert, 1971). The most common model is that of a discrete memoryless source, a source whose output is a sequence of letters (or messages), each letter being a selection from some fixed alphabet a,... The letters are taken to be random, statistically independent selections from the alphabet; the selection is made according to some fixed probability assignment $p(a)$,...

(Gallager, 1968). Without loss of generality, the code alphabet is assumed to be {0, 1} here. The modifications necessary for larger code alphabets are straightforward.

In addition, it is assumed that any cost associated with the code letters is uniform. This is a reasonable assumption, although it omits applications such as telegraphy where the code symbols are of different durations. The assumption is also important, as the problem of constructing optimal codes over unequal code letter costs is a significantly different and more difficult problem. Perl and Varn have developed algorithms for minimum-redundancy prefix coding in the case of arbitrary symbol cost and equal codeword probability (Perl et al., 1975; Varn, 1971). The assumption of equal probabilities mitigates the difficulty presented by the variable symbol cost. For the more general unequal letter costs and unequal probabilities model, Karp proposes an integer linear programming approach (Karp, 1961). Several approximation algorithms have been proposed for this more difficult problem (Krause, 1962; Cot et al., 1977; Mehlhorn, 1980).

When data are compressed, the goal is to reduce redundancy, leaving only the informational content. The measure of information of a source message x (in bits) is $-\log p(x)$ [log denotes the base 2 logarithm]. This definition has intuitive appeal; in the case where $p(x=1)$, it is clear that x is not at all informative since it had to occur. Similarly, the smaller the value of $p(x)$, the more unlikely x is to appear, hence, the larger its information content (for a longer, more elegant discussion of the legitimacy of this technical definition see Abramson, 1963).

The average information content over the source alphabet can be computed by weighting the information content of each source letter by its probability of occurrence, yielding the expression SUM{i = 1 to n} $[-p(a(i)) \log p(a(i))]$. This quantity is referred to as the entropy of a source letter, or the entropy of the source, and is denoted by H. Since the length of a codeword for message $a(i)$ must be sufficient to carry the information content of $a(i)$, entropy imposes a lower bound on the number of bits required for the coded message. The total number of bits must be at least as large as the product of H and the length of the source ensemble. As the value of H is generally not an integer, variable length codewords must be used if the lower bound is to be achieved. Given that message EXAMPLE is to be encoded one letter at a time, the entropy of its source can be calculated using the probabilities given in Fig. 2.9: $H = 2.894$, so that the minimum number of bits contained in an encoding of EXAMPLE is 116. The Huffman code given in Section 2.4.1.2 does not quite achieve the theoretical minimum in this case (Lelewer and Hirschberg, 1987).

Note: both definitions of information content given here are due to Shannon and Weaver (1949). A derivation of the concept of entropy as it relates to information theory is presented by Shannon and Weaver (1949). A simpler, more intuitive explanation of entropy is offered by Ash (1965).

2.4.4 Data Transmission

Data transmission refers to computer-mediated communication among system users and among systems. To consider data transmission functions, we must adopt a broad perspective. Data transmitted via computer may include words and pictures, as well as numbers. And the procedures for data transmission may take different forms for different system applications.

Data might be transmitted by transferring a data file from one user to another, perhaps with an accompanying message to indicate the file transfer has been initiated. Data might be transmitted by directly linking two display terminals, so that whatever data one user keys onto a display will be displayed to another user as well. More commonly, however, data are transmitted as "messages," which involves the creation of a specially formatted file with a formal header to specify the sender, recipient, addresses, etc. The word "message" is commonly used to denote data transmitted with a formal header. But the phrase "message handling" sometimes refers more generally to user participation in data transmission.

In a designed information system, data transmission by file transfer may be largely automatic, accomplished with little user involvement. Data transmission by direct linking is presumably rare, achieved only under operational constraints, as when a supervisor links to a subordinate's terminal to review or direct work.

In some applications, computer-mediated data transmission may be a discrete, task-defined activity. Perhaps a system used for planning/scheduling is later used to generate and transmit orders to implement a plan. In such a system, a user can "shift gears," first creating a plan and then transmitting it.

In other applications, data transmission may be a continuing, intermittent activity mixed with other tasks. As an example, air traffic controllers might use their computer facilities to exchange information (or to hand over responsibility), while they are performing other essential flight monitoring tasks (Smith and Mosier, 1986).

An even broader requirement for computer-based message handling is found in systems whose explicit, primary purpose is to support communication. In such applications, computer-mediated data transmission is sometimes called "electronic mail." In conjunction with other new technology, the development of electronic mail has brought forecasts of a "wired society" in which we become ever more dependent on computers for communication (Martin, 1978).

Effective communication is critically important in systems where information handling requires coordination among groups of people. This is true whether communication is mediated by computer or by other means. Computer-based message handling may offer a potential means of improving communication efficiency, but careful design of the user interface is needed to realize that potential.

In the interests of efficiency, much data transmission among computers is designed to be automatic, representing a programmed message exchange between one computer and another, with no direct user involvement. If a user does not participate in data transmission, there is obviously no need to include data transmission functions in user interface design. Only when the users themselves are involved in data transmission transactions will interface design guidelines be needed.

For the users of computer systems, data transmission can impose an extra dimension of complexity. A user not only must keep track of transactions with the computer, but also must initiate and monitor data exchange with other people. Users will need extra information to control data transmission, perhaps including status information about other systems, and the communication links with other systems. They will need feedback when sending or receiving data. They may need special computer assistance in composing, storing, and retrieving messages, as well as in actual data transmission. And they will wish to control the disposition of received messages, perhaps renaming a message and storing it with other related messages, and/or sending it on to other users.

When data transmission functions can be designed as an integral part of an information system, there is a clear opportunity for the interface designer to ensure the compatibility of the procedures. For example, if the system provides procedures for text editing, file storage, and retrieval in support of so-called "word processing" functions, those same procedures should serve to edit, store, and retrieve messages. Users will not have to learn a different set of commands or select from a different assortment of menu options.

In some current applications, however, data transmission functions have been grafted onto an existing system. There is a practical advantage in buying a separate package of message handling software for use in conjunction with an already existing system. The message handling functions do not have to be designed from scratch, and they can often be added without any fundamental redesign of the existing system capabilities (Smith and Mosier, 1986).

There is a potentially serious disadvantage, however, in trying to combine separately designed software packages; they will almost surely look different to a user, unless care is taken to design an integrated user interface. Differences in user interface logic can sometimes be "papered over" by the software designer, perhaps by providing a menu overlay that conceals inconsistencies of control logic (Goodwin, 1983, 1984). How well the user interface design has integrated the disparate software packages should be evaluated in operational use (Tammaro, 1983).

Whether the user interface to data transmission functions are designed from scratch or designed separately and then integrated with other system functions, some guidelines may be needed to help the interface designer. Recent studies of computer-based message handling are chiefly concerned with determining the functional capabilities required in data transmission

(Goodwin, 1980). There is already evidence, however, that the practical use of data transmission functions can be limited by deficiencies in user interface design (Goodwin, 1982).

The general objectives of user interface design in other functional areas are equally valid for data transmission functions. Procedures for data transmission should be consistent in themselves and also consistent with procedures for data entry and display. Interface design should minimize effort and memory load on the user and permit flexibility in user control of data transmission (Smith and Mosier, 1986).

Objectives:

- Consistency of data transmission,
- Minimal user actions,
- Minimal memory load on user,
- Compatibility with other information handling,
- Flexibility for user control of data transmission.

Data are transmitted over networks using signals, which are transformed, or encoded, by computers into the voice, video, graphics, and/or the print seen on computer screens. The signals used by computers to transmit data are either digital or analog (Nortel Networks, 2002).

- Analog signals are continuous signals that vary in strength. Sound is an example of an analog signal. Sound is actually a wave and is quite similar, or analogous, to electromagnetic waves, hence the name analog. Telephones have transmitters that encode sound waves into electromagnetic waves, which then travel over wires toward their destination. The receiving telephone decodes the electromagnetic waves back into sound waves. Our brains then decode the sound waves into the words we hear. Computer modems use the same principle. Analog signals can be represented digitally; for instance, a high electromagnetic voltage could be interpreted as 1 and low voltage as 0.
- Digital signals are discrete rather than continuous. Either there is a signal or there is not one. Telegraphs transmit data with discrete signals. We either hear a tap or we do not hear a tap. Discrete signals can be represented by on and off pulses. The duration of a discrete signal can be varied, as with dots and dashes in Morse code.

Discrete signals can also be represented digitally. The presence of a signal could be coded as a 1 and the absence of a signal coded as a 0. The digits 0 and 1 are used because computer circuitry is based on binary digital data. Codes are used to group a set number of bits together, with a group of bits representing a letter, number, or other character. The computer's brain, the central processing unit, transforms these codes of 0s and 1s into the voice, video, and data that are seen. One coding scheme, ASCII, codes an "a" as the binary number 0110-0001.

Digital data are based on two states, on or off. The binary numbering system uses only two digits, 0 and 1, so it makes sense to use the binary numbering system. One digit, 0, represents off; the other digit, 1, represents on. A single 0 or 1 is called a bit. One byte is equal to eight bits (also called an octet when discussing TCP/IP). In ASCII code, one octet is the equivalent of one alphabetic or numeric character. To appreciate how computers communicate over networks, we need to know how they encode information (Nortel Networks, 2002).

2.4.4.1 Connection-Oriented and Connectionless Transmissions

Data transmission may be

- Connection-oriented, or
- Connectionless.

In a connection-oriented transmission, the destination device acknowledges receipt; in a connectionless one, it does not.

In connection-oriented transmissions, the sending (source) device establishes a connection with the receiving (destination) device. The connection is continued until all data packets have been transmitted, and the source device receives notification that the data were received by the destination device and checked for errors.

A telephone conversation is an example of a connection-oriented transmission. When a call is made, the receiving party picks up the phone, and a conversation takes place, transmitting data across the phone line. The individual making the call knows that it arrived at the correct destination and are understood.

In a connectionless transmission, the source device transmits data, but the connection is not maintained. The source device does not wait for notification that the destination device has actually received the information accurately. This method is faster than connection-oriented transmission, but less reliable, as there is no notification of whether the data were received or not. It is more common to find connectionless transmissions on LANs.

To understand a connectionless transmission, think of a radio broadcast: A radio disc jockey tells his/her friends to be sure to listen to her/his program at 9:00 p.m. At that time, the disc jockey broadcasts a message to them. Did they receive the message? Although it is quite likely they turned their radios on, listened to the message, and understood it, the disc jockey cannot be sure (Nortel Networks, 2002).

2.4.4.2 Synchronous and Asynchronous Transmission

Computers need to know when to expect data and where a character begins and ends. When receiving data, the timing on both computer devices must be coordinated if they are to work together efficiently. This coordination is called clocking, timing, or framing. There are two protocols for the timing or coordination of data signals:

- Synchronous
- Asynchronous

When data are transferred, the transmitting and receiving nodes need to agree when the signal begins and ends, so the signals can be correctly measured and interpreted. This timing process is called bit synchronization, framing, or clocking.

Synchronization of data is very important. It is somewhat like timing in music. There are a specific number of beats expected per bar. When computer devices are synchronized, a specific number of signals or "beats" are expected within a set amount of time. Timing is important because it helps us be prepared. In many schools, every 50 min, a new class period starts. Students watch the clock and expect a signal. They are usually already prepared to leave the classroom when the clock sounds because they expected the signal.

Synchronous transmission requires the communicating devices to maintain synchronous clocks during the entire connection. The sending device transmits on a specific schedule, and the receiving device accepts the data on that same fixed schedule. The receiving device knows the timing of the sending device because the timing information is embedded in it. Synchronous transmissions are common in internal computer communications and usually are sent as entire frames. Synchronous transmission is common when large blocks of data are transferred because it is efficient and has a low overhead (number of bytes of data/control + data bytes).

Asynchronous data transmission does not involve synchronizing the clocks of the sending and receiving devices. Instead, start and stop bits are used for the synchronization of data signals. The start and stop bits tell the receiving device how to interpret the data. Asynchronous transmission sends one character at a time.

Data transmission may be half duplex, whereby data are transferred in only one direction at a time. An example of half duplex is a CB radio where only one person can talk at a time. Alternatively, data transmission may be full duplex, transmitted in two directions simultaneously. A telephone conversation is an example of full-duplex communication (Nortel Networks, 2002).

2.4.5 Data Transmission and Open Systems Interconnection Model

When transmitting data over networks, it is important to conform to the Open system interconnection (OSI) model. Data travel vertically through the seven OSI layers shown in Fig. 2.11. Data are encapsulated at each layer of the transmitting device from top to bottom and stripped at the receiving device in the reverse direction. The protocols of the OSI model are used to organize the data into packets, with headers and trailers (Nortel Networks, 2002).

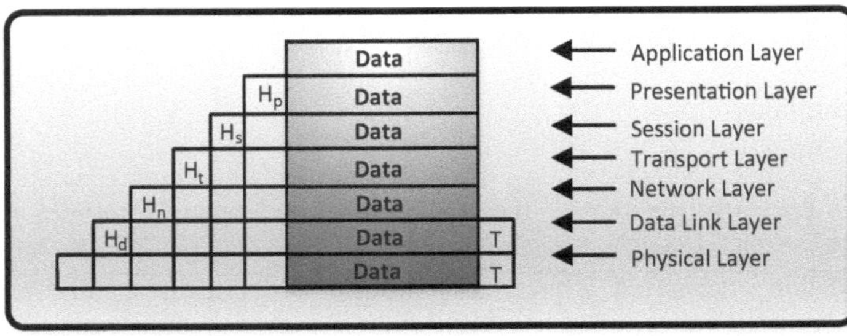

FIGURE 2.11 Open system interconnection model original data with headers and trailers. *Adapted from Borkar, V.R., Deshmukh, K., Sarawagi, S., 2000. Automatically extracting structure from free text addresses. IEEE Data Engineering Bulletin, 23.4, 27–32.*

Following the illustration in Fig. 2.11, OSI communication can be explained as follows:

- Each layer communicates with layers both immediately above and below it.
- Each layer from the sending (source) station also communicates with its peer layer at the receiving (destination) computer.
- Data start at the application layer of the source device and descend through the remaining layers before being transmitted to the destination device.
- As each layer receives the data from the layer above, it adds, in the form of headers, its own data; these data contain various protocols that enable communication.
- The original data, with the new header and the headers from the previous layers, are now sent to the next layer down.
- When the data reach the physical layer, they are transmitted across various media to the destination device.
- The destination device receives the entire frame and sends it up through the layers, one after the other in sequence.
- Each layer strips the header added by its peer layer at the source device.

Ethernet packets can contain approximately 1000 bytes. If the data being transmitted are larger than 1000 bytes, the computer breaks them down into packets. Each packet is transmitted and received separately. Packets are sequentially numbered, allowing the receiving computer to recreate the data in the correct order. Depending on the protocols used, packet size can change (Nortel Networks, 2002).

REFERENCES

Abiteboul, S., Clue, S., Milo, T., Mogilevsky, P., Simeon, J., 1999. Tools for data translation and integration. IEEE Data Engineering Bulletin 3–8 [26].

Abramson, N., 1963. Information Theory and Coding. McGraw-Hill, New York.

Ash, R.B., 1965. Information Theory. Interscience Publishers, New York.

Batini, C., Lenzerini, M., Navathe, S.B., 1986. A comparative analysis of methodologies for database schema integration. Computing Surveys 18 (4), 323–364.

Bell, T., Witten, I.H., Cleary, J.G., December 1989. Modelling for text compression. ACM Computing Surveys 21, 4.

Bentley, J.L., Sleator, D.D., Tarjan, R.E., Wei, V.K., April 1986. A locally adaptive data compression scheme. Communication of the ACM 29 (4), 320–330.

Bernstein, P.A., Dayal, U., 1994. An overview of repository technology. In: Proc. 20th VLDB.

Beskales, G., Ilyas, I.F., Golab, L., 2010. Sampling the repairs of functional dependency violations under hard constraints. PVLDB 3 (1–2), 197–207.

Bohannon, P., Fan, W., Flaster, M., Rastogi, R., 2005. A cost-based model and effective heuristic for repairing constraints by value modification. In: SIGMOD. ACM, pp. 143–154.

Borkar, V.R., Deshmukh, K., Sarawagi, S., 2000. Automatically extracting structure from free text addresses. IEEE Data Engineering Bulletin 23.4, 27–32.

Chalamalla, A., Ilyas, I.F., Ouzzani, M., Papotti, P., 2014. Descriptive and prescriptive data cleaning. In: SIGMOD, pp. 445–456.

Chaudhuri, S., Dayal, U., 1997. An overview of data warehousing and OLAP technology. ACM SIGMOD Record 26 (1).

Chu, X., Ilyas, I.F., Krishnan, S., Wang, J., June 26–July 01, 2016. Data cleaning: overview and emerging challenges. In: SIGMOD'16. San Francisco, CA, USA.

Chu, X., Ilyas, I.F., Papotti, P., 2013a. Discovering denial constraints. PVLDB 6 (13), 1498–1509.

Chu, X., Ilyas, I.F., Papotti, P., 2013b. Holistic data cleaning: putting violations into context. In: ICDE, pp. 458–469.

Cot, N., et al., 1977. Characterization and Design of Optimal Prefix Codes (Ph.D. dissertation). Computer Science Dept., Stanford Univ., Stanford, CA.

Doan, A.H., Domingos, P., Levy, A.Y., 2000. Learning source description for data integration. In: Proc. 3rd Intl. Workshop the Web and Databases (WebDB).

Do, H.H., Rahm, E., 2000. On Metadata Interoperability in Data Warehouses. Technical Report. Dept. of Computer Science, University of Leipzig.

Dwork, C., Feldman, V., Hardt, M., Pitassi, T., Reingold, O., Roth, A.L., 2015. Preserving statistical validity in adaptive data analysis. In: STOC, pp. 117–126.

FAO, May 18–30, 1998. Guidelines for the Routine Collection of Capture Fishery Data. FAO Fisheries Technical Paper 382. ISSN 0429-9345. FAO/DANIDA Expert Consultation, Bangkok, Thailand. http://www.fao.org/docrep/003/x2465e/x2465e00.htm#Contents.

Fayyad, U., 1998. Mining database: towards algorithms for knowledge discovery. IEEE Technical Bulletin Data Engineering 21 (1).

Fisher, T., April 28, 2016. Data Sanitization Methods. A List of Software Based Data Sanitization Methods. https://www.lifewire.com/data-sanitization-methods-2626133.

Gallager, R.G., 1968. Information Theory and Reliable Communication. Wiley, New York.

Galhardas, H., Florescu, D., Shasha, D., Simon, E., October 2000. Declaratively cleaning your data using AJAX. In: Journees Bases de Donnees.

Gartner, 2013. Forecast: The Internet of Things, Worldwide. https://www.gartner.com/doc/2625419/forecast-internet-things-worldwide-.

Gilbert, E.N., May 1971. Codes based on inaccurate source probabilities. IEEE Transactions on Information Theory 17 (3), 304–314.

Gokhale, C., Das, S., Doan, A., Naughton, J.F., Rampalli, N., Shavlik, J., Zhu, X., 2014. Corleone: hands-off crowdsourcing for entity matching. In: SIGMOD.

Golab, L., Karloff, H., Korn, F., Srivastava, D., Yu, B., 2008. On generating near-optimal tableaux for conditional functional dependencies. PVLDB 1 (1), 376–390.

Goodwin, N.C., 1980. A user-oriented evaluation of computer-aided message handling. In: Proceedings of the Human Factors Society 24th Annual Meeting. Human Factors Society, Santa Monica, CA, pp. 585–589.

Goodwin, N.C., 1982. Effect of interface design on usability of message handling systems. In: Proceedings of the Human Factors Society 26th Annual Meeting. Human Factors Society, Santa Monica, CA, pp. 69–73.

Goodwin, N.C., 1983. Designing a multipurpose menu driven user interface to computer based tools. In: Proceedings of the Human Factors Society 27th Annual Meeting. Human Factors Society, Santa Monica, CA, pp. 816–820.

Goodwin, N.C., 1984. Building a usable office support system from diverse components. In: Shackel, B. (Ed.), Human-Computer Interaction INTERACT'84. Elsevier Science Publishers, Amsterdam, pp. 577–581.

Graefe, G., Shapiro, L.D., April 1991. Data compression and database performance. Appeared Previously with Fewer Performance Results in Proc. ACM/IEEE-CS Symp. on Applied Computing, Kansas City, MO.

Haas, L.M., Miller, R.J., Niswonger, B., Tork Roth, M., Schwarz, P.M., Wimmers, E.L., 1999. Transforming heterogeneous data with database middleware: beyond integration. Data Engineering 31–36 [26].

Haas, D., Wang, J., Wu, E., Franklin, M.J., 2014. Clamshell: speeding up crowds for low-latency data labeling. PVLDB 9 (4), 372–383. December 2015.

Hellerstein, J.M., Stonebraker, M., Caccia, R., 1999. Independent, open enterprise data integration. IEEE Data Engineering Bulletin 43–49 [26].

Hernandez, M.A., Stolfo, S.J., 1998. Real-world data is dirty: data cleansing and the merge/purge problem. Data Mining and Knowledge Discovery 2 (1), 9–37.

Huffman, D.A., September 1952. A method for the construction of minimum-redundancy codes. Proceeding of the IRE 40 (9), 1098–1101.

Ingels, F.M., 1971. Information and Coding Theory. Intext, Scranton, PA.

Jarke, M., Lenzerini, M., Vassiliou, Y., Vassiliadis, P., 2000. Fundamentals of Data Warehouses. Springer.

Karp, R.M., January 1961. Minimum redundancy coding for the discrete noiseless channel. IRE Transactions on Information Theory 7 (1), 27–38.

Kashyap, V., Sheth, A.P., 1996. Semantic and schematic similarities between database objects: a context-based approach. VLDB Journal 5 (4), 276–304.

Krause, R.M., March 1962. Channels which transmit letters of unequal duration. Information and Control 5 (1), 13–24.

Krishnan, S., Patel, J., Franklin, M.J., Goldberg, K., 2014. A methodology for learning, analysing, and mitigating social influence bias in recommender systems. In: RecSys.

Krishnan, S., Wang, J., Wu, E., Franklin, M.J., Goldberg, K., 2015a. Active Clean: Interactive Data Cleaning While Learning Convex Loss Models.

Krishnan, S., Wang, J., Franklin, M.J., Goldberg, K., Kraska, T., 2015b. Stale view cleaning: getting fresh answers from stale materialized views. PVLDB 8 (12).

Lakshmanan, L., Sadri, F., Subramanian, I.N., 1996. SchemaSQL – a language for interoperability in relational multi-database systems. In: Proc. 26th VLDB.

Lelewer, D.A., Hirschberg, D.S., September 1987. Data compression. ACM Computing Surveys 19, 3. https://www.ics.uci.edu/~dan/pubs/DataCompression.html.

Lee, M.L., Lu, H., Ling, T.W., Ko, Y.T., 1999. Cleansing data for mining and warehousing. In: Proc. 10th Intl. Conf. Database and Expert Systems Applications (DEXA).

Li, W.S., Clifton, S., 2000. SEMINT: a tool for identifying attribute correspondences in heterogeneous databases using neural networks. Data and Knowledge Engineering 33 (1), 49–84.

Mahler, J., Krishnan, S., Laskey, M., Sen, S., Murali, A., Kehoe, B., Patil, S., Wang, J., Franklin, M., Abbeel, P., Goldberg, K.Y., 2014. Learning accurate kinematic control of cable-driven surgical robots using data cleaning and gaussian process regression. In: CASE.

Marcus, A., Parameswaran, A., 2013. Crowdsourced data management: industry and academic perspectives. Foundations and Trends in Databases 6 (1–2), 1–161.

Martin, J., 1978. The Wired Society. Prentice-Hall, Englewood Cliffs, NJ.

Mayfield, C., Neville, J., Prabhakar, S., 2010. ERACER: a database approach for statistical inference and data cleaning. In: SIGMOD.

Mehlhorn, K., September 1980. An efficient algorithm for constructing nearly optimal prefix codes. IEEE Transaction on Information Theory 26 (5), 513–517.

Milo, T., Zohar, S., 1998. Using schema matching to simplify heterogeneous data translation. In: Proc. 24th VLDB.

Monge, A.E., 2000. Matching algorithm within a duplicate detection system. IEEE Technical Bulletin Data Engineering 23 (4).

Mozafari, B., Sarkar, P., Franklin, M.J., Jordan, M.I., Madden, S., 2014. Scaling up crowd-sourcing to very large datasets: a case for active learning. PVLDB 8 (2).

Net 2000 Ltd., 2010. Data Sanitization Techniques. A Net 2000 Ltd. White Paper. http://www.DataMasker.com.

Nortel Networks, July 2002. Net-knowledge: Internetworking Fundamentals: LAN Configurations. Course 2890AEN. Student Guide 556-2890AEN-001.01.02. http://k-12.pisd.edu/currinst/network/if2_2st.pdf.

Parent, C., Spaccapietra, S., 1998. Issues and approaches of database integration. Communications of the ACM 41 (5), 166–178.

Perl, Y., Garey, M.R., Even, S., April 1975. Efficient generation of optimal prefix code: equiprobable words using unequal cost letters. Journal of the ACM 22 (2), 202–214.

Rahm, E., Hai Do, H., 2000. Data Cleaning: Problems and Current Approaches. University of Leipzig, Germany. Bulletin of the IEEE Computer Society Technical Committee on Data Engineering. Copyright 2000 IEEE.

Raman, V., Hellerstein, J.M., 1999. Potter's Wheel: An Interactive Framework for Data Cleaning. Working Paper.

Roth, T.T.M., Schwarz, P.M., 1997. Don't scrap it, wrap it, a wrapper architecture for legacy data sources. In: Proc. 23rd VLDB.

Rubin, F., November 1976. Experiments in text file compression. Communication of the ACM 19 (11), 617–623.

Rundensteiner, E., 1999. Special issue on data transformation. IEEE Technical Bulletin Data Engineering 22 (1).

Russo, D., Zou, J., 2015. Controlling Bias in Adaptive Data Analysis Using Information Theory. CoRR, abs/1511.05219.

Ruth, S.S., Kreutzer, P.J., September 1972. Data compression for large business files. Datamation 18 (9), 62–66.

Sapia, C., Höfling, G., Müller, M., Hausdorf, C., Stoyan, H., Grimmer, U., 1999. On supporting the data warehouse design by data mining techniques. In: Proc. GI-Workshop Data Mining and Data Warehousing.

Shannon, C.E., Weaver, W., 1949. The Mathematical Theory of Communication. University of Illinois Press, Urbana, IL.

Simpson, E.H., 1951. The interpretation of interaction in contingency tables. Journal of the Royal Statistical Society. Series B (Methodological) 13.

Smith, S., October 22, 2013. Digital Signal Processing: A Practical Guide for Engineers and Scientists. Newnes, 672 p.

Smith, S.L., Mosier, J.N., 1986. Guidelines for Designing User Interface Software. The MITRE Corporation, Bedford, MA, USA. ESD-TR-86-278.

Tammaro, S.G., 1983. An analysis of the use of computer-based office support tools: user characteristics and usage patterns. In: Proceedings of the Human Factors Society 27th Annual Meeting. Human Factors Society, Santa Monica, CA, pp. 882–886.

Varn, B., November 1971. Optimal variable length codes (arbitrary symbol cost and equal code word probability). Information and Control 19 (4), 289–301.

Volkovs, M., Chiang, F., Szlichta, J., Miller, R.J., 2014. Continuous data cleaning. In: ICDE, pp. 244–255.

Wargo, D., March 2016. Data Sanitization. University of Pennsylvania. https://www.med.upenn.edu/pmacs/assets/user-content/SecComm%203_16%20Data%20Sanitization.pdf.

Welch, T.A., June 1984. A technique for high-performance data compression. Computer 17 (6), 8–19.

Wiederhold, G., 1992. Mediators in the architecture of future information systems. Computer 25 (3), 38–49.

Yakout, M., Berti-Equille, L., Elmagarmid, A.K., 2013. Don't be scared: use scalable automatic repairing with maximal likelihood and bounded changes. In: SIGMOD.

Yakout, M., Elmagarmid, A.K., Neville, J., Ouzzani, M., Ilyas, I.F., 2011. Guided data repair. PVLDB 4 (5), 279–289.

Chapter 3

Preprocessing and Features

3.1 TIME AND FREQUENCY DOMAINS FOR DATA REPRESENTATION

A signal can be represented as a function of time; i.e., it varies with time. However, it can be also expressed as a function of frequency; i.e., a signal can be considered a composition of different frequency components. Thus, a signal has both time domain and frequency domain representations (Ray and Bhattacharya, 2016).

3.1.1 Time Domain Versus Frequency Domain

Time domain and frequency domain are two modes used to analyze data. Both are widely used in electronics, acoustics, telecommunications, and many other fields (MITx–Circuits and Electronics, 2012).

- Frequency domain analysis is used in conditions where processes such as filtering, amplifying, and mixing are required.
- Time domain analysis gives the behavior of a signal over time; this allows predictions and regression models for the signal.
- Frequency domain analysis is very useful in creating desired wave patterns such as binary bit patterns of a computer.
- Time domain analysis is used to understand data sent in such bit patterns over time.

3.1.1.1 Time Domain

Time domain analysis looks at data over a time period. Electronic signals, market behaviors, and biological systems are some of the functions analyzed using time domain analysis. For an electronic signal, time domain analysis is mainly based on the voltage–time plot. In a time domain analysis, the variable of interest is always measured against time. Several devices are used to analyze data on a time domain basis. The cathode ray oscilloscope is commonly used to analyze electrical signals in a time domain (MITx–Circuits and Electronics, 2012).

eMaintenance. http://dx.doi.org/10.1016/B978-0-12-811153-6.00003-8

A signal is continuous over a period if $\lim_{x \to a} s(t) = s(a)$ for all a; i.e., there is no break in the signal. A signal is *discrete* if it takes on a finite number of values. A signal is *periodic* if and only if $s(t+T) = s(t)$ for $-\alpha < t < \alpha$, where T is a constant, known as the *period*. The period is measured in seconds.

In other words, a signal is a *periodic signal* if it completes a pattern within a measurable time frame. A periodic signal is characterized by the following three parameters:

- Amplitude: the value of the signal at different instants of time, measured in volts.
- Frequency: the inverse of the time period; i.e., $f = 1/T$. The unit of frequency is Hertz (Hz) or cycles per second.
- Phase: a measure of the relative position in time of two signals within a single period, represented by ϕ in degrees or radians.

A sine wave, the most fundamental periodic signal, can be completely characterized by its amplitude, frequency, and phase. Examples of sine waves with different amplitude, frequency, and phase are shown in Fig. 3.1. The

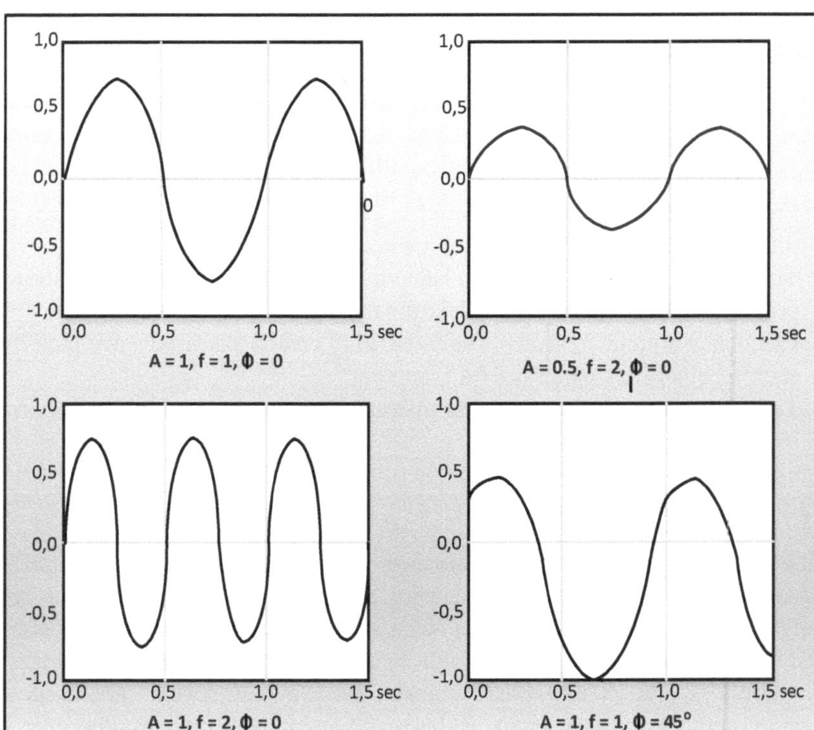

FIGURE 3.1 Examples of signals with different amplitude, frequency, and phase. *Adapted from Ray, A.K., Bhattacharya, B., 2016. Module 2-data communication fundamentals. Version 2 CSE IIT, course co-ordinated by: Kharagpur. National Programme on Technology Enhanced Learning (NPTEL). http://nptel.ac.in/courses/106105080/pdf/M2L1.pdf.*

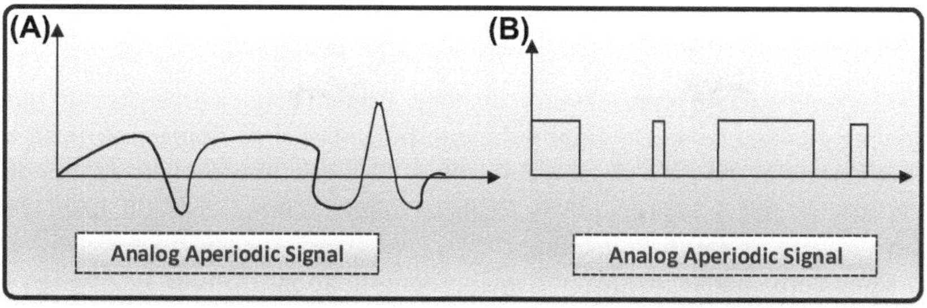

FIGURE 3.2 Examples of aperiodic signals. *Adapted from Ray, A.K., Bhattacharya, B., 2016. Module 2-data communication fundamentals. Version 2 CSE IIT, course co-ordinated by: Kharagpur. National Programme on Technology Enhanced Learning (NPTEL). http://nptel.ac.in/courses/106105080/pdf/M2L1.pdf.*

phase angle ϕ shown in Fig. 3.1A is the reference waveform (Ray and Bhattacharya, 2016).

An *aperiodic signal* or no periodic signal changes constantly without exhibiting a pattern or cycle that repeats over time, as shown in Fig. 3.2 (Ray and Bhattacharya, 2016).

3.1.1.2 Frequency Domain

Frequency domain is a method used to analyze data; in this method, a mathematical function or a signal is analyzed with respect to the frequency. It is widely used in fields such as control systems engineering, electronics, and statistics, mostly for signals or functions that are periodic over time. This does not mean frequency domain analysis cannot be used for signals that are not periodic.

The most important concept of frequency domain analysis is transformation. Transformation is used to convert a time domain function to a frequency domain function and vice versa. The most common transformation used in the frequency domain is the Fourier transformation. Because all signals have a frequency domain representation, Fourier argued that any real world waveform can be generated by the addition of sinusoidal waves. Therefore, the Fourier transformation is used to convert a signal of any shape into the sum of an infinite number of sinusoidal waves. As analyzing sinusoidal functions is easier than analyzing shaped functions, this method is very useful and widely used.

A number of mathematical transforms are used to analyze time functions; these are referred to as frequency domain methods. Signals can be transformed from the time to the frequency domain in a variety of ways. The signals can be processed within these domains, and each process in one domain has a corollary in the other (MITx–Circuits and Electronics, 2012). The following are some common transforms and the fields in which they are used:

- Fourier series—repetitive signals, oscillating systems.
- Fourier transforms—no repetitive signals, transients.

- Laplace transforms—electronic circuits and control systems.
- Z transform—discrete signals, digital signal processing.

The time domain representation displays a signal using a time-domain plot, showing changes in signal amplitude with time. The time-domain plot can be visualized with the help of an oscilloscope. The relationship between amplitude and frequency is provided by frequency domain representation, displayed with the help of spectrum analyzer.

Time domain and frequency domain representations of three sine waves of three different frequencies are shown in Fig. 3.3 (Ray and Bhattacharya, 2016).

Although simple sine waves help us understand the difference between the time domain and frequency domain representation, they are of little use in data communication, where composite signals made of many simple sine waves are common. Any composite signal can be represented by a combination of simple sine waves using Fourier analysis. For example, the signal shown in Fig. 3.4C is a combination of two sine waves with frequencies f_1 and $3f_1$, shown in Fig. 3.4A and B, respectively; it can be represented by $s(t) = \sin(\omega t) + (1/3)\sin 3(\omega t)$, where $\omega = 2\pi * f_1$.

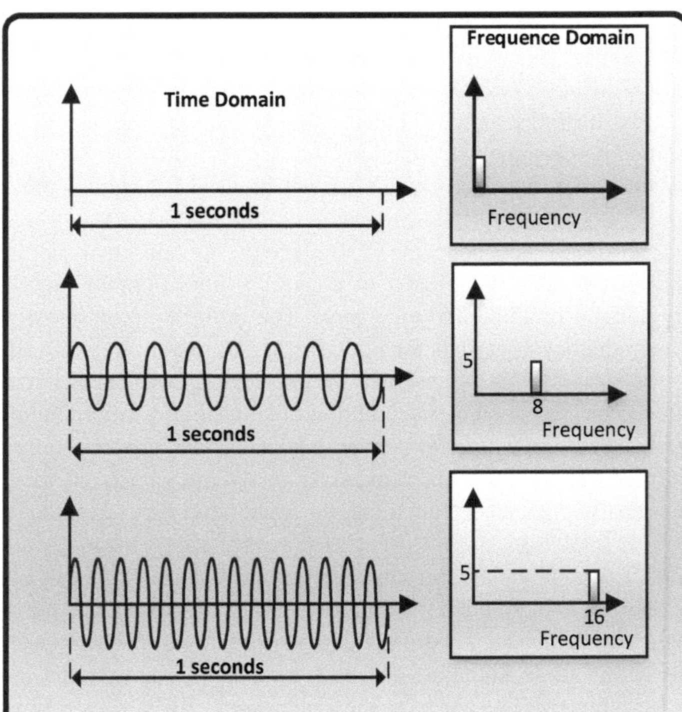

FIGURE 3.3 Time and frequency domain representations of sine waves. *Adapted from Ray, A.K., Bhattacharya, B., 2016. Module 2-data communication fundamentals. Version 2 CSE IIT, course co-ordinated by: Kharagpur. National Programme on Technology Enhanced Learning (NPTEL). http://nptel.ac.in/courses/106105080/pdf/M2L1.pdf.*

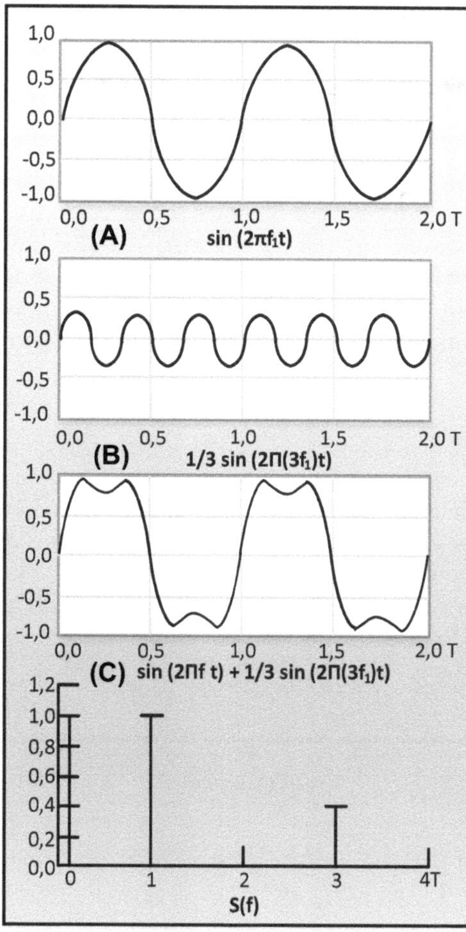

FIGURE 3.4 Time and frequency domain representations of a composite signal. *Adapted from Ray, A.K., Bhattacharya, B., 2016. Module 2-data communication fundamentals. Version 2 CSE IIT, course co-ordinated by: Kharagpur. National Programme on Technology Enhanced Learning (NPTEL). http://nptel.ac.in/courses/106105080/pdf/M2L1.pdf.*

The frequency domain function $s(f)$ specifies the constituent frequencies of the signal. The range of frequencies in a signal is known as the *spectrum*; it can be visualized with the help of a spectrum analyzer. The band of frequencies over which most of the energy of a signal is concentrated is called the *bandwidth* of the signal (Ray and Bhattacharya, 2016).

Many useful waveforms do not change in a smooth curve between maximum and minimum amplitude; they jump, slide, wobble, spike, and dip. But as long as these irregularities are consistent, cycle after cycle, a signal is still periodic and logically must be describable in the same terms as those used for sine waves. In fact, it can be decomposed into a collection of sine waves, each having a measurable amplitude, frequency, and phase (Ray and Bhattacharya, 2016).

3.1.1.3 Frequency Spectrum

Frequency spectrum of a signal is the range of frequencies contained by a signal. For example, a square wave is shown in Fig. 3.5A. It can be represented by a series of sine waves, $S(t) = 4A/\pi \sin(2\pi ft) + 4A/3\pi \sin(2\pi(3f)t) + 4A/5\pi \sin(2\pi(5f)t + \ldots)$ having frequency components f, $3f$, and $5f$, and amplitudes $4A/\pi$, $4A/3\pi$, and $4A/5\pi$ and so on. The frequency spectrum of this signal can be an approximation comprising only the first and third harmonics, as shown in Fig. 3.5B.

3.1.1.4 Bandwidth

The range of frequencies over which most of the signal energy of a signal is contained is known as the bandwidth or the effective bandwidth of the signal. The term "most" is arbitrary. It is usually defined in terms of its 3 dB cut-off frequency. The frequency spectrum and spectrum of a signal are shown in Fig. 3.6. Here, the f_l and f_h may be represented by 3 dB below ($A/\sqrt{2}$) the maximum amplitude.

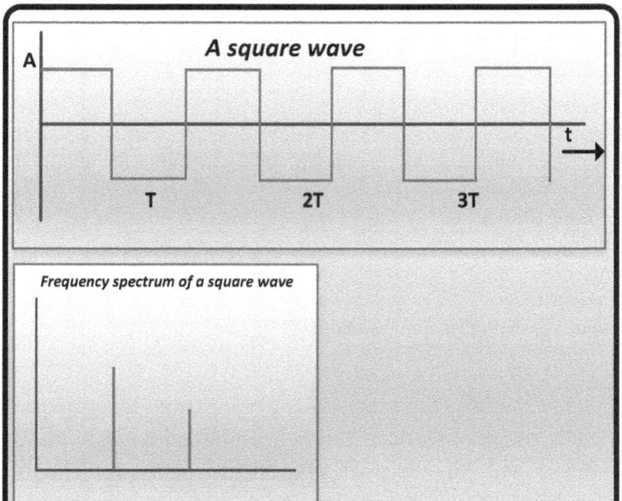

FIGURE 3.5 (A) A square wave, (B) frequency spectrum of a square wave. *Adapted from Ray, A.K., Bhattacharya, B., 2016. Module 2-data communication fundamentals. Version 2 CSE IIT, course co-ordinated by: Kharagpur. National Programme on Technology Enhanced Learning (NPTEL). http://nptel.ac.in/courses/106105080/pdf/M2L1.pdf.*

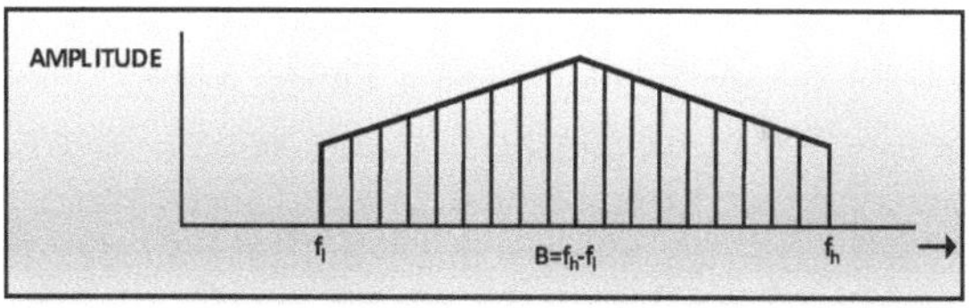

FIGURE 3.6 Frequency spectrum and bandwidth of a signal. *Adapted from Ray, A.K., Bhattacharya, B., 2016. Module 2-data communication fundamentals. Version 2 CSE IIT, course co-ordinated by: Kharagpur. National Programme on Technology Enhanced Learning (NPTEL). http://nptel.ac.in/courses/106105080/pdf/M2L1.pdf.*

3.1.2 Vibration Data Representation for Advanced Technology Facilities

Vibration control engineering of vibration-sensitive advanced technology facilities (ATFs) generally involves some combination of measurement and analysis for which data representation must be consistent. Applications include the following:

1. Design of structural and mechanical systems, including vibration isolation, and
2. Measurements in operating facilities to verify compliance with facility or equipment requirements.

The differences between the data representation used in these applications and those used for "typical" civil engineering dynamics applications (such as seismic analysis or construction vibration assessment) often lead to confusion within a design team. In some cases, generic-sounding terminology may be associated with concepts different from those in civil applications.

A fundamental difference between civil structural dynamics and ATF dynamic analysis is the reason for the analysis. In ATF analysis, we want to avoid the disruption of sensitive processes; in contrast, in civil structural dynamic analysis, we care about structural integrity. In other terms, the ATF dynamicist is concerned with defining and quantifying a "mean" or "mean + σ" environment, whereas the civil dynamicist is generally concerned with "maxima."

This section presents some of the fundamentals of signal processing and how they are applied to civil and structural engineering aspects of ATFs. As an increasing number of civil engineers are becoming involved with the design of these facilities, it is important to become conversant with the relevant analytical approaches (Amick, 1998).

3.1.2.1 Time Versus Frequency Domain in Vibrations

Vibration displacement, velocity, or acceleration can be stated in either time or frequency domains. Time domain data are representations of physical motion, wherein motion is quantified as a set of amplitudes as a function of time. Frequency domain data may be defined in a similar manner, except as a set of amplitudes as a function of frequency. It is common to refer to time domain data as a "time history" and frequency data as a "spectrum."

In the time domain, we can work with either instantaneous amplitude or an average, such as root-mean-square (RMS). Use of instantaneous amplitude requires consideration of the algebraic sign of the amplitude with respect to the "at-rest" position. The severity of instantaneous amplitude can be characterized by a maximum value over some period of time, either as 0-to-peak (the maximum absolute value) or peak-to-peak (the absolute sum of positive and negative peak amplitudes).

In frequency domain analysis, time domain data are transformed in some manner to spectra. Practitioners use several forms of spectral representations of the low-vibration environment, some based on fast Fourier transform (FFT) spectra and others based on digitally filtered signals, but response spectra are rarely used (if ever).

Spectra are defined by their frequency bandwidth and, in the context of vibration-sensitive facilities, are most commonly stated as follows:

1. Constant bandwidth (a.k.a. narrowband),
2. One-third-octave (a.k.a. proportional or percentage) bandwidth, or
3. Spectral density.

When working with measured vibrations, constant bandwidth and density spectra are typically obtained using FFT analysis, while one-third-octave band spectra are obtained by using either parallel filtering or a synthesis based on FFT analysis. All have evolved from the digital signal processing requirements of acoustics, physics, electrical engineering, and mechanical engineering. There is a large body of literature associated with spectral analysis of random and tonal vibrations (Amick, 1998).

3.1.2.2 Types of Data Signals

Measured vibration data are commonly acquired as analog time history signals produced by acceleration or velocity transducers, such as accelerometers or seismometers, respectively. From a data analysis viewpoint, Bendat and Piersol (1980) divide time history signals into two broad categories, each with two subcategories, as follows:

1. Deterministic data signals:
   - **a.** steady-state signals;
   - **b.** transient signals.
2. Random data signals:
   - **a.** stationary signals;
   - **b.** nonstationary signals.

3.1.2.2.1 Deterministic Data

Deterministic data signals are those for which we can, in theory, predict future time history values of the signal (within reasonable error) based on a knowledge of the applicable physics or past observations of the signal. A periodic signal is one that repeats itself after a constant time interval. The time history of the vibrations generated by one rotary mechanical system can take the form of either a single frequency (a pure tone) corresponding, for example, to the shaft rotation rate or a series of harmonics, the frequencies of which are integer multiples of the shaft rate. In the presence of a collection of independent (unsynchronized) periodic sources (such as the mechanical plant in a "fab"—a microelectronics fabrication facility), the collective time history may not be rigorously periodic—in this case it is referred to as "almost periodic." "Transient" deterministic data signals begin and end within a reasonable measurement time interval; they can include those from equipment start up and some types of well-controlled impacts (Bendat and Piersol, 1980).

3.1.2.2.2 Random Data

Random vibrations are broadly defined as those that are not deterministic, that is, where it is not theoretically feasible to predict future time history values based on knowledge of the applicable physics or past observations. In the time domain, the amplitude—including the peak amplitude—is random. There is no periodicity (Bendat and Piersol, 1980).

3.1.2.3 Instantaneous Versus Time-Averaged Representation

It is often convenient to represent an oscillating signal in a form that does not involve positive and negative amplitude. We cannot generally use a time-varying average, as commonly defined, because in many cases, the average of an oscillating signal is zero. We use the average energy, or RMS, as it is based on the square of the amplitude, and this is always positive. The process is usually carried out electronically by an analyzer of some sort.

For most purposes, a sinusoid can be completely characterized by its amplitude and frequency. Unfortunately, the situation of nondeterministic vibrations is not as straightforward. Random motion is typically characterized by average amplitude, the most common of which is the RMS value discussed above. In addition, there is no frequency directly associated with random vibration, and a spectrum must be used to define the frequency content.

If the average properties of the signals are time invariant, such random data are said to be "stationary." If the average properties of random signals vary with time, the signals are said to be "nonstationary." (An example of nonstationary random ground vibration is that associated with the passage of a train.) The property of ergodicity dictates that for stationary vibrations the energy average (RMS) amplitudes will be repeatable. This repeatability of the average amplitude—as opposed to the no repeatability of the peak amplitude—is a major reason why random vibrations are characterized by averages. The same repeatability can be found with the average RMS spectrum of a collection of steady-state sinusoids, allowing that RMS spectra to be used to quantify a wide variety of environments (Amick, 1998).

3.1.3 Time Series Data Representation

3.1.3.1 Piecewise Approximation

One simple and common approach to data representation is piecewise approximation. The algorithms of piecewise approximation generally segment the entire dataset into some number of nonoverlapping intervals over time and fit local models into the intervals. Formally, given time series $X = \{x_t | t = 1, 2, \ldots, N\}$, where t is the time index, the entire dataset is divided into k disjoint subsets ($k << N$) such that

$$\begin{aligned} X_1 &= \{x_t | t = 1, \ldots, b_1\} \\ X_2 &= \{x_t | t = b_1 + 1, \ldots, b_2\} \\ &\cdot \\ &\cdot \\ &\cdot \\ X_k &= \{x_t | t = b_{k-1} + 1, \ldots, N\}. \end{aligned} \tag{3.1}$$

where $b_1, \ldots, b_{k-1}$ ($b_i < b_{i+1}$, for all i) are breakpoints and $X_1 U \ldots U X_k = X$. In piecewise approximation, segmenting data over time and determining the local model are the main objectives. The length of segments or the number of the segments (k in Eq. 3.1) might be determined by a predefined constant number over time. Or the length of each segment may be determined based on the homogeneity of some properties of aggregate data, e.g., small variations or similar trends. In the latter case, the length of segments is often determined by identifying breakpoints where some property of interest of the local model significantly changes. Thus, this method may focus on identifying critical points, such as points at which the trend changes, while piecewise approximation with constant length for all segments can be more helpful to understand the overall trend of data over time.

The form of a local model for segments may be determined by some representative value or by a local parametric model. One simple local model is mean value. Using the mean value, the original data are represented as piecewise constant functions or step functions. Alternatively, linear line and polynomial models might be used along with the trend of piecewise aggregate data. Instead of using the mean, the sum of variation or volatility might be used as a representative value of the data points in each segment. The choice of segmenting and the form of local model, of course, must reflect the purpose of analysis (Seunghye, 2016).

3.1.3.2 Identification of Important Points

While piecewise approximation represents data by fitting local models or capturing some statistics of segments, data representation by identifying important points focuses on selecting a subset of points from the entire dataset. These selected data points contribute critically to the feature of the original data. Although the "importance" of the points can be defined depending on the feature the user would like to find in the data, many approaches to data reduction in the time domain attempt to find points that contribute to the shape of the original data, e.g., when there is a sudden jump or drop. If all data points are available before preprocessing, it is possible to analyze the overall data structure and choose the important points continuously for the entire dataset by the criterion of importance (batch processing). Otherwise, this criterion may be applied to a group of data points sequentially as new data are updated to identify critical points (online processing) (Seunghye, 2016).

3.1.3.3 Symbolic Data Representations

A common approach to time series representation is to convert numeric data into a finite number of discrete variables, usually symbolic variables. Converting numerical values into strings saves memory space and enables fast computation. One simple method of symbolic representation is to map numerical data within a certain range of values onto a symbol. Time series $X = \{x_i | x_i \in R,\ i = 1,\dots,\ N\}$ is mapped to symbolic series $S = \{s_i | s_i \in C,\ i = 1,\dots,\ N\}$ by symbolic representation, where C is a collection of symbols.

Another popular method is to discretize piecewise data and convert these piecewise data into strings. That is, the data representation consists of two steps: first, piecewise approximation and second, transformation of the reduced data obtained from the first step into symbols. The latter method enables data reduction, as well as memory space saving and more efficient computation, but the dimension of the original data does not change in the former method (Seunghye, 2016).

3.2 FEATURE SELECTION

Feature selection is also called variable selection or attributes selection. The automatic selection of attributes in data (such as columns in tabular data) is most relevant to predictive modeling.

A definition of feature selection is the process of selecting a subset of relevant features for use in model construction (Brownlee, 2014).

Enhancement of generalization, i.e., the performance of the learning algorithm over the test set, often motivates feature selection, i.e., detecting the relevant features and discarding the irrelevant ones. Feature selection has several advantages, such as:

- Improving the performance of the machine-learning algorithm;
- Data understanding, gaining knowledge of the process and perhaps visualizing it;

- Data reduction, limiting storage requirements and perhaps helping to reduce costs;
- Simplicity, possibility of using simpler models and gaining speed.

A unified view of feature selection is inspired by Liu and Motoda (1998). There are four aspects of feature extraction:

- Feature construction;
- Feature subset generation (or search strategy);
- Evaluation criterion definition (e.g., relevance index or predictive power); and
- Evaluation criterion estimation (or assessment method).

The last three are schematically summarized in Fig. 3.7 (Guyon et al., 2006).

Two main models deal with feature selection: filter methods and wrapper methods (Kohavi and John, 1997). While wrapper models involve optimizing a predictor as part of the selection process, filter models rely on the general

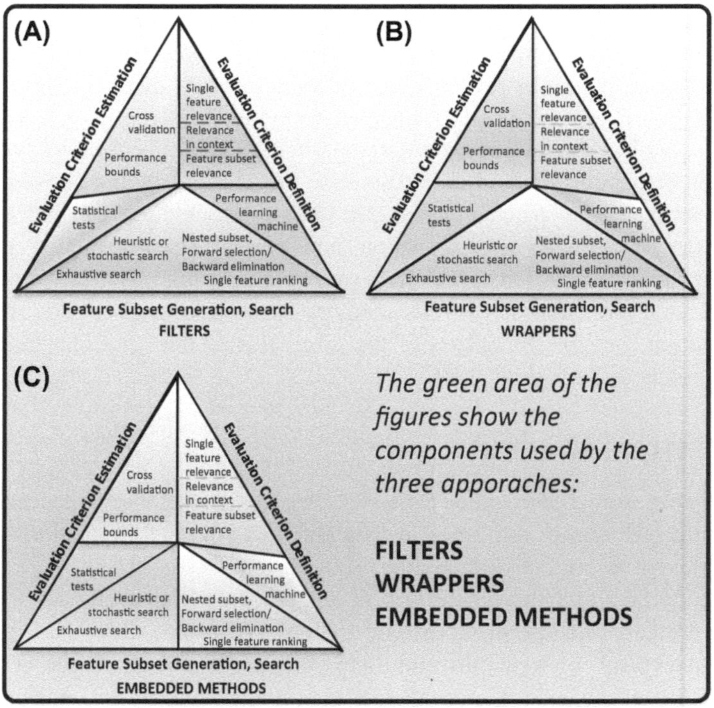

FIGURE 3.7 Three approaches to feature selection. The *shaded areas* show the components of each approach: filters, wrappers, and embedded methods. *Adapted from Guyon, I., Gunn, S., Nikravesh, M., Zadeh, L.A., 2006. Feature Extraction Foundations and Applications. Springer 2006. (pattern recognition).*

characteristics of the training data to select features independent of any predictor. Wrapper models tend to give better results, but filter models are usually computationally less expensive. When there are a large number of features, filter methods are indispensable; they can obtain a reduced set of features that can be treated by other more expensive feature selection methods (Sánchez et al., 2007). In fact, this is the basis for recent hybrid algorithms for feature selection. Such algorithms try to take advantage of the strong points of both approaches (Guyon et al., 2006). One problem is determining which filter gives the most relevant index for each case. This is not easy to solve. The many different approaches to relevance evaluation lead to a large number of indices for ranking and selection (Guyon et al., 2006). In this section, we apply several filter methods to several synthetic problems to test their effectiveness under different situations: increasing number of relevant features and samples, noisy output, and interaction between features (Sánchez et al., 2007).

3.2.1 Filter Methods Used for Feature Selection

As described above, filter methods perform the feature selection process as a preprocessing step with no induction algorithm. The general characteristics of the training data are used to select features (for example, distances between classes or statistical dependencies). This model is faster than the wrapper model and results in a better generalization because it acts independently of the induction algorithm. However, it tends to select subsets with a high number of features (even all the features), so a threshold is required to choose a subset (Sánchez et al., 2007).

Some filter methods are described in the following subsections.

3.2.1.1 Relief

The original Relief algorithm (Kira and Rendell, 1992) estimates the quality of attributes according to how well their values distinguish between instances that are near to each other. For this purpose, given a randomly selected instance, $x_i = \{x_{1i}; x_{2i}, \ldots, x_{ni}\}$, Relief searches for its two nearest neighbors: one from the same class, called nearest hit H, and the other from a different class, called nearest miss M. It then updates the quality estimate for all the features, depending on the values for x_i, M, and H.

The original Relief can deal with discrete and continuous features but is limited to two-class problems. An extension, ReliefF (Kononenko, 1994), not only deals with multiclass problems but is also more robust and capable of dealing with incomplete and noisy data. ReliefF has been adapted for continuous class (regression) problems, resulting in the RReliefF algorithm (Robnik-Sikonja and Kononenko, 2003). The Relief family of methods is especially attractive because it may be applied in all situations, has low bias, includes interaction among features, and may capture local dependencies that other methods miss (Sánchez et al., 2007).

3.2.1.2 Correlation-Based Feature Selection

Correlation-based feature selection (CFS) is a simple filter algorithm that ranks feature subsets according to a correlation-based heuristic evaluation function (Hall, 1999). The bias of the evaluation function is toward subsets containing features that are highly correlated with the class and uncorrelated with each other. Irrelevant features should be ignored because they will have low correlation with the class. Redundant features should be screened out, as they will be highly correlated with one or more of the remaining features. The acceptance of a feature will depend on the extent to which it predicts classes in areas of the instance space not already predicted by other features. CFS's feature subset evaluation function is expressed as

$$M_S = \frac{k\overline{r_{cf}}}{\sqrt{k + k(k-1)\overline{r_{ff}}'}} \tag{3.2}$$

where M_S is the heuristic "merit" of a feature subset S containing k features, $\overline{r_{cf}}$ is the mean feature-class correlation ($f \in S$), and $\overline{r_{ff}}$ is the average feature–feature intercorrelation. The numerator of this equation can be thought of as providing an indication of how predictive of the class a set of features is and the denominator of how much redundancy there is among the features (Sánchez et al., 2007).

3.2.1.3 Fast Correlated-Based Filter

The fast correlated-based filter method (Yu and Liu, 2003) is based on symmetrical uncertainty (SU) (Press et al., 1988), defined as the ratio between the information gain (IG) and the entropy (H) of two features, x and y:

$$SU(x,y) = 2\frac{IG(x/y)}{H(x) + H(y)} \tag{3.3}$$

where IG is defined as:

$$IG(x/y) = H(x) + H(y) + H(x,y) \tag{3.4}$$

where $H(x)$ and $H(x,y)$ are the entropy and joint entropy, respectively. This method was designed for high-dimensional data and has been effective in removing both irrelevant and redundant features. However, it fails to take into consideration the interaction between features (Sánchez et al., 2007).

3.2.2 Wrapper Method Approach

In the wrapper approach, shown in Fig. 3.8, feature subset selection uses the induction algorithm as a black box (i.e., no knowledge of the algorithm is needed, just the interface). The feature subset selection algorithm conducts a search for a good subset using the induction algorithm itself as part of the evaluation function. The accuracy of the induced classifiers is estimated using accuracy estimation techniques (Kohavi, 1995).

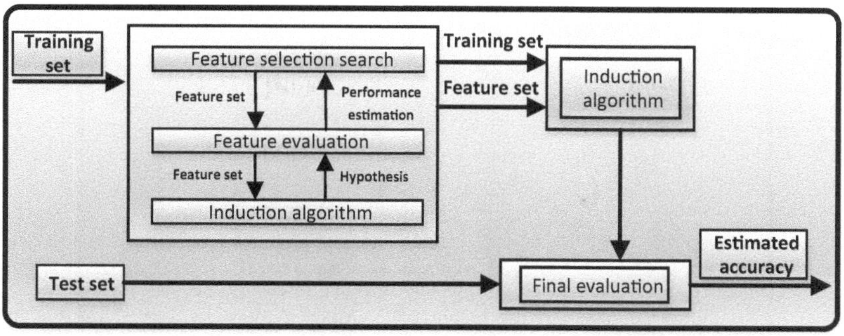

FIGURE 3.8 Wrapper approach to feature subset selection. *Adapted from Kohavi, R., John, G., 1997. Wrappers for feature subset selection. Artificial Intelligence Journal, Special Issue on Relevance 97 (1–2), 273–324.*

The wrapper approach conducts a search for possible parameters. A search requires a state space, an initial state, a termination condition, and a search engine (Ginsberg, 1993; Russell and Norvig, 1995).

In the search-space organization we have chosen, each state represents a feature subset. For n features, there are n bits in each state, and each bit indicates whether a feature is present (1) or absent (0). Operators determine the connectivity between the states, and we use operators that add or delete a single feature from a state, corresponding to the search space commonly used in stepwise methods in statistics. Fig. 3.9 shows the state space and operators for a four-feature problem. The size of the search space for n features is $0(2^n)$, making it impractical to search the whole space exhaustively, unless n is small (Kohavi and John, 1997).

The goal of the search is to find the state with the highest evaluation, using a heuristic function to guide it. As we do not know the actual accuracy of the induced classifier, we use accuracy estimation as both the heuristic function and the evaluation function. Our evaluation function is a fivefold cross-validation (Fig. 3.10) repeated multiple times. The number of repetitions is determined on the fly by looking at the standard deviation of the accuracy estimate, assuming they are independent. If the standard deviation of the accuracy estimate is above 1% and five cross-validations have not been executed, we execute another cross-validation run. While this is only a heuristic, it seems to work well in practice and avoids multiple cross-validation runs for large datasets (Kohavi and John, 1997).

This heuristic has the nice property of forcing the accuracy estimation to execute cross-validation more times on small datasets than on large datasets. Because small datasets require less time to learn, the overall accuracy estimation time, the product of the induction algorithm running time, and the cross-validation time do not grow too quickly. Thus, we conserve "hardness" using this heuristic: small datasets will be cross-validated many times to overcome the high variance resulting from small amounts of data. For much larger datasets, we could switch to a holdout heuristic to save even more time (a factor of five).

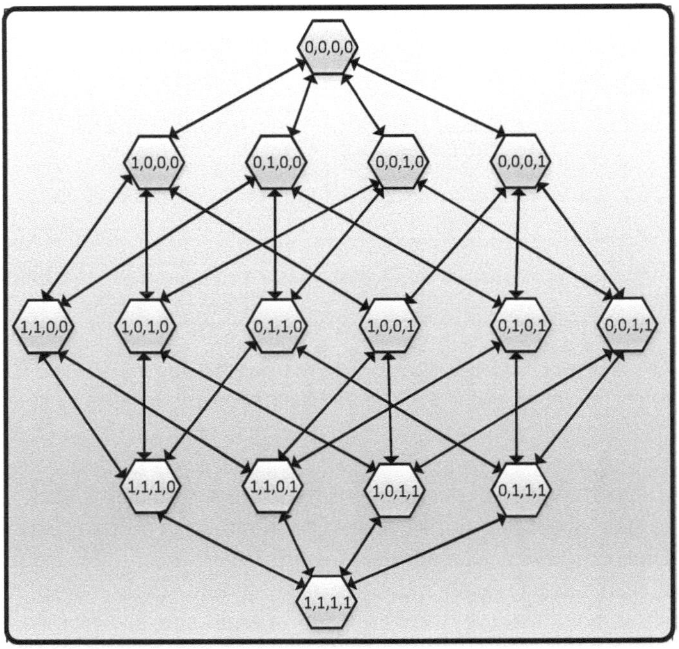

FIGURE 3.9 State space search for feature subset selection. *Adapted from Kohavi, R., John, G., 1997. Wrappers for feature subset selection. Artificial Intelligence Journal, Special Issue on Relevance 97 (1–2), 273–324.*

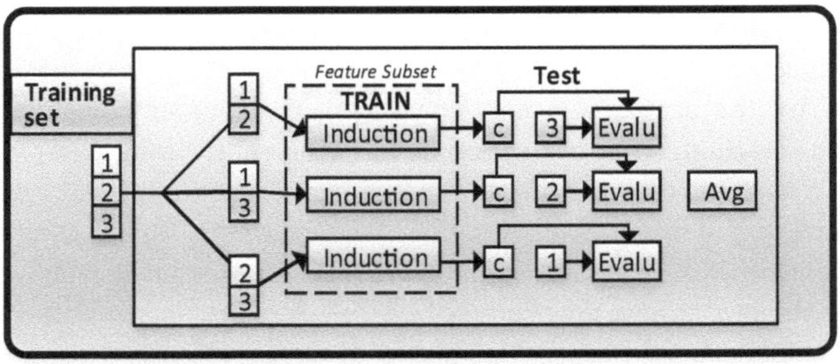

FIGURE 3.10 Cross-validation method for accuracy estimation (threefold cross-validation). *Adapted from Kohavi, R., John, G., 1997. Wrappers for feature subset selection. Artificial Intelligence Journal, Special Issue on Relevance 97 (1–2), 273–324.*

The term *forward selection* refers to a search that begins at the empty set of features; the term *backward elimination* refers to a search that begins at the full set of features (Devijver and Kittler, 1982; Miller, 1990). The initial state we use in most of our experiments is the empty set of features; hence, we are using

TABLE 3.1 Summary of the Instantiation of the Search Problem

State	A Boolean vector, one bit per feature
Initial state	The empty set of features (0,0,0,... 0)
Heuristic/evaluation	Fivefold cross-validation repeated multiple times with a small penalty (0.1%) for every feature
Search algorithm	Hill-climbing or best-first search
Termination condition	Algorithm dependent (see below)

Adapted From Kohavi, R., John, G., 1997. Wrappers for feature subset selection. Artificial Intelligence Journal, Special Issue on Relevance 97 (1–2), 273–324.

a forward selection approach. The main reason is computational: building classifiers is much faster when there are few features in the data. Although in theory, going backward from the full set of features may capture interacting features more easily, the method is extremely expensive when only the add-feature and delete-feature operators are used. Table 3.1 summarizes the instantiation of the search problem.

Note that it is possible to add a complexity penalty to the evaluation function, penalizing feature subsets with many features so as to break ties in favor of smaller subsets. For example, in our work, we set the penalty at 0.1%, quite small compared to the standard deviation of the accuracy estimation, aimed at being below 1%. No attempts are made to set this value optimally for the specific datasets. It is simply added to pick the smaller of two feature subsets with the same estimated accuracy.

3.2.3 A Statistical View of Feature Selection

The problem of feature selection can be viewed in the following, abstract way. Assume that a relevance index $r(x,y)$ has been defined to provide a quantitative assessment of the relevance of a candidate variable x for modeling a quantity y; the latter may be either real (for regression problems) or categorical (for classification problems). A vector x of measured values of x and a vector y of measured values of y are available; their components are modeled as realizations of independent and identically distributed random variables X and Y. Hence, $r(x,y)$ can be viewed as a realization of a random variable $R(X,Y)$.

Fig. 3.11 shows the probability distributions of R for relevant features and for irrelevant features (both distributions are unknown, of course). In that framework, feature selection consists of setting a threshold r_0 and making the decision that all candidate features with relevance smaller than r_0 should be discarded. The probabilities of keeping a feature although it is irrelevant (false

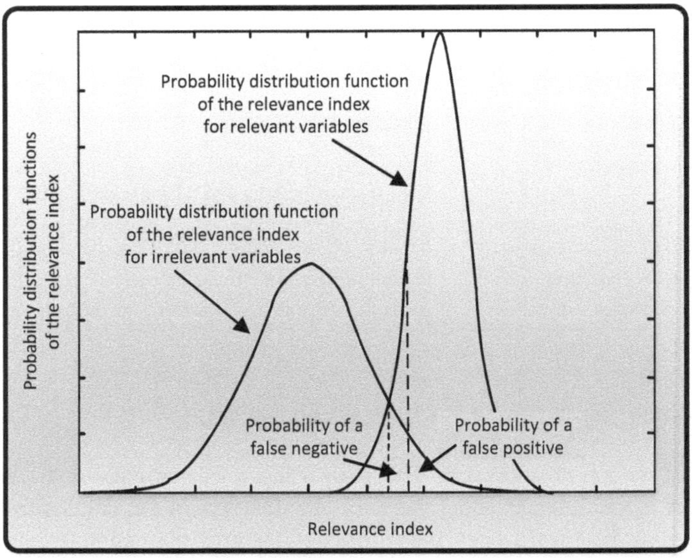

FIGURE 3.11 Probability distribution functions of the relevance index for relevant and irrelevant features. *Adapted from Guyon, I., Gunn, S., Nikravesh, M., Zadeh, L.A., 2006. Feature Extraction Foundations and Applications. Springer (pattern recognition).*

positive) and the probability of discarding a feature although it is relevant (false negative) are displayed in Fig. 3.11 (Guyon et al., 2006).

Although it is conceptually simple, the above view of the problem is abstract for the following reasons (Guyon et al., 2006):

- The relevance index must be defined, and
- Both probability distribution functions are unknown.

3.2.4 A Machine Learning View of Feature Selection

In traditional approaches to modeling, the main issue is the estimation of the parameters of a model assumed to be "true"; i.e., it is assumed that the family of functions within which the model is sought contains the target function itself. In practice, an analysis of the phenomenon of interest leads to a mathematical relation (a function or a differential equation) that contains parameters; the latter have a physical meaning and must be estimated from measurement data. The validity of the estimated parameter values must be assessed; e.g., confidence intervals for the parameters must be computed, but the validity of the mathematical function whose parameters are estimated is not questioned at the level of the statistical analysis itself.

In contrast, in machine learning approaches, the family of functions within which the model is sought has no specific a priori justification, so the model is generally not true. Models may differ in various respects:

- Sets of variables or features,
- Model complexity: number of support vectors for support vector machines (SVMs), number of hidden neurons for neural nets, number of monomials for polynomial models, etc., and,
- Initial parameter values or stopping criteria.

The choice of the model itself is part of the statistical analysis: it is the purpose of the model selection step of the modeling process. As a consequence, the parameters of the model have no specific meaning, so the accuracy of the estimation of the parameters is not an issue per se. The real issue is the generalization ability of the model, i.e., its ability to provide accurate estimates of the entity of interest on fresh data (data not used to estimate the parameters).

3.2.4.1 General Principles of Validation and Cross-Validation

In this section, we introduce the vocabulary, notations, and basic concepts of learning, generalization, overfitting, validation, and cross-validation.

In its simplest setting, model selection generally involves two datasets: a training set for parameter estimation and a validation set for model selection (Stone, 1974). Model selection can usually not be performed by comparing performances on training data: adjusting a model very accurately to the training examples is always possible if the number of adjustable parameters is large enough and comparatively few training data points are available. If the learning problem is thus underdetermined and/or if the data are noisy, fitting the training data very accurately may result in a model with poor predictive performance on new data points (poor generalization.) That effect is known as overfitting. The validation set provides a means of comparing the generalization ability of various models and detecting overfitting: if the performance on the validation set is significantly poorer than the performance on the training set, the model exhibits overfitting. If many models are compared with the validation set, the performance of the best selected model on the validation set is an optimistic estimate of the prediction performance on new data; in a sense, learning is then performed on the validation set. A third set called test set, distinct from the training and validation set, is then needed to assess the final performance of the model.

In the following, we assume the observed quantities can be modeled as the realization of a random variable Y such that $Y = f(x) + \Omega$, where Ω is the "noise," a random variable whose expected value is zero. A parameterized model $g(x,\theta)$, with vector of parameters θ, is sought, such that $g(x,\theta)$ is as close as possible to the unknown function $f(x)$. In that context, overfitting means fitting both the deterministic part of the data—the function of interest—and the noise present in the data. Fitting the noise means that the model is very sensitive to the specific realization of the noise present in the data and, hence will generalize poorly (Guyon et al., 2006).

In the risk minimization framework, the generalization error is called the expected risk, expressed as

$$R[g] = \int l(y, g(x, \theta))dP(x, y) \tag{3.5}$$

where $l(\cdot)$ is a loss function and P is the data distribution. The goal is to minimize the expected risk.

In process or signal modeling, the squared loss $(y - g(x,\theta))^2$ is commonly used; for classification problems, the 0/1 loss inflicts a penalty of 1 for classification errors and 0 otherwise. When the data distribution in unknown but m data samples are given, we can compute the empirical risk:

$$R_{emp}[g] = \frac{1}{m}\sum_{k=1}^{m} l(y_k, g(x_k, \theta)) \tag{3.6}$$

Thus, the performance on the training set is the empirical risk measured on the m_T training examples, and the performance on the validation set is the empirical risk measured on the m_V validation examples.

For the squared loss, the performance on the training set is usually called the training mean square error, expressed as $T_{MSE} = (1/m_T)\sum_{k=1}^{m_T} r_k^2$, with $r_k = y^k - g(x^k - \theta)$. Here, r_k is the modeling error on example k of the training set. Similarly, the performance on the validation set may be expressed by the validation mean square error, $V_{MSE} = (1/m_V)\sum_{k=1}^{m_V} r_k^2$. Assuming m_V and $m_T >> 1$, the ideal model, $g(x,\theta) = f(x)$, would yield values of the T_{MSE} and V_{MSE} that are equal to the estimated variance of the noise, $(1/m - 1)\sum_{k=1}^{m} (y^k - f(x^k))^2$. The "best" model will be the model that comes closest to that ideal situation, given the available data, the chosen features, and the model g.

For small datasets, a single split of the data into a training set and a validation set may provide a very inaccurate estimate of the generalization error. A preferred strategy consists of splitting the data into D subsets, training on $D - 1$ subsets, validating on the last subset, and repeating that D times with D different splits between training and validation data. The technique is known as "D-fold cross-validation." Each example of the dataset is present once, and only once, in the validation subset. The cross-validation performance is the average of the results on all the splits. For instance, the cross-validation mean square error is given by $CV_{MSE} = (1/m)\sum_{k=1}^{m} \left(r_k^V\right)^2$, where m is the total number of examples and r_k^V is the modeling error on example k when the latter is in the validation subset. Variants of the scheme are considered in the next section (Guyon et al., 2006).

Learning theory has addressed the problem of bounding the error made by approximating the expected risk with an empirical risk (see, e.g., by Vapnik, 1998). The bounds are of the type:

$$R[g] \leq R_{emp}[g] + \in \tag{3.7}$$

The right hand side of the inequality is called "guaranteed risk."

Such bounds may be used for various purposes:

- To avoid splitting the training data between a training set and a validation set: models are trained using all the available training data; the training examples are used to compute $R_{emp}[g]$; models are then selected by comparing the guaranteed risks. Alternatively, training and model selection may be performed in a single optimization process, directly minimizing the guaranteed risk. Here, the ratio of the complexity of the model and the number of training examples often appears in ε.
- To avoid reserving an additional test set to compute the final prediction performance: the validation error is used for $R_{emp}[g]$, and the bound provides a worst-case guaranteed generalization performance. Here, the ratio of the log of the number of models and the number of validation examples appears in ε (Guyon et al., 2006).

3.2.4.2 Cross-Validation and Its Variants

This section describes some useful variants of the D-fold cross-validation technique (Guyon et al., 2006).

3.2.4.2.1 Leave-One-Out

Leave-one-out is a special case of cross-validation, whereby the number of data splits is equal to the number of examples: in other words, each example is withdrawn in turn from the training set and:

- Training is performed on all examples except the left-out example and
- The modeling error on the left-out example is computed.

After m repetitions of the procedure (where m is the number of available examples), the "leave-one-out" score is computed as the average of the modeling error on the left-out examples. For instance, for the squared loss, we obtain

$$LOO_{MSE} = \frac{1}{m} \sum_{k=1}^{m} \left(k_k^{(-k)}\right)^2 \tag{3.8}$$

where $k_k^{(-k)}$ is the modeling error on example k when the latter is withdrawn from the training set. Vapnik (1982) shows that the leave-one-out score is an unbiased estimator of the generalization error for a training set of $(m-1)$ examples. However, it has a large variance and is computationally expensive, as the number of trainings is equal to the number of available examples. This last drawback is alleviated by the *virtual leave-one-out* technique, discussed in the next section (Guyon et al., 2006).

3.2.4.2.2 Virtual Leave-One-Out

The virtual leave-one-out technique relies on the assumption that, in general, the withdrawal of a single example from the training set will yield a model that is not substantially different from the model obtained by training on the whole set of available data. Therefore, the output of the model trained without example k can be written as a first-order Taylor expansion with respect to the parameters:

$$g\left(x, \theta^{(-k)}\right) \approx g(x, \theta) + Z\left(\theta^{(-k)} - \theta\right) \tag{3.9}$$

where:

- $g(x,\theta)$ is the m-vector of the predictions of model g when its input vector is x and its parameter vector, after training on the whole dataset, is θ,
- $g(x,\theta^{(-k)})$ is the m-vector of the predictions of the model trained when example k is withdrawn from the training set, with parameter vector $\theta^{(-k)}$,
- Z is the jacobian matrix of the model, i.e., the matrix whose columns are the values of the gradient of the model output with respect to each parameter of the model; hence, Z is an (m,q) matrix, where q is the number of parameters and where x^j denotes the vector of variables pertaining to example j.

This means the approximate model is linear with respect to its parameters and results from linear modeling hold approximately, insofar as the Taylor expansion is accurate enough (Seber and Wild, 1989).

For the squared loss, the predicted residual sum of squares statistic can be used to assess the leave-one-out error:

$$r_k^{(-k)} \cong \frac{r_k}{1 - h_{kk}} \tag{3.10}$$

where $r_k^{(-k)}$ is the modeling error in example k when the latter is left out of the training set, r_k is the modeling error in example k when the latter belongs to the training set, and h_{kk} is the k-th diagonal element of matrix $H = Z(Z^T Z)Z^T$. The quantity h_{kk} is the leverage of example k. The above relation is exact for linear-in-their-parameters models (Allen, 1974), such as kernel machines, and approximate otherwise (Monari and Dreyfus, 2002, Oussar et al., 2004). Therefore, the leave-one-out score can be expressed as the virtual leave-one-out score:

$$VLOO_{MSE} = \frac{1}{m} \sum_{k=1}^{m} \left(\frac{r_k}{1 - h_{kk}}\right)^2 \tag{3.11}$$

which is exact for linear-in-their-parameters machines and approximate otherwise; it is obtained at a computational cost that is m times smaller than the cost of computing LOO_{MSE} from Eq. (3.9).

Ideally, the leave-one-out score should be of the order of the variance of the noise present in the measurements (assuming homoscedasticity, i.e., that the variance of the noise is identical for all measurements.) Therefore, for sound model selection, some estimate of the noise should be available, for instance, by performing repeated measurements.

For classification, an approximate leave-one-out score for SVM classifiers (Opper and Winther, 2000) assumes that, on withdrawal of an example from the training set,

1. nonsupport vectors will remain nonsupport vectors,
2. margin support vectors will remain margin support vectors, and
3. misclassified patterns will remain misclassified.

When several candidate models with different features and/or different structures have virtual leave-one-out scores of the same order of magnitude, the significance of the differences between the estimated performances of the models must be assessed by statistical tests (see, for instance, Anders and Korn, 1999).

3.2.4.3 Performance Error Bars and Tests of Significance

Given the widespread use of the generalization error estimates provided by cross-validation and its variants, it would be desirable to have an estimator of the variance of those estimates. Unfortunately, no unbiased estimator of the variance of cross-validation has been developed (Bengio and Grandvalet, 2003).

Conversely, estimators can be found for simple validation methods where data points are used only once. For instance, for classification problems, if error rate E is computed with a single validation set (or test set) of m examples, the standard error of E is

$$stderr(E) = \sqrt{\frac{E(1-E)}{m}} \tag{3.12}$$

This error bar can be used to determine the size of a test set needed to obtain good error rate estimates (see, e.g., Guyon et al., 1998). If we prefer that the error bar does not exceed $0.1E$, assuming $E << 1$, the number of test examples should be larger than $100/E$.

Determining an optimum size for the validation set is a more challenging task. We have to monitor the trade-off between biasing the model unfavorably by withholding too many training examples and adding variance to the performance estimate by reserving too few validation examples. The complexity of the models and the number of models to be compared should also be taken into account, but this remains an open problem (Guyon et al., 2006).

3.2.4.4 Bagging

When data are sparse, "bagging" (short for "bootstrap aggregating") can be used to advantage for modeling and performance assessment. The idea underlying bagging is that averaging predictions over an ensemble of predictors will usually improve the generalization accuracy (Breiman, 1996). The ensemble of predictors is trained from an ensemble of training sets obtained by bootstrap from the initial training data (Efron and Tibshirani, 1993): subsets are drawn randomly with replacement, so that an example may be altogether absent from some training sets, but present several times in others. Different predictors are trained from those datasets; for process modeling, the predictions are averaged, while classification uses majority voting.

The generalization performance of bagged predictors can be assessed on out-of-bag samples: for each example i, the loss $l\left(y_i, g^{j(-i)}(x_i)\right)$ of each predictor $g^{j(-i)}$ whose training set does not contain example i is computed; the generalization performance estimate is $\left(1/N_{j(-i)}\right)\sum_{j(-i)} l\left(y_i, g^{j(-i)}(x_i)\right)$, where $N_{j(-i)}$ is the number of training sets that do not contain example i (Wolpert and Macready, 1999).

In the same spirit, a given number M of D-fold cross-validation splits can be generated. In each cross-validation split, a given example appears $D - 1$ times in the training sets and one time in the validation sets. Overall, each example appears M times in the validation sets and $M(D - 1)$ times in the training sets; a total of MD models are trained. For each example, the average prediction (or classification) error made by the D models not trained on that example is computed. The procedure is repeated for each example, and the average error is computed (Guyon et al., 2006).

3.2.4.5 Model Selection for Feature Selection

In this section we describe some aspects of model selection more specific to feature selection. Recall that techniques of feature selection involving the performance of a learning machine are known as "wrappers" (Guyon et al., 2006).

3.2.4.5.1 Cross-Validation for Feature Selection

Since cross-validation provides an estimate of the generalization ability of models, it is frequently used for feature selection. Actually, the prediction ability of models depends both on the features used and on the complexity of the model. Therefore, several cross-validation loops may be nested when feature selection and machine selection are performed simultaneously.

Section 3.2.1 shows the number of features can be selected from statistical principles, such as upper bounding the number of false positives, or the false discovery rate, irrespective of the machine used. Cross-validation can then be used for model selection (e.g., find the best kernel, the best number of hidden neurons, the best regularization hyperparameter, etc.), with the features selected in the previous step.

A natural but possibly dangerous alternative would be to rank all features with the whole dataset using any appropriate ranking algorithm and performing feature selection by cross-validation or leave-one-out. The procedure is the following:

- Rank candidate features using all training data and construct nested feature subsets.

  For each candidate feature subset, perform *D*-fold cross-validation or leave-one-out.
- Select the feature subset that provides the best error estimate.

However, we caution against this method; it is prone to overfitting because the validation data are used in the feature selection process (Ambroise and McLachlan, 2002).

It is more appropriate to include the feature selection in the cross-validation loop, determine the best number of features n_0, and then select n_0 features using the whole training set to produce the final model. In the end, the model must be tested on an independent test set. The procedure can be summarized as follows:

- Split data into training and validation sets.
- For each training subset:
  - Rank all features and construct nested feature subsets.
  - For each feature subset:
    - Train a model and compute its validation error.
- Average the validation errors of the subsets of the same size.
- Find the number n_0 of features that provides the smallest average validation error.
- Rank the features on all training data, select the top n_0 features, and train the final model.
- Test the final model with an independent test set.

For model selection, an inner cross-validation (or leave-one-out or virtual leave-one-out) loop may be inserted in the "for each feature subset" loop above; alternatively, model selection by cross-validation, leave-one-out or virtual leave-one-out, can be performed before and/or after feature selection (Guyon et al., 2006).

3.2.4.5.2 Guaranteed Risk and Feature Selection

Performance bounds of the type given in Eq. (3.12) are useful in several respects for feature selection. Embedded methods include feature selection algorithms derived from performance bounds. Kernel methods such as SVMs lend themselves particularly well to this methodology. In principle, reserving a validation set for feature selection may be avoided altogether, as the approach consists of minimizing a "guaranteed risk"; this incorporates a penalization for model complexity that may also penalize large feature set sizes. In practice, the number of features is sometimes determined by cross-validation (Guyon et al., 2006).

Performance bounds yield insight into the relative statistical complexity of feature selection strategies. When feature selection is performed in the context of machine learning using the performance of a model on a validation set, comparing many feature subsets comes at a price: the confidence interval for the generalization error will suffer. For instance, for classification problems, if N models (here corresponding to N feature subsets) are being compared with a validation set of size m_V, the error bar on the prediction increases monotonically with $\sqrt{Ln\, N/m_V}$ (see Vapnik, 1982; Theorem 6.1). Therefore, if all possible combinations of n_c candidate features are considered, $N = 2^{n_c}$. To keep the confidence interval from growing, the number of validation examples needed will have to scale with the number of candidate features. However, for feature ranking and nested feature subset strategies, only $N = n_c$ feature subsets are considered; only a number of validation examples scaling with $Ln(n_c)$ will be needed. Other algorithms with a "logarithmic sample complexity" are described by Langley (1994) and Ng (1998).

3.2.4.5.3 Bagging and Feature Selection

Ensemble methods such as bagging can be used for feature ranking: features can be ranked for each model of the bag and a "bagged feature ranking" performed based on the frequency of occurrence of each feature at or above a given rank in the individual rankings (Guyon et al., 2006).

3.2.5 Cross-Validation Versus Overfitting

The goal of feature selection is to choose a subset X_s of the complete set of input features $X = \{x_1, x_2,\ldots, x_M\}$ such that the subset X_s can predict the output Y with accuracy comparable to the performance of the complete input set X and with a great reduction of the computational cost.

First, let us clarify how to evaluate the performance of a set of input features. In this chapter, we propose a very conservative form of feature set evaluation to avoid overfitting. This is important. Even if feature sets are evaluated by test set cross-validation or leave-one-out cross-validation (LOOCV), an exhaustive search of possible feature sets is likely to find a misleadingly well-scoring feature set by chance. To prevent this, we use the *cascaded cross-validation* procedure in Fig. 3.12; the procedure selects from increasingly large sets of features (and, thus, from increasingly large model classes). The score for the best feature set of a given size is computed by an independent cross-validation from the score for the best size of feature set (Deng, 1998).

1. Shuffle the data set and split into a training set of 70% of the data and a testset of the remaining30%.

2. Let vary among feature-set sizes: j = (0, 1, 2, ..., m)

a) Let fs_j = best feature set of size,j where "best" is measured as a minimizer of the leave-one-out cross-validation error over the training set.
b) Let Testscore j= the RMS prediction error of feature set

b) Let Testscore = the RMS prediction error of feature fs on the test set

End of loop of (j).

3. Select the feature set fs for which the test-set score is minimized.

FIGURE 3.12 Cascaded cross-validation procedure to find the best set of up to m features. *RMS*, root-mean-square. *Adapted from Deng, K., 1998. OMEGA: On-Line Memory-Based General Purpose System Classifier (Doctoral thesis): CMU-RI-TR-98–33. The Robotics Institute, School of Computer Science, Carnegie Mellon University, Pittsburgh.*

We should make two comments about the procedure in Fig. 3.12. First, the choice of 70/30 split for training and eating is somewhat arbitrary but is empirically a good practical ratio according to more detailed experiments. Second, Fig. 3.12 does not describe how we search for the best feature set of size j in Step 2a. This is the subject of Section 3.2.6.

Evaluating the performance of a feature selection algorithm is more complicated than evaluating a feature set because to evaluate an algorithm, we must first ask the algorithm to find the best feature subset. Moreover, to give a fair estimate of how well the feature selection algorithm performs, we should try the first step on different datasets. Therefore, evaluating the performance of a feature selection algorithm, described in Fig. 3.13, has two layers of loops. The inner loop uses an algorithm to find the best subset of features. The outer loop evaluates the performance of the algorithm using different datasets (Deng, 1998).

3.2.6 Feature Selection Algorithms

This section introduces the conventional feature selection algorithm: forward feature selection. It explores three greedy variants of the forward algorithm to improve computational efficiency without sacrificing too much accuracy.

3.2.6.1 Forward Feature Selection

The forward feature selection procedure begins by evaluating all feature subsets consisting of only one input attribute. In other words, it starts by measuring the LOOCV error of one-component subsets, $\{X_1\}$, $\{X_2\}$,..., $\{X_M\}$, where M is the input dimensionality, allowing us to find the best individual feature, $X_{(1)}$.

```
1. Collect a training data set from the specific domain.
2. Shuffle the data set.
3. Break it into P partitions. (say P = 20)
4. For each partition (i = 0, 1,...., P-1)
        a) Let OuterTrainset (i) = all partitions except i.
        b) Let Outertestset (i) = the i'th partitions
        c) Let InnerTrain (i) = randomly chosen 70% of the
        OuterTrainset (i).
        d) Let InnerTest (i) = the remaining 30% of the
        OuterTrainset (i).
        e) For j = 0, 1,....., m
                Search for the best feature set with j
                components, fs using leave-out on
                InnerTrain (i).
           End loop of (j).
        f) Select the fs with the best inner test score.
        g) Let OuterScore = RMS score of the selected feature
        set on OuterTestset (i).
     End of loop of (i).
5. Return the mean Outer Score.
```

FIGURE 3.13 Full procedure for evaluating feature selection of up to m attributes. *RMS*, root-mean-square. *Adapted from Deng, K., 1998. OMEGA: On-Line Memory-Based General Purpose System Classifier (Doctoral thesis): CMU-RI-TR-98–33. The Robotics Institute, School of Computer Science, Carnegie Mellon University, Pittsburgh.*

Forward selection now finds the best subset consisting of two components: $X_{(1)}$ and one other feature from the remaining $M - 1$ input attributes. This yields a total of $M - 1$ pairs. We assume $X_{(2)}$ is the other attribute in the best pair besides $X_{(1)}$.

The input subsets with three, four, and more features are evaluated. In forward selection, the best subset with m features is the m-tuple consisting of $X_{(1)}, X_{(2)}, \ldots, X_{(m)}$, while the best feature set overall is the best of all the M steps. Assuming the cost of an LOOCV evaluation with i features is $C_{(i)}$, then the computational cost of forward selection searching for a feature subset of size m out of M total input attributes will be

$$MC(1) + (M-1)C(2) + \cdots + (M-m+1)C(m) \qquad (3.13)$$

For example, the cost of one prediction with one-nearest-neighbor as the function approximator, using a kd-tree with j inputs, is $O(j \log N)$, where N is the number of data points. Thus, the cost of computing the mean leave-one-out error which involves N predictions is $O(j\,N \log N)$. The full cost of feature selection using the above formula is $O(m^2\,MN \log N)$ (Deng, 1998).

To find the best input feature set overall, we can also employ an exhaustive search. The search begins by searching for the best one-component subset of the input features, just as in the forward selection algorithm. It goes on to find the best two-component feature subset, which may consist of any pairs of the input features. It then moves to find the best triple out of all the combinations of any three input features. The cost of an exhaustive search is the following:

$$MC(1) + \binom{M}{2} C(2) + \cdots + \binom{M}{m} C(m) \tag{3.14}$$

Forward selection is much cheaper than an exhaustive search, but it may suffer because of its greediness. For example, if $X_{(1)}$ is the best individual feature, this does not guarantee that either $\{X_{(1)}, X_{(2)}\}$ or $\{X_{(1)}, X_{(3)}\}$ must be better than $\{X_{(2)}, X_{(3)}\}$. Therefore, a forward selection algorithm may select a feature set different from that selected by exhaustive searching. With a bad selection of input features, the prediction Y_q of a query $X_q = \{x_1, x_2, \ldots, x_M\}$ may be significantly different from the true Y_q (Deng, 1998).

3.2.6.2 Three Variants of Forward Selection

This subsection investigates two questions about empirical analysis using real world datasets mixed with artificially designed features:

1. How likely is the greediness of forward selection to lead to a bad selection of input features?
2. If the greediness of forward selection does not have a significantly negative effect on accuracy, how can we modify the forward selection algorithm to be greedier to improve the efficiency even further?

We propose three greedier feature selection algorithms whose goal is to select no more than m features from a total of M input attributes, with a tolerable loss of prediction accuracy: the super greedy algorithm, greedy algorithm, and restricted forward selection (Deng, 1998).

3.2.6.2.1 Super Greedy Algorithm

First, calculate all the one-attribute LOOCV calculations, sort the individual features according to their LOOCV mean error, and then take the m best features as the selected subset. Next, calculate all M computations involving one feature and one computation involving m features. If nearest neighbor is the function approximator, the cost of a super greedy algorithm *is* $O((M + m)$ $N \log N)$ (Deng, 1998).

3.2.6.2.2 Greedy Algorithm

Calculate all the one-attribute LOOCVs, sort the data, take the best two individual features and evaluate their LOOCV error, then take the best three individual features, and so on, until m features have been evaluated. This algorithm may conclude with a subset whose size is smaller than m but whose inner test set error is smaller than that of the m component feature set. Hence, the greedy algorithm may end up with a better feature set than the super-greedy one. The cost of the greedy algorithm for nearest neighbor is $O((M + m^2)\, N \log N)$ (Deng, 1998).

3.2.6.2.3 Restricted Forward Selection

Calculate all the one-feature set LOOCV errors, and sort the features according to the corresponding LOOCV errors. Suppose the feature ranking from the most important to the least important is $x_{(1)}, x_{(2)}, \ldots, x_{(M)}$.

Now calculate the LOOCVs of two-feature subsets, which consist of the winner of the first round, $X_{(1)}$, along with another feature, either $X_{(2)}$ or $X_{(3)}$, or any other one, until $X_{(M/2)}$. There are $M/2$ of these pairs. The winner of this round will be the best two-component feature subset chosen by restricted forward selection (RFS).

Next calculate the LOOCV errors of $M/3$ subsets, which consist of the winner of the second round, along with the other $M/3$ features at the top of the remaining rank. In this way, RFS will select its best feature triple.

Continue this procedure, until RFS has found the best m-component feature set.

By the following of these four steps, RFS finds m feature sets whose sizes range from 1 to m. By comparing their LOOCV errors, RFS can find the best overall feature set.

The difference between RFS and conventional forward selection (FS) is that at each step to insert an additional feature into the subset, FS considers all the remaining features, while RFS only tries those that seem more promising. The cost of RFS for nearest neighbor is $O(M\, m\, N \log N)$ (Deng, 1998).

3.3 FEATURE EXTRACTION

Feature construction and feature selection are two potential problem areas of feature extraction.

3.3.1 Feature Extraction From the Time Domain

3.3.1.1 State-of-the-Art Features From the Time Domain

Failure diagnosis is measured numerically using features. The existing failure diagnosis for rolling element machinery focuses on features in the frequency domain, e.g., using Fourier transform, or the time-frequency domain, e.g., using wavelet transform. In the early stage of failure development, damage is not likely significant, and the defect signal can be masked by noise in the acquired signal. When analyzed in the frequency domain, the periodicity of the signal is not significant. Therefore, spectral analysis may not be effective. However, as the periodicity is definitely significant, using the time domain feature is also recommended; normal and defect signals differ in their statistical characteristics in the time domain so combined usage of time domain features with other domains can improve diagnosis accuracy.

Table 3.2 lists some domain features used in the literature. Of these, Kurtosis is an important and popular feature used in rolling element machines. Kurtosis defines the peakedness of the amplitude in the signal. When the amplitude follows normal distribution, the Kurtosis value is fixed. Beta distribution is used for both shape and scale parameters. Beta distribution is a flexible distribution, and most signals can fit it. Since the parameters in Beta distribution for normal and defect signals differ, these parameters can be used to diagnose failure (Heng and Nor, 1998). However, Heng and Nor (1998) say the Beta method does not show significant advantages over Kurtosis or Crest factor for rolling element bearings (Fuqing et al., 2011).

Of the 10 features listed in the table, Kurtosis, Crest, and Impulse have nondimensional magnitudes, making them independent of the magnitude of the signal power. In contrast, RMS, peak, standard deviation, and normal negative likelihood (NNL) are fully dependent on the power of the signal. Poor quality sensors or the location where they are mounted may negatively

TABLE 3.2 State-Of-The Art Time Domain Features

	Feature	Definition		Feature	Definition
1	Peak value	Pv = (1/2)[max(xi) − min(xi)]	6	Clearance factor	$\mathrm{Clf} = \frac{\mathrm{Pv}}{\left(\frac{1}{n}\sum_{i=1}^{n}\sqrt{\lvert x_i\rvert}\right)^2}$
2	RMS	$\mathrm{RMS} = \sqrt{\frac{1}{n}\sum_{i=1}^{n}(x_i)^2}$	7	Impulse factor	$\mathrm{Imf} = \frac{\mathrm{Pv}}{\frac{1}{n}\sum_{i=1}^{n}\lvert x_i\rvert}$
3	Standard deviation	$\mathrm{Std} = \sqrt{\frac{1}{n}\sum_{i=1}^{n}(x_i - \overline{x})^2}$	8	Shape factor	$\mathrm{Shf} = \frac{\mathrm{RMS}}{\frac{1}{n}\sum_{i=1}^{n}\lvert x_i\rvert}$
4	Kurtosis Value	$\mathrm{Kv} = \frac{\frac{1}{n}\sum_{i=1}^{n}(x_i - \overline{x})^4}{\mathrm{RMS}^4}$	9	Normal negative likelihood Value	$\mathrm{NNLV} = -\ln L;$ $L = \prod_{i=1}^{N} f(x_i, u, \sigma)$
5	Crest factor	Crf = Pv/RMS	10	Beta Parameter	The parameters in beta function

RMS, root-mean-square.
Adapted from Fuqing, Y., Kumar, U., Galar, D., 2011. Feature Extraction From Time Domain and Automatic Failure Detection Using Support Vector Machine for Rolling Element Machine. Division of Operation and Maintenance Engineering, Luleå, SE-971 87, Sweden.

influence the power and quality of the acquired signal. The main advantage of the nondimensional features, such as Kurtosis, Crest, and Impulse, is that they are immune from these nuisance factors. Nevertheless, RMS is an important feature in signal processing. It measures the power of the signal and can be used to normalize the signal. Many other features have been used in the past, for example, Beta-Kurtosis (Wang et al., 2001), Weibull negative likelihood value (Abbasion, 2007; Sreejith et al., 2008), Kurtosis Ratio (Vass, 2008), etc. All have been discarded because they do not really work, are too complex for calculation, or make nonrelevant contributions to existing features (Fuqing et al., 2011).

3.3.1.2 Normal Negative Likelihood

NNL has been used by some researchers as a feature to diagnose failure (Sreejith et al., 2008). In NNL, the amplitudes of the signal are assumed to follow normal distribution. The parameters u and σ are calculated using the maximum likelihood estimator method. As the following makes clear, the performance of NNL is equivalent to another much cheaper feature in terms of complexity of calculation (Fuqing et al., 2011).

Let the amplitudes of the signal be denoted by a series x_1, x_i,..., x_n discretely. When parameters u and σ are known, the negative likelihood function of this series is

$$f(x_1, x_2, \ldots, x_n | \mu, \sigma^2) = \prod_{i=1}^{n} f(x_i | \mu, \sigma^2) = \left(\frac{1}{2\pi\sigma^2}\right)^{n/2} \exp\left(-\frac{\sum_{i=1}^{n}(x_i - \mu)^2}{2\sigma^2}\right) \tag{3.15}$$

The maximum likelihood estimator of u and σ is

$$\widehat{u} = \overline{x} = \sum_{i=1}^{n} x_i / n \text{ and } \widehat{\sigma}^2 = \frac{\sum_{i=1}^{n}\left(x_i - \widehat{\mu}\right)}{n} \tag{3.16}$$

Substituting the above formula (3.16) into formula (3.15) and simplifying it, the follow equation is obtained:

$$f(x_1, x_2, \ldots, x_n | \mu, \sigma^2) = \left(\frac{1}{2\pi}\right)^{n/2} \exp(-n/2)\sigma^{-n} \tag{3.17}$$

Thus, the negative likelihood can be rewritten as

$$Ln = \ln\left(\left(\frac{1}{2\pi}\right)^{n/2} \exp(-n/2)\right) - n \ln \sigma \tag{3.18}$$

It can be concluded from the above that NNL is essentially equivalent to ln σ. Obviously, ln σ is not a nondimensional feature. By normalizing it using the signal energy, a new feature can be derived as

$$\text{NNNL} = \ln\frac{\sigma}{\text{RMS}} \tag{3.19}$$

The above feature is called normalized normal negative likelihood (NNNL) (Fuqing et al., 2011).

3.3.1.3 Mean Variance Ratio

The distribution of the amplitude differs in the signals of normal and defective bearings. The density function of a normal signal without defect is dominated by some noise signals; therefore, the shape of the signal tends to be a peak. The distribution of the defect signal has a wide amplitude so the variance is bigger than that of a normal signal. The left image in Fig. 3.14 denotes the normal signal of a bearing, and the right image is a defect signal. It is evident that the defect signal differs from the normal one; it has a much wider variance (Fuqing et al., 2011).

The mean and variance ratio (MVR) feature may find both defective and normal condition signals. The definition of MVR is

$$\text{MVR} = \frac{\frac{1}{n}\sum_{i=1}^{n} x_i}{\sqrt{\frac{1}{n}\sum_{i=1}^{n}(x_i - \overline{x})^2}} \tag{3.20}$$

MVR implies the degree of scatter for the distribution of signal amplitude. Obviously, MVR is also a nondimensional feature, independent of signal power (Fuqing et al., 2011).

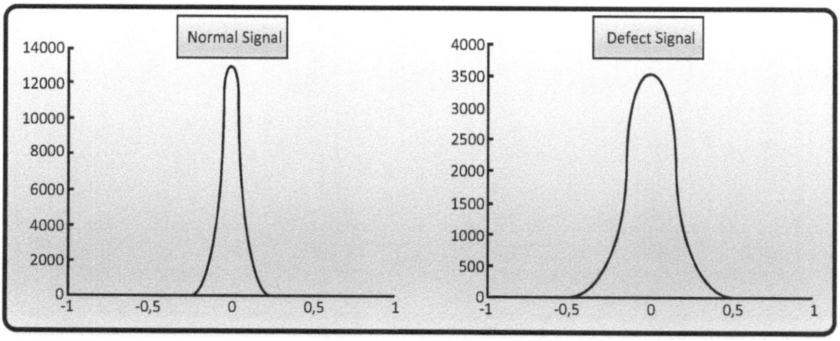

FIGURE 3.14 Normal and defect signals. *Adapted from Fuqing, Y., Kumar, U., Galar, D., 2011. Feature Extraction From Time Domain and Automatic Failure Detection Using Support Vector Machine for Rolling Element Machine. Division of Operation and Maintenance Engineering, Luleå, SE-971 87, Sweden.*

3.3.1.4 Symbolized Shannon Entropy

All the features listed in Table 3.2 are statistical features because they consider statistical characteristics of the amplitude distribution. However, all these features ignore information on the space patterns of the amplitude. For example, when rolling elements have a defect, the amplitude tends to be periodic. This periodicity cannot be reflected in the statistical features. Fig. 3.15 shows a simple example to verify this argument. This figure is comprised of 100 samples. The amplitude of each sample is {1, 2, 3, 4}, where each value has an identical probability of appearance. The upper figure and the lower figure have an identical number of samples but with different space distribution. In the upper periodic signal, the amplitude is distributed deterministically with a sequence of 1234, 1234,... iteratively. In the random signal, the amplitude is randomly distributed (Fuqing et al., 2011).

The signals in Fig. 3.15 are obviously different, but their statistical feature values listed in Table 3.1 are the same; i.e., the statistical features are not able to discriminate them.

The Shannon entropy can measure the uncertainty of a random process. Rolling element machinery without failure tends to generate a more random signal, and the machine with failure usually tends to have a more deterministic

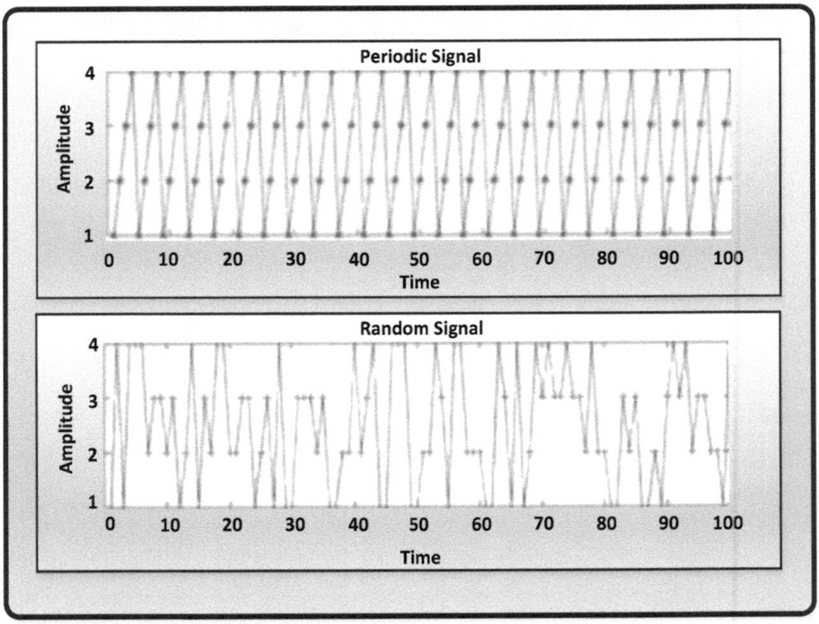

FIGURE 3.15 Periodic and random signals. *Adapted from Fuqing, Y., Kumar, U., Galar, D., 2011. Feature Extraction From Time Domain and Automatic Failure Detection Using Support Vector Machine for Rolling Element Machine. Division of Operation and Maintenance Engineering, Luleå, SE-971 87, Sweden.*

signal; i.e., the Shannon entropy will be different. To extract the periodicity in the signal, a feature named symbolized Shannon entropy (SSE) is proposed. In this feature, the signal is first symbolized, and the Shannon entropy is then used. SSE has been used to detect weak signals (Tang, 1995; Finney et al., 1998). The procedure for SSE calculation is as follows:

1. Discretize the signal and predefine a threshold. The amplitude below the threshold is coded as 0 and the above is coded as 1. Thus, the signal is discretized into a binary sequence denoted by

$$b_1, b_2, b_3, \ldots, b_i, \ldots$$

2. Segment the binary signal into units of equal length L. For example, segment the binary sequence 110010010 into 110, 010, and 010 with length $L = 3$. Calculate the decimal value of each segment. We are using "6," "2," and "2" in this example.
3. Calculate the probability of each segment using the frequency of each segment. For "6" in this example, it is 1/3, and for "2" it is 2/3.
4. Calculate the entropy using the following Shannon entropy formula:

$$H = -\frac{1}{\log N} \sum_i p_i \log p_i \tag{3.21}$$

where N is the total number of observed events, and p_i is the probability of the i event.

In a periodic signal, some sequences will occur frequently so the Shannon entropy will be lower. As the Shannon entropy values vary with different acquired data, it can be used as feature to measure the characteristics of a signal. For random data, the Shannon entropy value is 1. For deterministic signals, it is between 0 and 1 (Finney et al., 1998). Using the above procedure, for the example of Fig. 3.15, the SSE of the periodic signal is 0 and that of the random signal is 0.905. These two signals can be significantly discriminated. Similarly, different faulty bearings can be diagnosed by attending to the randomness in the acquired signals (Fuqing et al., 2011).

3.3.1.5 *Simulation of Feature Performance*

Following Heng and Nor (1998), this section simulates various signals to test the performance obtained with different features (Fuqing et al., 2011) (Fig. 3.16).

Table 3.3 shows a number of time domain features. As the table shows, the feature values differ for different signals. This means the features can segregate the different waves considered in a test. The entropy parameter shows small values for deterministic signals, such as square, sine, and triangle. For random signals, such Gaussian random and uniform random, the entropy values are above 0.9 (Fuqing et al., 2011).

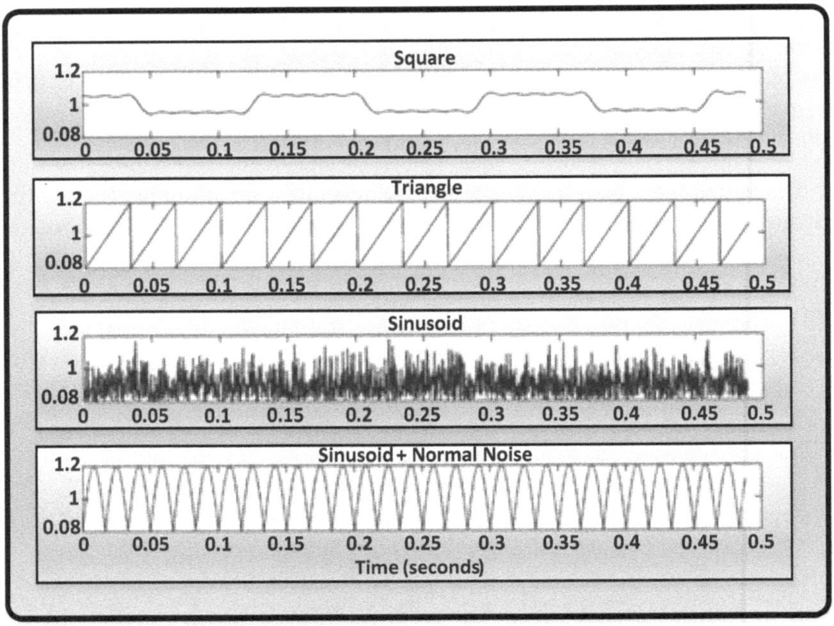

FIGURE 3.16 Simulated signals. *Adapted from Fuqing, Y., Kumar, U., Galar, D., 2011. Feature Extraction From Time Domain and Automatic Failure Detection Using Support Vector Machine for Rolling Element Machine. Division of Operation and Maintenance Engineering, Luleå, SE-971 87, Sweden.*

TABLE 3.3 State-Of-The Art Time Domain Features

Signals	Kurtosis	Crest Factor	NNNL	MVR	Entropy	Shape	Clear
Square	1.0866	0.059	3.02	20.42	0.12	1.00	0.06
Triangle	1.8	0.87	0.69	1.73	0.19	1.15	1.12
N(0,1)	3.91	2.08	0.50	1.32	0.90	1.25	3.07
U(0,1)	1.79	0.86	0.70	1.74	0.91	1.15	1.12
Sin	1.93	0.71	0.83	2.07	0.26	1.11	0.86
Sin + *N*(0,0.1)	1.97	0.95	0.81	2.02	0.46	1.11	1.17
Sin + *N*(0,1)	3.33	1.78	0.52	1.35	0.90	1.24	2.60

MVR, mean and variance ratio; *NNNL*, normalized normal negative likelihood.
Adapted from Fuqing, Y., Kumar, U., Galar, D., 2011. Feature Extraction From Time Domain and Automatic Failure Detection Using Support Vector Machine for Rolling Element Machine. Division of Operation and Maintenance Engineering, Luleå, SE-971 87, Sweden.

The table also shows two trials done with a sinusoidal signal (bottom two lines). Two white Gaussian noises, $N(0,0.1)$ and $N(0,1)$, are added to the trials respectively. Further calculation shows all the time domain features differ, suggesting this domain is extremely sensitive to noise (Fuqing et al., 2011).

3.3.2 Other Types of Feature Extraction Methods

If a signal contains frequency components emerging and vanishing in certain time intervals, both time and frequency localization are required. The traditional method proposed for such an analysis is the short-time Fourier transform (STFT) or Gabor transform. The STFT enables the time localization of a certain sinusoidal frequency but with an inherent limitation of the Heisenberg's uncertainty principle, which states that resolution in time and frequency cannot be arbitrarily small, because their product is lower bounded by

$$\Delta t \Delta f \geq \frac{1}{4\pi} \tag{3.22}$$

To overcome the resolution limitation of the STFT, a decomposition of square integrable signals $L^2(R)$ has recently been developed are termed wavelets. These families of functions $h_{a,b}$

$$h_{a,b}(t) = |a|^{\frac{1}{2}} h(t-b)a, b \in \_, \quad a \neq 0 \tag{3.23}$$

are generated from a single function $h(t)$ by the operation of dilations and translations. The wavelet transform of a continuous signal $x(t)$ can be defined as

$$CWT_x(b,a) = \left\langle x(t), |a|^{-\frac{1}{2}} h * ((t-b)/a) \right\rangle = |a|^{-\frac{1}{2}} \int x(t) h * ((t-b)/a) dt \tag{3.24}$$

where * represents the complex conjugation, and <> represents the inner product.

Eq. (3.23) is interpreted as a multiresolution decomposition of the signal into a set of frequency channels having the same bandwidth in a logarithmic scale (i.e., constant Q or constant relative bandwidth frequency analysis by octave band filters). For the STFT, the phase space is uniformly sampled, whereas in wavelet transform, the sampling in frequency is logarithmic, enabling us to analyze higher frequencies in shorter windows and lower frequencies in longer windows in time (Micheli-Tzanakou, 2000).

3.3.2.1 Wavelets

Wavelet analysis has become a ubiquitous tool in image and signal processing. Two basic properties, space and frequency localization and multiresolution analysis, make it very attractive. Unlike the complex sine and cosine basis functions of the Fourier transform, the basic functions of the wavelet transform are localized in both space and frequency.

Taking the wavelet transform of an image involves convolving a pair of filters, one high pass and one low pass, with the image. This is followed by decimation by two, repeated for as many octaves as desired. This algorithm is depicted in Fig. 3.17. The figure shows a one-octave decomposition of an image into four components: low pass rows, low pass columns (LP–LP); high pass rows, low pass columns (HP–LP); low pass rows, high pass columns (LP–HP); and high pass rows, high pass columns (HP–HP). These will later be referred to as components 0 through 3, respectively. For the purposes of decomposing images, subsequent octaves are created by transforming the LP–LP component of the previous octave (Micheli-Tzanakou, 2000).

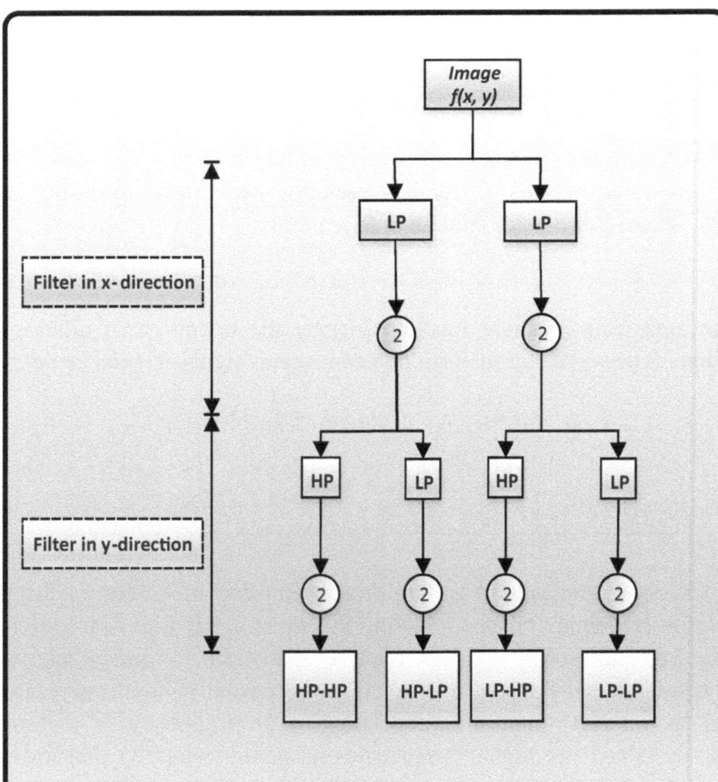

FIGURE 3.17 Wavelet transform algorithm. *LP–HP*, low pass rows, high pass columns; *LP–LP*, low pass rows, low pass columns; *HP–HP*, high pass rows, high pass columns; *HP–LP*, high pass rows, low pass columns. *Adapted from Micheli-Tzanakou, E., 2000. Supervised and Unsupervised Pattern Recognition. In: David Irwin, J. (Series Ed.: Auburn University), Feature Extraction and Computational Intelligence. CRC Press, Boca Raton, London, New York, Washington, DC.*

FIGURE 3.18 Wavelet transform of Lena.bmp. *Adapted from Micheli-Tzanakou, E., 2000. Supervised and Unsupervised Pattern Recognition. In: David Irwin, J. (Series Ed.: Auburn University), Feature Extraction and Computational Intelligence. CRC Press, Boca Raton, London, New York, Washington, DC.*

The result of the wavelet transform on a test image is shown in Fig. 3.18. Fig. 3.18A shows the original image Lena.bmp. Fig. 3.18B shows the first octave wavelet decomposition using a 4-tap Daubechies (1988) filter bank (see also Press et al., 1996) enhanced to show the high-pass components. From left to right, top to bottom is LP−LP (component 0), HP−LP (component 1), LP−HP (component 2), and HP−HP (component 3). Notice that component 1 accentuates the vertically oriented details, component 2 accentuates the horizontally oriented details, and component 3 accentuates the 45 and 135 degrees diagonal details (Micheli-Tzanakou, 2000).

While there are many practical wavelet filters applicable to a wide variety of problems, the filter used in our analysis is the Daubechies 8 tap filter. These filter coefficients are as follows:

- High-pass coefficients = {0.010597, 0.032883, −0.030841, −0.187035, 0.027984, 0.630881, −0.714847, 0.230378}
- Low-pass coefficients = {0.230378, 0.714847, 0.630881, −0.027984, −0.187035, 0.030841, 0.032883, −0.010597}

3.3.2.1.1 Discrete Wavelet Series

Although various ways of discretizing time-scale parameters are possible, the conventional scheme is dyadic grid sampling, where time remains continuous, but time-scale parameters are sampled by choosing $a = 2^i$ and $b = k\ 2^i\ i$, $k \in Z$. The wavelets in this case are given by

$$h_{ik}(t) = 2^{\frac{i}{2}} h\left(2^{-i}t - k\right) \tag{3.25}$$

A wavelet series decomposes a signal $x(t)$ onto a basis of continuous-time wavelets or synthesis wavelets $\alpha_{i,k}(t)$ as shown by

$$x(t) = \sum_{i \in Z} \sum_{k \in Z} C_{i,k} \alpha_{i,k}(t) \tag{3.26}$$

The wavelet coefficients are defined as

$$C_{i,k} = \int x(t) h^*_{i,k}(t) dt \tag{3.27}$$

The signal decomposition may be done using orthogonal wavelets, in which case, the analysis and synthesis wavelets are identical (Micheli-Tzanakou, 2000).

3.3.2.1.2 Discrete Wavelet Transform

Discrete wavelet transform (DWT) is very close to a wavelet series, but it applies to discrete time signals $x[n]$. It achieves a multiresolution decomposition of $x[n]$ on I octaves labeled $i = 1,\ldots, I$, given by

$$x[n] = \sum_{i=I}^{\infty} \sum_{k \in Z} a_{i,k} g_i\left[n - 2^i k\right] + \sum_{k \in Z} b_{I,k} h_I\left[n - 2^I k\right] \tag{3.28}$$

DWT computes wavelet coefficients $a_{i,k}$ for $i = 1,\ldots, I$ and scaling coefficients $b_{i,k}$, given by

$$\text{DWT}\{x[n]; 2^I, k2^I\} = a_{i,k} = \sum_n x[n] g^*_i\left[n - 2^i k\right] \tag{3.29}$$

and

$$b_{i,k} = \sum_n x[n] h^*_I\left[n - 2^I k\right] \tag{3.30}$$

where $g_i[n - 2^i k]$ are the discrete wavelets, and $h_I[n - 2^I k]$ are the scaling sequences (Micheli-Tzanakou, 2000).

3.3.2.1.3 Spline Wavelet Transform

The attractiveness of the Gabor representation of a signal comes from its optimal time–frequency localization (Gabor, 1946). However, the use of fixed window size, redundancy, and nonorthogonality are some major limitations. The use of B-spline wavelets has near optimal time–frequency localization (Unser et al., 1991, 1992). They are not orthogonal like the Battle/Lemarie polynomial spline wavelets used by Mallat (1989), which exponentially decay; however, they are semiorthogonal and have compact support (Micheli-Tzanakou, 2000).

The B-spline wavelet transform is used to construct a sequence of embedded polynomial spline function spaces $\{S^n_{(i)}\ i \in Z\}$ of order n such that $S_{(i)}{}^n \supset S_{(i+1)}{}^n$ for $i \in Z$ where Z is the set of integers (Unser et al., 1992). $S_{(i)}{}^n$ is the subset of functions in $L_2(R)$ that are of class C^{n-1}, i.e., continuous functions with continuous derivatives up to order $(n-1)$ and are equal to a polynomial of degree n in intervals $[k2^i, (k+1)2^i]$ with $k \in Z$ (Micheli-Tzanakou, 2000).

Hence,

$$S^n_i = \phi^n_i(x) = \sum_{-\infty}^{\infty} C_i(k)\beta^n_{2^i}\left(x - 2^i k\right) \tag{3.31}$$

where

$$\beta^n_{2^i}(x) = \beta^n\left(x/2^i\right) \tag{3.32}$$

The basis function $b^n(x)$ is the B-spline of order n. For the spline basis functions of order n, the wavelet sequence q^n is

$$q^n(k-1) = (-1)^k b^{2n+1}(k) * p^n(k) \tag{3.33}$$

where $b^{2n+1}(k) = \beta^{2n+1}(k)$, and $p^n(k)$, the scaling sequence, is the binomial kernel expressed as

$$p^n(k) = (1/2^n)\binom{n+k}{k}, \quad k = 0, \ldots, n+ \tag{3.34}$$

When the wavelet and the scaling sequences are determined, the wavelet function $\underline{b}^n_2(x)$ can be constructed by a scaling function $\beta^n(x)$ by solving a two-scale equation (a dilation equation) given as

$$\underline{\beta}^n_2(x) = \underline{\beta}^n(x/2) = \sum_{K=-\infty}^{\infty} q^n(k)\beta^n(x-k) \tag{3.35}$$

Given a function $\phi^n(x)$ is obtained, the B-spline representation at the finest resolution level is defined as level (0) using

$$\phi_{(0)}(x) = \sum_{K=-\infty}^{\infty} C_{(0)}(k)\beta^n(x-k) \tag{3.36}$$

The essence of the wavelet transform is to decompose the above expression using basis functions expanded by a factor of two:

$$\phi^{(0)}_n(x) = \sum_{K=-\infty}^{\infty} d_{(1)}(k)\underline{\beta}^n_2(x-2k) + \sum_{K=-\infty}^{\infty} C_{(1)}(k)\beta^n_2(x-2k) \tag{3.37}$$

The B-spline wavelet is a polynomial compact support with the property $\underline{\beta}_2^n \perp \beta_2^n(x-2k)\ k \in Z$. This means the first term of the right-hand side of Eq. (3.35) is the projection of $\phi_n^{(0)}$ on $S_{(1)}^n$, and the second is represents the residual error.

The decomposition can be implemented iteratively up to level I to yield the wavelet representation

$$\phi_n^{(0)}(x) = \sum_{i=1}^{I} \sum_{K=-\infty}^{\infty} d_{(i)}(k)\beta_{2^i}^n\left(x - 2^i k\right) + \sum_{K=-\infty}^{\infty} C_I(k)\beta_{2^I}^n\left(x - 2^I k\right) \quad (3.38)$$

where the coefficients $\left\{d_{(i)}, \ldots, d_{(I)}\right\}$ are wavelet coefficients ordered from fine to coarse; the sequence $\left\{C_{(I)}\right\}$ characterizes the lower resolution signal at level (I) (Micheli-Tzanakou, 2000).

3.3.2.1.4 Discrete B-Spline Wavelet Transform

It is possible to regard a wavelet function as a band-pass filter and represent it as a combination of a low-pass and a high-pass filter. In this case, the wavelet analysis becomes a multiresolution analysis. The low-pass and the high-pass filters for nth order spline wavelet multiresolution decomposition may be computed as

$$h(k) = \frac{1}{2}\left[b^{2n+1}\right]^{-1} \uparrow_2 * b^{2n+1}(k) * p^n(k) \quad (3.39)$$

$$g(k+1) = \frac{1}{2}\left[b^{2n+1}\right]^{-1} \uparrow_2 * (-1)^k p^n(k) \quad (3.40)$$

where $\uparrow_2$ indicates up-sampling by 2 (Micheli-Tzanakou, 2000).

3.3.2.1.5 Design of Quadratic Spline Wavelets

For the analysis of the waveforms used in our examples, the quadratic spline wavelets ($n = 2$) are designed for their antisymmetric property, conforming to the morphological character of the signal. The quadratic spline and wavelet functions are shown in Figs. 3.19 and 3.20. The low-pass $h(n)$ and the high-pass $g(n)$ filter kernels for the quadratic spline wavelet are

$$h(k) = \frac{1}{2}\left[\left(b^5\right)^{-1}\right](k) \uparrow_2 * b^5(k) * p^2(k) \quad (3.41)$$

$$g(k+1) = \frac{1}{2}\left[b^5(k)\right]^{-1} \uparrow_2 * (-1)^k p^2(k) \quad (3.42)$$

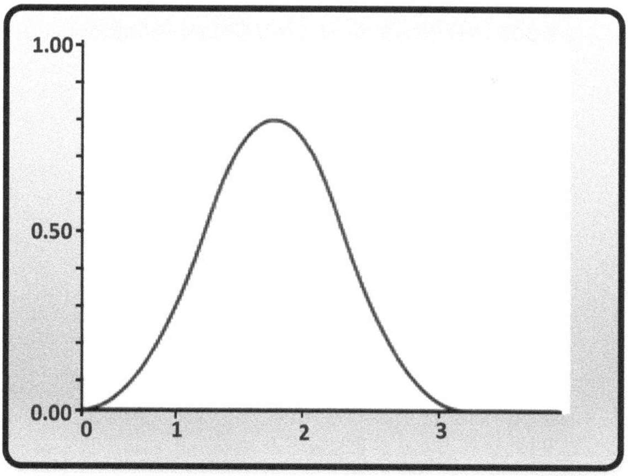

FIGURE 3.19 Quadratic B-spline functions. *Adapted from Micheli-Tzanakou, E., 2000. Supervised and Unsupervised Pattern Recognition. In: David Irwin, J. (Series Ed.: Auburn University), Feature Extraction and Computational Intelligence. CRC Press, Boca Raton, London, New York, Washington, DC.*

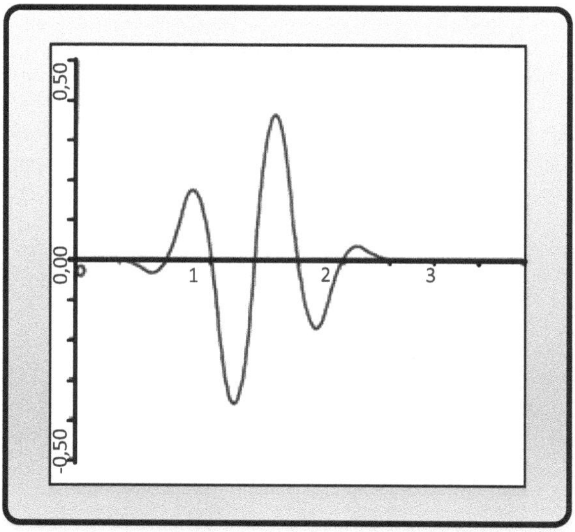

FIGURE 3.20 Quadratic B-spline wavelet. *Adapted from Micheli-Tzanakou, E., 2000. Supervised and Unsupervised Pattern Recognition. In: David Irwin, J. (Series Ed.: Auburn University), Feature Extraction and Computational Intelligence. CRC Press, Boca Raton, London, New York, Washington, DC.*

where

$$\left[\left(b^5\right)^{-1}\right](k) = Z^{-1}\left\{\frac{120}{z^2 + 26z + 66 + 26z^{-1} + z^{-2}}\right\} \tag{3.43}$$

and

$$p^2(k) = \frac{1}{2^2}Z^{-1}\left[1 + 3z^{-1} + 3z^{-2} + z^{-3}\right] \tag{3.44}$$

Further, $\left[b^5(k)\right]^{-1}$ can be expressed in factorized form by

$$\left[b^5(k)\right]^{-1} = Z^{-1}\left[\frac{\alpha_1\alpha_2}{(1-\alpha_1 z^{-1})(1-\alpha_1 z)(1-\alpha_2 z^{-1})(1-\alpha_2 z)}\right] \tag{3.45}$$

where $\alpha_1 = -0.04309$ and $\alpha_2 = -0.43057$.

$$\left[\left(b^5\right)^{-1}\right](k) = \left[\begin{array}{c}\dfrac{\alpha_1\alpha_2}{\left(1-\alpha_1^2\right)\left(1-\alpha_2^2\right)(1-\alpha_1\alpha_2)(\alpha_1-\alpha_2)} \\ \left(\alpha_1\left(1-\alpha_2^2\right)\alpha_1^{|k|} - \alpha_2\left(1-\alpha_1^2\right)\alpha_2^{|k|}\right)\end{array}\right] \tag{3.46}$$

The coefficient values for these filter kernels *h*(*n*) and *g*(*n*) are given in Table 3.4 for the quadratic spline wavelets (Micheli-Tzanakou, 2000).

3.3.2.1.6 Fast Algorithm

The initial step in the derivation of a fast algorithm is to find the B-spline coefficients *C*(*k*) at the resolution level 0. This is efficiently implemented using

$$C^+(k)f(k) + b_1C^+(k-1), \quad (k = 2, \ldots, K) \tag{3.47}$$

$$C^-(k)f(k) + b_1C^-(k+1), \quad (k = K-1, \ldots 1) \tag{3.48}$$

$$C(k) = b_0(C^+(k) - f(k)) \tag{3.49}$$

where $b_0 = -8a/(1-a^2)$ $b_1 = a = \surd 8 - 3$, with $C^+(1) = C^-(K) = 0$. The wavelet coefficients are then computed iteratively for *I* octaves by filtering and decimating by a factor of 2:

$$C_{i+1}(k) = [h * C_i]\downarrow_2(k) \tag{3.50}$$

$$d_{i+1}(k) = [g * C_i]\downarrow_2(k), \quad i = 0, 1, 2, \ldots, I-1 \tag{3.51}$$

where $\downarrow_2$ indicates down-sampling by 2, and where *h* and g are the low-pass and the high-pass filter kernels for decomposition, respectively, computed as

$$h(k) = \frac{1}{2}\left[b^{2n+1}\right]^{-1}\uparrow_2 * b^{2n+1}(k) * p^n(k) \tag{3.52}$$

TABLE 3.4 Coefficients of Truncated Decomposition Filters h, g (IIR), and Reconstruction Filters p^2, q^2 (FIR) for Quadratic Spline Filters

K	H(k)	g(k)	$p^2(k)$	$Q^2(k)$
−10	+0.00157	−0.00388		
−9	+0.01909	−0.03416		
−8	−0.00503	+0.00901		
−7	−0.04440	+0.07933		
−6	+0.01165	−0.02096		
−5	+0.10328	−0.18408		
−4	−0.02593	+0.04977		+1/480
−3	−0.24373	+0.42390		−29/480
−2	+0.03398	−0.14034	+1/4	+147/480
−1	+0.65523	−0.90044	+3/4	−303/480
0	+0.65523	+0.90044	+3/4	+303/480
1	+0.03398	+0.14034	+1/4	−147/480
2	−0.24373	−0.42390		+29/480
3	−0.02593	−0.04977		−1/480
4	+0.10328	+0.18408		
5	+0.01165	+0.02096		
6	−0.04440	−0.07933		
7	−0.00503	−0.00901		
8	+0.01909	+0.03416		
9	+0.00157	+0.00388		

Adapted from Micheli-Tzanakou, E., 2000. Supervised and Unsupervised Pattern Recognition. In: David Irwin, J. (Series Ed.: Auburn University), Feature Extraction and Computational Intelligence. CRC Press, Boca Raton, London, New York, Washington, DC.

$$g(k) = \frac{1}{2}\left[b^{2n+1}\right]^{-1} \uparrow_2 * (-1)^k p^n(k) \tag{3.53}$$

where $\uparrow_2$ indicates up-sampling by 2. The basic computational block diagram for the I octave wavelet decomposition is given in Fig. 3.21. The input to the block diagram is the signal at resolution level (0). The $H(z)$ and $G(z)$ denote the z-transform of the low-pass and high-pass filter kernels h and g, respectively. The DWT by the spline wavelets is performed by these two filters, in the

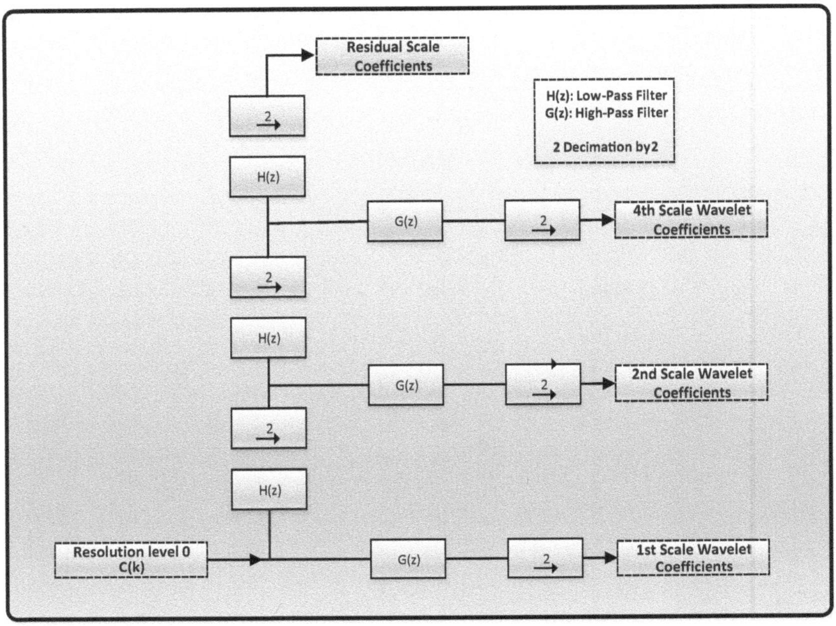

FIGURE 3.21 Basic computational block diagram of one octave wavelet decomposition algorithm. *Adapted from Micheli-Tzanakou, E., 2000. Supervised and Unsupervised Pattern Recognition. In: David Irwin, J. (Series Ed.: Auburn University), Feature Extraction and Computational Intelligence. CRC Press, Boca Raton, London, New York, Washington, DC.*

form of multiresolution decomposition. In each resolution level, the output of the high-pass filtering yields the wavelet coefficients for that resolution (i), and the output of the low-pass filtering yields the coarser input to the next resolution level ($i + 1$). For the purpose of scaling, the discrete sequences' decimation by two is performed after each application of filtering. Rioul and Duhamel (1992) have proposed various fast algorithms for the implementation of the recursive filtering and decimation schemes for wavelet decomposition (Micheli-Tzanakou, 2000).

The wavelet transform is a sequential band-pass filtering operation using filters with logarithmically ordered band-pass characteristics in the frequency domain. In the time domain, this operation corresponds to projecting a signal onto a subspace spanned by the dilated and translated versions of prototype functions called wavelet and scaling functions. Dilation yields logarithmically ordered filters, and translation yields the information about where in time any frequency band activity in the signal occurs.

In the multiresolution scheme, the number of wavelet coefficients halves from one scale to the next; this implies that longer time windows for lower frequencies and shorter time windows for higher frequencies are employed in the analysis. The wavelet coefficients can be used to investigate the amplitude

TABLE 3.5 Frequency Band for Scale for 1 KHz

Scale	Frequency (Hz)
First scale	250–500
Second scale	125–250
Third scale	62.5–125
Fourth scale	31.3–62.5
Fifth scale	15.6–31.3
Sixth scale	7.8–15.6
Residual scale	0.0–7.8

Adapted from Micheli-Tzanakou, E., 2000. Supervised and Unsupervised Pattern Recognition. In: David Irwin, J. (Series Ed.: Auburn University), Feature Extraction and Computational Intelligence. CRC Press, Boca Raton, London, New York, Washington, DC.

and phase characteristics of oscillations in various frequency bands forming the waveform or the image under consideration.

For the exploration of different oscillatory components in a waveform using multiresolution analysis, the waveforms are decomposed into six or more scales by the quadratic spline wavelets. As an example, if the sampling rate is 1 KHz, each scale covers the frequency band shown in Table 3.5.

REFERENCES

Abbasion, S., 2007. Rolling element bearings multi-fault classification based on the wavelet denoising and support vector machine. Mechanical Systems and Signal Processing 21 (7), 2933–2945.

Allen, D.M., 1974. The relationship between variable selection and prediction. Technometrics 16, 125–127.

Ambroise, C., McLachlan, G.J., 2002. Selection bias in gene extraction on the basis of microarray gene-expression data. PNAS 99, 6562–6566.

Amick, H., Colin Gordon & Associates, 1998. Vibration data representation for advanced technology facilities. In: Proceedings of 12th ASCE Engineering Mechanics Conference, La Jolla, California, May 17–20, 1998, pp. 306–309.

Anders, U., Korn, O., 1999. Model selection in neural networks. Neural Networks 12, 309–323.

Bendat, J.S., Piersol, A.G., 1980. Engineering Application of Correlation and Spectral Analysis, New York. Wiley-Interscience.

Bengio, J., Grandvalet, Y., 2003. No unbiased estimator of the variance of K-fold cross-validation. Journal of Machine Learning Research 5, 1089–1105.

Breiman, L., 1996. Bagging predictors. Machine Learning 24, 123–140.

Brownlee, J., October 6, 2014. An Introduction to feature selection. Machine Learning Process. http://machinelearningmastery.com/an-introduction-to-feature-selection/.

Deng, K., 1998. OMEGA: On-Line Memory-Based General Purpose System Classifier (Doctoral thesis): CMU-RI-TR-98–33. The Robotics Institute, School of Computer Science, Carnegie Mellon University, Pittsburgh.

Daubechies, I., 1988. Orthonormal bases of compactly supported wavelets. Communications on Pure and Applied Mathematics 41 (7).

Devijver, P.A., Kittler, I., 1982. Pattern Recognition: A Statistical Approach. Prentice-Hall, Englewood, Cliffs, NJ.

Efron, B., Tibshirani, R.J., 1993. Introduction to the Bootstrap. Chapman and Hall, New York.

Finney, C.E.A., Green, J.B., Daw, C.S., 1998. Symbolic Time-Series Analysis of Engine Combustion Measurements. Society of Automotive Engineers. Finney, Green.

Fuqing, Y., Kumar, U., Galar, D., 2011. Feature Extraction from Time Domain and Automatic Failure Detection Using Support Vector Machine for Rolling Element Machine. Division of Operation and Maintenance Engineering, Luleå, SE-971 87, Sweden.

Gabor, D., 1946. Theory of communication. Journal of IEE (London) 93, 429.

Ginsberg, M.L., 1993. Essentials of Artificial Intelligence. Morgan Kaufmann, Los Altos, CA.

Guyon, I., Gunn, S., Nikravesh, M., Zadeh, L.A., 2006. Feature Extraction Foundations and Applications. Springer 2006 (pattern recognition).

Guyon, I., Makhoul, J., Schwartz, R., Vapnik, V., 1998. What size test set gives good error rate estimates? IEEE Transactions on Pattern Analysis and Machine Intelligence 20, 52–64.

Hall, M.A., 1999. Correlation-Based Feature Selection For Machine Learning (Ph.D. thesis). University of Waikato, Hamilton, New Zealand.

Heng, R.B.W., Nor, M.J.M., 1998. Statistical analysis of sound and vibration signals for monitoring rolling element bearing condition. Applied Acoustics 53 (1–3), 211–226.

Kira, K., Rendell, L., 1992. A practical approach to feature selection. In: Proceedings of the Ninth International Conference on machine learning, pp. 249–256.

Kohavi, R., 1995. A study of cross-validation and bootstrap for accuracy estimation and model selection. In: Mellish, C.S. (Ed.), Proceedings IJCAI-95, Montreal, Que. Morgan Kaufmann, Los Altos, CA, pp. 1137–1143.

Kohavi, R., John, G., 1997. Wrappers for feature subset selection. Artificial Intelligence Journal, Special Issue on Relevance 97 (1–2), 273–324.

Kononenko, I., 1994. Estimating attributes: analysis and extensions of RELIEF. In: European Conference on Machine Learning, pp. 171–182.

Langley, P., 1994. Selection of Relevant Features in Machine Learning.

Liu, H., Motoda, H., 1998. Feature Extraction, Construction and Selection: A Data Mining Perspective. Kluwer Academic.

Mallat, S., 1989. A theory for multiresolution signal decomposition: the wavelet representation. IEEE Transactions on Pattern Analysis and Machine Intelligence 11 (7), 674.

Micheli-Tzanakou, E., 2000. Supervised and Unsupervised Pattern Recognition. In: David Irwin, J. (Series Ed.: Auburn University), Feature Extraction and Computational Intelligence. CRC Press, Boca Raton, London, New York, Washington, DC.

Miller, A.J., 1990. Subset Selection in Regression. Chapman and Hall, London.

MITx–Circuits and Electronics, May 01, 2012. Time Domain Versus Frequency Domain Analysis. https://www.edx.org/course/circuits-electronics-1-basic-circuit-mitx-6-002-1x-0.

Monari, G., Dreyfus, G., 2002. Local overfitting control via leverages. Neural Computation 14, 1481–1506.

Ng, A.Y., 1998. On feature selection: learning with exponentially many irrelevant features as training examples. In: In 15th International Conference on Machine Learning. Morgan Kaufmann, San Francisco, CA, pp. 404–412.

Opper, M., Winther, O., 2000. Advances in Large Margin Classifiers, Chapter Gaussian Processes and Support Vector Machines: Mean Field and Leave-One-Out. MIT Press, pp. 311–326.

Oussar, Y., Monari, G., Dreyfus, G., 2004. Reply to the comments on "local overfitting control via leverages" in "Jacobian conditioning analysis for model validation". Neural Computation 16, 419–443.

Press, W.H., Flannery, B.P., Teukolsky, S.A., Vetterling, W.T., 1988. Numerical Recipes in C. Cambridge University Press, Cambridge.

Press, W.H., Flannery, B.P., Teukolsky, S.A., Vettering, W.T., 1996. Numerical Recipes, second ed. Cambridge Univ. Press, Cambridge (Chapter 13).

Ray, A.K., Bhattacharya, B., 2016. Module 2-data communication fundamentals. In: Version 2 CSE IIT, Course Co-ordinated by: Kharagpur, National Programme on Technology Enhanced Learning (NPTEL). http://nptel.ac.in/courses/106105080/pdf/M2L1.pdf.

Rioul, O., Duhamel, P., 1992. Fast algorithms for discrete and continuous wavelet transforms. IEEE Transactions on Information Theory 38 (2), 569.

Robnik-Sikonja, M., Kononenko, I., 2003. Theoretical and empirical analysis of ReliefF and RReliefF. Machine Learning 53, 23–69.

Russell, S.J., Norvig, P., 1995. Artificial Intelligence: A Modern Approach. Prentice Hall, Englewood Cliffs, NJ.

Sánchez, N., Betanzos, A., Tombilla, M., 2007. Filter Methods for Feature Selection—A Comparative Study. University of A Coruña, Department of Computer Science, A Coruña, Spain.

Seber, G.A., Wild, C.J., 1989. Nonlinear Regression. John Wiley and Sons, New York.

Sreejith, B., Verma, A.K., Srividya, A., 2008. Fault Diagnosis of Rolling Element Bearing Using Time-Domain Features and Neural Networks.

Seunghye, J.W., 2016. Data Representation for Time Series Data Mining: Time Domain Approaches, vol. 9. January/February 2017. © 2016 Wiley Periodicals, Inc.

Stone, M., 1974. Cross-validatory choice and assessment of statistical predictions. Journal of the Royal Statistical Society: Series B 36, 111–147.

Tang, X.Z., 1995. Symbol sequence statistics in noisy chaotic signal reconstruction. Physical Review E 51 (5), 3871–3889.

Unser, M., Aldroubi, A., Eden, M., 1991. Fast B-spline transforms for continuous image representation and interpolation. IEEE Transactions on Pattern Analysis and Machine Intelligence 13 (3), 277.

Unser, M., Aldroubi, A., Eden, M., 1992. On the asymptotic convergence of B-spline wavelets to Gabor functions. IEEE: Transactions on Information Theory 38 (2), 864.

Vapnik, V.N., 1982. Estimation of Dependencies Based on Empirical Data. Springer, New York.

Vapnik, V.N., 1998. Statistical Learning Theory. John Wiley & Sons, New York.

Vass, J., 2008. Avoidance of speckle noise in laser vibrometry by the use of Kurtosis ratio: application to mechanical fault diagnostics. Mechanical Systems and Signal Processing 22 (3), 647–671.

Wang, W.Q., Ismail, F., Golnaraghi, M.F., 2001. Assessment of gear damage monitoring techniques using vibration measurements. Mechanical Systems and Signal Processing 15 (5), 905–922.

Wolpert, D., Macready, W.G., 1999. An efficient method to estimate Bagging's generalization error. Machine Learning 35 (1), 41–55.

Yu, L., Liu, H., 2003. Feature selection for high-dimensional data: a fast correlation-based filter solution. In: Proceedings of the Twentieth International Conference on machine learning, ICML, pp. 856–863.

Chapter 4

Data and Information Fusion From Disparate Asset Management Sources

4.1 ONLINE AND OFF-LINE CONDITION MONITORING INFORMATION

4.1.1 Condition Monitoring Data and Automatic Asset Data Collection

Condition monitoring (CM) involves comparing online or off-line data with expected values; it should be able to generate alerts, if necessary based on preset operational limits. Health assessment determines if the health of the monitored component or system has degraded and conducts fault diagnostics, leading to the development of prognostics. The primary tasks of prognostics are calculating the component or system's future health and estimating its remaining useful life (RUL). This sounds simple enough, but in reality, it can be difficult to perform reliable and effective condition-based maintenance (CBM). First, initiating CBM is costly. The cost of instrumentation can be quite large, especially if the goal is to monitor equipment that is already installed. Therefore, it is important to decide whether the equipment is important enough to warrant the cost.

Implementing CBM requires the setting of an information system able to meet the following basic requirements:

- Collecting and processing a large quantity of information not previously available on the condition of each part of a machine.
- Initiating corrective maintenance actions within an established lead-time (the period between the off-limits condition and an emergency shutdown). Two different situations are possible:
  - The condition of machine is not yet close to breakdown. In this case, the normal procedure of the maintenance planning section is followed.
  - The condition of machine is already well within the lead-time (near to breakdown). In this situation, the information is passed to maintenance supervision for emergency corrective maintenance actions.

eMaintenance. http://dx.doi.org/10.1016/B978-0-12-811153-6.00004-X

To operate a CBM program correctly, the maintenance personnel should consider the following:

- Condition of machine
- Part of machine that is probably defective
- Probable defect
- Time during which failure must be repaired

By scrutinizing and correlating the diagnosis with the actual findings during repair work, it is possible to:

- Control the training of maintainers
- Improve the correlation between parameters chosen for condition measurement and actual defects found
- Obtain severity curves specific to each component or system

Making the potential of CM a reality requires that large amounts of data be collected, monitored, filtered, and turned into actionable information. The cheaper and more ubiquitous the computerized monitoring hardware becomes, the greater the volume of data and the more challenging it becomes to manage and interpret. The vast amount of diagnostic data produced by today's smart field devices can be a very important source for accurate documentation of maintenance activities. But the sheer volume and complexity of such information can be daunting and difficult for maintenance personnel to manage. What is required is an effective means of compiling and organizing the data for day-to-day utilization by maintenance staff, while preserving and recording significant events for future reference.

Data are becoming more and more available. However, these data may not be used because they are either of bad quality or have been improperly stored for the following reasons (Soibelman and Kim, 2002):

- Project managers do not have sufficient time to analyze the computerized data so they do not care about proper storage.
- The complexity of the data analysis process is beyond the capabilities of the relatively simple maintenance systems commonly used.
- There is no well-defined automated mechanism to extract, preprocess, and analyze the data and summarize the results so stored data are not reliable.

Maintenance personnel not only cope with large amounts of field-generated data, they also turn that information to their advantage in a number of ways. Real-time condition monitoring (RTCM) systems produce warnings, alarms, and reports that can be used by maintenance people for many purposes. Such systems permit the most important issues to be identified and handled quickly.

The ultimate goal is to fully integrate RTCM data with computerized maintenance management systems (CMMSs) to generate work orders as needed. This would provide true automation from the time a field device begins to show signs of reduced performance until a work order is printed in

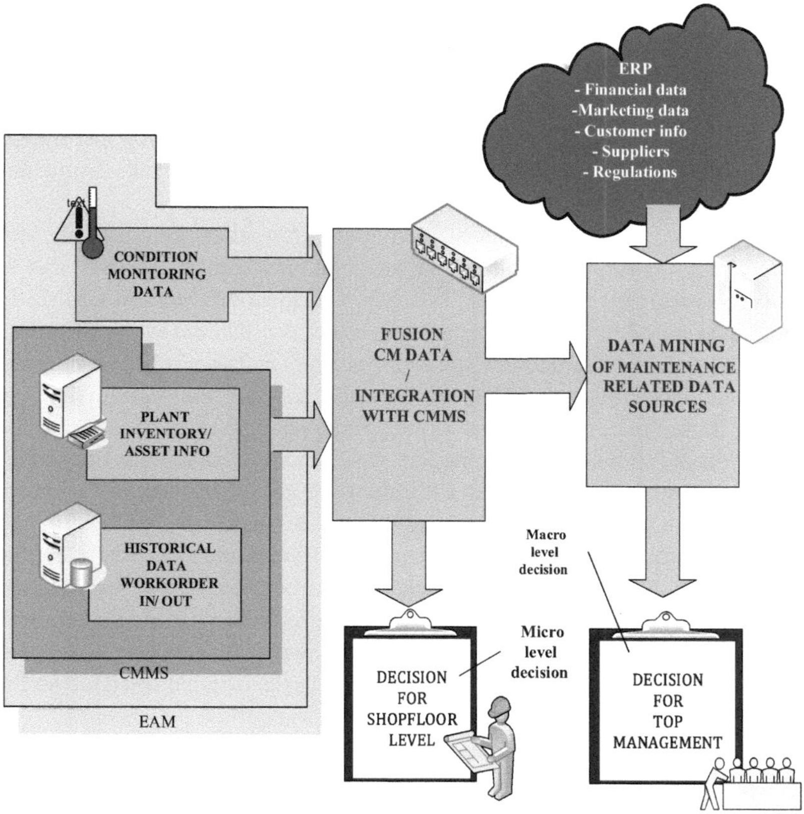

FIGURE 4.1 Two-step integration of real-time condition monitoring and computerized maintenance management system (CMMS) databases. *CM*, condition monitoring; *EAM*, enterprise asset management; *ERP*, enterprise resource planning. *Adapted from Galar, D., Parida, A., Kumar U., 2012b. e-Maintenance Decision Making Based On Different Types of Data Fusion. Division of Operation and Maintenance Engineering. Luleå University of Technology, Luleå, Sweden, 97187.*

the maintenance department, and a technician is dispatched to the scene. Fig. 4.1 shows the automation of work order dispatching.

With the evolution of IT, this level of integration of CMMS and CM has become possible. The development of open communication protocols allows the information accumulated by smart field devices to be captured by asset management software. It is no longer necessary for technicians to carry handheld communicators or laptops into a plant to evaluate the condition of instruments, some of which are inaccessible or in hazardous areas, to be followed by manually documenting test results and current device status (Galar et al., 2012b).

Since all records are date- and time-stamped, users can easily determine when and by whom a particular device has been changed or tested, including "as found/as left" notations. With this information available in a database, it should never be necessary for technicians to spend time searching for historical information on a device. Since events can also be recorded manually, users can document unusual occurrences affecting the entire plant, such as a lightning strike or power outage, or individual events like device inspections. This decision level is extremely useful for technicians to take immediate action.

In addition, a vast amount of available information can produce new knowledge if it is exploited with proper artificial intelligence (AI) tools that are physically integrated into the database type and location. Modern CMMS information is stored in very large relational, or tabular, databases. This format is appropriate for an integrated investigation, as many software tools are now able to query and investigate the tables. For historical analysis, only certain fields are required, thus allowing sensitive data to be removed or filtered. The data subset still contains a full history of component faults and related actions, providing a comprehensive maintenance history profile while alleviating security concerns.

Importing CM data into these relational databases is challenging but possible, as each type of sensor generates different data classes, sampling rates, and number of compiled indicators. Furthermore, each manufacturer stores the collected information in unique proprietary formats, requiring platform-specific importation software to be written. However, most CM software allows the CM data to be exported from the original interface so that it can be expanded and generalized.

Although CMMS and CM data now coexist within a single database where they can be queried and explored, automating the discovery of linked events requires additional processing. Relating a given maintenance fault or action, which is textual, to sensor data, which are some arbitrary data class type, can only be accomplished through the compilation of overlapping metadata (Tianhao et al., 1992). The fields that are generated characterize the location and significance of events, creating a quantified set of parameters by which the disparate data can be compared. Metadata for CM records are generated differently depending on the data class involved. One-dimensional and dimensionless quantities can be assigned rarity parameters through statistical distribution analysis, and higher dimensional data require using neural networks to identify anomalies. Determining rarity is often accomplished through simple single variable statistical analysis, while severity is typically derived from developers' recommended threshold values. More complex domain types require more advanced, though typically well-understood, analyses, such as neural networks which can isolate anomalous points from multidimensional data. Some predict that through the integration process, more advanced metrics and indicators can be discovered to implement previously unexplored relationships in the data, such as multiparameter trending. This new knowledge can help maintenance personnel determine the RUL of a system to better schedule operation and maintenance processes. This information affects replacement of assets, shutdown of the plant, overhauls, etc., so it constitutes the second decision level displayed in Fig. 4.1. While it is strongly related to business goals, however, it is useless for immediate interventions (Galar et al., 2012b).

4.1.2 Fusion of Maintenance and Control Data

Process control systems, such as those used in the oil and gas industry, pulp and paper industry, or other process industries, typically include one or more centralized or decentralized process controller communicatively coupled to at least one host or operator workstation and to one or more process control and instrumentation device, such as field devices. These devices, for example, valve, switches, and sensors, perform functions within the process, such as opening or closing valves and measuring process parameters.

The process controller receives signals indicative of process measurements or process variables made by or associated with the field devices and/or other information pertaining to the field devices. It uses this information to implement a control routine and then generates control signals which are sent over one or more of the buses to the field devices to control the operation of the process. Information from the field devices and the controller is typically made available to one or more applications executed by an operator workstation; this enables the operator to perform desired functions with respect to the process, such as viewing the current state of the process, modifying the operation of the process, etc. While a typical process control system has many process control and instrumentation devices, such as valves, transmitters, sensors, etc., connected to one or more process controllers with software that controls these devices during the operation of the process, many other supporting devices are necessary for or related to process operation.

These additional devices include power supply equipment, power generation and distribution equipment, rotating equipment, etc., located at numerous places in a typical plant. This equipment does not necessarily create or use process variables and, in many instances, is not controlled or even coupled to a process controller for the purpose of affecting the process operation. Nevertheless, it is important to and ultimately necessary for proper operation of the process. In the past, however, process controllers were not necessarily aware of these other devices or simply assumed they were operating properly. Fig. 4.2 shows a fan monitored by two accelerometers in such a way that the information generated by vibrations can close the loop and couple the device into the control system (Galar et al., 2012a).

Many process plants have other software systems, which execute applications related to business functions (enterprise resource planning, ERP) or maintenance functions (CMMS). In fact, many process plants, especially those using smart field devices, have applications to monitor and maintain the plant's devices regardless of whether they are process control and instrumentation devices or other types.

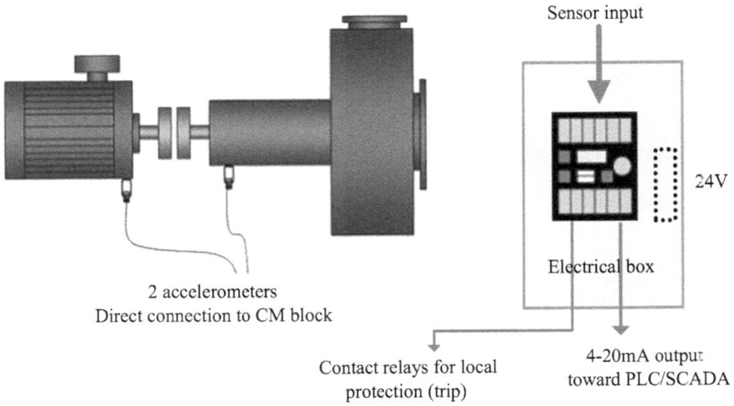

FIGURE 4.2 Monitoring a fan and its integration in the control loop. *CM*, condition monitoring; *PLC*, programmable logic controller; *SCADA*, supervisory control and data acquisition. *Adapted from Galar, D., Kumar, U., Juuso, E., Lahdelma, S., 2012a. Fusion of maintenance and control data: a need for the process. In: 18th World Conference on Nondestructive Testing, 16–20 April 2012, Durban, South Africa.*

The integration of maintenance information, management, and monitoring is essential to close the loop of the process. That is why CMMS systems have evolved. Enterprise asset management (EAM) software is more sophisticated than CMMS (Fu et al., 2002). Such solutions usually enable communication with field devices and store data pertaining to them, allowing the operating state of the field devices to be tracked. In some instances, the EAM application may be used to communicate with a device to change parameters within that device, to cause it to run applications on itself, such as self-calibration routines or self-diagnostic routines, or to obtain information about its status or health, etc.

The information may be stored and used by a maintenance person to monitor and maintain these devices. Other types of applications are used to monitor other types of devices, such as rotating equipment and power generation and supply devices. These other applications are sometimes available to maintenance persons and used to monitor and maintain the devices within a process plant. In many cases, however, outside service organizations perform services related to monitoring process performance and equipment. They acquire the data they need, run typically proprietary applications to analyze the data and merely provide results and recommendations to the process plant personnel. While this is helpful, the plant personnel have little or no ability to view the raw data or to use the analysis data in any other manner. Fig. 4.3 shows a flow diagram of the information produced when CM is outsourced and a final report is recorded in the system (Galar et al., 2012a).

In typical plants, the functions associated with process control activities, device and equipment maintenance and monitoring activities, and business

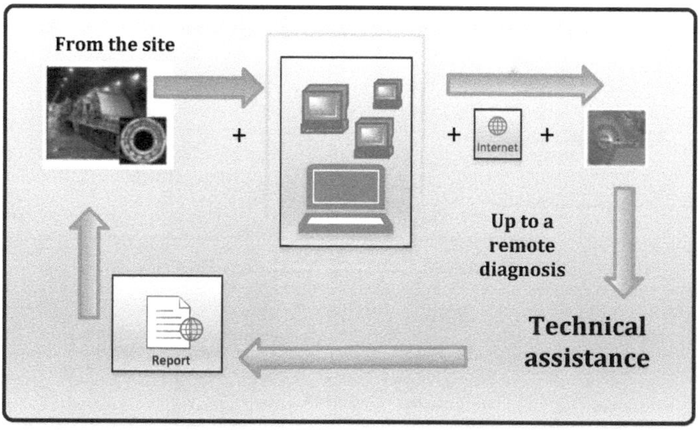

FIGURE 4.3 Typical process of condition monitoring outsourcing. *CM*, condition monitoring; *CMMS*, computerized maintenance management system. *Adapted from Galar, D., Kumar, U., Juuso, E., Lahdelma, S., 2012a. Fusion of maintenance and control data: a need for the process. In: 18th World Conference on Nondestructive Testing, 16–20 April 2012, Durban, South Africa.*

activities such as process performance monitoring are separate, both in the location where they take place and in the personnel who typically perform them. Furthermore, the people involved in these various functions generally use different tools, such as different applications run on different computers. In many instances, these different tools collect or use different types of data associated with or collected from the different devices within the process and are set up differently to collect the data they need. However, there should be cooperation among all departments in an enterprise and between experts in their respective domain knowledge if a maintenance policy is to succeed (Yu et al., 2004).

Process control operators generally oversee the day-to-day operation of a process and are primarily responsible for assuring the quality and continuity of its operation. They typically affect the process by setting and changing set points within it, tuning its loops, scheduling its operations, etc. Maintenance personnel are primarily responsible for assuring that the actual equipment within the process is operating efficiently and for repairing and replacing malfunctioning equipment. They use tools, such as maintenance interfaces, the EAM application discussed above, and many other diagnostic tools, to get information about the operating states of the devices within the process. Maintenance persons also schedule maintenance activities, which may require portions of the plant to be shut down. For many newer types of process devices and equipment, for example, smart field devices, the devices themselves may include detection and diagnostic tools which automatically sense problems with the operation of the device and automatically report them to a maintenance person via a standard maintenance interface.

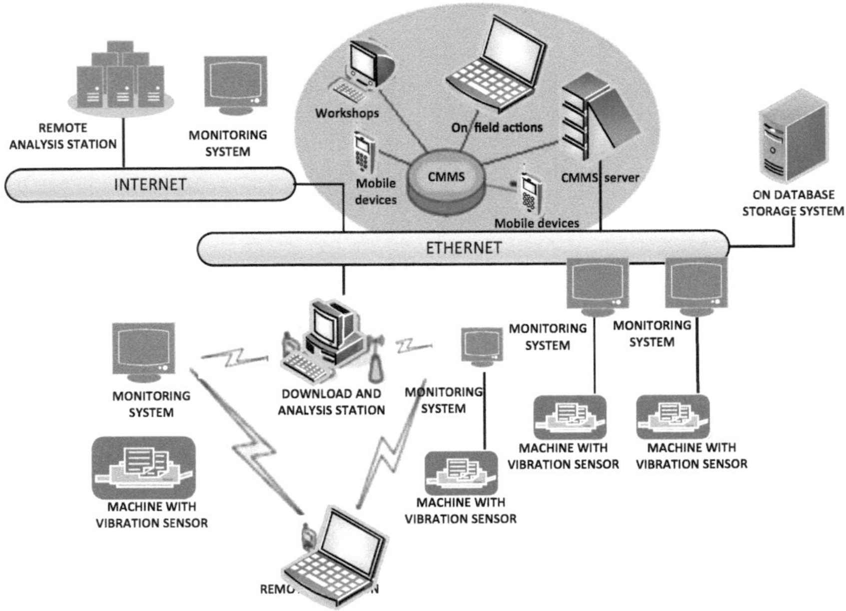

FIGURE 4.4 Typical architecture of maintenance information system. *Adapted from Galar, D., Kumar, U., Juuso, E., Lahdelma, S., 2012a. Fusion of maintenance and control data: a need for the process. In: 18th World Conference on Nondestructive Testing, 16–20 April 2012, Durban, South Africa.*

Maintenance interfaces and maintenance personnel comprise a huge data network (Davies and Greenough, 2000), generally located apart from process control operators, as shown in Fig. 4.4. This is not always the case, however. In some process plants, process control operators perform the duties of maintenance persons or vice versa, or the different people responsible for these functions may use the same interface (Galar et al., 2012a).

Many applications are used to perform the different functions within a plant, i.e., process control operations, maintenance operations, and business operations. They are not integrated and, thus, do not share data or information. In many cases, tasks, such as monitoring equipment, testing the operation of devices, determining if the plant is running in an optimal manner, etc., are performed by outside consultants or service companies who measure the data needed, perform an analysis, and provide only the results of the analysis to the plant personnel. In these cases, the data are typically collected and stored in a proprietary manner and rarely made available to the plant personnel for other uses.

Given the abundance of data analysis and other detection and diagnostic tools available in the process control environment, either in the plant itself or via outside service companies or consultants, a great deal of information about the health and performance of devices is available to maintenance personnel which could also be helpful to process operators or business officers. Similarly, information available to the process operator about the current operational status of the process control loops and other routines might be helpful to the maintenance person. And information generated by or used in the course of performing business functions could be helpful to the maintenance crew or the process control operator.

However, in the past, because these functions were separated, the information generated or collected in one functional area was not used at all or not used very well in other functional areas, leading to an overall suboptimal use of the assets in process plants (Galar et al., 2012a).

4.1.3 Data Fusion: A Need for Maintenance Processes

A process control system includes a data collection and distribution system that collects and stores data from various data sources, each of which may use its own proprietary manner of acquiring or generating the data in the first place. The data collection and distribution system then makes the stored data available to other applications associated with or provided in the process control system or to applications associated with the data sources themselves for use in any desired manner. In this manner, applications may use data from vastly different data sources to provide a better view of or insight into the operational status of a plant, to make better or more complete diagnostic or financial decisions about the plant, etc. (Galar and Morant, 2012).

The diagnostic applications may generate measurement, control and device indexes pertaining to nonprocess variables, such as the health of a device. These equipment performance indexes may be determined by models calculating key performance variables, such as efficiency or cost of production (Galar et al., 2012a).

Thus, applications may combine or use data from previously disparate collection systems such as process control monitoring systems, CM systems, and process performance models to get a better overall view of the state of a process control plant, to diagnose problems, and to take or recommend actions in production planning and maintenance.

This information may be sent to a process operator or maintenance person to inform him/her of a current or future problem. It may be used by the process operator to correct a current problem within a loop or to make a change.

A process control expert may use the measurement, control and device indexes, along with process variable data, to optimize operation of the process. Process variable data and nonprocess variable data may also be combined using the data collection and distribution system.

The detection of a device problem, for example, one requiring shutdown of the process, may cause business software to automatically order replacement parts or alert the business officer that chosen strategic actions will not produce the desired results because of the actual state of the plant. Likewise, the change of a control strategy performed within the process control function may cause business software to automatically order new or different raw materials.

There are, of course, many other types of applications to which the fused data related to process control, equipment monitoring, and performance monitoring data can be an aid by providing different and more complete information about the status of the assets within a process control plant to all areas of the plant (Hall and Llinas, 1997). However, because the functions are so different, and the equipment and personnel used to oversee these functions are different, there is often little or no meaningful data sharing between them (Galar and Morant, 2012).

As shown in Fig. 4.4, a typical process control plant includes a number of business and other computer systems interconnected with a number of control and maintenance systems by one or more communication networks. The operator interfaces may store and execute tools available to the process control, including control optimizers, diagnostic experts, neural networks, tuners, etc.

Further, maintenance systems, such as computers executing the CMMS/EAM application or any other device or equipment monitoring and communication applications may be connected to the process control systems or to the individual devices therein to perform maintenance and monitoring activities.

Some plants may connect various assets, such as turbines, motors, etc., to a maintenance computer via some permanent or temporary communication link (such as a bus, a wireless communication system, or a handheld device connected to the equipment to take readings and then removed). The maintenance computer may store and execute applications to diagnose, monitor, and optimize the operating state of the assets.

In the case of rotating equipment, maintenance personnel use the applications to maintain and oversee performance, to locate problems, and to determine when and if the equipment must be repaired or replaced. Outside consultants or service organizations may temporary acquire or measure data on the equipment and then analyze these data to detect problems, poor performance, or other issues effecting the equipment. The computers running these analyses may not be connected to the rest of the system via any communication line or may be connected only temporarily.

Similarly, with power generation and distribution systems, outside consultants or service organizations may temporarily acquire or measure data on the equipment and analyze these data to detect problems, poor performance, or other issues effecting the equipment. Once again, the computers running the analyses may not be connected to the rest of the system via any communication line or may be connected only temporarily.

In the past, the various process control systems, power generating and maintenance systems have not been interconnected in a manner that enables them to share data in a useful manner. As a result, each of the different functions has operated on the assumption that the other equipment within the plant which may be affected by or have an effect on that particular function is operating perfectly which, of course, is almost never the case. However, because the functions are so different and the equipment and personnel used to oversee these functions are different, there is generally little or no meaningful data sharing between the various functional systems (Galar et al., 2012a). The solution is the "asset cloud" (Dasarathy, 2001, 2003).

4.1.4 Cloud Computing

Cloud computing has been defined as "a model for enabling convenient, on-demand network access to a shared pool of configurable computing resources (e.g., networks, servers, storage, applications, and services) that can be rapidly provisioned and released with minimal management effort or service provider interaction" (Mell and Grance, 2009). The cloud is a set of hardware, networks, storage, services, and interfaces that enable the delivery of computing as a service. It provides the means through which everything from computing power to computing infrastructure, applications, and business processes to personal collaboration can be delivered as a service wherever and whenever it is required (Mell and Grance, 2009 Amrhein and Quint, 2009; Rhoton, 2010). Cloud services include the delivery of software, infrastructure, and storage over the Internet or Intranet (either as separate components or a complete platform) based on user demand. Unlike other approaches, the cloud is as much about the business model as it is about the technology (Gustafson and Galar, 2012).

For asset management, the cloud seems to solve the problem of dispersed data stored in different repositories. The data collection and distribution applications may be dispersed throughout the network, with data collection accomplished at distributed locations. The collected data may be converted to a common format at the distributed locations and sent to one or more central databases for subsequent distribution. These distributed databases constitute the asset cloud. The end users (maintenance personnel and operators) do not have to know anything about the underlying technology.

The asset cloud was designed to acquire data from various sources of data and format these data to a common data format, or structure and provide the data, as needed, to applications run by a computer system or disbursed between workstations. As it can fuse or integrate data from previously disparate and separate systems, the asset cloud provides better measurement, viewing, control and understanding of the operation (Dasarathy, 2001, 2003).

Generally speaking, data collection routines should be created to collect data from disparate sources and to input these data in a common or consistent format into the cloud. The cloud operates as a data and information base to coordinate the distribution of data or information from one functional area, such as the maintenance area, to other functional areas, such as the process control or the business areas. As a result, the cloud may use the collected data

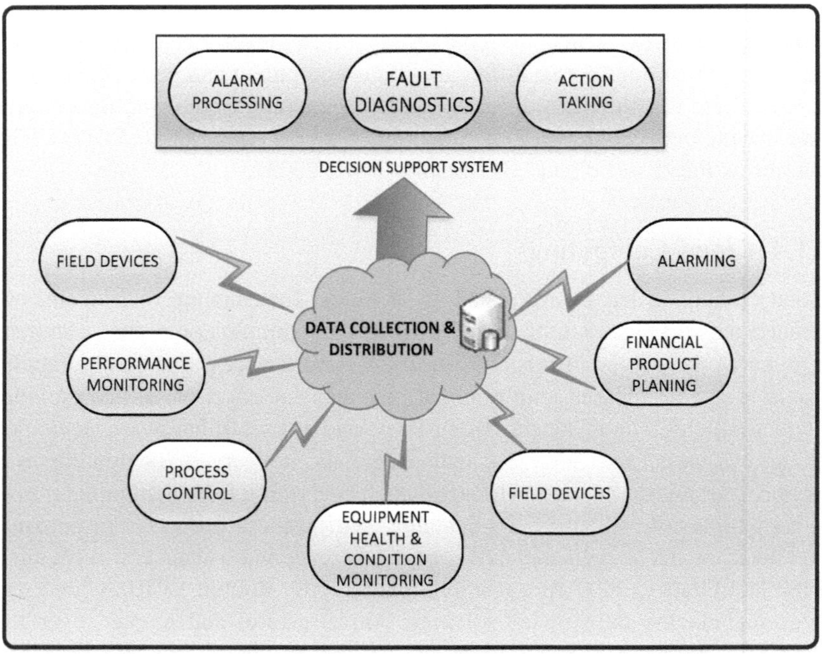

FIGURE 4.5 Services provided by the asset cloud (Galar et al., 2012a).

to generate new information or data which can be distributed to one or more of the computer systems associated with the different functions; it may also execute or oversee the execution of other applications that use the collected data to generate new types of data (Gustafson and Galar, 2012).

Fig. 4.5 is a simplified functional block diagram of the data flow and communication associated with the asset cloud. The data collection and distribution system portrayed here receive data from numerous sources: a process control data source, an equipment or process health data source, a financial or production data source and a performance monitoring data source. The cloud collects the data from the data sources in a common format or converts those data, once received, into a common format for storage and later use by other elements, devices, or applications in the process control system. Once received and converted, the data are stored in a database in some accessible manner and made available to applications or users within the asset cloud. Applications related to process control, fault diagnostics, predictive maintenance, financial planning, optimization, etc. may use, combine, and integrate the data from one or more data source. This ensures better performance than in the past, when such data were not combined (Gustafson and Galar, 2012).

4.2 COMPUTERIZED MAINTENANCE MANAGEMENT SYSTEMS

A CMMS is a type of management software that performs functions to support management and track Operation and Maintenance (O&M) activities.

4.2.1 Computerized Maintenance Management System Needs Assessment

To determine the need for a CMMS, managers should assess their current mode of operation. Key questions to ask include the following:

- Do you have an effective way to generate and track work orders? How do you verify the work was done efficiently and correctly? What is the notification procedure upon completion?
- Are you able to access historical information on the last time a system was serviced, by whom, and for what condition?
- How are your spare-parts inventories managed and controlled? Do you have excess inventory or are you consistently waiting for parts to arrive?
- Do you have an organized system to store documents (electronically) related to O&M procedures, equipment manuals, and warranty information?
- When service staff are in the field, what assurances do you have that they are compliant with health and safety issues and are using the right tools/equipment for the task?
- How are your assets, i.e., equipment and systems, tracked for reporting and planning?

If the answers to these questions are not well defined or even entirely lacking, it may be worth investigating the benefits of a well-implemented CMMS (Sullivan et al., 2010).

4.2.2 Computerized Maintenance Management System Capabilities

CMMS systems automate most of the logistical functions performed by maintenance staff and management. They come with many options and have numerous advantages over manual maintenance tracking systems. Depending on the complexity of the system, typical CMMS functions include the following:

- Work order generation, prioritization, and tracking by equipment/component.
- Historical tracking of all work orders generated, sorted by equipment, date, person responding, etc.
- Tracking of scheduled and unscheduled maintenance activities.
- Storing of maintenance procedures, as well as all warranty information, by component.

- Storing of all technical documentation or procedures by component.
- Availability of real-time reports of ongoing work activity.
- Calendar-based or run time–based preventive maintenance (PM) work order generation.
- Capital and labor cost tracking by component, as well as shortest, median, and longest times to close a work order by component.
- Complete parts and materials inventory control with automated reorder capability.
- Digital interfaces to streamline input and work order generation.
- Outside service calls/dispatch capabilities.

Many CMMS programs can now interface with existing energy management and control systems, as well as ERP or even supervisory control and data acquisition (SCADA) systems for interaction from the managerial to the shop floor level in maintenance decisions. Coupling these capabilities allows condition-based monitoring and component use profiles (Sullivan et al., 2010).

While CMMS can go a long way toward automating and improving the efficiency of most maintenance programs, there are some common pitfalls (Sullivan et al., 2010). These include the following:

- *Improper selection of a CMMS*. This is a site-specific decision. Time should be taken to evaluate needs and look for the proper match.
- *Inadequate training of staff on proper use of the CMMS*. Staff need dedicated training on input, function, and maintenance of the CMMS.
- *Lack of commitment to properly implement the CMMS*. A commitment needs to be in place for the startup/implementation of the CMMS.
- *Lack of commitment to persist in CMMS use and integration*. While a CMMS provides significant advantages, it needs to be maintained.

However, CMMS is an information repository whose outcomes strongly depend on the quality of the data stored in the system. What happens if the data in the CMMS are wrong, or there are no data at all? If managers do not have the right kind of data or enough of it, they must make decisions without all the facts. The wrong decision could be costly.

To maximize CMMS use, managers need to continually revisit the system's features and functions, review and upgrade training for users, and reinforce among users the need for a steady stream of reliable maintenance and repair data (Hounsell, 2008).

4.2.3 Computerized Maintenance Management System Benefits

One of the greatest benefits of a CMMS is the elimination of paperwork and manual tracking activities, permitting maintenance staff to become more productive. In addition, the maintenance cost can be tracked, so the tool can be used as a cost control. It should be noted that the functionality of a CMMS lies in its ability to collect and store information in an easily retrievable format. A CMMS does not make decisions; rather, it provides the manager with the best information to make maintenance decisions (Sullivan et al., 2010).

Benefits of implementing a CMMS include the following:

- Detecting impending problems before a failure occurs, resulting in fewer failures and customer complaints.
- Achieving a higher level of planned maintenance activities to enable more efficient use of staff resources.
- Improving inventory control by enabling better spare parts forecasting to eliminate shortages and minimize existing inventory.
- Maintaining optimal equipment performance to reduce downtime and increase equipment life.
- Keeping a detailed record of maintenance costs and investments, tracking the consequences, and meeting the annual budget. Some CMMSs may perform life cycle costing calculations for asset renovation, overhaul, and replacement.

4.2.4 Computerized Maintenance Management System Resources

The Internet provides a great resource to find CMMS vendors. A simple search under "CMMS" provides links to the many vendors and the resources they offer.

4.2.5 The Role of Computerized Maintenance Management System

CMMSs have evolved over the last 3 decades, starting as elementary asset trackers with some PM functionality to becoming maintenance information systems. Hundreds of vendors provide solutions on a variety of platforms. State of the art systems give users the comprehensive ability to facilitate the flow of maintenance information and to check the health of the maintenance organization at a glance.

As the maintenance, repair, and operations (MRO) software market continues to expand, vendors have developed solutions that focus on specific segments of asset and work management. Systems described as enterprise asset management (EAM), asset life cycle management, asset performance management, asset or enterprise reliability management and CM are all focused on achieving the same goal; increasing equipment availability and performance, increasing product quality, and reducing maintenance expenses. When a CMMS is implemented to facilitate an established process and standards, those goals can be realized (Crain, 2003).

4.2.5.1 Does Your Organization Need a Computerized Maintenance Management System?

The vendor hype for business software is only rivaled by the number of unsuccessful implementations, and maintenance management software is no exception. If you are considering a maintenance management implementation, you need to start by identifying the goals and vision of the maintenance department.

If you just want a list of assets and parts for accounting, CMMS is overkill. Say your goal is to reduce unproductive time spent going through filing cabinets trying to answer questions like the following:

- When was the last time we worked on this asset? And what was the problem?
- Where did we buy this part last time? Who has their phone number?
- Is there a substitute part in the storeroom? Where is it?
- Is this equipment under warranty? Have we performed the required warranty maintenance task?
- How much have all the repairs on this asset cost? Should we replace it?
- When was the last time a part was used? Is it obsolete?
- What did we spend on PM last year? Reactive maintenance?

If those questions sound familiar, you probably could benefit from a CMMS (Crain, 2003).

4.2.5.2 Process First

Far too often organizations will purchase CMMS or EAM software with the expectation that their maintenance business will instantly operate more efficiently. As with everything else in life, CMMS can only provide what you put into it. By defining and documenting existing process requirements and establishing a method for measuring performance, you can define the role of the CMMS and determine what functionality is required to fulfill the maintenance vision:

- Define the basic processes, from purchasing inventory for a planned maintenance activity to preparing and following up on an emergency breakdown work request.
- Establish a team that represents all facets of the maintenance organization (planning, storeroom, purchasing, engineering, etc.).
- Diagram existing processes and determine possible bottle necks that can be eliminated and parts of the process that can be automated.

Sample maintenance processes include the following:

- Reactive or corrective work (with and without materials)
- Planned maintenance (with and without materials)
- Preventive maintenance (with and without materials)

- Engineering/maintenance projects
- Maintenance, repair, and operations (MRO) parts reorder
- MRO parts cycle count
- MRO parts receiving

It is a good idea to establish a continuous process improvement team to evaluate and add maintenance processes on a fixed basis, such as quarterly (Crain, 2003).

4.2.5.3 Measuring the Process

To evaluate the success of maintenance processes and their implementation as a whole, key performance indicators (KPIs) need to be defined. The KPIs will allow a maintenance team to identify and further optimize the processes, determine training requirements and quantify the return on investment (Crain, 2003).

Common measurements include the following:

- Planned versus reactive maintenance (cost and hours)
- Percent of preventive maintenance in total planned maintenance (cost and hours)
- Percent of a particular maintenance project in total planned maintenance (cost and hours)
- Percent of predictive maintenance in total planned maintenance (cost and hours)
- Inventory turnaround
- Maintenance cost as a percentage of equipment replacement cost
- Cost by equipment class
- Schedule compliance
- Percent equipment availability
- Loss of opportunity costs due to equipment failure
- Loss of opportunity costs due to equipment PM

4.2.5.4 Defining the Requirements

Once the primary processes and measurements are defined, the information requirements need to be determined.

- Identify what data are required, and what data are "nice to have."
- Evaluate how labor is scheduled and cost is tracked.
- Determine the types of work orders and how they will be prioritized.
- Determine what types of analysis will be performed and what supporting data are required.
- Develop standard nomenclature for items such as locations, assets, bins, and parts.
- Identify what entities will be tracked and what information needs to be tracked for each.

Evaluate the information requirements for other systems:

1. Will an interface be required to accounting, HR, timekeeping, purchasing or project management software?
2. Will other specialty systems need to be integrated such as control and CM software?

Identify requirements from the IT department, including the preferred operating system and database platforms:

- Does IT want a web or client server deployment?

All these requirements should be documented as a formal functional requirements document.

Take this opportunity to review the organization chart:

- Are positions defined in the work process filled? A full-time maintenance planner may need to be added or perhaps additional storeroom personnel are required.
- Are job descriptions modified to incorporate process requirements?
- Who will be responsible for supporting users from the maintenance group and from IT? (Crain, 2003).

4.2.5.5 Common Components of a Computerized Maintenance Management System

Once requirements have been defined, a checklist can be compiled to rate the functionality of systems being evaluated. Most CMMS applications are comprised of baseline modules that support basic maintenance functions of asset management, work management, PM and materials management. Some applications have "add-on" modules available for additional cost that address specialty requirements, such as fleet, mobile work order, barcode, or work request (Crain, 2003).

The following subsections list common modules and specific fields and functionality found in most CMMS applications (Crain, 2003).

4.2.5.5.1 Labor

- The ability to keep a listing of employees and certifications
- The ability to assign an employee to a specific area
- The ability to assign an employee to a specific crew and/or shift
- The ability to assign standard and overtime rates to an employee

4.2.5.5.2 Asset

- Nameplate data (manufacturer, model, serial)
- Asset criticality
- Material list
- Location or position
- Hierarchy for rolling up cost from the asset to area and facility

- Asset assemblies
- Configurable equipment class and/or type
- Cost codes
- Asset cost
- Warranty information
- Meters (used to trigger PMs)
- Safety procedures
- Documents or images can be printed with the work order
- User-defined fields

4.2.5.5.3 Work Management

- Availability of work order number
- Availability of work order description
- Availability of work order priority
- Ability to configure work order type (PM, corrective, safety, etc.)
- Availability of problem, component, cause and remedy codes
- Ability to reserve materials planned on work orders
- Ability to add materials recorded against work orders to the asset material list
- Ability to plan shifts and employees based on work orders
- Ability to record asset downtime
- Ability to record planned versus actual labor, materials and cost
- Ability to support work orders or projects
- Availability of safety procedures
- Ability to attach documents or images printed with work orders
- Availability of user defined fields

4.2.5.5.4 Task or Procedures

- Can create predefined procedures associated with a PM or work order, including the following:
  - Checklist of tasks to be performed
  - Estimated hours
  - Resource assignments
  - Parts list
  - Safety procedures

4.2.5.5.5 Preventive Maintenance

- Ability to associate multiple assets with one PM record
- Ability to trigger PMs based on time or usage based events
- Ability to trigger PMs on fixed or completion based intervals
- Ability to sequence PM work orders (i.e., monthly, quarterly, annually)
- Ability to assign seasonal or blackout dates
- Ability to batch generate PM work orders

4.2.5.5.6 Materials Management

- Has multiple store rooms
- Has bin locations
- Can discretely track parts by a unique item number
- Can support multiple manufactures and manufacturer part numbers associated with one inventory item number
- Has primary vendor and associated vendor contact information
- Has inventory class and commodity codes
- Has the ability to postinventory cost to work orders using "average" or "last price"
- Can set min and max reorder points
- Can see parts stored in multiple store rooms
- Can support cycle counting
- Can support work order issues/returns
- Can support storeroom transfers
- Can support nonstock items issued to work orders
- Can support repairable spares

4.2.5.5.7 Purchasing

- Purchase requisition generated automatically when parts are issued based on min max levels
- Consolidation of line items from multiple request into one PO
- Ability to track purchase order items on order, received or on back order
- Approval hierarchy

4.2.5.5.8 Plug-ins or Add-ons

- Interfaces with widely used ERP software
- Interfaces with plant automation systems
- Interfaces with customer relationship management and customer information systems
- Ad hoc reporting
- Fleet integration
- Integration of PDA clients (Pocket PC or Palm OS)
- Bar-coding integration
- Integration of work requests and dispatches
- Integration with CAD programs
- Integration with GIS
- Integration with project management systems
- e-Procurement

4.2.6 Computerized Maintenance Management System Implementation

Once the requirements have been established and a system selected, the implementation process begins. Unfortunately, it is not as easy as flicking a switch. To get the CMMS to provide useful information it must be configured, key data collected and entered, and employees must be motivated to use the system.

The team who identified the business processes and requirements is a good choice for the initial configuration. The best first step is to have them participate in an extensive overview of the selected application in consultation with the vendor or a consultant familiar with the selected CMMS. During the overview, the team needs to ensure that the system meets all requirements. It should also identify all the components of the system that are required to achieve the planned level of functionality (i.e., labor, value list, cost/GL codes, assets, bins, inventory, etc.).

As the data collection efforts begin, the system needs to be "sold" to end users by doing the following:

- Engage them in the process of collecting and "owning" the information entered into the system.
- Take advantage of assemblies like safety or tailgate meetings to provide updates on the progress of the implementation.
- Express how the system will benefit the maintenance users, management and the company as a whole.
- Ask senior management team to address employees on the significance of the system and the importance of their contributions.
- Promote question and answer sessions, possibly even short product demonstrations.
- Emphasize that the goal of the system is not to monitor employees, but to move the organization from a reactive to a proactive mode (Crain, 2003).

4.2.6.1 Consulting Services

Many vendors and consultants provide maintenance management system implementation support. Some focus on the business side of the system while others are systems integrators with an information technology focus. If you are considering the services of a consultant, these questions may prove useful:

1. Can your staff remove themselves enough from the existing way of business to evaluate ways to make the old processes more efficient?
2. Do you have someone on your maintenance staff with enterprise software implementation experience? If not, does the project schedule allow for the learning curve?
3. Does your maintenance and IT staff have the time to add this project to their existing workload?

If your answer is no to any of the above, or if you plan on your system being fully functional in under 6–12 months, you should consider a consultant. As well as contributing years of lessons learned to the implementation team, a seasoned consultant will help develop a project plan and guide teams through the implementation process.

If you want a turnkey type implementation, it is critical to have your team evaluate and approve key deliverables, such as business processes, nomenclature, customizations, data and the final system configuration. Regardless of the consultants' role, during testing, training and going live, the existing resources should be visibly leading the effort (Crain, 2003).

4.2.6.2 System Configuration

The bulk of the work that goes into a CMMS implementation is the collection of data, data migration, system configuration and testing. The data entered in the system must follow consistent naming standards. The identification numbers should be recognizable. If you're implementing a new naming convention, you need to label assets in the field and provide hierarchy or plot plan diagrams next to maintenance workstations. The data should be as complete and accurate as possible. Incorrect or inconsistent data represent the quickest road to frustration for maintenance users.

The system should be configured to provide the most intuitive user interface available. Work request and work order entry screens should take advantage of as many "value lists" as possible. Use default values if supported by the system. If the CMMS supports hiding unused fields, take advantage of that feature as well. To ease future upgrades, keep customizations to the application and custom database triggers to a minimum. When customizations can't be avoided, make sure they are well documented.

The system should be well tested before training. To this end, develop test scripts that model your work process and recruit computer savvy end users to participate in testing. They will also be able perform data quality spot checks. Do not start training until you are confident the system will perform "as advertised" (Crain, 2003).

4.2.6.3 Training

Training is where the "rubber meets the road." End users are getting their first hands-on experience with the system and developing first impressions. "Canned" data should never be used for training. Instead, the training environment should mirror the production database. The format of training should be role and process based. A work requestor is going to be overly confused if he/she is trained on all the functionalities of a work order instead of on the one screen and few fields required to create a work request. The following techniques are also useful:

- Make sure that one or two implementation team members are available to help users who fall behind during training.

- Follow up with those users during "go live" and provide auxiliary training.
- Provide a "play" environment where users can exercise and reinforce what they learned in training without corrupting production data (Crain, 2003).

4.2.6.4 Go Live

Schedule "go live" when you will have plenty of support for the users and lower than average work order volume. Have a backup plan to manage the flow of information in case the system goes down unexpectedly. If users have to fall back to the paper system for a few hours, have them create work orders in the CMMS at the end of their shift. Have quick reference or "how to" sheets available, as well as diagrams, to help them enter the proper values into the system.

At the end of everyday, review what went right and where changes are required. Reviewing the work orders daily is an excellent way of seeing how successfully each maintenance user is interacting with the system, or who needs extra help. Update and re-post the work processes the following day. Track and prioritize configuration issues and system "bugs." Provide a daily update to vendors or consultants, if applicable (Crain, 2003).

4.2.6.5 Postimplementation

Following "go live" but before the transition to normal operation, review all the defined requirements and determine whether they have been fulfilled. Insure that the mechanisms to facilitate and measure business processes are in place. If a vendor or consultant has been used, make sure all his/her deliverables have been completed.

Develop a support and issues escalation path. Have users go to their supervisor or resident "power user" with issues. If the issue cannot be resolved, escalate it to a site or company administrator. The company administrator should be the sole point of contact between the organization and the vendor. The administrator should be responsible for logging and tracking all open issues and enhancement requests (Crain, 2003).

4.2.7 Maintenance Knowledge Management Fusing Computerized Maintenance Management System and Condition Monitoring

The production and process industry has been transforming and improving for the last couple of decades because of global competition coupled with the advancement of information and communication technology (ICT). With the increased competition, the business scenario is focusing on meeting customers' needs, thus creating a need for enhanced value and improved asset management. More specifically, today's need for prognostic business solutions is compelling industries to minimize production and service downtime by reducing machine performance degradation. Proactive maintenance strategies are called for if a company wants to provide optimized and continuous process performance with minimized system breakdowns and maintenance—giving customers value for their money, in other words. This has led to the implementation of solutions from the business world, such as e-intelligence, e-factory, e-automation, eMaintenance, e-marketing, and e-service.

eMaintenance provides an organization with intelligent tools to monitor and manage assets (machines, plants, products, etc.) proactively through ICT, focusing on health degradation monitoring and prognostics, instead of fault detection and diagnostics (i.e., reactively). Maintenance effectiveness depends on the quality, timeliness, accuracy, and completeness of information, including knowledge and earlier experiences of machine degradation state and support processes, based on which decisions are made. This translates into a number of key requirements: preventing data and information overload, being able to differentiate and prioritize data and actions (during collection and reporting), preventing, as far as possible, information islands, and communicating status and vital information to relevant actors. The integration of maintenance knowledge management (MKM) into the processes and infrastructures of eMaintenance creates the foundation for a more comprehensive approach to ICT-based maintenance solutions, otherwise termed i-Maintenance ("intelligence-based maintenance"). The i-Maintenance approach not only aims at integrating MKM into solutions but also at integrating collaborative environments, remote computational services, ontologies for effective tagging of resources, solutions, etc. It is designed to be used across different levels of an organization and between organizations.

CMMS and CM are the most popular repositories of information in maintenance, and most of the deployed technology is installed in them. Unfortunately, isolated information islands are frequently the result.

While using CMMS and CM technology as isolated systems can achieve maintenance goals, combining the two into one seamless system can have exponentially more positive effects on maintenance performance than either system alone might achieve. The combination of the strengths of a top-notch CMMS (PM scheduling, automatic work order generation, maintenance inventory control, and data integrity) with the wizardry of a leading-edge CM system (multiple-method CM, trend tracking, and expert system diagnoses) allows work orders to be generated automatically based on information provided by CM diagnostic and prognostic capabilities. Fifteen years ago, linking CMMS and CM technology was just a vision, easily dismissed as infeasible or, at best, too expensive to warrant much investigation. Now, the technology available in CMMS and CM solutions has made it possible to achieve such a link relatively easily and inexpensively.

Integrating an MKM component with the CMMS and CM environments introduces the risk of creating additional information islands and complexities, if the integration is not properly designed, developed, and implemented. One promising approach to the integration of MKM and CMMS and CM data into overall solutions is the use of service-oriented architectures (SOAs) in combination with software agents. The danger of trying to intimately integrate MKM, collaborative structures, CMMS, and CM into one unified solution is that the overall solution will likely be suboptimized if not planned and implemented properly. This is especially dangerous when implementing solutions with a high rate of change in structure and processes—MKM is one of these. At the moment, the most viable route is to integrate at the end-user level, using a combination of application servers with end-user environments to allow the modular integration of different information sources, services, and functions.

A high-specification CMMS can perform a wide variety of functions to improve maintenance performance. It is the central organizational tool for world-class maintenance and is designed to facilitate a shift in emphasis from reactive to PM. It achieves this shift by allowing maintenance professionals to set up automatic PM work order generation. A CMMS can also provide historical information that is then used to adjust a PM system to minimize unnecessary repairs, while avoiding run-to-failure repairs. PMs for a given piece of equipment can be set up on a calendar schedule or a usage schedule based on measurements and readings. A fully featured CMMS also includes inventory tracking, workforce management, and purchasing, in a package that stresses database integrity to safeguard vital information. The final result can be optimized equipment uptime, lower maintenance costs, and better overall plant efficiency, depending on the ability of the maintenance staff to use the systems and processes, and these are highly dependent on the staff's knowledge level and experience.

A good CM system will accurately monitor real-time equipment performance and alert maintenance professionals to any changes in performance trends. A CM package tracks a variety of measurements, such as vibration, oil condition, temperature, operating and static motor characteristics, pump flow, and pressure output. These measurements are taken using monitoring tools, including ferrographic wear particle analysis, proximity probes, triaxial vibration sensors, accelerometers, lasers, and multichannel spectrum analyzers. An expert CM system can analyze the measurements and diagnose faults. Expert system analysis can place maintenance procedures on hold until absolutely necessary, thus extracting maximum equipment uptime. In addition, expert systems should offer diagnostic fault trending whereby individual machine fault severity is observed over time.

MKM allows the effective dissemination of experience, manuals, collaborative structures for access by internal and external specialists and knowledge resources. In the context of support for CMMS, MKM can integrate management of documents, instructions, access to remote servicing, decision support, and experience capture and management. In the context of support for CM, MKM can support the analysis processes of CM data by accessing collaborative structures for internal and external specialists. In addition, it can provide access to external CM analysis tools/computational engines and permit interaction with external vendors' specialized support structures. The MKM component can be instrumental in the ability of the maintenance staff to properly interpret the results of measurements and computations.

Both CMMS and CM systems have strong suits that make them indispensable to maintenance operation improvements. CMMS is a great organizational tool, but cannot directly monitor equipment conditions. A CM system excels at monitoring those equipment conditions, but is not suited to organizing the overall maintenance operation. The logical conclusion, then, is to combine CMMS and CM technologies into a seamless system that avoids catastrophic breakdowns, but eliminates needless repairs to equipment that is running satisfactorily. The MKM environment can play a strong role in improving the accuracy and quality of the analysis and the resulting decisions. MKM can also minimize the risk of mistakes and human error in the implementation of the decisions and the quality of the maintenance work performed on the work floor or in the field. It also leads to the improved quality of decisions made over time because of the more accurate feedback of experiences of and observations made by the maintenance operators (Björling et al., 2013).

Technology providers are trying to develop advanced tools but maintenance departments often struggle with daily problems of implementing, integrating, and operating such systems. MKM systems can play a vital role in speeding up effective implementation of the more advanced ICT-centric solutions by integrating competence resource structures into the solutions, allowing different types of end users to have support and instructions optimized for their own work contexts. MKM systems can supply infrastructures for experience capture and management, thus supporting maintenance departments in the management of ever-increasing complexities in assets and process flows.

Existing maintenance systems are often a heterogeneous combination of methods and systems in which the integrating factor of the information and business processes is the maintenance personnel, and those personnel often cannot utilize the full functionality of the underlying systems. The information in the maintenance systems goes through human minds, forming an organizational information system and creating a high reliance on the expertise of the maintenance staff. Thus, increasing support for the human component in the overall system will improve the overall maintenance processes and the effectiveness of operations. In this context, MKM has an important role (Rasovska et al., 2008).

With the emergence of intelligent sensors to measure and monitor the health state of a component or system and the gradual implementation of information and communication technologies (ICT) in organizations, eMaintenance is becoming a reality (Parida and Kumar, 2004). While eMaintenance techniques can provide benefits to an organization, the seamless integration of ICT into the industrial environment remains a challenge. It is necessary to address the requirements and constraints from the maintenance and ICT standpoints in parallel (Björling et al., 2013).

4.2.7.1 An Integrated Approach to Asset Management

Two main maintenance information sources need to be merged: CMMS and CM.

CMMS uses context-specific textual data to record information such as asset load and usage, component failures, servicing or repairs, and inventory control. Although for a given platform, there may be several different implementations, the underlying structure is typically heavily regulated, allowing a large base of consistently structured data. These systems are the core of traditional scheduled maintenance practices and rely on bulk observations from historical data to make modifications to regulated maintenance actions.

CM systems collect component-specific quantitative data to assist maintenance crews in the identification of failures that are imminent or have already occurred. There is typically no standardization in the way data are collected across platforms or vendors, because the technology is still in its infancy. There is still a need to investigate and debate the type of information required for asset health diagnostic.

At this point, CMMS environments do not support client platforms. More advanced resources, such as multimedia and integrated collaborative environments, are useful for effective remote support and interaction with internal and external specialists. Core functionalities for experience capture are not present in current CMMS environments—the main challenge is the lack of metadata management.

CMMS, CM, and MKM need to be linked. The measurements and analysis implementations supported by MKM and provided by a CM package must be available to maintenance planners who work with a CMMS to schedule predictive and other types of work orders. These maintenance planners are also supported by MKM.

In the past, maintenance organizations using both CMMS and CM technologies linked them by inputting CM data manually into the CMMS. While this is an acceptable way to transfer data for the purpose of scheduling predictive maintenance work orders, it is time-consuming. Another CM data transfer method used more recently is passive data exchange; this involves writing pertinent CM data to a specified local or network directory. Relevant data to be exchanged include equipment identification, date and time stamps, repair priority, repair recommendations, and observations (Björling et al., 2013). The CMMS program routinely checks this directory, and if a transfer file is found, CMMS reads it and imports it into the CMMS database. Historically, this method of data transfer has been very specific to formal cooperation between various manufacturers of CM and CMMS software. The passive data transfer method is better than manual data entry, but falls well short of the total automation and instant access to information that is possible when CMMS and CM are totally integrated.

Integration of MKM solutions can support effective training and learning of the staff responsible for the data capture (increasing the quality), and the staff responsible for the analysis and information/experience exchange between internal and external specialists. In future scenarios, CM analysis will be performed remotely by different companies with different specializations. An MKM environment can aid the internal analysis staff to select the correct CM analysis and modeling vendor/analysis approaches and handle the eventual training efforts. To this point, integration has been addressed this far largely from the point of view of representing the collected information to the end user (operator or manager) in an effective manner, i.e., bridging the gap between information collected from plants and equipment and the ERP platforms. A major initiative has been the development of information integration specifications to enable open, industry-driven, integrated solutions for asset management.

Some of the efforts to standardize the eMaintenance platforms currently underway are: Machinery Information Management Open Systems Alliance (MIMOSA) (Kahn and Klemme-Wolf, 2004), GEM@WORK (Wang et al., 2004), CASIP (Baptiste, 2004) and *PROTEUS* (Thomas et al., 2004). Such platforms provide an information schema at the application level and an application programming interface to communicate with the underlying protocol stack. Existing communication technologies are not well-suited for reliable and timely delivery of appropriate data between distributed end systems in industrial environments; this remains a critical missing link in the vision of seamless integration. Added to this is the current lack of effective integration of knowledge management structures into the overall maintenance environments. Such integration will soon become vital, given the increasing complexity of processes and products, not to mention the increased competition for knowledgeable staff and specialists. The main challenge for future maintenance systems is to support the end users of different types in their efforts to manage a complex work environment and complex and advanced assets and processes to decrease stress and risks (Björling et al., 2013).

4.2.7.1.1 Integration of Data Sources

The first step of integrating CMMS, CM, and MKM packages into an automatic system is setting up a way for the systems to communicate. In the case of CMMS and CM technologies, the first step can be setting up consistent data in each system so they can communicate using a common base of information. Next, there must be a system of data cross-references between the sensors, meter tags, or other measurement tools in a CM system and the appropriate module in the CMMS that associates readings in one system with readings in the other. Meter readings or alarm triggers that are out of the acceptable range setup in the CMMS should trigger a predefined work order. Any discrepancy in this cross-reference for a piece of equipment will nullify the link for that piece of equipment, making the ability to predict problems problematic. This makes the initial planning of data entry rules and database setup a critical part of the preintegration process. The third step is to provide a direct link between the systems' data tables. This is referred to as an "active exchange" of data. In today's environment, CMMS databases feature open architecture such SQL, Oracle, and others.

Integrating MKM systems and environments with CMMS and CM has not been a factor in early implementations because of the specialization of CMMS and CM system vendors. Knowledge management has been seen as a separate area with its own markets and usage contexts; it has not been considered viable to integrate MKM into CMMS and CM environments.

The most obvious obstacle in the integration of CMMS, CM, and MKM data and information is the disparate nature of the data types involved, and attempts to remedy this problem have met with inconsistent implementation and limited scalability. The first step is to assign the qualitative CMMS data with quantitative indexing, allowing CM data to be separated into discreet maintenance states. It is the responsibility of the maintainer to correctly insert the appropriate fault or work code into the maintenance logs, which to date has not been done with sufficient accuracy or consistency to be deemed reliable. Fig. 4.6 demonstrates an integrated end-user environment supporting a modular and work context-centric approach to information management for

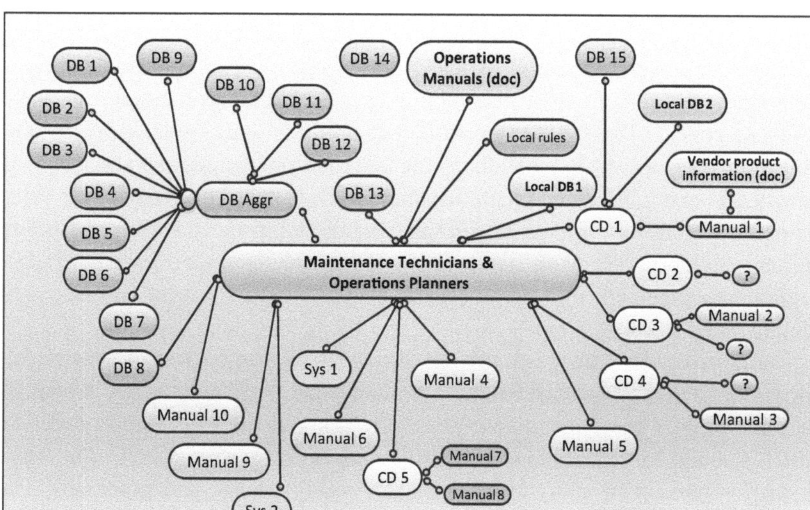

FIGURE 4.6 Integration of condition monitoring/computerized maintenance management system/maintenance knowledge management and support for multiple work contexts. Courtesy of Enviro Data, Sweden. *Adapted from Björling, S-E., Galar, D., Baglee, D., Singh, S., Kumar, U., 2013. Maintenance knowledge management with fusion of CMMS and CM. In: DMIN 2013 International Conference on Data Mining, 22–25 Jul 2013, Las Vegas, Nevada, USA.*

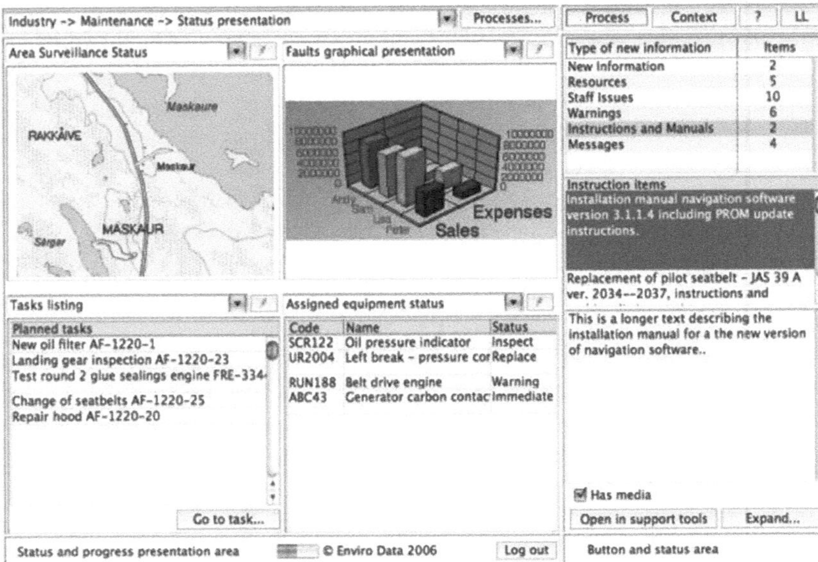

FIGURE 4.7 Example of an information sources map for a maintenance operations planner. *Adapted from Björling, S-E., Galar, D., Baglee, D., Singh, S., Kumar, U., 2013. Maintenance knowledge management with fusion of CMMS and CM. In: DMIN 2013 International Conference on Data Mining, 22–25 Jul 2013, Las Vegas, Nevada, USA.*

multiple organizational layers in an i-Maintenance structure. It has the ability to support end-user adaptation of the overall work environment depending on work roles, deployment environment, and user access control structures. It also presents a concept for end-user controlled mash-up (systems able to present and manage information from many disparate information sources and services—all presented in one unified end-user environment). It is based on research covering end-user environments for qualified maintenance of advanced technical systems (military fighters) allowing integration of knowledge/practices/experiences/standards management into a unified and integrated environment (Björling and Kumar, 2009).

The above example responds to information logistics and knowledge support challenges for maintenance planners and staff supporting advanced systems. Fig. 4.7 presents an anonymized schema of the information and data sources needed for maintenance planning and work for this system. The same scenario is present in many other contexts, organizations, and processes. The main challenge is to create productive and effective end-user environments to allow management and handling of these disparate information and competence resource sources and access methods. This is a good example of the challenges involved in integrating CMMS, CM, and MKM resources and, at the same time, allowing effective collaboration (Björling et al., 2013).

4.2.7.1.2 Definition of Integration/Relation Process

Although there have been many recent efforts to collect and maintain large repositories of CMMS and CM data, relatively few studies try to identify the ways these two data sets could be related. There are even fewer studies of how the CMMS, CM, and MKM resources can be linked and managed in work contexts. It is only logical to assume that written histories of maintenance records are linked to the measurements of on-board sensors, and it is in the interest of CBM research to develop a means by which these data sets can be consistently and reliably merged. At the same time, relevant competence resources connected to an asset must be managed and updated in end-user environments that do not introduce additional complexities for the users in their most prevalent work contexts and processes.

A full integration of CMMS and CM data sets requires a more advanced form of interfacing which more appropriately models the real-world relationships between observed maintenance and sensor data. Case studies to date have been generated by individuals who identify related events based upon their knowledge of the systems involved. For example, an abrupt change in a vibration sensor on a gearbox is assumed to be related to a recorded replacement of a nearby part. Taking an analytical approach to this decision-making process is complex, since the determination of causality and dependence is often performed through a highly subjective process. MKMs can, in this case, increase the accuracy of the analysis by supplying earlier experiences, faults and conclusions and, at the same time, speed up the processes by allowing faster access to external specialists and competence resources of different types. An MKM environment can also support wider and more complex analysis of overall process dependencies and more complex logging of faults and corrective actions than are allowed by the current selection of CMMSs.

The overall goal of enhanced interfacing require the automation of the complex process of linking events from different data sets—preferably using SOA to guarantee the security of the data. Developing this system begins with a four-step investigation: historical data collection, importation into a single database, data abstraction, and data analysis. Using a wealth of historical information in combination with knowledge of system components, software agents under development are attempting to bridge the gap between data types by allowing the proximity, severity, and rarity of events across data sets to be evaluated. Fig. 4.8 describes a process using direct integration of data sets, not an SOA-based interaction. When CM and CMMS are integrated, it is possible to regularly identify instances where CM data are reflected by real-world events. This allows an objective determination of asset parts prone to failure and an evaluation of CM effectiveness in monitoring those regions. Based upon these evaluations, feedback can be given to CMMS and CM developers to refine the means by which data are collected, and a strategy for the next generation of fully integrated CBM systems can be devised. This can be one of the tasks for MKM environments—acting as a collaborative platform for collecting, evaluating, and analyzing earlier experiences, designing mitigation

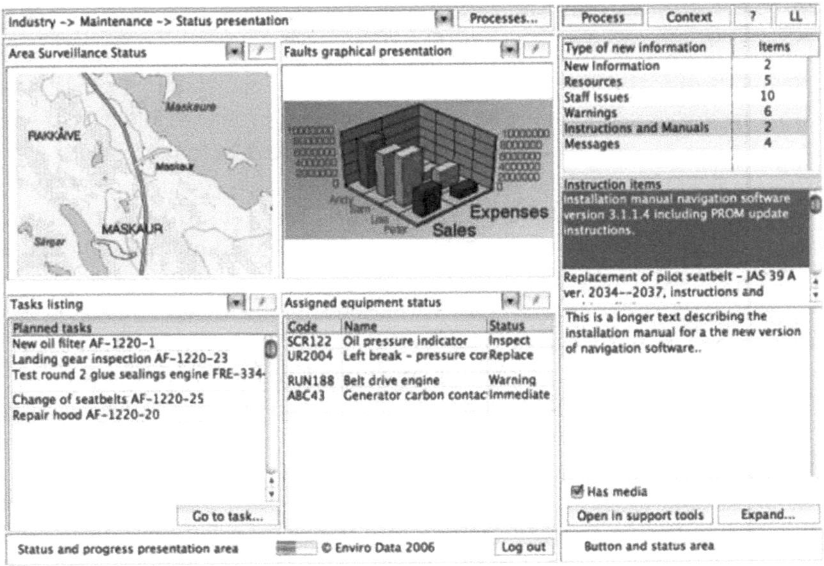

FIGURE 4.8 Four-stage integration process. *Adapted from Björling, S-E., Galar, D., Baglee, D., Singh, S., Kumar, U., 2013. Maintenance knowledge management with fusion of CMMS and CM. In: DMIN 2013 International Conference on Data Mining, 22–25 Jul 2013, Las Vegas, Nevada, USA.*

methods and processes, and providing support for the development of internal practices (Björling et al., 2013).

4.2.7.1.3 Data Source Collection

The history of maintenance management systems predates the information age, and it has traditionally been delegated to the unit level for implementation. Collecting data has required the permission of various units, thus limiting the scale of CMMS research to date. As a result, efforts to centralize CMMS data have been slow to materialize, and data collection for early integration studies only hints at the future capabilities of a centralized CBM system. In contrast, CM developers have relied on automated data centralization to evaluate and validate their systems since their inception.

To minimize the risk of unintended sabotage of the measurement data caused by faulty installation routines and a lack of experience handling the measurement equipment, an organization can use MKM structures to integrate training efforts and eventual certification processes into the overall installation and maintenance of the measurement equipment (Björling et al., 2013).

4.2.7.1.4 Importation of Relational Data

Modern CMMS information is stored in large relational, or tabular, databases. This format is appropriate for an integration investigation, as there are many software tools available to query and investigate the tables. For historical analysis, only certain fields are required, thus allowing sensitive data to be removed or filtered. The data subset will still contain a full history of component faults and related actions, providing a comprehensive maintenance history profile while alleviating security concerns.

Importing CM data into a relational database is more challenging, as each type of sensor generates different data classes, sampling rates, and number of compiled indicators. Furthermore, each manufacturer stores the collected information in unique proprietary formats, requiring platform-specific importation software to be written. This software allows the CM data to be exported from the original interface so that it can be expanded and generalized. Once this is accomplished, the benefits are tremendous: multiple manufacturer and cross-platform data can be viewed as though they are generic data classes (Björling et al., 2013).

4.2.7.1.5 Preprocessing and Data Abstraction

Even when CMMS and CM data coexist in a single database where they can be queried and explored, automating the discovery of linked events requires additional processing. In their original form, the data sets only have two fields in common: asset identification and date. Relating a given maintenance fault or action, which is textual, to sensor data, which is some arbitrary data class type, can only be accomplished through the compilation of overlapping metadata. The fields that are generated characterize the location and significance of events, creating a quantified set of parameters by which the disparate data can be compared.

Since CMMS is textual, the data are processed using AI tools applied to language processing (LP) (Manning and Schutze, 1999). LP is a subfield of both AI and linguistics and has a large variety of applications. It covers topics ranging from machine translation to speech recognition and often focuses on a computer's ability to interpret and respond to natural human languages. A recent success in LP is auto summarization and information extraction. Given the specific context of maintenance management data in which descriptions of assets' faults and performed actions are stored, the lexical domain is highly restricted, and text CMMS fields can be analyzed separately to create a set of interpreters which extract key information from the fault or action description. The AI LP tool outputs which component the record refers to, along with a list of other descriptor keywords. Categorical statistical analyses are performed to characterize the rarity of a given record, and a preprogrammed scoring chart assigns each record a severity based on the available keywords.

An additional level of abstraction of data for CM records is generated differently depending on the data class involved. Anomalies can be identified using statistical distribution analysis and, in the case of multidimensional data, neural networks to identify which components a particular sensor or indicator is monitoring are predefined by the CM manufacturer.

Automatic semantic analysis is dependent on the in-data; if the users and untrained and are not using coherent strategies in naming and characterizing the information, there is much more potential for error. MKM can give support in educating users and offer channels for end-user input on eventual modifications to the ontological and semantic definitions (Björling et al., 2013).

4.2.7.1.6 Analysis and Correlation

Metadata are extracted from all the available records and put into a single events table containing asset identification, component name, event time, a rarity parameter, and a severity parameter. The simplest method of determining event-relatedness is accomplished by a proximity study of the metadata. The results can be categorized by component to identify parts of assets where CM devices have a high success rate in identifying faults or reflecting maintenance actions. Known problematic subsystems that do not have a high count of related CMMS and CM events indicate a need for revision to the sensing strategy or changes in indicator definition. For these components, further analysis can be performed on the raw data to discover new algorithms for condition indicator computation. However, this computation is extremely complex because of the multilocation structure of many companies, with many information systems related to maintenance in each location (CMMS, CM, phones, PDA, laptops, SCADA, ERP, etc.). Replication of all data to perform this correlation is not feasible and would require enormous computation resources; the concept of cloud computing is seen as the answer to creating these metadata from all available information.

The analysis and correlation steps demand high levels of competence and experience—this is tricky when there is a lot of equipment, a large number of variants, and not enough units of a specific kind to locally create expertise on the units in question. An MKM environment can offer structures for collaboration between the specialists and maintenance staff covering the fault modes, mitigating efforts, experiences, and practices. It can also offer collaborative environments for collective assistance in fault analysis and CM analysis (Björling et al., 2013).

4.3 SUPERVISORY CONTROL AND DATA ACQUISITION AND AUTOMATION DATA FROM PROGRAMMABLE LOGIC CONTROLLERS AND SIMILAR DEVICES

Supervisory control and data acquisition (SCADA) is a software system that works with coded signals over communication channels. These coded signals are combined with a data acquisition system via communication channels to get information about the status of remote equipment for display or for recording functions (Boys, 2009). The SCADA system is mainly used to control the production of power, infrastructure and facility processes, including water treatment, distribution space stations, energy consumption, siren systems in civil defense etc.

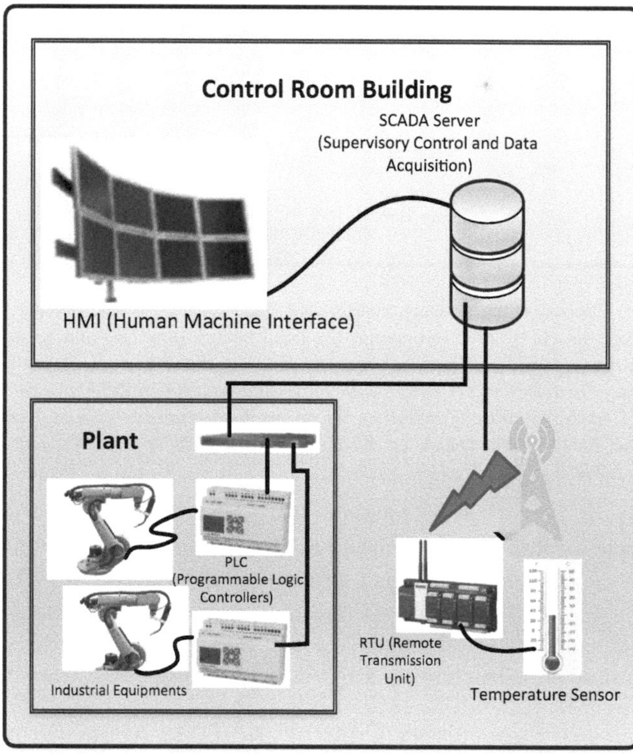

FIGURE 4.9 Supervisory control and data acquisition (SCADA) system. *Adapted from Arora, N., 2016. Supervisory control and data acquisition (SCADA) - BackBone of modern industry - a review supervisory control and data acquisition (SCADA). In: Conference Proceeding of 4th International Conference on Recent Innovation in Science, Technology and Management (ICRISTM-16) at YMCA, Jai Singh Road, Delhi, India on 3rd December 2016, ISBN: 978-81-932712-9-2.*

Shalini and Kumar (2013) have reported on the real-time implementation of SCADA in power distribution networks. The electrical distribution system consists of several substations; these substations have multiple controllers, sensors, and operator-interface points, as shown in Fig. 4.9 (Arora, 2016). The issues faced by SCADA are not the same as those of the past; more importance is given to integration, functionality, communication, network technologies, security, reliability, and so on. SCADA in the power system network increases the system's reliability and stability for integrated grid operation.

4.3.1 Supervisory Control and Data Acquisition System

In general, programmable logic controllers (PLCs), circuit breakers, and power monitors are used to control and monitor a substation in real time. Data are transmitted from the PLCs and other devices to a computer-based

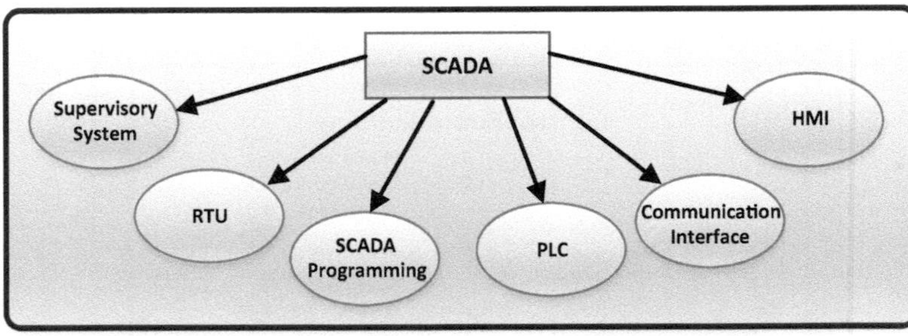

FIGURE 4.10 Basics of supervisory control and data acquisition (SCADA) system. *HMI*, human–machine interface; *PLC*, programmable logic controller; *RTU*, remote terminal unit. *Adapted from Arora, N., 2016. Supervisory control and data acquisition (SCADA) - BackBone of modern industry - a review supervisory control and data acquisition (SCADA). In: Conference Proceeding of 4th International Conference on Recent Innovation in Science, Technology and Management (ICRISTM-16) at YMCA, Jai Singh Road, Delhi, India on 3rd December 2016, ISBN: 978-81-932712-9-2.*

SCADA node located at each substation. One or more computers are located at different centralized control and monitoring points. The SCADA system, as shown in Fig. 4.10, consists of different blocks: human–machine interface (HMI), supervisory system, remote terminal units (RTUs), PLCs, communication infrastructure, and SCADA programming (Arora, 2016).

4.3.2 Basics of Supervisory Control and Data Acquisition

4.3.2.1 Human–Machine Interface

HMI is an input–output device that presents the process data to be controlled by a human operator. It is linked to the SCADA system's software programs and databases to provide management information, including the scheduled maintenance procedures, detailed schematics, logistic information, and trending and diagnostic data for a specific sensor or machine. HMI systems allow the operating personnel to see the information graphically (Arora, 2016).

4.3.2.2 Supervisory System

A supervisory system is used as a server for communicating between the equipment of the SCADA system, such as RTUs, PLCs, and sensors, etc., and the HMI software used in the control room workstations. A master station or supervisory station comprises a single PC in smaller SCADA systems; in larger SCADA systems, a supervisory system comprises distributed software applications, disaster recovery sites, and multiple servers.

4.3.2.3 Remote Terminal Units

Physical objects in SCADA systems are interfaced with microprocessor controlled electronic devices called RTUs. These units are used to transmit telemetry data to the supervisory system and receive messages on controlling the connected objects from the master system. Hence, they are also called remote telemetry units.

4.3.2.4 Programmable Logic Controllers

In SCADA systems, PLCs are connected to the sensors collecting the sensor output signals to convert the sensor signals into digital data. PLCs are used instead of RTUs because of their advantages, including flexibility, configuration, versatile, and affordability.

4.3.2.5 Communication Infrastructure

Generally, a combination of radio and direct wired connections is used for SCADA systems, but in large systems such as power stations and railways SONET/SDH is frequently used. Among the very compact SCADA protocols used in SCADA systems, a few communication protocols standardized and recognized by SCADA vendors send information only when the supervisory station polls the RTUs.

4.3.2.6 Supervisory Control and Data Acquisition Programming

SCADA programming is used to create maps and diagrams which will give important situational information in case of an event failure or process failure. Standard interfaces are used for programming most commercial SCADA systems. SCADA programming can be done using derived programming language or C language.

4.3.3 Architecture of Supervisory Control and Data Acquisition

The SCADA system usually includes the following components: local processors, operating equipment, PLCs, instruments, RTU, intelligent electronic device, and a master terminal unit or host computers.

The block diagram shown in Fig. 4.11 represents the basic SCADA architecture. The figure depicts an integrated SCADA architecture which supports TCP/IP, UDP and other IP-based communication protocols, as well as industrial protocols such as Modbus TCP, Modbus over TCP or Modbus over UDP. These all work over cellular, private radio, or satellite networks.

Complex SCADA architectures have a variety of wired and wireless media and protocols to get data back to the monitoring site. This allows the implementation of powerful IP-based SCADA networks over landline, mixed cellular, and satellite systems. SCADA communications can use a diverse range of wired and wireless media.

The choice of the existing communication depends on the characterization of a number of factors: remoteness, available communications at the remote sites, existing communications infrastructure, polling frequency and data rates. These factors impact the final decision on SCADA architecture.

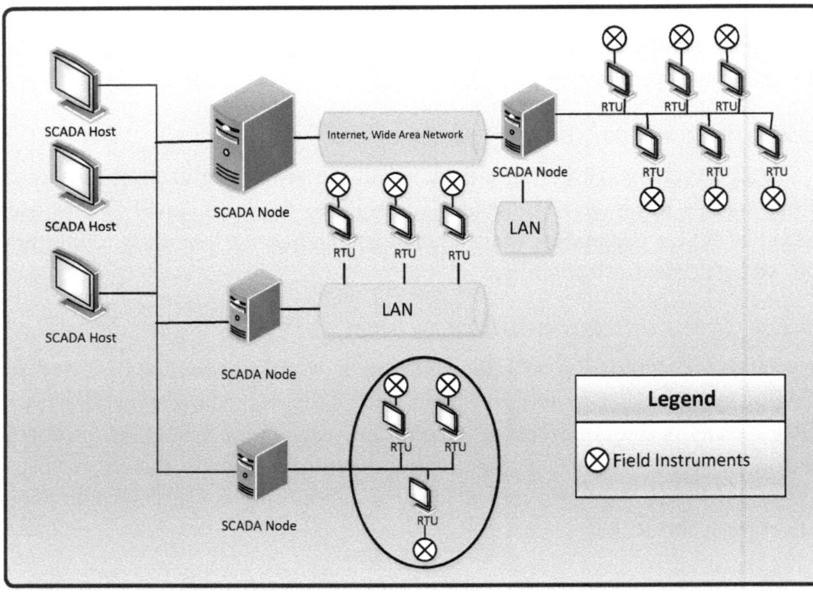

FIGURE 4.11 Architecture of supervisory control and data acquisition (SCADA). *LAN*, local area network; *RTU*, remote terminal unit. *Adapted from Arora, N., 2016. Supervisory control and data acquisition (SCADA) - BackBone of modern industry - a review supervisory control and data acquisition (SCADA). In: Conference Proceeding of 4th International Conference on Recent Innovation in Science, Technology and Management (ICRISTM-16) at YMCA, Jai Singh Road, Delhi, India on 3rd December 2016, ISBN: 978-81-932712-9-2.*

4.3.4 Types of Supervisory Control and Data Acquisition Systems

There are many different types of SCADA systems, and SCADA architectures have four different generations:

1. First generation: Monolithic or early SCADA systems,
2. Second generation: Distributed SCADA systems,
3. Third generation: Networked SCADA systems, and
4. Fourth generation: Internet of Things technology SCADA systems.

4.3.5 Applications of Supervisory Control and Data Acquisition

SCADA systems are used to monitor a variety of data, such as flows, currents, voltages, pressures, temperatures, water levels, etc., in various industries. If the system detects any abnormal conditions from any monitoring data, the alarms at central or remote sites will be triggered to alert the operators through the human-machine interface (HMI).

There are numerous applications of SCADA systems, but some frequently used ones are the following:

- Manufacturing industries
- Wastewater treatment and distribution plants
- Power systems

4.3.5.1 Supervisory Control and Data Acquisition in Manufacturing Industries

In manufacturing industries, the SCADA system can be applied to regular processes such as running production systems to meet productivity targets, checking the number of units produced, counting the completed stages of operations at various stages of the manufacturing process, and so on.

4.3.5.2 Supervisory Control and Data Acquisition in Wastewater Treatment and Distribution Plants

Wastewater treatment plants include surface-water treatment and well water treatment systems. SCADA systems are used to control the automatic operation of the equipment, such as backwashing filters based on the hours worked or amount of water flow through the filters. This is shown in Fig. 4.12. In distribution plants, the water tank levels, pressure of the system, temperature of the plant, sedimentation, filtration, chemical treatment, and other parameters or processes are controlled using SCADA applications, such as PLCs or PC-based workstations connected to each other using local area networks (LANs) such as Ethernet.

4.3.5.3 Supervisory Control and Data Acquisition in Power Systems

Power systems can be defined as consisting power generation, transmission, and distribution. All need to be monitored regularly to improve system efficiency. The

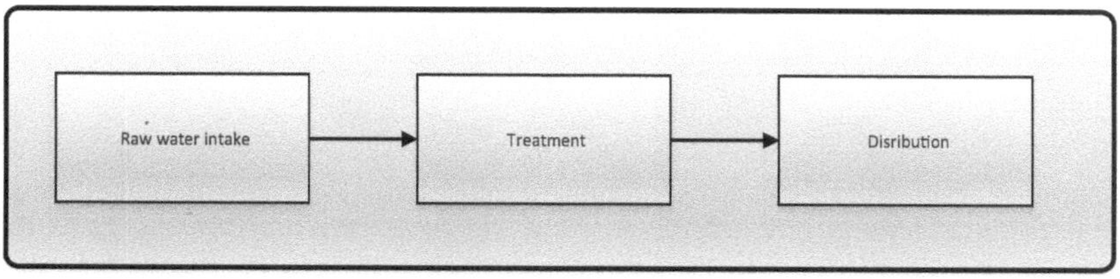

FIGURE 4.12 Wastewater treatment and distribution plants. *Adapted from Arora, N., 2016. Supervisory control and data acquisition (SCADA) - BackBone of modern industry - a review supervisory control and data acquisition (SCADA). In: Conference Proceeding of 4th International Conference on Recent Innovation in Science, Technology and Management (ICRISTM-16) at YMCA, Jai Singh Road, Delhi, India on 3rd December 2016, ISBN: 978-81-932712-9-2.*

application of SCADA improves the overall efficiency of the system by providing supervision and control over the generation, transmission, and distribution systems. SCADA in the power system network increases the system's reliability and stability for integrated grid operation; see Fig. 4.13 for an illustration.

4.3.5.4 Wireless Supervisory Control and Data Acquisition

In large-scale industries, such as power plants, steel plants, and so on, many processes and operations, such as the movement of conveyer belts for coal or product transport, boiler heat temperature, etc., must be monitored continuously, and it is necessary to control the factors affecting these parameters. The application of wireless SCADA will provide better control of the required control systems and operations; see Fig. 4.14.

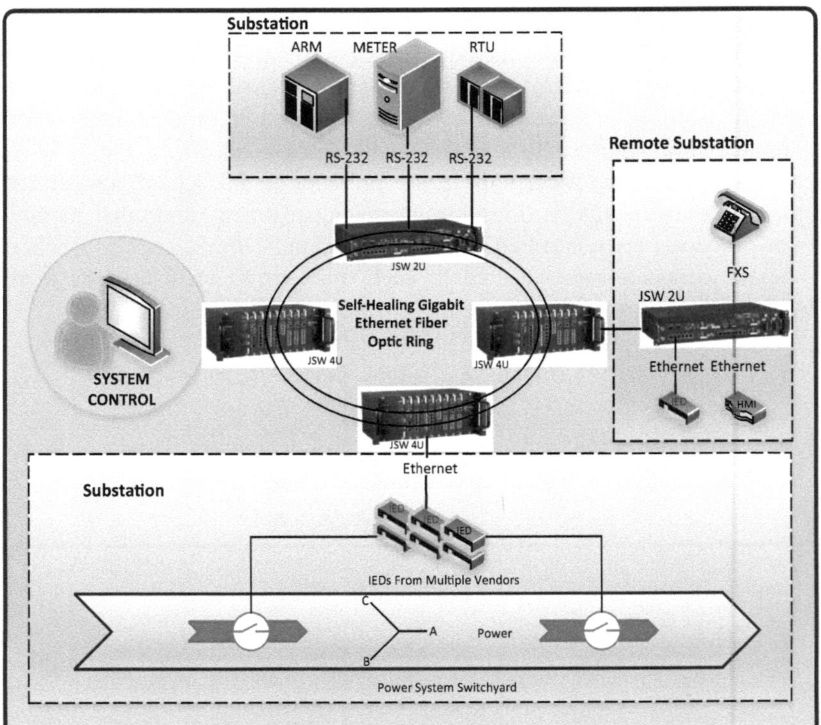

FIGURE 4.13 Supervisory control and data acquisition in power system. *Adapted from Arora, N., 2016. Supervisory control and data acquisition (SCADA) - BackBone of modern industry - a review supervisory control and data acquisition (SCADA). In: Conference Proceeding of 4th International Conference on Recent Innovation in Science, Technology and Management (ICRISTM-16) at YMCA, Jai Singh Road, Delhi, India on 3rd December 2016, ISBN: 978-81-932712-9-2.*

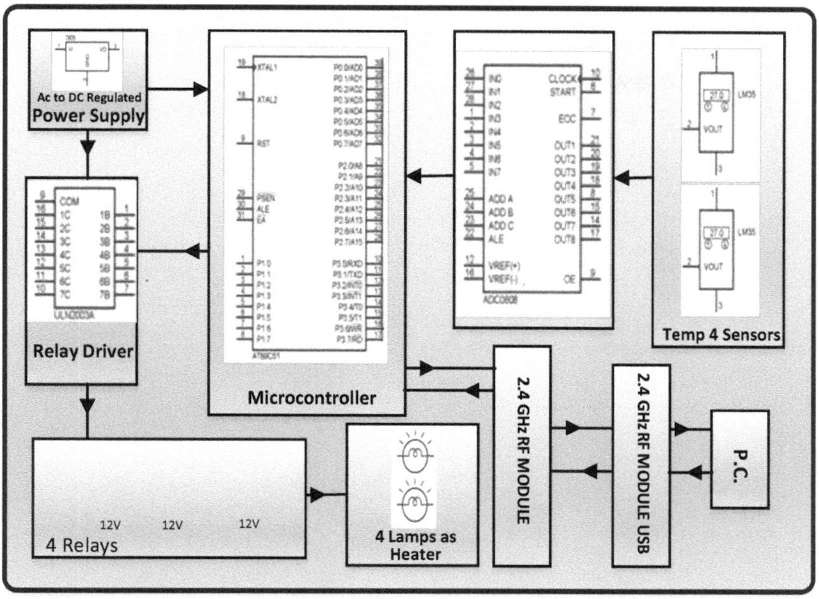

FIGURE 4.14 Wireless supervisory control and data acquisition system. *Adapted from Arora, N., 2016. Supervisory control and data acquisition (SCADA) - BackBone of modern industry - a review supervisory control and data acquisition (SCADA). In: Conference Proceeding of 4th International Conference on Recent Innovation in Science, Technology and Management (ICRISTM-16) at YMCA, Jai Singh Road, Delhi, India on 3rd December 2016, ISBN: 978-81-932712-9-2.*

4.3.6 Understanding Supervisory Control and Data Acquisition

To sum up the preceding discussion, SCADA refers to a class of industrial control systems (ICSs), i.e., computer systems to monitor and control industrial, infrastructure, or facility-based processes. A simplified ICS schematic is shown in Fig. 4.15. Relevant industrial processes include those of manufacturing, production, power generation, fabrication, and refining. Infrastructure processes may be public or private utilities, water or waste treatment, oil or gas distribution, and the power grid. Facility processes generally monitor and control HVAC and lighting, particularly in building energy management systems.

A SCADA system comprises the following elements: an HMI, a supervisory system, and RTUs, which interface to particular sensors and PLCs that can interface with sensors, actuators, and indicators. The communication infrastructure, which may be wireless, connects the supervisory system to the RTUs or PLCSs (Greenland, 2012). Fig. 4.16 gives a schematic representation of a SCADA system.

4.3.7 Programmable Logic Controller Programming

Programmable logic controllers (PLCs) used to be programmed in ladder logic, which is similar to relay logic. Today, PLCs are usually programmed in any one

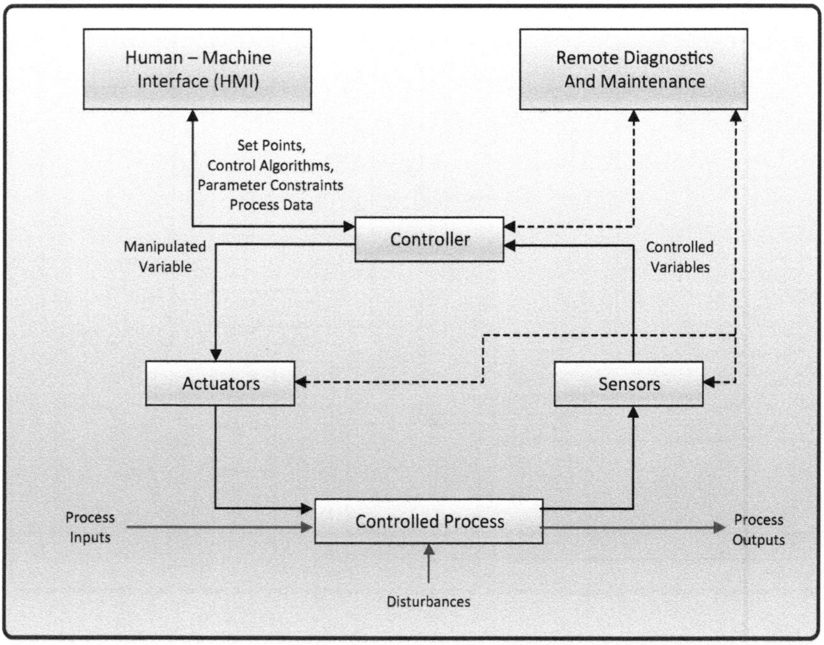

FIGURE 4.15 Simplified industrial control systems block schematic. *Adapted from Greenland, P., November 2012. Understanding Industrial System Structures Is Key to Powering Them. Exclusive Technology Feature. Cupertino, Calif.*

of several languages, ranging from ladder logic to Basic or C. The program is typically written in a development environment on a personal computer and then downloaded onto the PLC directly through a cable connection. A PLC diagram is shown in Fig. 4.17.

International standard IEC 61131-3 has recently become popular; it lists five programming languages for programmable control systems: FBD (function block diagram), LD (ladder diagram), ST (structured text, Pascal type lang), IL (instruction list), and SFC (sequential function chart) (Kadam et al., 2008).

4.3.7.1 Programmable Logic Controller Operation and Working Principles

A PLC works by continuously scanning a program. The scan cycle consists of three important steps. There are typically more than three but these are the most important:

- *Check input status*: First, the PLC looks at each input to determine if it is on or off. In other words, is the sensor connected to the first input on? How about the second input? The PLC continues to check all the connected inputs in this fashion and records the data into its memory to be used during the next step.

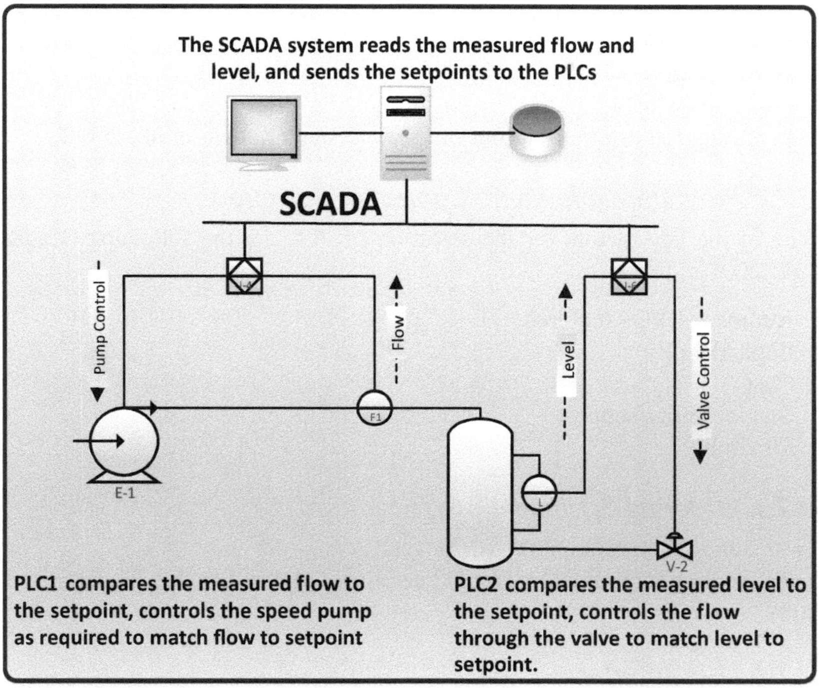

FIGURE 4.16 SCADA schematic example. *Adapted from Greenland, P., November 2012. Understanding Industrial System Structures Is Key to Powering Them. Exclusive Technology Feature. Cupertino, Calif.*

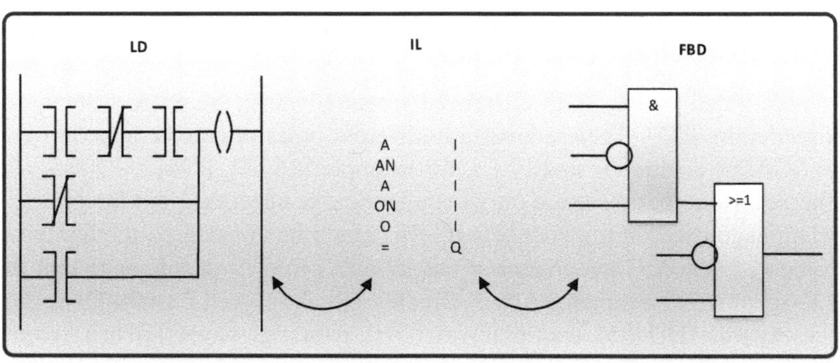

FIGURE 4.17 Programmable logic controller block diagram. *FBD*, function block diagram; *IL*, instruction list; *LD*, ladder diagram. *Adapted from Kadam, D.P., Sonwane, P.M., Landge, S.C., Dhote, V.P., Thorat, S.S., 2008. PLC & SCADA, a case study: autoclave, automation. Proceedings of International Conference-PSACO-2008, 13–15th March 2008. Andhra University, Visakhapatnam, AP, India.*

- *Execute program*: Next, the PLC executes the program, one instruction at a time.
- *Update output status*: Finally, the PLC updates the status of the outputs based on which inputs were on during the first step and the results of executing the program during the second step (Kadam et al., 2008).

4.3.7.2 Programmable Logic Controller Selection

Some of the key factors for the selection of PLC are the following (Kadam et al., 2008):

- Number of I/Os required
- Expandability
- Cost
- Serviceability/support
- Flexibility

4.3.7.3 Ordering Specifications

Power Supply: AC/DC, 220V AC/24 V DC.
Configuration: Total input and output details; digital and analog (Kadam et al., 2008).

4.3.7.4 User Interface

PLCs may need to interact with people for configuration, alarm reporting or everyday control. A HMI is employed for this purpose. A simple system may use buttons and lights to interact with the user. Text displays are available, as well as graphical touch screens. Most modern PLCs can communicate over a network to some other system, such as a computer running a SCADA system (Kadam et al., 2008).

4.3.8 Connections and Protocols

As mentioned earlier, a PLC is used frequently in a SCADA system, for example, the RTU. There are several different types of connections between the SCADA computers and PLCs, for example, RS232, RS422, RS485, and Ethernet. Note that these are only the electrical connections, not the protocol or language used on the connection. This could be considered the hardware protocol. Each PLC manufacturer has its own protocol or language that the PLC speaks. For example, Allen–Bradley (Rockwell) PLCs talk DF1, data highway 485 (DH485), data highway (DH), data highway + (DH+), remote I/O (RIO), Devicenet, Controlnet, and a few others. Modicon PLCs speak Modbus, Modbus Plus, and now Modbus TCP/IP (Modbus over Ethernet). Automation Direct PLCs speak Directed and K-sequence protocols, but some of their PLCs still communicate using their old protocols. The old general electrics PLC protocol was continuity check message but it now uses simple network protocol and variants such as simple network protocol extended. Siemens PLCs work mainly with the Profibus protocol, considered more of an open than a proprietary protocol.

The most common "standard" protocols are DF1, Modbus, and Profibus. DF1 and Modbus are popular and have been around longer than many of the people using them. Profibus is popular because it combines simplicity (daisy chain, twisted pair), industrial ruggedness (RS485), speed (12 MBaud), adaptability (users can select from many different speeds), scalability (from simple COM port up to communications controllers), and function (most automation devices can talk Profibus). All PLC manufacturers typically have a PLC module that will talk these protocols.

Now that Ethernet is becoming the most "hyped" capability in automation, there are several protocols over Ethernet. But even though Allen–Bradley and Siemens PLCs are both put on the same Ethernet cable, they cannot talk to each other because they do not use the same protocol (Kadam et al., 2008).

4.3.8.1 Supervisory Control and Data Acquisition Configuration

Configuration in SCADA should be done sequentially using the following sequence (Kadam et al., 2008):

1. Protocol configuration
2. Port configuration
3. Device configuration
4. Tag configuration
5. Role configuration

4.4 ENTERPRISE RESOURCE PLANNING AND OTHER COOPERATIVE INFORMATION RELATED TO THE ASSET

ERP is software that attempts to integrate all departments and functions across a company onto a single computer system that can serve all those departments' particular needs (Koch, 2005). Evolving out of the manufacturing industry, ERP implies the use of packaged software rather than proprietary software written by or for one customer. ERP modules may be able to interface with an organization's own software with varying degrees of effort, and, depending on the software, ERP modules may be alterable via the vendor's proprietary tools or via proprietary or standard programming languages (Han et al., 2005).

As we all know, a company consists of many function departments, such as finance, HR, purchasing, manufacturing, logistics, etc. Because the application of information technologies is increasingly popular, each department typically has its own computer system optimized for the particular ways the department does its work, not only for office automation, but also to help people analyze data and make the right decisions. ERP combines all software together into single, integrated software program that runs off a single database so that the various departments can easily share information and communicate with each other (Han et al., 2005).

4.4.1 Understanding Enterprise Resource Planning

There are two flows across the supply chain: product flow and information flow. In the past, information system tended to be islands, each with a different function within a company. For instance, when orders came from customers, they were processed and recorded by the sales department; sales then transferred the information to manufacturing. After the production department made a master schedule, the logistics department learned the distribution requirements and planned the delivery. Finally, accounting was able to bill customers. This business process led to numerous problems, such as delays, lost orders, input errors, and long lead time. The process is illustrated by the chart in Fig. 4.18 (Han et al., 2005).

Ideally, everyone should have access to the same real-time data through some interface when they need these data. This would require a single-point-of-contact system. That is one of the original ideas of ERP. To guarantee the effectiveness of ERP implementation, all the functions of the chain need to be linked seamlessly. As indicated above, there are walls between departments; the system needs to smash those walls (Han et al., 2005). The post-ERP situation is shown in Fig. 4.19.

4.4.2 Core Components of Enterprise Resource Planning

ERP totally changed the old computer systems, i.e., one for each separate department and replaced them with a single unified software program that can be divided into software modules. The modules roughly approximate the old stand-alone systems. Since the 1990s, ERP software has become flexible enough that some modules can be installed without buying the whole package. Many companies install only certain ERP modules, such as finance, HR, material management, or production planning, and leave the rest for the future (Han et al., 2005).

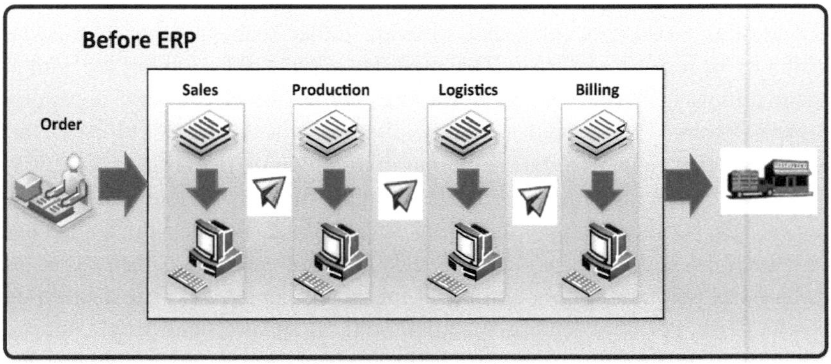

FIGURE 4.18 Order fulfillment before enterprise resource planning (ERP). *Adapted from Andrea, R., Sieber, S., Valor, J., 2003. Introduction of ERP.*

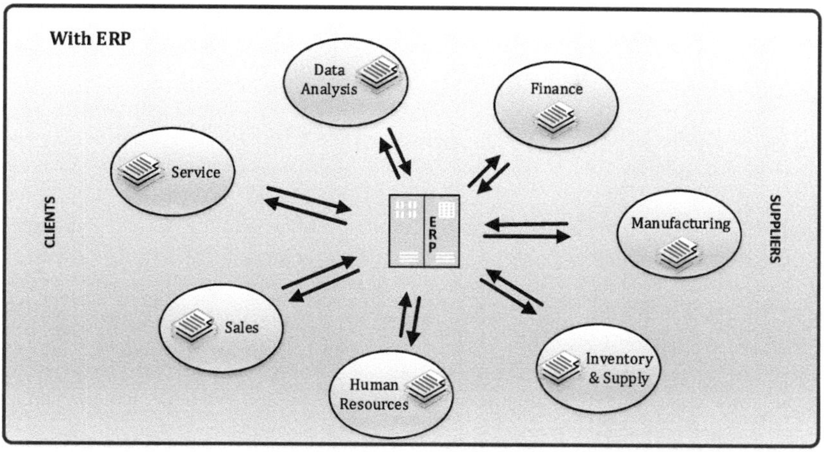

FIGURE 4.19 Order fulfillment after enterprise resource planning (ERP). *Adapted from Andrea, R., Sieber, S., Valor, J., 2003. Introduction of ERP.*

To enable easy handling of the system, ERP is divided into the following core subsystems:

- Sales and marketing
- Master scheduling
- Material requirement planning
- Capacity requirement planning
- Billing for materials
- Purchasing
- Shop floor control
- Accounts payable/receivable
- Logistics
- Asset management
- Financial accounting

4.4.3 Why Use Enterprise Resource Planning?

4.4.3.1 Key Motivating Factors

When asked to identify the motivating factors behind the ERP decision, decision-makers at three companies mentioned similar factors (Bozarth, 2006):

- ERP can standardize supply chain practices across multiple sites.
- ERP can simplify the software environment by replacing multiple, highly customized legacy systems with a single "fleet" solution.
- ERP can support corporate-level visibility and control of key supply chain processes, such as procurement and production scheduling.

TABLE 4.1 Tangible and Intangible Benefits

Tangible Benefits	Intangible Benefits
Inventory reduction	Information visibility
Personnel reduction	New/improved processes
Productivity improvements	Customer responsiveness
Order management improvements	Cost reductions
Financial cycle improvements	Integration
Information technology cost reduction	Standardization
Procurement cost reduction	Flexibility
Cash management improvement	Globalization
Revenue/profit increase	Supply/demand chain
Transportation/logistics cost reduction	Business performance
Maintenance reductions	Dismantling inefficient legacy systems
On-time delivery improvements	

Adapted from Gargeya, V.B., Brady, C., 2005. Success and failure factors of adopting SAP in ERP system implementation. Business Process Management Journal 11(5), 501–516.

4.4.3.2 Tangible and Intangible Benefits

Benefits of ERP are both tangible, i.e., obvious financial factors, and intangible, i.e., nonfinancial factors. Nonfinancial benefits include information visibility and flexibility. A more complete listing of tangible and intangible benefits is provided in Table 4.1 (Gargeya and Brady, 2005).

4.4.4 Enterprise Resource Planning Implementation Process

To understand the process of ERP adoptions, a number of researchers have developed conceptual ERP life cycle frameworks or process models. Ehie and Madsen (2005) suggest a five-stage ERP implementation process: project preparation, business blueprint, realization, final preparation, "go live" and support; see Fig. 4.20 (Seo, 2013).

Project preparation refers to a comprehensive planning phase; a project team with a leadership role is formed, budget targets are set, and project objectives and plans are defined. In the business blueprint phase, the current business process is analyzed in detail so that an appropriate ERP system can be selected. A project team is trained on the functionality and configuration of the selected ERP system. An understanding of the selected ERP system allows a project team to gain insight into reengineering its business processes.

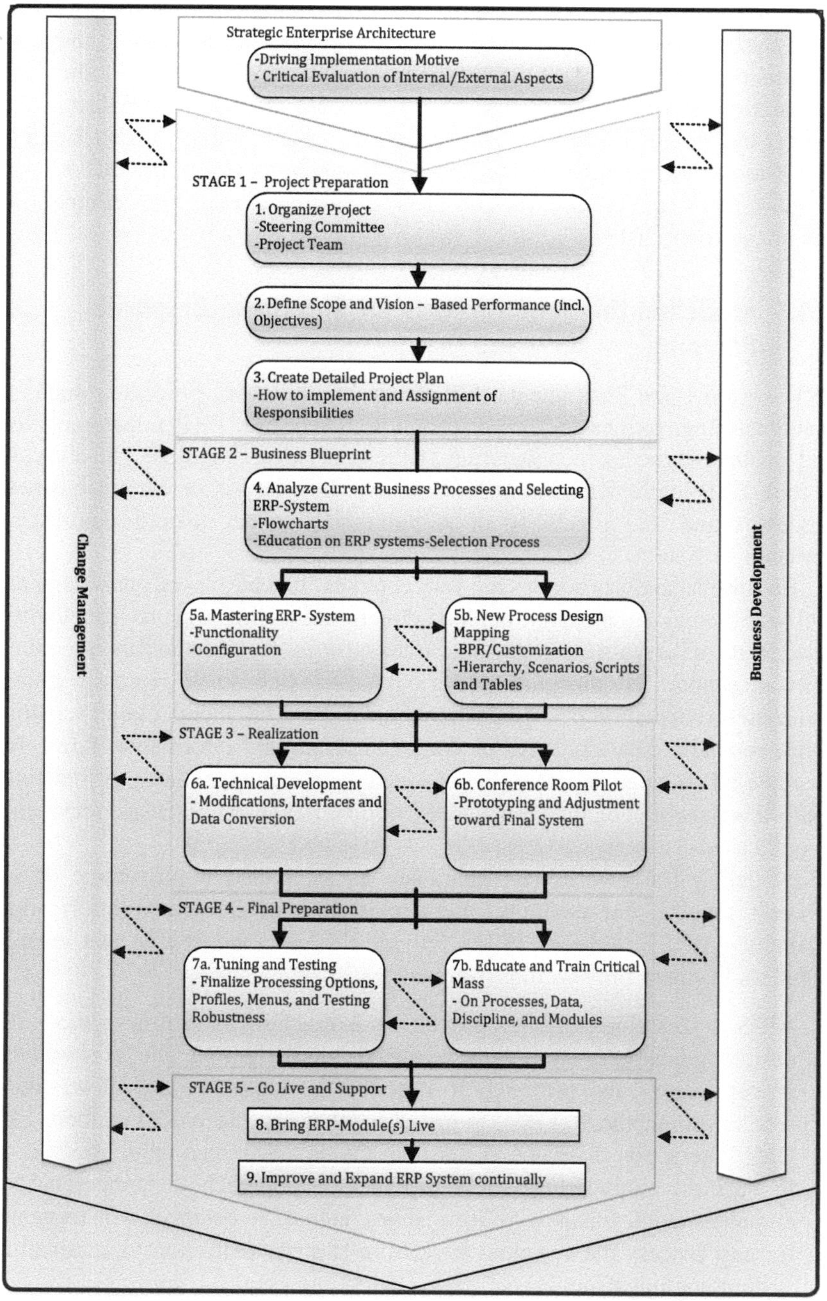

FIGURE 4.20 Five-stage enterprise resource planning (ERP) implementation process. *BPR*, business process reengineering. *Adapted from Ehie, I.C., Madsen, M., 2005. Identifying critical issues in enterprise resource planning (ERP) implementation. Computers in Industry 56(6), 545–557.*

In the realization phase, a project team concentrates on implementing an ERP system, including modification, development of interfaces, and data conversion. At the same time, each process design is tested. In the final preparation phase, the entire process is fully integrated and tested throughout the organization with full data and various scenarios. End users are trained in this phase as well. Finally, in the "go live" and support phase, the ERP system is constantly assessed and may be given extensions to ensure a competitive advantage (Seo, 2013).

4.4.5 Modeling the Requirements for Enterprise Resource Planning

ERP systems are instruments for improving business processes such as manufacturing, purchasing, or distribution. Therefore, ERP implementation and business process reengineering (BPR) activities should be closely connected. ERP implementation should involve the analysis of current business processes and the possibility of reengineering, rather than designing, an application system.

Because business processes are very complex, in many cases, analysis of the real-world application cannot be done directly. Modeling aims at reducing the complexity of the reality to better understand business processes and their required software support. For this purpose, various aspects of business processes can be considered (interaction of organizational units involved, input–output relationships, semantic data structures or communication, and information flows, for example). This creative process is supported by modeling methods. A modeling method is a set of components and a description of how to use these components to build a model.

Modeling frameworks can guarantee an integrated improvement of all relevant aspects. For example, the Architecture of Integrated Information Systems (ARIS) (Scheer, 1998) provides three applications that enable seamless business process improvement:

- ARIS is an architecture that provides and structures modeling methods for designing business processes from various views, for example, event-driven process chains and petri nets to describe the logical process flow, entity relationship methods (ERM) to model the data aspects, or OO methods like UML interaction diagrams to illustrate the organizational units involved.
- In business engineering, ARIS provides an approach to comprehensive computer-aided business process management. It comprises instruments for any process management step, from strategy definition to application system configuration.
- The ARIS software system toolkit supports the entire ARIS methodology (Scheer, 1999). It is fully integrated with SAP's "Implementation Management Guide" (IMG) (Walldorf, 1999), enabling model-based business improvement up to the customization of the ERP system's parameters.

When business processes are being improved, reference models may be useful. Reference models provided by ERP software vendors or consultant companies benefit the customer by using business process knowledge and best practices, providing the opportunity to compare business software solutions, and/or pinpointing positive or negative implementation issues. Probably the most comprehensive business reference model is SAP's R/3 software documentation, developed largely in accordance with the event-driven process chain (EPC) method.

4.4.6 Model-Based Customization

If business processes are properly redesigned and business process models document the improved situation, ERP systems can be configured by these conceptual models. However, this requires that the process models are linked with the repository of their respective ERP systems.

Direct interaction between the modeling tool, the conceptual models, and the application system modifies the implementation strategy of ERP software. Rigid-phase concepts used to be customary, with as-is analysis pinpointing any potential weak spots. Then ERP software independent target concepts were developed and implemented, using customizing techniques. Now, these phases are increasingly executed simultaneously and interactively. This lets business and implementation staff work in close collaboration, leading to simultaneous addressing of business and implementation issues. SAP's Accelerated SAP, Baan's Dynamic Enterprise Modeling, and Oracle's Fast Forward are all examples of ERP implementation strategies based on business process models.

Fig. 4.21 shows an example: for clarification purposes, the four windows are shown separately. In a real-world application, these windows would be displayed on one screen, providing the user with all the information at once. The upper right window shows the excerpt of the business process model in the ARIS modeling tool, illustrating the part of the standard software process that can be hidden. An additional process branch, not contained in the standard software, must be added (Scheer and Habermann, 2000).

The function "create inquiry" asks users which screen is being used in SAP R/3. Using the process model as a modeling tool and by clicking on the function (or starting a command), users can seamlessly invoke SAP R/3. This screen is shown at the bottom left.

SAP's customizing tool IMG is activated to adapt the function—the upper left window in the figure depicts the available function parameters. With the modeling tool, results of discussions, parameter decisions, unresolved issues, and the like can be stored in the function, as depicted in the bottom right window. This enables detailed documentation of business and IT-specific business process engineering. This documentation can be used at a later

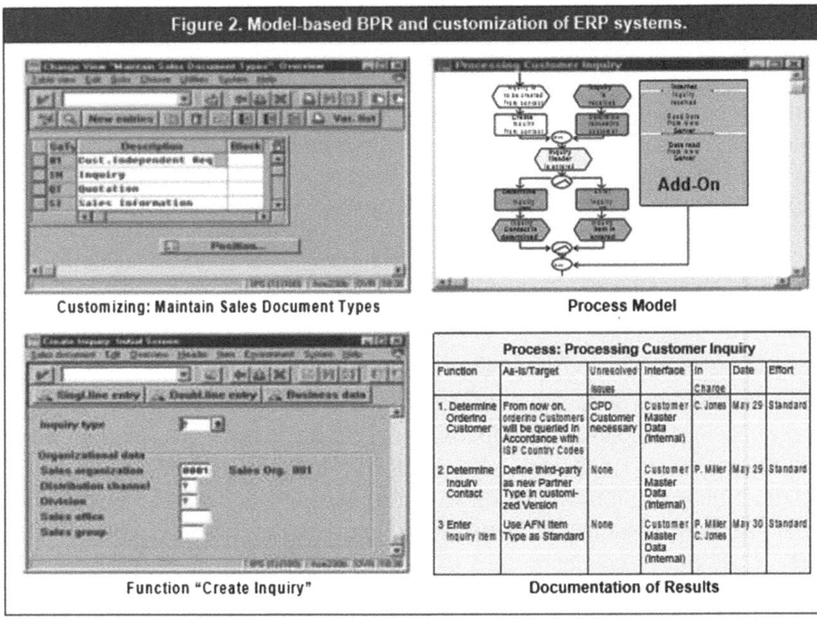

FIGURE 4.21 Model-based business process reengineering and customization of enterprise resource planning systems. *Adapted from Scheer, A-W., Habermann, F., April 2000. Making ERP a success. Communications of the ACM 43(4).*

point in time to monitor a project or clarify any questions; it can also be applied to following projects. Thus, model-based ERP implementation enables knowledge management for continuous process improvement (Scheer and Habermann, 2000).

4.4.7 Modeling the Future: Enterprise Resource Planning Goes e-Business

In the mid-1990s, the BPR debate drew attention from isolated business activities to entire value chains. Yet "entire" process management still mostly focused on the information flow within departmental, corporate, or national boundaries. Obstacles within those areas appeared difficult to cope with, so interorganizational communication and cooperation were seldom seriously considered on the improvement agenda. Consequently, ERP systems were restricted to intraorganizational process support.

Now, after various BPR and ERP lessons have been learned, companies seem better prepared for business scope redefinition. They better understand the limitations of intraorganizational improvement versus the urge to play an active role in the global e-business community. This means not only creating a company's Website but redesigning office processes according to the new requirements (Scheer and Habermann, 2000).

Obviously, this attitude has effects on business application systems. With companies on their way to new business dimensions, implemented ERP systems cannot remain inside organizational boundaries. On the technical side, ERP vendors are—as many other software vendors—forced to move from a traditional client/server to a browser/Web server architecture to deliver e-business capabilities. Hence, for their first-generation e-business solutions, almost all big ERP vendors are using a mixed Java/XML strategy. On the conceptual side, they are facing the even bigger challenge of providing instruments to cope with the increasing complexity of e-business. Business process models appear to be particularly useful in this context. While e-business process models are fundamentally the same as intraorganizational process models, integration and coordination mechanisms, such those being examined by MIT's Center for Coordination Science, are becoming even more important (Malone, 1999):

- With increasing globalization, e-business almost inevitably means international—sometimes even intercultural—business cooperation. While ERP systems were multilingual from the very first, the human understanding of foreign business terms, process management concepts, legal restrictions and cultural individualities is much more limited. Since models consist of graphic symbols that can be used according to formal or semiformal grammars, they have the means to reduce or even overcome those problems.
- Many improvement plans fail because of a lack of transparency in business processes and structures. If people do not realize the reengineering needs and benefits, they will not take part in a BPR project or accept the proposed or instituted changes. In geographically distributed cooperation, such as international business mergers or virtual organizations, the problem of transparency becomes worse. Yet to establish a successful partnership, business processes have to be designed at a very high level of detail. Thus, business process models can help to clearly define the goals, the process interfaces and the organizational responsibilities of interorganizational cooperation.
- At an operational level of e-business, models are very useful. In business-to-business applications, such as supply chain management, the ERP systems of all business partners must be connected. This means fighting not only with the thousands of parameters of a single ERP system but with twice as many (or even more). Another likely scenario is that the business partners involved in an interorganizational supply chain are using ERP systems of different vendors. In this case, a business process model can be used to define the conceptual value chain. Then a conceptual model that is independent of a particular software vendor can be linked to the repositories of the different ERP systems to customize an integrated software solution.

Business process models have the potential to play a major role in the success of e-business. Some ERP vendors like SAP understand this and are already implementing their first model-based e-business applications (Scheer and Habermann, 2000).

4.4.8 Conclusion: Enterprise Resource Planning Versus Computerized Maintenance Management System

- CMMS is a forward-looking technical management system, addressing practical, real-time maintenance needs in the workplace. It is an essential tool, adding increased effectiveness and efficiency of work management and asset care. It drives quality output, regulatory compliance and optimal operating costs.
- ERP is a backward-looking financial management system. It is concerned with enterprise overview and business process integration—human resources, payroll, inventory, financial management, and supplier collaboration.
- The systems are very different and, by and large, they are mutually exclusive (Shire Systems, 2007).

REFERENCES

Amrhein, D., Quint, S., 2009. Cloud Computing for the Enterprise: Part 1: Capturing the Cloud. Developer Works, IBM.

Andrea, R., Sieber, S., Valor, J., 2003. Introduction of ERP.

Arora, N., 2016. Supervisory control and data acquisition (SCADA) - backbone of modern industry - a review supervisory control and data acquisition (SCADA). In: Conference Proceeding of 4th International Conference on Recent Innovation in Science, Technology and Management (ICRISTM-16) at YMCA, Jai Singh Road, Delhi, India on 3rd December 2016, ISBN 978-81-932712-9-2.

Baptiste, J., 2004. A case study of remote diagnosis and e-maintenance information system. In: e-Proceedings of Intelligent Maintenance System. Arles, France, pp. 15–17.

Björling, S.-E., Galar, D., Baglee, D., Singh, S., Kumar, U., 2013. Maintenance knowledge management with fusion of CMMS and CM. In: DMIN 2013 International Conference on Data Mining, 22–25 July 2013, Las Vegas, Nevada, USA.

Björling, S.-E., Kumar, U., 2009. ICT Concepts for Managing Future Challenges in e-Maintenance. COMADEM 2009, San Sebastian, Spain.

Boys, W., August 18, 2009. Back to Basics: SCADA. Automation TV: Control Global - Control Design.

Bozarth, C., 2006. ERP implementation efforts at three firms. International Journal of Operations & Production Management 26 (11), 1223–1239. http://dx.doi.org/10.1108/01443570610705 83.

Crain, M., 2003. The Role of CMMS. A white paper on the selection and implementation of Computerized Maintenance Management Systems (CMMS). Industrial Technologies, Northern Digital, Inc.

Dasarathy, B.V., 2001. Information fusion. What, where, why, when, and how? Information Fusion 2, 75–76.

Dasarathy, B.V., 2003. Information fusion as a tool in condition monitoring. Information Fusion 4, 71–73. http://www.data-fusion.org/article.php?sid=70.

Davies, C., Greenough, R.M., 2000. The use of information systems in fault diagnosis. In: Proceedings of the 16th National Conference on Manufacturing Research. University of East London.

Ehie, I.C., Madsen, M., 2005. Identifying critical issues in enterprise resource planning (ERP) implementation. Computers in Industry 56 (6), 545–557.

Fu, C., Ye, L.Q., Cheng, Y.C., Liu, Y.Q., Iung, B., 2002. Mas-based model of intelligent control maintenance management system (ICMMS) and its application. In: Proceedings of 2002 International Conference on Machine Learning and Cybernetics, vol. 1, pp. 376–380.

Galar, D., Kumar, U., Juuso, E., Lahdelma, S., 2012a. Fusion of maintenance and control data: a need for the process. In: 18th World Conference on Nondestructive Testing, 16–20 April 2012, Durban, South Africa.

Galar, D., Parida, A., Kumar, U., 2012b. e-Maintenance Decision Making Based On Different Types of Data Fusion. Division of Operation and Maintenance Engineering. Luleå University of Technology, Luleå, Sweden, 97187.

Galar, D., Morant, A., 2012. Integration of Production Data in CM for Non-stationary Machinery: A Data Fusion Approach. Division of Operation and Maintenance Engineering. Luleå University of technology, Luleå, Sweden.

Gargeya, V.B., Brady, C., 2005. Success and failure factors of adopting SAP in ERP system implementation. Business Process Management Journal 11 (5), 501–516.

Greenland, P., November 2012. Understanding Industrial System Structures Is Key to Powering Them. Exclusive Technology Feature. Cupertino, Calif.

Gustafson, A., Galar, D., 2012. Fusion of production, operation and maintenance data for underground mobile mining equipment. In: The Ninth International Conference on Condition Monitoring and Machinery Failure Prevention Technologies. Division of Operation and Maintenance Engineering. Luleå, Sweden.

Hall, D.L., Llinas, J., 1997. An introduction to multisensor data fusion. Proceedings of the IEEE 85 (1), 6–23.

Han, J., Liu, R., Swanner, B., Yang, S., 2005. Enterprise Resource Planning.

Hounsell, D., September 2008. Justify Budgets with a CMMS. Facility Maintenance Decisions. http://www.facilitiesnet.com/software/article/Justify-Budgets-with-a-CMMS-Facility-Management-Software-Feature–9600.

Kadam, D.P., Sonwane, P.M., Landge, S.C., Dhote, V.P., Thorat, S.S., 2008. PLC & SCADA, a case study: autoclave, automation. In: Proceedings of International Conference-PSACO-2008, 13–15th March 2008. Andhra University, Visakhapatnam, AP, India.

Kahn, J., Klemme-Wolf, H., 2004. Overview of MIMOSA and the open system architecture for enterprise application integration. In: Proceedings of the 17th European Maintenance Congress, 333–341.

Koch, C., 2005. The ABCs of ERP. http://wikifab.dimf.etsii.upm.es/wikifab/images/d/da/The_ABCs_of_ERP.pdf.

Manning, C., Schutze, H., 1999. Foundations of Statistical Natural Language Processing. Press, Cambridge, MA.

Malone, T.W., 1999. Tools for Inventing Organizations: Toward a Handbook of Organizational Processes. MIT Center for Coordination Science, 21st Century Initiative Working Paper No. 5.

Mell, P., Grance, T., 2009. The NIST Definition of Cloud Computing, Version 15. National Institute of Standards and Technology (NIST), Information Technology Laboratory.

Parida, A., Kumar, U., 2004. Managing information is the key to maintenance effectiveness. In: e-Proceedings of Intelligent Maintenance System. Arles, France, 15–17.

Rasovska, I., Chebel-Morello, B., Zerhouni, N., 2008. A mix method of knowledge capitalization in maintenance. Journal of Intelligent Manufacturing 19, 347–359.

Rhoton, J., 2010. Cloud Computing Explained: Implementation Handbook for Enterprises. Recursive Press, London.

Scheer, A.-W., 1998. ARIS—business Process Frameworks, second ed.

Scheer, A-W., 1999. IDS. ARIS Toolset. Ed. Saarbruecken 1999.

Scheer, A.-W., Habermann, F., April 2000. Making ERP a success. Communications of the ACM 43 (4).

Seo, G., May 2013. Challenges in Implementing Enterprise Resource Planning (ERP) System in Large Organizations: Similarities and Differences Between Corporate and University Environment. Working Paper CISL# 2013-07. Massachusetts Institute of Technology, Cambridge, MA 02142.

Shalini, J., Kumar, S.J., 2013. Birtukan Teshome, Samrawit Bitewlgn Muluneh, Bitseat Tadesse Aragaw. International Journal of Advanced Research in Electrical, Electronics and Instrumental Engineering 2.

Shire Systems, 2007. ERP Asset Management Systems Versus Best of Breed CMMS. Shire Systems Limited. http://www.shiresystems.co.uk/downloads/slideshows/erpversuscmms.pdf.

Soibelman, L., Kim, H., 2002. Data preparation process for construction knowledge generation through knowledge discovery in databases. Journal of Computing in Civil Engineering 16 (1), 39–48.

Sullivan, G.P., Pugh, R., Melendez, A.P., Hunt, W.D., August 2010. Operations & Maintenance Best Practices. A Guide to Achieving Operational Efficiency, Release 3.0. Chapter 4: Computerized Maintenance Management System. http://energy.gov/sites/prod/files/2013/10/f3/OM_4.pdf.

Tianhao, W., Khan, F.M., Fisher, T.A., Shuler, L.A., Pottenger, W.M., 1992. Posting act tagging using transformation-based learning. In: ICDM'02 Proceedings of the Workshop on Foundations of Data Mining and Discovery, IEEE International Conference on Data Mining.

Thomas, B., Denis, R., Jacek, S., Jean-Pierre, T., Noureddine, Z., 2004. PROTEUS – an integration platform for distributed maintenance systems. In: e-Proceedings of Intelligent Maintenance System (IMS' 2004). Arles, France, pp. 15–17.

Walldorf, 1999. Accelerated SAP: Driving Rapid Implementations for Rapid Results. SAP.

Wang, X., Liu, C., Lee, J., 2004. Intelligent maintenance based on multi-sensor data fusion to web-enabled automation systems. In: e-Proceedings of Intelligent Maintenance System. Arles, France, pp. 15–17.

Yu, R., Ye, L., Fu, C., 2004. A multi-agent based, remote maintenance support and management system. In: Proceedings. IEEE/WIC/ACM International Conference on Intelligent Agent Technology (IAT 2004), pp. 496–499.

Chapter 5

Diagnosis

5.1 GOALS OF DETECTION, IDENTIFICATION, AND LOCALIZATION OF FAILURES

Fault diagnosis is the process of tracing a fault by identifying its symptoms, applying knowledge, and analyzing test results. Accurate diagnosis of faults in complex engineering systems requires acquiring the information through sensors, processing the information using advanced signal processing algorithms, and extracting required features for efficient classification or identification of faults. Identification of faults and subsequent remedial action can increase productivity and reduce maintenance costs in various industrial applications. Machine learning (ML) methods involving feature extraction, feature selection, and classification of faults offer a systematic approach to fault diagnosis and can be used in automated or unmanned environments. These are increasingly used in industrial sectors, such as manufacturing, automotive, marine, and aerospace, to maximize equipment uptime and minimize maintenance and operating costs (Vijayaraghavan, 2010).

Fault diagnosis has been the subject of numerous investigations over past decades. Researchers in such diverse disciplines as medicine, engineering, science, business, and finance have developed methodologies to detect faulty (failure) or anomaly conditions, pinpoint or isolate which component/object in a system/process is faulty, and decide on the potential impact of a failing or failed component on the health of the system. The diversity of application domains is matched only by the plurality of enabling technologies that have surfaced in attempts to diagnose detrimental events while meeting certain performance specifications. More recently, research and development (R&D) activities have increased with the need to maintain legacy equipment, improve their reliability, and make them available when needed. This section focuses on some important historical events in the development of condition-based maintenance/prognostics and health management (CBM/PHM) and addresses machine-related fault phenomena.

Many years ago now, the US Navy initiated an integrated diagnostic support system, which included adaptive diagnostics, feedback analysis, fault precursors, and some prognostic capability (Rosenberg, 1989). Its CBM

eMaintenance. http://dx.doi.org/10.1016/B978-0-12-811153-6.00005-1

program was aimed at enhancing the safety and affordability of Navy and Marine Corps platforms and vehicles. An integrated system that processes sensor data from shipboard machinery, assesses machine condition, and provides early fault detection and prediction of time to failure was developed, implemented, and tested successfully on a shipboard chiller by an R&D research team (Hadden et al., 1999a,b).

More recently, the open systems architecture (OSA) CBM initiative suggests seven levels of activity to integrate data acquisition, signal processing, condition monitoring (CM), health assessment, prognosis, decision support, and presentation. Fig. 5.1 depicts the seven-layered OSA CBM architecture. An OSA CBM XML scheme has been proposed for networking and communication capabilities. The proposed decision-support level integrates information to support a "decision to act" based on information from the prognostic layer, external constraints, goals, financial incentives, etc. In parallel, International Standards Organization (ISO) standards have attempted to capture the conceptual framework for processing models aimed at providing the basic CBM/PHM modules from data acquisition and analysis to health assessment, prognostic assessment, and advice generation (Vachtsevanos et al., 2006).

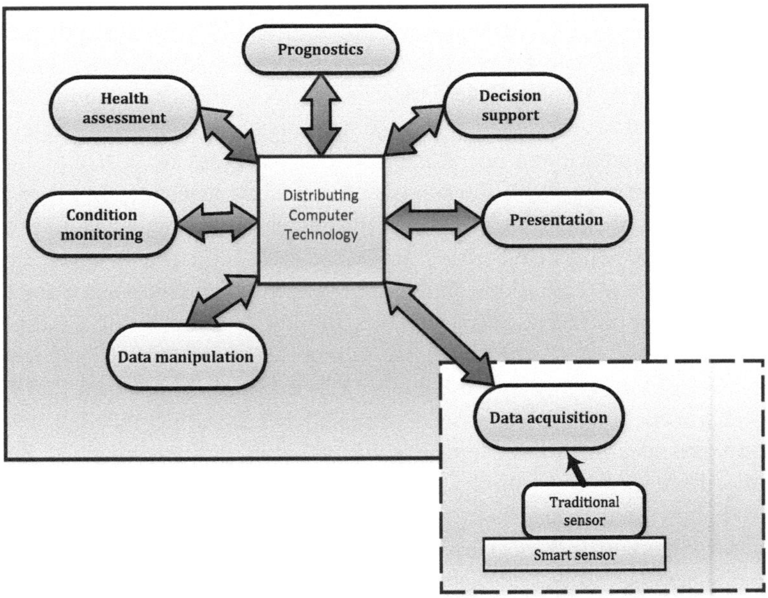

FIGURE 5.1 Seven-layered open systems architecture condition-based maintenance architecture. *Adapted from Vachtsevanos, G., Lewis, F., Roemer, M., Hess, A., Wu, B., 2006. Intelligent Fault Diagnosis and Prognosis for Engineering Systems, John Wiley & Sons, Inc., Hoboken, New Jersey.*

Fig. 5.2 depicts the ISO 13374-1 processing model. R&D activities by several major companies (GE, Boeing, Lockheed, and Honeywell, among many others) have produced promising results, but these focus on specific application domains and lack widespread acceptance as technology standards.

Statistical distribution, Bayesian networks, stochastic autoregressive moving-average models, state-estimation techniques, knowledge-intensive expert systems, finite state machines, Markov chain models, soft computing (SC) tools (neural nets, fuzzy logic, and genetic algorithms (GAs)) have been suggested as possible ways to perform mechanical system diagnosis/prognosis. Fault diagnosis, or fault detection and identification (FDI), is a mature field, with contributions ranging from model-based techniques to data-driven configurations that capitalize on SC and other "intelligent" tools (Konrad and Isermann, 1996; Mylaraswamy and Venkatasubramanian, 1997).

Modern diagnostic algorithms require massive databases that collect data from multiple and diverse sensor suites. Information typically is extracted from the databases in the form of features or condition indicators and used as input to diagnostic routines whose goal is to declare impending failure

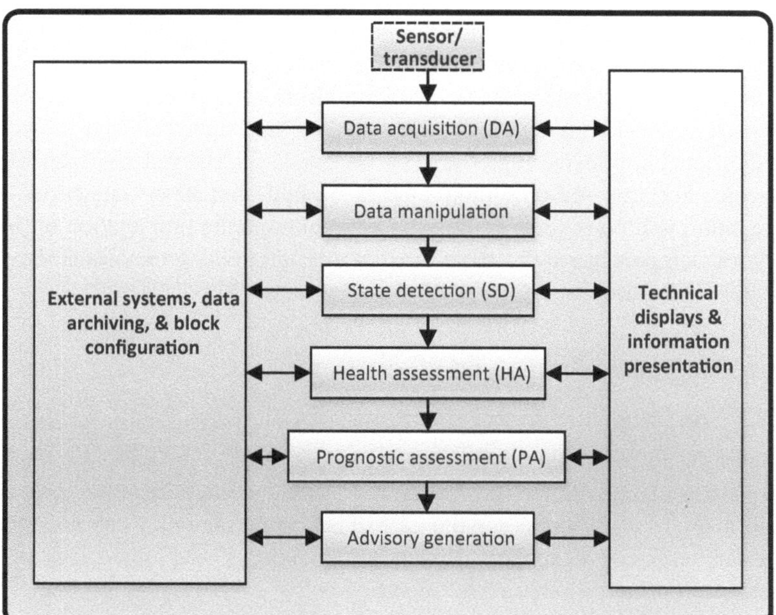

FIGURE 5.2 ISO 13374-1 processing model. *Adapted from Vachtsevanos, G., Lewis, F., Roemer, M., Hess, A., Wu, B., 2006. Intelligent Fault Diagnosis and Prognosis for Engineering Systems, John Wiley & Sons, Inc., Hoboken, New Jersey.*

conditions. Machine diagnosis and prognosis for CBM involves an integrated system architecture, with the following main modules:

1. A sensor suite and suitable sensing strategies to collect and process data of critical process variables and parameters;
2. A failure modes effect and criticality analysis (FMECA) module that determines and prioritizes according to the frequency of occurrence and the severity of possible failure modes and systematically catalogs effect–root-cause relationships;
3. An operating-mode identification routine that determines the current operational status of the system and correlates fault-mode characteristics with operating conditions;
4. A feature extractor that selects and extracts from raw data the features or condition indicators to be used by the diagnostic module;
5. A diagnostic module—the diagnostician—that assesses, through online measurements, the current state of critical machine components;
6. A prognostic module—the prognosticator—that estimates the remaining useful life (RUL) of a failing component/subsystem;
7. A maintenance scheduler whose task is to schedule maintenance operations without adversely affecting the overall system functionalities; note that the machine in question is only one of many constituent elements.

The chapter focuses on fundamental technologies that enable the accurate and timely diagnosis of faulty or incipient-failure conditions. Fault prognosis issues are addressed in Chapter 6. Elements of fault diagnosis and prognosis and their enabling technologies are, by necessity, interwoven and difficult to delineate. For this reason, some aspects of fault initiation, detection, and propagation will be covered in this chapter, although the propagation or time-evolution component may be more directly relevant to the prognosis module of the CBM/PHM architecture (Vachtsevanos et al., 2006).

5.1.1 Diagnostic Framework

5.1.1.1 Introduction and Definitions

This section reviews definitions and basic notions of fault diagnosis. Definitions have evolved with the incorporation of machine fault diagnosis tools into the CBM/PHM framework, and the CBM/PHM community has not yet agreed on a standardized vocabulary. However, the military and other sectors are converging on the following definitions:

- *Fault diagnosis.* Detecting, isolating, and identifying an impending or incipient failure condition; the affected component (subsystem, system) is still operational, albeit in a degraded mode.
- *Failure diagnosis.* Detecting, isolating, and identifying a component (subsystem, system) that has ceased to operate.

To estimate the RUL or time to failure of a failing component, that is, to conduct fault prognosis, and maximize uptime through CBM, it is essential to determine accurately and without false alarms impending or incipient failure conditions, i.e., faults. The designer's objective is to achieve the best possible performance of a given diagnostic routine by minimizing false positives (FPs) and false negatives (FNs) while reducing the time delay between the initiation and detection/isolation of a fault event (Vachtsevanos et al., 2006).

5.1.1.2 Basic Definitions

The basic notions presented in this subsection follow the International Federation of Automatic Control (IFAC) Technical Committee—SAFEPROCESS (Fault Detection, Supervision, and Safety for Technical Processes)—terminology in the field (Isermann and Ballé, 1997) as used by Chen and Patton (1999).

5.1.1.2.1 Fault Diagnosis: Fault-Tolerant Control

A fault represents an unexpected change of system function, although it may not represent a physical failure. The term *failure* indicates a serious breakdown of a system component or function that leads to the significantly deviated behavior of the whole system. In contrast, the term *fault* indicates a malfunction that does not significantly affect the normal behavior of the system.

An incipient (*soft*) fault represents a small and often slowly developing continuous fault. Its effects on the system are, in the beginning, almost unnoticeable. A fault is called hard or abrupt if its effects on the system are larger and are bringing the system very close to the limit of acceptable behavior.

A fault is called *intermittent* if its effects on the system are hidden for discontinuous periods of time (Isermann, 1997). Although a fault may be tolerable at the moment it occurs, it must be diagnosed as early as possible, as it may lead to serious consequences over time.

A fault diagnosis system is a monitoring system used to detect faults and diagnose their location and significance in a system. The system performs the following tasks:

- Fault detection—To indicate if a fault has occurred in the system, or not,
- Fault isolation—To determine the location of the fault, and
- Fault identification—To estimate the size and nature of the fault.

The first two tasks—fault detection and isolation—are considered the most important, so fault diagnosis is often simply called fault detection and isolation.

A *fault-tolerant control system* is a controlled system that continues to operate acceptably following faults in the system or in the controller. An important feature of such a system is automatic reconfiguration, once a malfunction is detected and isolated. The contribution of fault diagnosis to a fault-tolerant control system is the detection and isolation of faults, so users can decide how to perform reconfiguration (Palade and Bocaniala, 2006).

5.1.1.2.2 Diagnosis Based on Analytical Models

Model-based fault diagnosis has been defined as "the determination of the faults in a system by comparing available system measurements with a priori information represented by the system's analytical/mathematical model, through generation of residuals quantities and their analyses. A residual is a fault indicator that reflects the faulty condition of the monitored system" (Chen and Patton, 1999).

The problem that occurs when using an analytical model for a given system is that it cannot perfectly model uncertainties because of disturbances and noise. Nonmodeled dynamics and other uncertainties result in differences between the analytical model output and the system output. These differences may cause the residuals to indicate erroneous faults. A robust FDI scheme provides satisfactory sensitivity to faults, while being robust (insensitive or even invariant) to modeling uncertainties (Frank, 1991; Patton and Chen, 1996) and to noise. One of the main challenges in designing a robust FDI scheme is to make it able to diagnose incipient faults. The effects of an incipient fault on a system are almost unnoticeable in the beginning; thus, the effects of uncertainties on the system could hide these small effects.

A fault diagnosis task has two main stages: residual generation and decision-making (Chow and Willsky, 1980) (Fig. 5.3). Residual generation is a procedure for extracting fault symptoms from the system, using available input and output information. A residual generator is an algorithm used to generate residuals (Chen and Patton, 1999). Decision-making refers to examining the residual signals to establish if a fault has occurred and, if so, to isolate it (Palade and Bocaniala, 2006).

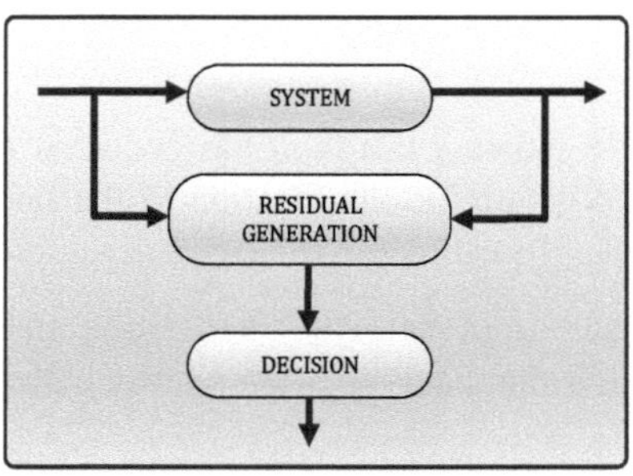

FIGURE 5.3 Two main stages of fault diagnosis. *Adapted from Palade, V., Bocaniala, C.D., 2006. Computational Intelligence in Fault Diagnosis. ©Springer-Verlag London Limited.*

5.1.1.3 Fault Diagnosis Requirements and Performance Metrics

CBM/PHM system designers must adhere to design requirements imposed by the system user and conform to overall system objectives. Although it may be possible to define a generic set of requirements associated with diagnostic routines, specific quantitative requirements definitions are system and application-dependent. In general, a CBM/PHM system must meet the following requirement:

- Ensure enhanced maintainability and safety while reducing operational and support cost,
- Be designed as an OSA,
- Closely control PHM weight,
- Meet reliability, availability, maintainability, and durability requirements,
- Meet monitoring, structural, cost, scalability, power, compatibility, and environmental requirements.

More specifically, fault diagnosis requirements include the following:

- The PHM system must detect no less than x percent of all system or subsystem (component) failures being monitored and no less than y percent of system failures specifically identified through reliability-centered maintenance (RCM) and FMECA studies.
- The PHM system must demonstrate a fault-isolation rate to one (or x) organizational removable assembly units for all faults detected by PHM.
- Fault-identification requirements may be defined based on the impact of a particular fault on the system's operating characteristics.
- For false alarms, the PHM must have a demonstrated mean operating period between false alarms of x.
- If built-in testing capability is available, the PHM diagnosis built-in testing must complete an end-to-end test within an x-min test period.
- Additional requirements may be defined in terms of allowable FPs, FNs, and the time window between fault initiation and fault detection.

Specific numerical quantities must be prescribed by the end user. It should be pointed out that there are no accepted standards for CBM/PHM requirements. The community is currently debating the issues as the technological basis continues to advance and systems are developed and implemented.

Fault-diagnostic algorithms must have the ability to detect system performance, degradation levels, and faults (failures) based on physical property changes, i.e., detectable phenomena. The faults detected are typically a subset of those identified through FMECAs. Furthermore, such algorithms must be able to identify the specific subsystem and/or component within the system being monitored that is failing, as well as the specific failure mechanism that has occurred. Knowing the potential impact of the fault on the operational integrity of the system is another desirable outcome of diagnosis. Since the ultimate objective of a CBM/PHM system is to predict the time progression of a fault and estimate the time to failure of the failing component, it is essential that faults be detected and isolated as early as possible after their initiation stage.

Reducing false alarms, FPs, and FNs is a major objective of a diagnostic system. This important task can be facilitated by:

- Data validation,
- Optimal feature selection/extraction,
- Optimal choice of fault detection/identification algorithms,
- Combining (fusing) evidence (Dempster–Shafer theory),
- Conflict resolution and ignorance representation,
- Fusion of fault-classification algorithms, and
- Declaring a fault only when confidence (certainty) exceeds a set threshold.

Enabling technologies for each of these means are discussed in sections of the present and other chapters of this book (see also Vachtsevanos et al., 2006).

5.1.1.4 Fault Monitoring and Diagnosis Framework

Fault detection and isolation strategies have been developed in recent years and have found extensive utility in a variety of application domains. The term FDI is used here to imply fault detection and the determination of a specific fault component/subsystem because it is prevalent in the technical literature. A general FDI structure is depicted in Fig. 5.4. In this figure, a residual signal that represents a deviation from standard operating conditions is generated by comparing, for example, a model's output with the actual system output. Based on this signal, we can make decisions about the operating condition of the machinery. Methods to generate the residual are given in this chapter and can be based on time signal statistics, system identification, frequency domain techniques, etc. (Vachtsevanos et al., 2006).

The enabling technologies typically fall into two major categories: model-based and data-driven, as shown in Fig. 5.5. Model-based techniques rely on an accurate dynamic model of the system and are capable of detecting even unanticipated faults. They take advantage of the actual system and model outputs to generate a "discrepancy" or residual (innovation), as it is known,

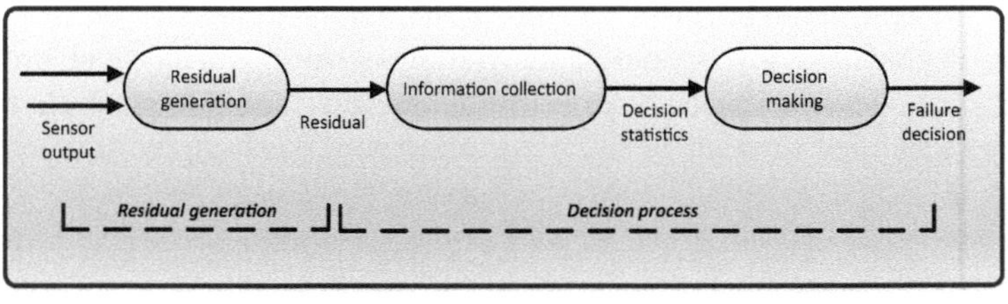

FIGURE 5.4 General fault detection and isolation structure. *Adapted from Vachtsevanos, G., Lewis, F., Roemer, M., Hess, A., Wu, B., 2006. Intelligent Fault Diagnosis and Prognosis for Engineering Systems. John Wiley & Sons, Inc., Hoboken, New Jersey.*

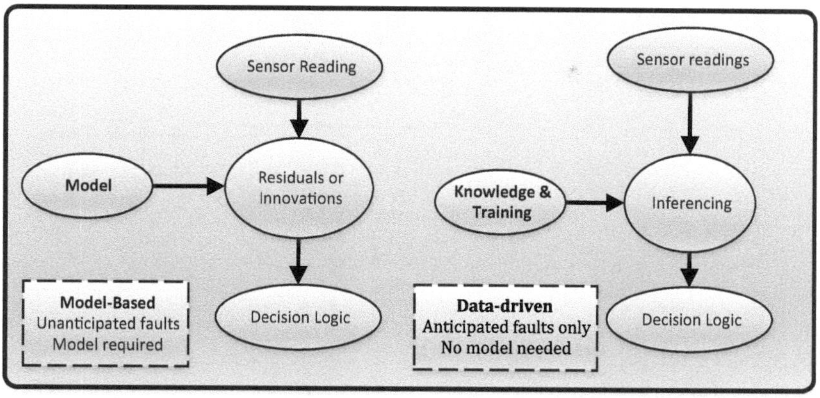

FIGURE 5.5 Model-based and data-driven fault diagnosis techniques. *Adapted from Vachtsevanos, G., Lewis, F., Roemer, M., Hess, A., Wu, B., 2006. Intelligent Fault Diagnosis and Prognosis for Engineering Systems. John Wiley & Sons, Inc., Hoboken, New Jersey.*

between the two outputs that is indicative of a potential fault condition. In one method, a bank of filters is used to pinpoint the identity of the actual fault component. Data-driven techniques often address only anticipated fault conditions, with a fault "model" construct or a collection of constructs, such as neural networks (NNs), expert systems, etc., that must be trained with known prototype fault patterns (data) and then employed online to determine the faulty component's identity (Vachtsevanos et al., 2006).

A general scheme for online fault monitoring and diagnosis is shown in Fig. 5.6 (Vachtsevanos et al., 2006).

The figure shows the entire online procedure required for monitoring and diagnosis and serves as a focus for the following discussion. The major components of this procedure are collecting data through sensor outputs and then building a feature vector with enough information about the current machine operating condition to allow fault identification and classification. The information contained in the feature vector can be obtained by three means:

1. Model-based methods (e.g., identify physical model parameters using Kalman filtering or recursive least squares),
2. Data-driven methods (e.g., vibration signal statistical moments, vibration frequency spectrum information), and
3. Statistical regression and clustering techniques on existing historical legacy data.

Once the current feature vector has been obtained, it is used as input to a fault-classification block that can contain various types of decision-making algorithms. If more information is needed before a decision can be reached, probing signals can be sent back to the machine (e.g., injecting a test vibration signal into a machinery mount to test for loose fittings) (Vachtsevanos et al., 2006).

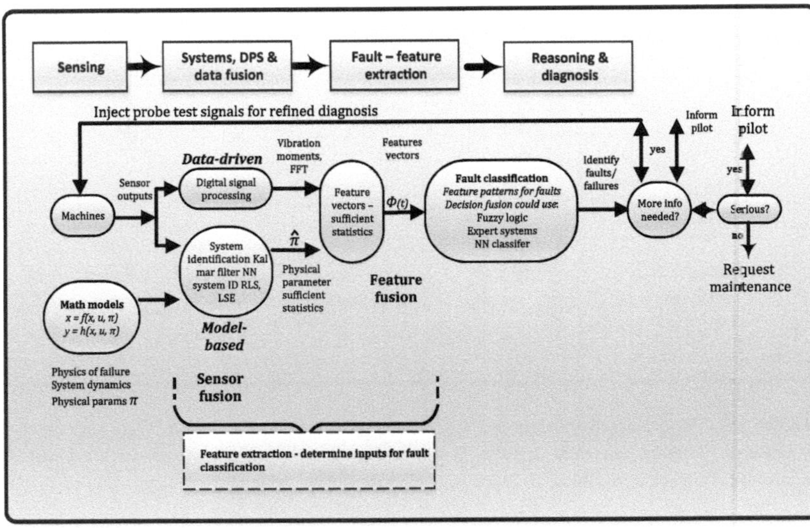

FIGURE 5.6 Online fault monitoring and diagnosis. *FFT*, fast fourier transform; *LSE*, least squares estimator; *NN*, neural network; *RLS*, recursive least squares. *Adapted from Vachtsevanos, G., Lewis, F., Roemer, M., Hess, A., Wu, B., 2006. Intelligent Fault Diagnosis and Prognosis for Engineering Systems. John Wiley & Sons, Inc., Hoboken, New Jersey.*

5.1.1.4.1 Fault-Feature Vector

The fault-feature vector is a set of data useful for determining the current fault status of the machinery. Feature vectors contain a variety of information about the system, including system parameters relating to fault conditions (bulk modulus, leakage coefficient, temperatures, pressures, etc.) and vibration and other signal analysis data (energy, kurtosis, discrete Fourier transform, etc.). Feature vector components are selected using physical models and/or historical legacy data such as that available in the USN SH-60 HIDS study (Hardman et al., 1999). Physical models might show that components such as bulk modulus and leakage coefficient should be included, and historical failure data might show the importance of vibration signature energy, kurtosis, etc. Different feature vectors are needed to diagnose different subsystems (Vachtsevanos et al., 2006).

In probabilistic terms, the feature vector $\varphi[x(t)]$ is a sufficient statistic for the full state and history $x(t)$ of the system to determine the fault/failure condition $z(t)$. That is, the fault likelihood function probability density function (PDF) can be decomposed as

$$f[z(t)/x(t)] = g[x(t)]h\{\varphi[x(t)], z(t)\} \tag{5.1}$$

so that knowledge of $\varphi[x(t)]$ is sufficient to diagnose the fault state $z(t)$. The data component factor $g[x(t)]$ depends only on $x(t)$ and is not needed for diagnosis. It is not necessary to know the full system state and history $x(t)$.

The feature vectors are extracted using system identification and digital signal-processing techniques. $\varphi[x(t)]$ is monitored continuously and is a time-varying vector. At each time, the fault classification may be determined by comparing $\varphi[x(t)]$ to a library of stored fault patterns or to predetermined alarm conditions. $\varphi[x(t)]$ contains the *symptoms* of the fault status, whereas the fault pattern library (FPL) contains *diagnostic information in terms of known fault patterns*. Many techniques have been used to create the FPL for fault classification, including physical modeling, statistical techniques, historical fault-test data, rule-based expert systems, reasoning systems, fuzzy logic, NNs, etc. (Vachtsevanos et al., 2006).

5.1.2 Artificial Intelligence–Based Machine Condition Monitoring and Fault Diagnosis

In the recent past, artificial neural networks (ANNs), fuzzy systems, and neuro-fuzzy systems have been extensively used for speed, torque estimation, and solid state drive control of both DC and AC machines (Vas, 1999). However, they are particularly suited for AC applications where the relationships between motor current and speed are nonlinear.

Artificial intelligence (AI) techniques are increasingly used for CM and fault detection of machines (Schoen et al., 1995; Filippetti et al., 1996; Vas, 1999). An NN–based fault diagnosis system using the stator current spectra is described in Fig. 5.7. The preprocessor extracts the frequency components of the sampled current data. Using the rule-based frequency filters, these frequency components are classified into four categories with decreasing level of importance. Based on these rules, an NN, trained for all possible operating condition of the machine, is used to classify the incoming data. A spectral signature that falls outside the trained clusters is marked as a potential motor fault. To prevent false diagnosis, the postprocessor sends an alarm only when fault signatures are observed persistently. This function is performed by

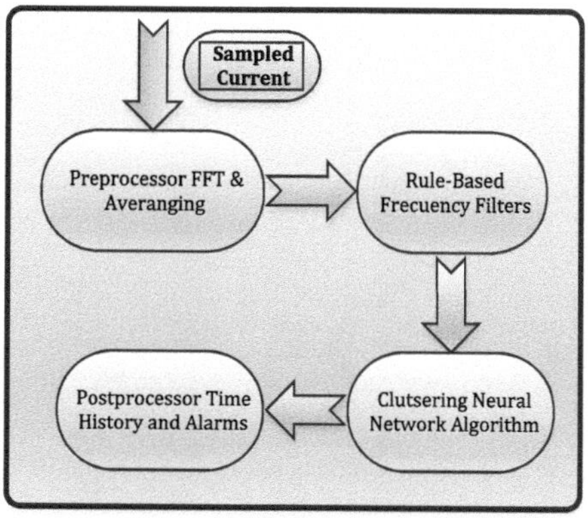

FIGURE 5.7 Artificial neural networks–based fault diagnosis. *FFT*, fast fourier transform; *NN*, neural network. *Adapted from Nandi, S., Toliyat, H.A., 1999. Condition Monitoring and Fault Diagnosis for Electrical Machines – A Review. Electric Machines & Power Electronics Laboratory. Department of Electrical Engineering. Texas A&M University, College Station, TX, pp. 77843–83128.*

maintaining a time history of the motor being monitored. Such a scheme has been successfully implemented by Schoen et al. (1995) to diagnose bearing and unbalanced rotor faults of induction motors (Nandi and Toliyat, 1999).

Similarly, fuzzy logic–based systems have been used by Vas (1999) to classify broken bar–related faults by categorizing the two sideband components around the fundamental component of induction motor line current using a set of nine rules. Consider the following example. We denote the sidebands as A1 and A2, the two inputs of the system, and *n*, the number of broken bars, as the output of the system. Therefore, under these rules, we might say "If A1 is small and A2 is large, *n* equals one broken bar." This fuzzy logic system is the Mamdani type; the fuzzy inference is performed by using the fuzzy implication min–max method and the centroid defuzzification technique. The membership functions for A1 and A2 are small, medium, and large (Nandi and Toliyat, 1999).

5.1.3 Neural Network Alternatives

When used for fault diagnosis, NNs use sensor measurements and process alarms as their inputs and particular fault types appear in their outputs (see Fig. 5.8). Sometimes, when facing novel fault classes, NN processing might result in incorrect diagnosis, mainly because of NN's generalization limitations.

Here, to cope with the previously mentioned drawbacks, two different types of NNs, i.e., extended dynamic back propagation and the radial basis function (RBF), are considered and applied (Dalianis et al., 1993).

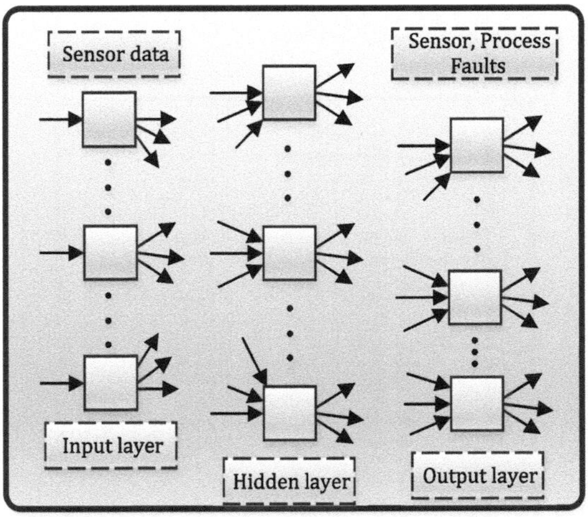

FIGURE 5.8 Multilayer neural network for fault recognition. *Adapted from Tzafestas, S.G., Dalianis, P.J., 1994. Fault diagnosis in complex systems using artificial neural networks. In: 3rd IEEE CCA. Intelligent Robotics and Control Unit (IRCU). Computer Science Division. Electrical & Computer Engineering Department. National Technical University of Athens, Zografou 15773, Athens, Greece.*

5.1.3.1 Extended Dynamic Back Propagation Networks

The extension of the backpropagation (BP) algorithm presented by Dalianis et al. (1993) provides a new energy function. As is well known, too many weights (degrees of freedom) in a BP network can overfit the network to the training data, possibly leading to poor generalization. This algorithm provides the best generalization with the minimum network size. Experiments start with rather large networks (i.e., considering the number of nodes in the hidden layer) for the given problem, and during training weight-node eliminations take place, until the best possible performance is reached.

The BP network has one hidden layer. Fig. 5.9 shows the structure. The inputs are on the bottom; above this, we see first layer, the hidden layer, and the output one. The node transition function is $\sigma(x) = 1/(1 + e^{-x})$, one of the most commonly used in such networks. This algorithm converges much faster than the standard BP algorithm (Tzafestas and Dalianis, 1994).

The new energy function proposed by Dalianis et al. (1993), is

$$E = c_1E_c + c_2E_0 + c_3E_w = c_1\sum_p E_p + c_2\sum_p \frac{n_o * \sum_1 o_{ip}^2}{\sum o_{ip}^2 + n_ok_1} + c_3\frac{n_w * \sum_1 w_{ij}^2}{\sum w_{ij}^2 + n_wk_2} \quad (5.2)$$

where o_{ip} is the output of node i for pattern p, n_o is greater than or equal to the number of nodes in hidden layers, and n_w is greater than or equal to the

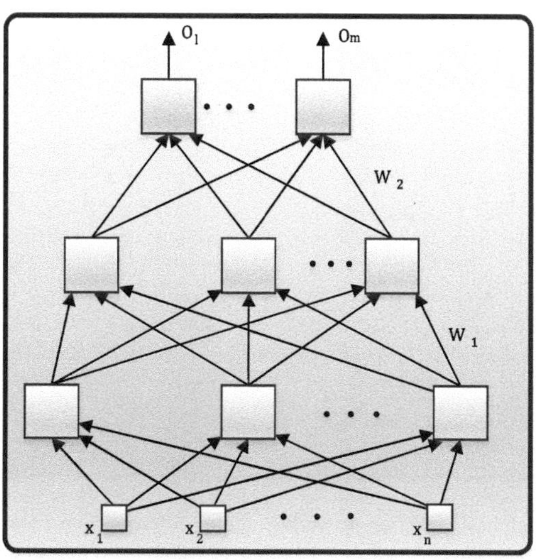

FIGURE 5.9 Back propagation neural network structure. *Adapted from Tzafestas, S.G., Dalianis, P.J., 1994. Fault diagnosis in complex systems using artificial neural networks. In: 3rd IEEE CCA. Intelligent Robotics and Control Unit (IRCU). Computer Science Division. Electrical & Computer Engineering Department. National Technical University of Athens, Zografou 15773, Athens, Greece.*

number of weights. In this case, the minimization of E is equivalent to the minimization of $E_c = \sum E_p$, under the constraints $E_o = 0$, $E_w = 0$, which reduce the network dimension, creating dependencies among o_i and w_{ij}. The new algorithm reduces the number of weights and nodes by minimizing some of them to zero. For suitable k, n_w, and n_o, the E_w, E_o are equal to the number of weights and nodes. Hence, their minimizations leave the network with the optimal number of parameters.

This approach allows us to build more complicate approximators, guaranteeing better performance even with incomplete training sets, providing stability and a high degree of generalization, and offering faster training times (Tzafestas and Dalianis, 1994).

5.1.3.2 Radial Basis Function Networks

RBF networks have characteristics similar to those of BP networks. They do a nonlinear, albeit localized, transformation and a weighted linear combination. However, RBF networks can usually be trained much faster than BP ones.

An RBF network implements an input–output mapping using a linear combination of radially symmetric functions (Moody and Darken, 1989). It has three layers, with feedforward connections between the nodes, as shown in Fig. 5.10 (Tzafestas and Dalianis, 1994).

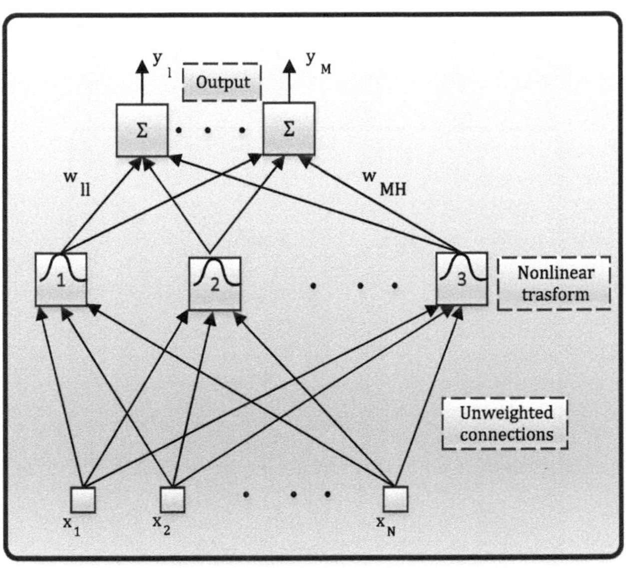

FIGURE 5.10 Radial basis function network structure. *Adapted from Tzafestas, S.G., Dalianis, P.J., 1994. Fault diagnosis in complex systems using artificial neural networks. In: 3rd IEEE CCA. Intelligent Robotics and Control Unit (IRCU). Computer Science Division. Electrical & Computer Engineering Department. National Technical University of Athens, Zografou 15773, Athens, Greece.*

In the first layer, weightless connections pass inputs to the hidden layer. The hidden nodes are radial units. Theoretical studies and practical results both suggest the choice of the nonlinearity is not crucial to the performance of an RBF network. The multivariate Gaussian density function is chosen for the hidden node transfer function and expressed as

$$a_{hk} = \exp\left(- \frac{\|x_h - c_k\|^2}{\sigma_h^2} \right) \tag{5.3}$$

where a_{hk} is the activation of the hth unit in the hidden layer (H in total) with input x_k, $\| \bullet \|$ denotes the Euclidean norm, and c_k is the RBF unit center determined by clustering. Each node in the hidden layer has N+1 parameters (N, the dimension of x, and a bias factor). σ_h is a distance scaling parameter, determining over what distance in the input space the unit will have a significant influence. Connections in the second layer are weighted as usual, and the output nodes are linear summation units. The output y_m at node m is given by

$$y_m = \sum_{h=1}^{H} w_{mh} a_h \qquad m = 1, \ldots, \text{M number of classes} \tag{5.4}$$

The extension of RBF to indicate novelty in fault classes may permit the estimation of the probability density of the training data. A comparison of the RBF network to the classical BP network mentions improvements in the former that allow it to identify situations where a novel class appears close to a test case of the original training data or to determine the time required for diagnosis (Tzafestas and Dalianis, 1994).

5.1.4 Supervising the Diagnostic Neural Network

To benefit from the advantages of the two different approaches, i.e., NNs and expert systems, to the diagnostic task, an integrated (hybrid) diagnostic strategy is suggested. The sensor processing capabilities of NNs can be integrated with the reasoning capabilities of expert systems using a number of different approaches. One is shown in Fig. 5.11.

The NN is embedded into the expert system as a module that interprets the data passed to it and acts as a candidate generator. The expert system interacts with the user, retrieves the sensor measurements, makes the required data preprocessing, interprets the NN results, and performs a postprocessing diagnostic reasoning, before passing the diagnosis to the system user or the automatic control units.

The suggested supervising expert system allows us to automate the preprocessing of the data fed to the NN, to control the NN dynamics, to interface the system with the user, and to provide necessary information on the diagnostic system's operation and its behavior under different cases (Tzafestas and Dalianis, 1994).

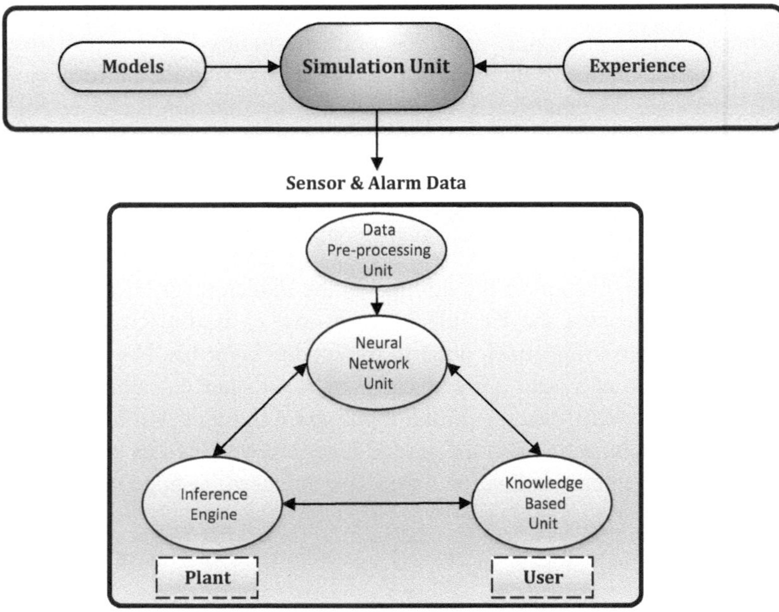

FIGURE 5.11 Combining expert system techniques with neural networks for fault diagnosis. *Adapted from Tzafestas, S.G., Dalianis, P.J., 1994. Fault diagnosis in complex systems using artificial neural networks. In: 3rd IEEE CCA. Intelligent Robotics and Control Unit (IRCU). Computer Science Division. Electrical & Computer Engineering Department. National Technical University of Athens, Zografou 15773, Athens, Greece.*

5.1.5 Other Methods for Fault Diagnosis

This section summarizes other fault diagnostic methods. Each addresses a particular diagnostic system requirement, but all, as well as others we cannot cover here, present a unique technological advantage that is of interest to CBM/PHM designers (Vachtsevanos et al., 2006).

5.1.5.1 Statistical Change Detection

B. Eugene Parker et al. of Barron Associates, Inc. (Parker et al., 2000), introduced the statistical change detection (SCD) technique used for early detection of subtle system changes. The algorithm seeks to detect situations in which a given model that describes the initial behavior of a time series eventually fails to describe that time series accurately. Given a sequence of random observations, the SCD technique aims at both online and off-line fault detection. Until some (unknown) time, t_p, the measurements are generated by a statistical model, M_p. Thereafter, they are generated by another model, M_i. The goal of online SCD is to detect changes from M_p to M_i as quickly as possible after they occur (i.e., minimize detection delay for a given mean time between false alarms). The goal of off-line SCD is to detect a change from M_p to M_i

with minimum sensitivity (i.e., maximize the power of the test subject to the constraint of a fixed false-alarm probability).

For diagnostic applications, the emphasis of the SCD approach is on "quickest detection." In some cases, early fault detection benefits the prognosis of the RUL of a failing component because it allows sufficient time for prediction and follow-up corrective action when the failing component is endangering the operational integrity of the system. SCD takes advantage of the bispectral domain, thus avoiding the challenges posed by its spectral counterpart, such as low signal-to-noise ratio (SNR) at characteristic defect frequencies. Higher-order spectra are sensitive to non-Gaussian vibration signatures only. It is well known that third- and higher-order cumulants vanish for Gaussian processes, thus enhancing SNR for fault signatures. In the SCD scheme, failure modes involve sets of phase-coupled harmonics and sidebands because cumulants are sensitive to phase coupling among different frequencies. Barron Associates used a time-local bispectrum and the magnitude-squared bicoherence function as higher-order statistics for fault detection.

The SCD approach was applied to accelerometer data obtained from a CH-47D helicopter combiner transmission spiral bevel input pinion gear that exhibited a classic tooth-bending fatigue failure as a result of test-cell overloading. SCD detected the fault clearly at the same time as did special purpose Steward Hughes, Ltd., Mechanical Systems Diagnostic Analysis. Details of the SCD method can be found in Parker et al., 2000 (Vachtsevanos et al., 2006).

5.1.5.2 State-Based Feature Recognition

George D. Hadden et al. at Honeywell International introduced a state-based feature-recognition software technique using temporally correlated features from multiple data streams to perform anomaly detection and diagnosis (Hadden et al., 1994). In the technique, features are detected using software structures called *finite state machines or just state machines*. Unique advantages of this approach are as follows:

- Hierarchical pattern recognition can be built on recognition of simpler patterns.
- It requires only 100–300 bytes per individual state machine.
- Dynamically downloadable sensors can take a "closer look."
- Parallelizable state machines operate independently, yet can use output from each other.

Failure modes are modeled as state machines whose structure simulates a state register, local and global variables, and time stamps; the transition list entails state-to-state transitions, conditions, and actions. Fig. 5.12 shows a general state machine structure. Honeywell applied state-based feature-recognition concepts to engine and chiller fault diagnosis, among others (Hadden et al., 1994; Hadden and Nelson, 1991).

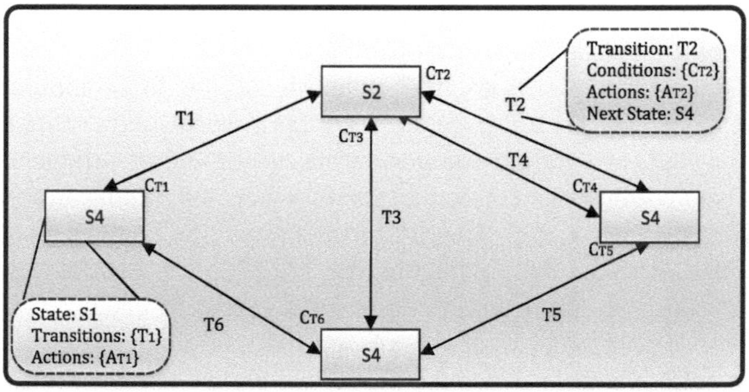

FIGURE 5.12 General state machine. *Adapted from Vachtsevanos, G., Lewis, F., Roemer, M., Hess, A., Wu, B., 2006. Intelligent Fault Diagnosis and Prognosis for Engineering Systems, John Wiley & Sons, Inc., Hoboken, New Jersey.*

5.1.5.3 Bayesian Networks

Bayesian networks, or belief networks, knowledge maps, and probabilistic causal networks, as they are sometimes known, are helpful tools to model multifault, multisymptom dependency relations (Jensen, 2001). Probabilistic reasoning is a significant tool in situations where causality seems to play a role but our understanding of the actual cause–effect is incomplete. The Bayesian philosophy's foundation is Bayes' theorem, where the posterior condition is derived as the ratio of the likely times the prior is divided by the evidence. Applying Bayes' theorem to multiple variables requires chain rules, for example

$$P(A, B, C, D) = P(A|B, C, D)P(B|C, D)P(C|D)P(D) \qquad (5.5)$$

As the number of variables increases, the distributions required for the joint distribution of all variables increase exponentially! For example, for n binary random variables, we need $2^n - 1$ joint distribution.

Bayesian networks offer a way to compactly describe distributions. Formally, they are directed acyclic graphs whose nodes are random variables, such as causes, effects, and pieces of evidence. The network's arcs specify the dependence assumption between random variables. In designing a Bayesian network, we typically specify the dependence of variables, the prior probabilities of "root" nodes, and the conditional probabilities of nonroot nodes, given only their direct predecessors. The simple schematic in Fig. 5.13 illustrates the computational advantage of this approach. The network shows that A causes B and C, and C and D cause E. The quantities required are the probabilities $P(A)$ and $P(D)$ and the conditional probabilities $P(B|A)$, $P(C|A)$, and $P(E|D, C)$. If A, B, C, D, and E are binary (true/false type), direct representation, $P(A, B, C, D, E)$, needs $2^5 - 1 = 31$ distributions, whereas the Bayesian network needs only 5 (Vachtsevanos et al., 2006).

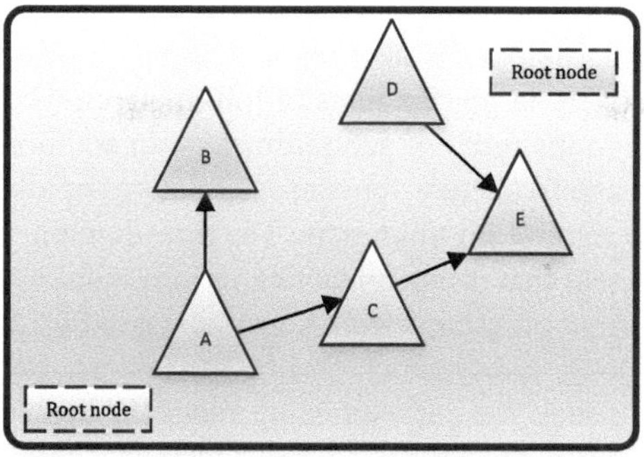

FIGURE 5.13 Simple Bayesian network. *Adapted from Vachtsevanos, G., Lewis, F., Roemer, M., Hess, A., Wu, B., 2006. Intelligent Fault Diagnosis and Prognosis for Engineering Systems, John Wiley & Sons, Inc., Hoboken, New Jersey.*

Advantageous properties of such network configurations include the following:

1. If the specified distributions are consistent, all probabilities are consistent.
2. The network always uniquely defines a distribution; that is, the joint distribution for the Bayesian network is uniquely defined by each random variable.

Evaluating a Bayesian network is an arduous task because it is generally an NP-hard problem. Exact solutions are available only for singly connected networks. It is necessary, therefore, to convert the network from multiply connected to singly connected by clustering or using approximate methods. Among the latter methods, we may randomly position values for some nodes and then use them to pick values for other nodes. Alternatively, statistics can be maintained on values that the nodes take; we can arrive at answers from those statistics.

The superiority of Bayesian reasoning for fault diagnosis stems from the likelihood principle; that is, our inferences depend on probabilities assigned to the data that have been received, not on data that might have occurred but did not. This is in sharp contrast to classic statistical inference. In our diagnosis problems, diagnosis is based on reasoning about cause—effect or fault symptom relations. Probabilistic reasoning remains the only viable alternative in situations where one symptom may be caused by several faults and many fault symptoms are not deterministic or well understood (Vachtsevanos et al., 2006).

5.1.5.4 Mechanical Face Seal and Drive Shaft Fault Diagnosis

Itzhak Green et al. introduced a methodology to diagnose faults in mechanical face seals and drive shafts using eddy-current proximity sensors and xy orbit plots (Zou and Green, 1997; Zou et al., 2000). The eddy-current sensors allow us to observe directly relative shaft motion, for example, inside clearance of fluid-film bearings. They are commonly employed in high-speed turbomachinery and often supplied as standard equipment on these machine types. In shaft monitoring, deflections are relatively large (fractions of millimeters), whereas clearances in seal monitoring are of the order of submillimeters. In addition to the usual fault features in the time and frequency domains, angular orbit plots are used to detect seal absolute and/or relative misalignment. The time domain, frequency domain, and orbit plots of a seal that is noncontacting or in normal operation, as derived from three eddy-current proximity probes, are shown in Fig. 5.14 (Vachtsevanos et al., 2006).

For a contacting seal, the corresponding plots are depicted in Fig. 5.15.

The plots show that noncontacting behavior and contacting behavior are distinctly different and can clearly be identified. Features can be derived in all three domains, including the tilt orbit plot, to suggest contacting seal behavior. Furthermore, cracked-shaft dynamic behavior is distinctly different (having a dominant second-order harmonic component) from a seal failure in contact (having higher harmonic oscillations). Hence, multifaults (in this case, a dual fault) can be detected using a single monitoring system. This maximizes the return on the investment. Theoretical models for seal and shaft dynamic behavior and experimental results are shown by Zou and Green (1997) and Zou et al. (2000) to agree well (Vachtsevanos et al., 2006).

5.1.5.5 Hidden Markov Models

Hidden Markov models (HMMs) have been studied extensively since their introduction in the late 1960s and early 1970s. An excellent tutorial on this subject can be found in Rabiner (1989). Because of their rich mathematical structure, they have become increasingly popular in the last several years, with a wide range of applications. A problem of fundamental interest to fault diagnosis is characterizing signals in terms of signal models. A system may be described at any time as being in one of several distinct states. A typical Markov chain with five states and selected state transitions is shown in Fig. 5.16 (Vachtsevanos et al., 2006).

At regularly spaced, discrete times, the system undergoes a change of state (possibly from normal to a faulty condition) according to a set of probabilities associated with the state and estimated from actual measurements. HMMs are well suited to modeling quasistationary signals, such as those derived from vibration monitoring, and can be used to perform fault detection and estimation (Bunks et al., 2000). Similar to the notional description earlier, HMMs are composed of two main components: the state-transition matrix, which probabilistically describes how the model moves from one stationary type of behavior to another, and the probability distribution associated with each state, which describes the statistics of the particular stationary model. Details of application of the HMM approach to a Westland helicopter gearbox data for gearbox fault detection and isolation can be found in Bunks et al. (2000). If sufficient data are available to estimate the parameters of an appropriate probability density for each state (i.e., load level, fault mode, etc.) and a priori

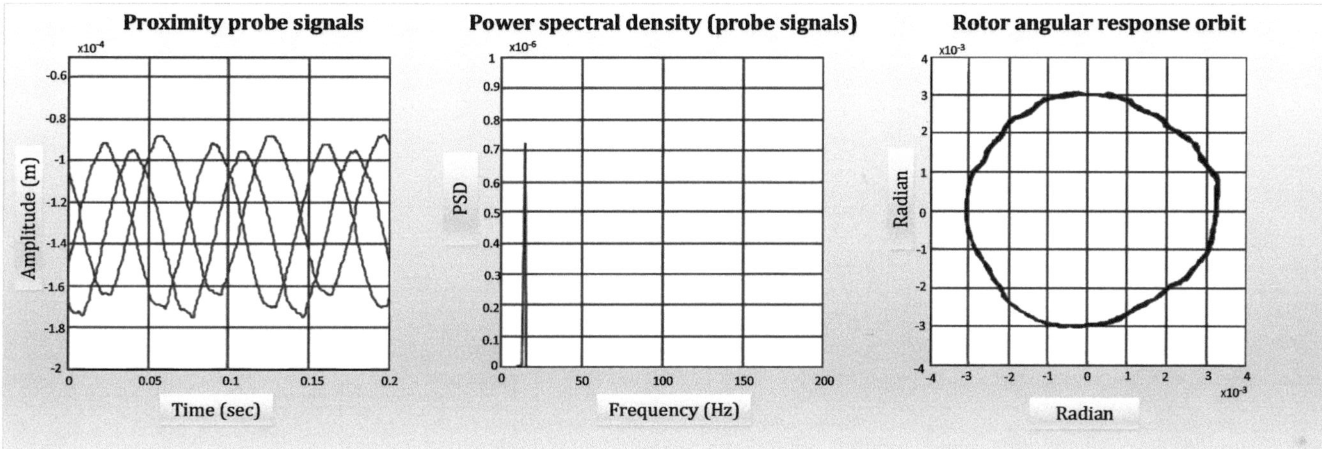

FIGURE 5.14 Noncontacting face seal operation. *Adapted from Vachtsevanos, G., Lewis, F., Roemer, M., Hess, A., Wu, B., 2006. Intelligent Fault Diagnosis and Prognosis for Engineering Systems, John Wiley & Sons, Inc., Hoboken, New Jersey.*

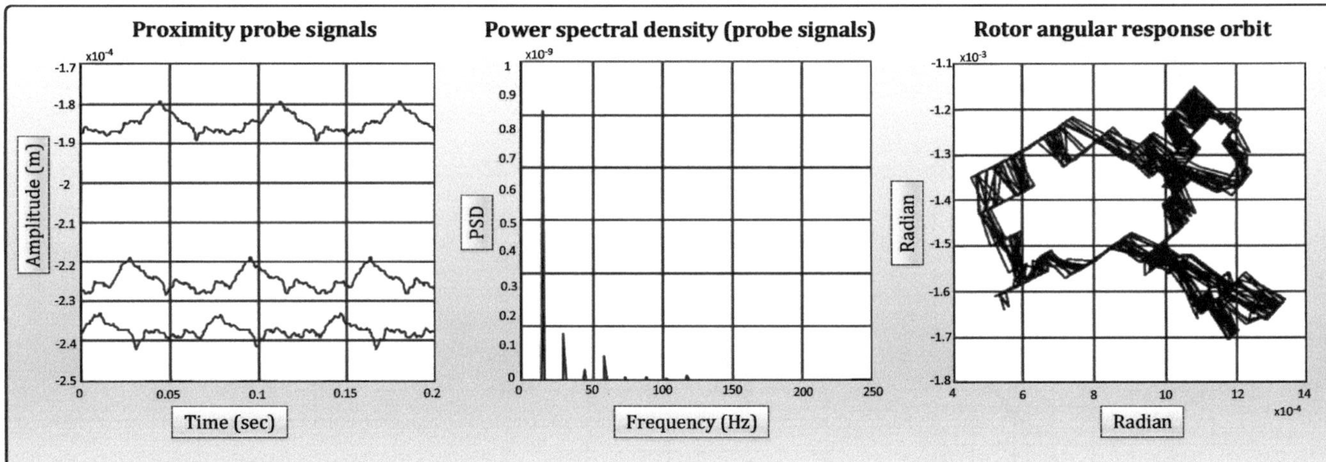

FIGURE 5.15 Contacting seal operation. *Adapted from Vachtsevanos, G., Lewis, F., Roemer, M., Hess, A., Wu, B., 2006. Intelligent Fault Diagnosis and Prognosis for Engineering Systems, John Wiley & Sons, Inc., Hoboken, New Jersey.*

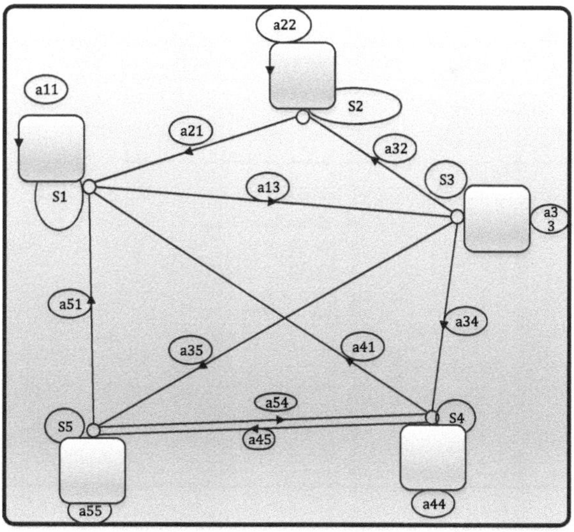

FIGURE 5.16 Markov chain with five states and selected state transitions. *Adapted from Vachtsevanos, G., Lewis, F., Roemer, M., Hess, A., Wu, B., 2006. Intelligent Fault Diagnosis and Prognosis for Engineering Systems, John Wiley & Sons, Inc., Hoboken, New Jersey.*

evidence is given of the frequency of occurrence of each fault mode and the average time spent at each load level, then the HMM is a flexible and powerful tool, although technical challenges remain in terms of suitable data sets, training of the HMM, the quasistationarity assumption, etc. (Vachtsevanos et al., 2006).

5.2 DATA-DRIVEN VERSUS PHYSICAL MODELS

5.2.1 Introduction

CBM, as illustrated in Fig. 5.17, is a cost-effective maintenance strategy in which maintenance schedules are predicted based on the results provided from diagnostics and prognostics. Diagnostics characterizes the status of damage through detection, isolation, and identification using collected data from structural health monitoring. Based on the diagnostics information, prognostics predicts the future behavior of damage and the RUL, i.e., the remaining time/cycles before maintenance is required. The collected data that are important for both diagnostics and prognostics can be classified into two types:

1. Event data or run-to-failure data, and
2. CM data.

The latter provide information on damage, such as wear volume, crack size, vibration signal, oil debris, and thermography, to name a few. Some CM

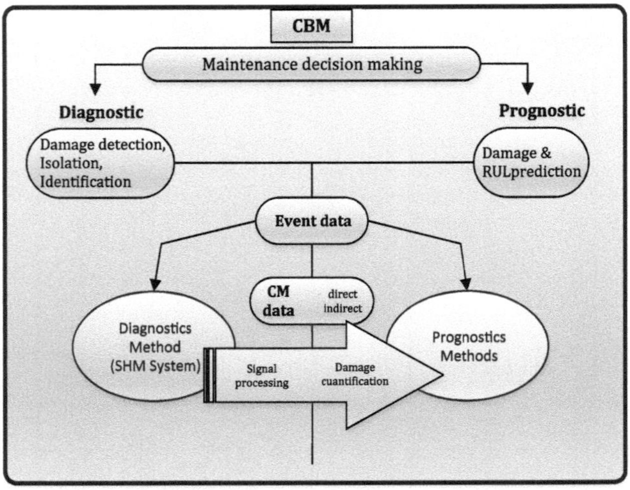

FIGURE 5.17 Building blocks of condition-based maintenance (CBM). *CM*, condition monitoring; *RUL*, remaining useful life; *SHM*, structural health monitoring. *Adapted from An, D., Kim, N.H., Choi, J.-H., 2015. Practical options for selecting data-driven or physics-based prognostics algorithms with reviews. Reliability Engineering and System Safety 133, 223–236. Journal homepage: www.elsevier.com/locate/ress.*

data are direct, such as crack size, while others are indirect, such as oil debris. In the case of indirect CM data, a damage quantification process is required for diagnostics and prognostics (An et al., 2015).

There are several reviews of diagnostics methods and a number of publications on diagnostics, signal processing, and damage quantification. For example, Jardine et al. (2006) provide an overall review of diagnostics and prognostics in terms of implementing CBM and decision-making. Most reviews provide a simple comparison of different methods for a specific application only. In addition, most focus on an algorithm, rather than interpreting the algorithm in the context of prognostics. In what follows, we provide a practical review of prognostic methods. We introduce prognostics algorithms, give their attributes, and enumerate their pros and cons in terms of model definition, model parameter estimation, and the ability to handle noise and bias in data.

In general, prognostics methods can be categorized into data-driven, physics-based, and hybrid approaches, as illustrated in Fig. 5.18. Data-driven approaches use information from previously collected data (training data) to identify the characteristics of the currently measured damage state and to predict the future trend. Physics-based approaches assume that a physical model describing the behavior of damage is available; they combine the physical model with measured data to identify model parameters and predict future behavior. There are two main differences between data-driven and physics-based approaches:

1. Availability of a physical model, and
2. Use of training data to identify the characteristics of the damage state.

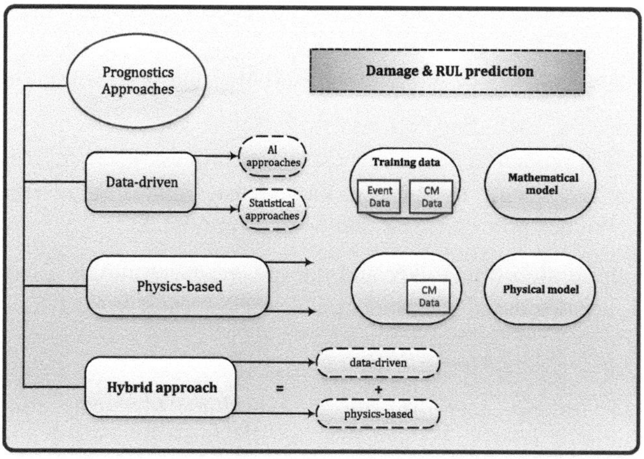

FIGURE 5.18 Categorization and definition of prognostics methods. *AI*, artificial intelligence; *CM*, condition monitoring; *RUL*, remaining useful life. *Adapted from An, D., Kim, N.H., Choi, J.-H., 2015. Practical options for selecting data-driven or physics-based prognostics algorithms with reviews. Reliability Engineering and System Safety 133, 223–236. Journal homepage: www.elsevier.com/locate/ress.*

Hybrid approaches combine the above methods to improve the prediction performance (Liao and Köttig, 2014; Xu and Xu, 2011; Xing et al., 2011.). Such approaches are not mature, however, and are not discussed in this section.

Data-driven and physics-based approaches have different properties that must be reflected in their respective algorithms. Providing a standard to select the best algorithm for different conditions is important for users. In what follows, we consider the number of data sets, the level of noise and bias in data, the availability of loading conditions and physical models, and model complexity. We test typical prognostics algorithms to show the influence of each algorithm's attributes, as well as the pros and cons of the results. As an example, we use a fatigue crack growth because a physical model is available, and this allows us to compare prognostics methods. However, our conclusions are not limited to the crack growth problem and can be interpreted in a general context (An et al., 2015).

5.2.2 Data-Driven Approaches

Data-driven approaches use information from observed data to identify the characteristics of damage progress and predict the future state without using any particular physical model. Instead, they use mathematical models and/or weight parameters, determined from the training data obtained under various usage conditions. Since data-driven approaches depend on the trend of data, which often show a distinct characteristic near the end of life, they are powerful in predicting near-future behaviors, especially toward the end of life (An et al., 2015).

Data-driven approaches are generally divided into two categories:

1. AI approaches, including NN and fuzzy logic, and
2. Statistical approaches, including Gaussian Process (GP) regression, relevance/support vector machine (SVM), least squares regression, gamma process, Wiener processes, hidden Markov model, etc.

Among these algorithms, NN and GP are most commonly used for prognostics and, are discussed in detail in Chapter 6 (An et al., 2015).

5.2.2.1 Data-Driven Modeling

Hydrological models can be characterized as physical, mathematical (including lumped conceptual and distributed physically based models), and empirical. The latter class of models, in contrast to the first two, involves mathematical equations that are not derived from physical processes in the catchment but from analysis of time series data. Examples include the unit hydrograph method, linear regression, and artificial intelligence in medical applications (ARIMA) models. Recent developments in computational intelligence, especially in the area of ML, have greatly expanded the capabilities of empirical modeling. The field encompassing these new approaches is called Data-Driven Modeling (DDM). As the name suggests, DDM is based on analyzing the data about a system, in particular finding connections between the system state variables (input, internal, and output variables) without explicit knowledge of the physical behavior of the system. These methods represent large advances on conventional empirical modeling and include contributions from the following overlapping fields:

- Artificial intelligence (AI), the overarching study of how human intelligence can be incorporated into computers.
- Computational intelligence (CI), including NNs, fuzzy systems, and evolutionary computing, as well as other areas within AI and ML.
- SC, close to CI, but with special emphasis on fuzzy rule-based systems (FRBS) induced from data.
- Machine learning (ML), once a subarea of AI that concentrates on the theoretical foundations used by CI and SC.
- Data mining (DM) and knowledge discovery in databases (KDD), often focusing on very large databases and associated with applications in banking, financial services, and customer resources management. DM is seen as a part of a wider KDD. Methods are mainly from statistics and ML.
- Intelligent data analysis, with a focus on data analysis in medicine and research and incorporating methods from statistics and ML.

DDM uses CI and ML methods to build models to complement or replace physically based models. A machine-learning algorithm is used to determine the relationship between a system's inputs and outputs using a training data set representative of all the behavior found in the system (see Fig. 5.19) (Abrahart et al., 2008).

Once the model is trained, it can be tested using an independent data set to determine how well it can generalize to unseen data (Abrahart et al., 2008).

5.2.2.2 Data-Driven Modeling Techniques

5.2.2.2.1 Neural Network

The NN algorithm is a representative data-driven method in which a network model learns a way to produce a desired output, such as the level of degradation or lifespan, by reacting to given inputs, such as time and usage conditions. Once the network model learns enough about the relationship between inputs and output, it can be used for diagnosis and prognosis. A typical architecture, feed-forward neural network (FFNN) (Svozil et al., 1997), is illustrated in Fig. 5.20. The circles represent nodes (also called neurons or units), and each set of nodes in the same column is called a layer. The nodes in the input and output layers represent the input and output variables, respectively. The number of nodes in the hidden layer should be determined to correctly express the mechanism between the input and output, receiving signals from the input layer and forwarding them to the output layer. The learning process is equivalent to determining weight parameters, such that the network model accurately represents the relationship between inputs and outputs. Once the network model learns enough, the model is functionalized using transfer functions and weight parameters. The transfer functions characterize the relationship between two adjacent layers, where several types of transfer functions are available, such as the sigmoid, inverse, and linear functions (Duch and Jankowski, 1999). The weight parameters include weights and biases (Liu et al., 2010; Firth et al., 2003). The weights, shown as rectangles in Fig. 5.20, are usually multiplied by the value at the previous

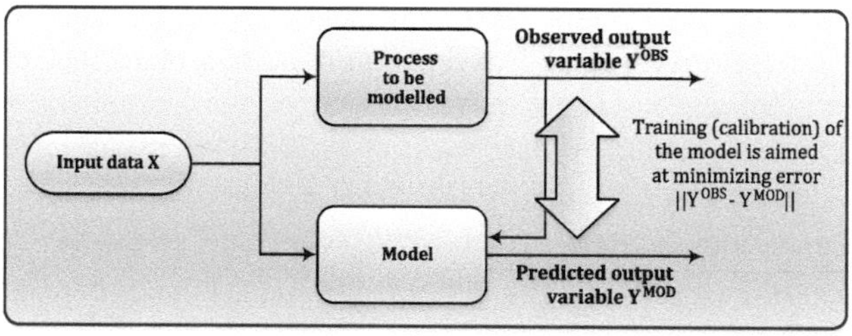

FIGURE 5.19 General approach to modeling. *Adapted from Abrahart, R.J., See, L.M., Solomatine, D.P., 2008. Practical hydroinformatics. Computational intelligence and technological developments in water applications. In: Singh, V.P. (Ed.), Water Science and Technology Library, vol. 68. Springer (Texas A&M University, College Station, U.S.A. ISBN: 978-3-540-79880-4).*

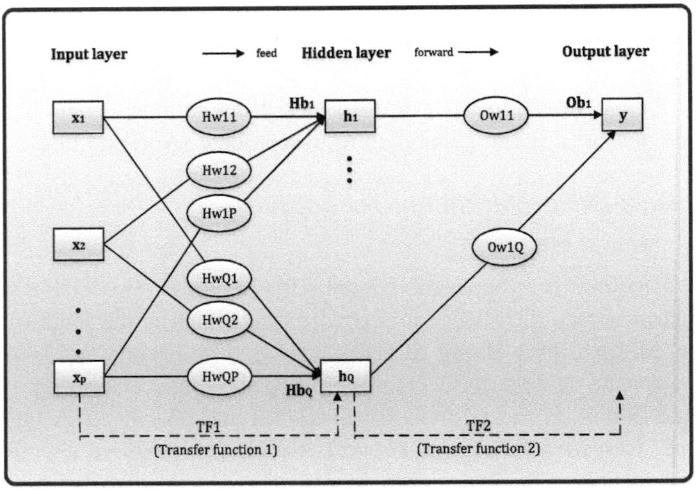

FIGURE 5.20 Feed-forward neural network. *Adapted from An, D., Kim, N.H., Choi, J.-H., 2015. Practical options for selecting data-driven or physics-based prognostics algorithms with reviews. Reliability Engineering and System Safety 133, 223–236. Journal homepage: www.elsevier.com/locate/ress.*

nodes; the biases shown as ellipses are then added to the sum of the results to be inputs to the transfer function. The process of finding the optimal weight parameters is called training, or learning, and requires many sets of training data (An et al., 2015).

FFNN is often called the back-propagation neural network (BPNN) because the weight parameters are determined through a learning/optimization algorithm by the backward propagation of errors between the training data and the output from the network model (Rumelhart et al., 1986). In other words, FFNN and BPNN are used, respectively, to calculate the response forward and to update the weight parameters backward, based on the response. Recurrent NN (RNN) (Bodén, 2002) is the second most common architecture; its concept is not much different from FFNN, except that it has local feedback connections between the input and hidden/output nodes. Since the Levenberg–Marquardt backpropagation algorithm (Wilamowski et al., 2001) is usually employed for the RNN's learning algorithm, RNN and Levenberg–Marquardt NNs are sometimes used interchangeably. In addition to FFNN and RNN, there are other NNs, such as fuzzy-neural, wavelet, associative-memory, modular, and hybrid NNs (An et al., 2015).

5.2.2.2.2 Gaussian Process Regression

Gaussian Process (GP) is a commonly used regression-based data-driven approach to prognostics, whose concept is illustrated in Fig. 5.21. The GP

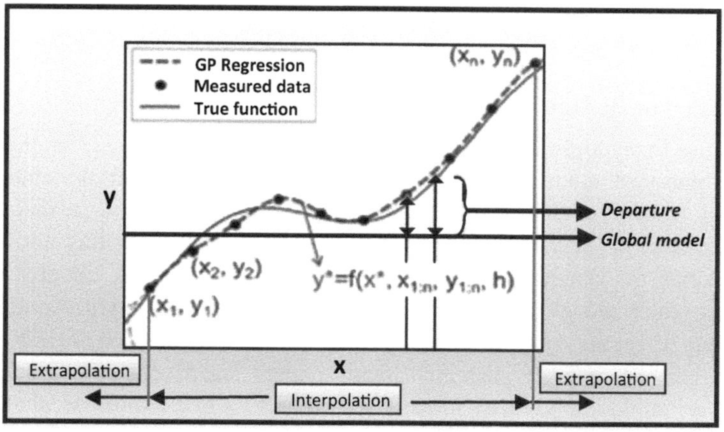

FIGURE 5.21 Gaussian process regression network. *Adapted from An, D., Kim, N.H., Choi, J.-H., 2015. Practical options for selecting data-driven or physics-based prognostics algorithms with reviews. Reliability Engineering and System Safety 133, 223–236. Journal homepage: www.elsevier.com/locate/ress.*

regression model is composed of a global model and its local departure. An outstanding feature of GP is that the simulated output, denoted by the blue dashed curve in Fig. 5.21, passes through a set of measured data (training data) $(x_{1:n}; y_{1:n})$. When the prediction point (a new input), x^*, is at a measurement point (e.g., point A), $x_i(i = 1, 2,\ldots, n)$, the magnitude of departure is the same as the difference between measured data, $y_i(i = 1, 2,\ldots, n)$ and the global model; hence, the prediction output y^* becomes measured data, y_i. When the prediction point deviates from a measurement point (e.g., point B), the magnitude of departure is changed based on the correlation between B and the measurement points. As a result, if the prediction point is between measurement points, the simulated output smoothly interpolates the measured data. If, however, the prediction point moves out of the measurement points, i.e., extrapolation, the influence of this departure is reduced as correlation decreases and the GP becomes closer to the global model. Therefore, the case of extrapolation is not so different from an ordinary least-squares regression (Coppe et al., 2011). For interpolation, correlation between data points is an important factor to characterize GP's attribute; this is determined by selecting an appropriate type of covariance function and estimating its scale parameters (or hyperparameters). The scale parameters are usually determined using an optimization algorithm based on the measured data (training data) (An et al., 2015).

GP regression can also be described in stochastic terms to account for the uncertainty in prediction by assuming that the measured output follows multivariate normal distribution, with the mean and its covariance being the global model and its departure, respectively. We can obtain the GP model as the mean (composed of a global model and departure), and the variance is used to represent uncertainty in simulated GP outputs (An et al., 2015).

5.2.2.2.3 Fuzzy Rule-Based Systems

FRBS use fuzzy logic for inference. Fuzzy logic is based on fuzzy set theory in which binary set membership has been extended to include partial membership ranging from 0 to 1 (Zadeh, 1965). Fuzzy sets, in contrast to their crisp counterparts, have gradual transitions between defined sets; this allows the uncertainty associated with these concepts to be modeled directly. After defining each model variable with a series of overlapping fuzzy sets, the mapping of inputs to outputs can be expressed as a set of IF-THEN rules, which can be entirely specified from expert knowledge, or from data. However, unlike NNs, fuzzy models are prone to a rule explosion; that is, as the number of variables or fuzzy sets per variable increases, there is an exponential increase in the number of rules, making it difficult to specify the entire model from expert knowledge alone (Kosko, 1997). Various automated methods for optimizing fuzzy models are now available (Wang, 1994), including NNs and GAs.

The fuzzy sets and rules are called the fuzzy model knowledge base. Crisp inputs to the model are first fuzzified via this knowledge base, and a fuzzy inference engine is used to process the rules in parallel via a fuzzy inference procedure such as max-min or max-product operations (Jang et al., 1997). The fuzzy solution surface resulting from the execution of the rule base is defuzzified to produce the system output(s). Fuzzy IF-THEN rules can also be comprised of functional consequents, usually of a linear or polynomial form, in a formulation referred to as a Takagi–Sugeno–Kang model (Takagi and Sugeno, 1985; Sugeno and Kang, 1988). The crisp inputs are fuzzified according to the fuzzy set definitions, combined via the inference engine, and the functional consequents are weighted by the memberships that result from the execution of the rules. The overall result is a weighted average of the equations, as more than one rule can fire positively during a single pass of the rule base (Abrahart et al., 2008).

Fuzzy logic has found multiple applications, mainly in control theory (see, e.g., Kosko, 1997). FRBSs can be built by interviewing human experts or by processing historical data and forming a data-driven model. The basics of the latter approach and its use in a number of water-related applications can be found in Bárdossy and Duckstein (1995). FRBS have effectively been used for drought assessment (Pesti et al., 1996), prediction of precipitation events (Abebe et al., 2000a), analysis of groundwater model uncertainty (Abebe et al., 2000b), control of water levels in polder areas (Lobbrecht and Solomatine, 1999), and modeling rainfall-discharge dynamics (Vernieuwe et al., 2005).

5.2.2.2.4 Genetic Algorithms in Model Optimization

Genetic algorithms (GAs) or, more widely, evolutionary algorithms, are nonlinear search and optimization methods inspired by the biological processes of natural selection and survival of the fittest (Goldberg, 1989). They do not belong to the class of data-driven models, but as they are widely used in optimizing models, we are considering them here. Genetic (evolutionary) algorithms are typically attributed to the area of computational intelligence.

Unlike other methods such as hill climbing and simulated annealing, a GA, like other randomized search algorithms such as adaptive cluster covering (Solomatine, 1999), exhibits implicit parallelism, considering many points at once during the search process, thereby reducing the chance of converging to a local optimum. GAs also use probabilistic rules in the search process, and they can generally outperform conventional optimization techniques on difficult, discontinuous, and multimodal functions. Despite their unique and adaptive search capabilities, there is no guarantee that GAs will find a global solution; however, they can often find an acceptable one quite quickly. A detailed introductory survey appears in Reeves and Rowe (2003).

The basic unit of a GA is the gene, which, in biological terms, represents a given characteristic of an individual, such as eye color. In a GA, a gene represents a parameter that is being optimized. An individual or chromosome is simply the combined set of all the genes, i.e., all the parameters needed to generate the solution. To start the search, a population of these individuals or strings is randomly generated. Each string is then evaluated by a fitness or objective function according to some measure of performance. This represents the success of the solution and is analogous to the survival ability of an individual within the population. To evolve better performing solutions, the fittest members of the population are selected and exposed to a series of genetic operators, which produce offspring for the next generation. The least fit solutions will die out through natural selection as they are replaced by new, recombined individuals.

The main genetic operator is crossover; in crossover, a position along the bit string is randomly chosen to cut two parent chromosomes into two segments, which are then swapped. The new offspring are comprised of a segment from each parent, thereby inheriting bits from both. The occurrence of crossover is determined probabilistically; when crossover is not applied, offspring are simply duplicates of the parents, giving each individual a chance of passing on a pure copy of its genes into the gene pool. The second main genetic operator is mutation, which is applied to each of the offspring individually after crossover. Mutation can alter the bits in a string but with an extremely low probability. Crossover allows the GA to explore new areas in the search space and gives it the majority of its searching power, while mutation exploits existing areas to find a near optimal solution and provides a small amount of random search to ensure that no point in the search space has a zero probability of being examined. The newly generated offspring are placed back into the population, and the exercise is repeated for many generations until a set of user-specified termination criteria are satisfied, such as exceeding a preset number of generations, or until no improved solution is found after a given period of time. Over many generations, a whole new population of possible solutions, possessing a higher proportion of the characteristics found in the fitter members of the previous generation, is produced (Abrahart et al., 2008).

GAs are useful tools for handling difficult problems when conventional techniques cannot cope, or alternatively, they can be used to improve existing methods through hybridization. For example, fuzzy logic rule-based models can be entirely optimized by a GA in a completely inductive approach, or expert knowledge can be used to specify the rules or membership functions, leaving the GA to optimize only the unknown parts of the model. See Cordón and Herrara (1995) and Karr (1991) for more details of fuzzy logic model optimization using a GA. GAs can also be used to optimize other data-driven models such as NNs (Yao and Liu, 1997).

5.2.2.2.5 Other Approaches

In addition to NNs and FRBS, other data-driven methods have been used to solve hydrological problems, but these methods are less well-known than ANN or FRBS. Methods currently attracting research attention are described below:

- *Genetic programming* is a method for evolving equations by taking various mathematical building blocks such as functions, constants, and arithmetic operations and combining them into a single expression. Genetic programming was originally proposed by Koza (1992). Evolutionary regression is similar to genetic programming, but the goal is to find a regression equation, typically a polynomial regression, where the coefficients are determined through an evolutionary approach such as a GA. Examples of hydrological applications include Khu et al. (2001) application of genetic programming to real-time runoff forecasting for a catchment in France, and Giustolisi and Savic's (2006) use of evolutionary regression for groundwater and river-temperature modeling (Abrahart et al., 2008).
- *Classification* is a method for partitioning data into classes and attributing data vectors to these classes. The output of a classification model is a class label, not a real number as in regression models. The classes are typically created such that they are far from one another in attribute space, but the points within a class are as tightly clustered around the center point as possible. Examples of classification techniques include k-nearest neighbor (*k*-NN), Bayesian classification, decision tree classification (Witten and Frank, 2000), and SVMs (Vapnik, 1998). There are many examples of applying

classification methods in hydrology. For example, Frapporti et al. (1993) use fuzzy c-means clustering to classify shallow Dutch groundwater sites, and Hall and Minns (1999) use an (self organizing feature map) SOFM to classify catchments into subsets on which ANNs are applied to model regional flood frequency. Hannah et al. (2000) use clustering to find groups of hydrographs on the basis of their shape and magnitude; clusters are then used for classification by experts. Harris et al. (2000) apply clustering to identify the classes of river regimes. Velickov et al. (2000) use self-organizing feature maps as clustering methods and SVM as a classification method in aerial photograph interpretation with the purpose of subsequent construction of flood severity maps. Solomatine et al. (2007) uses decision trees and *k*-NN to classify river flow levels according to their severity in a flood forecasting problem in Nepal. Zhang and Song (2006) use a combination of (Special Operations Forces) SOF and (adaptive resonance theory) ART networks for special pattern identification of soil moisture (Abrahart et al., 2008).

- In *Instance-Based Learning* (IBL), classification or prediction is made by combining observations from the training data set that are close to the new vector of inputs (Mitchell, 1997). This is a local approximation and works well in the immediate neighborhood of the current prediction instance. The nearest neighbor classifier approach classifies a given unknown pattern by choosing the class of the nearest example in the training set as measured by some distance metric, typically Euclidean. Generalization of this method is the *k*-NN method. For a discrete-valued target function, the estimate will be the most common value among k training examples nearest to x_q. For real-valued target functions, the estimate is the mean value of the *k*-NN examples. Locally weighted regression is a further extension in which a regression model is built on *k*-nearest instances. Applications of IBL in water-related problems mainly use the simplest method, *k*-NN. Karlsson and Yakowitz (1987) show the use of this method in hydrology, focusing on (single variate) time series forecasts. Galeati (1990) demonstrates the applicability of the *k*-NN method (with the vectors composed of the lagged rainfall and flow values) for daily discharge forecasting and favorably compares it to the statistical ARX model. Shamseldin and O'Connor (1996) use the *k*-NN method to adjust the parameters of the linear perturbation model for river flow forecasting. Toth et al. (2000) compare the *k*-NN approach to other time series prediction methods in a problem of short-term rainfall forecasting. Solomatine et al. (2007) consider IBL in a wider context of ML and test their applicability in short-term hydrologic forecasting (Abrahart et al., 2008).
- *Chaos theory* and nonlinear dynamics can be used for time series prediction when the time series data are of sufficient length and carry enough information about the behavior of the system (Abarbanel, 1996). The main idea is to represent the state of the system at time *t* by a vector in *m*-dimensional state space. If the original time series exhibits chaotic

properties, its equivalent trajectory in phase space has properties allowing accurate prediction of future values of the independent variable. Hydrological examples include the work by Solomatine et al. (2000) and Velickov et al. (2003); both use chaos theory to predict a surge in water level in the North Sea. For two-hourly predictions, the error is as low as 10 cm and is at least on par with the accuracy of hydrodynamic models. Babovic et al. (2000) use a chaos theory—based approach to predict water levels in the Venice lagoon, while Phoon et al. (2002) uses it to forecast hydrologic time series (Abrahart et al., 2008).

- *Support vector machine (SVM)* is a relatively new important method based on the extension of the idea of identifying a line (or a plane or some surface) that separates two classes in classification. It is based on statistical learning theory initiated by Vapnik in the 1970s (Vapnik, 1998). This classification method has been extended to solving prediction problems and used in hydrology-related tasks. Dibike et al. (2001) and Liong and Sivapragasam (2002) report using SVMs for flood management and to predict water flows and stages (Abrahart et al., 2008).
- *Machine-learning methods* for clustering and classification are often used to support regression methods, and methods of IBL are used for both classification and regression (Abrahart et al., 2008).

5.2.3 Physics-Based Approaches

Physics-based approaches combine a physical damage model with measured data to predict future behavior of degradation or damage and to predict RUL, as illustrated in Fig. 5.22. The behavior of a physical model depends on the parameters obtained from laboratory tests or estimated in real time, based on measured data up to the current time. RUL is predicted by advancing the damage state until it reaches a certain threshold, as indicated by the dashed curves in Fig. 5.22.

Note that the following subsections explain the issues of physics-based approaches, not their algorithms (An et al., 2015).

5.2.3.1 Physical Model Adequacy

As physics-based approaches employ a physical model describing the behavior of damage, they have advantages in predicting long-term behaviors of damage. However, model validation should be carried out first, as most models contain assumptions and approximations. There is a great deal of literature on model validation using statistical methods, such as the hypothesis test or the Bayesian method (Oden et al., 2013; Rebba et al., 2006; Sargent, 2013; Ling and Mahadevan, 2013). In general, as the model complexity increases, the number of model parameters increases, and the estimation of parameters becomes more difficult. Recently, Coppe et al. (2012) have showed that model accuracy can be relieved by identifying equivalent parameters from a simpler model.

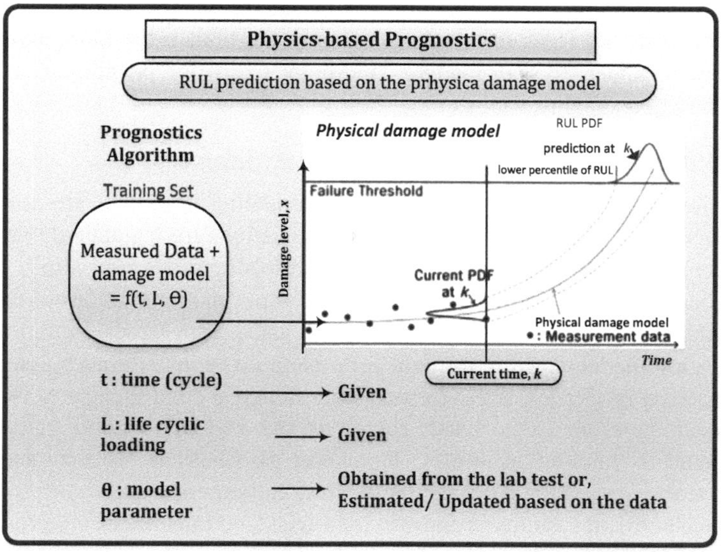

FIGURE 5.22 Physics-based prognostics. *PDF*, probability density function; *RUL*, remaining useful life. *Adapted from An, D., Kim, N.H., Choi, J.-H., 2015. Practical options for selecting data-driven or physics-based prognostics algorithms with reviews. Reliability Engineering and System Safety 133, 223–236. Journal homepage: www.elsevier.com/locate/ress.*

Coppe uses a simple Paris model with an assumed stress intensity factor to predict crack growth of complex geometries; the model parameters are adjusted to compensate for the error in the stress intensity factor. Although this analysis is limited to the case of similar damage behavior in the simple and complex model, additional efforts to validate model accuracy may not be necessary (Vachtsevanos et al., 2006).

5.2.4 Physical Model–Based Methods

Model-based methods rely on the availability of a good dynamic model of the system. Such models can be derived using physical modeling principles, or the model can be determined using online parameter estimation and system identification methods. Often the two approaches can be combined, as shown in the following examples.

The structural integrity of critical systems, such as aircraft, has been attracting major interest from the CBM/PHM community in recent years. Attention has focused primarily on crack initiation and propagation models. Crack initiation models, used as fault diagnostic tools, have been studied extensively by materials engineering experts, and an excellent review of the crack initiation constitutive equations is found in Sehitoglu (1997). Three-dimensional finite-element models, such as ANSYS, FRANC-3D, and

FASTRAN (Newman, 1992), are making inroads in the quest for effective and efficient modeling tools that capture robustly the fault behavior of structural components under stress or loading conditions. Corrosion phenomena are also being targeted (Vachtsevanos et al., 2006).

5.2.4.1 Fatigue and Crack Propagation Models

Crack initiation models based on physical modeling principles are now well understood and can be used to account for deviations from standard operating conditions. Crack models must account for load profiles and stress ratios, variations in temperature, sustained load hold time, and interactions of damage mechanisms, such as fatigue, environment, and creep. Crack initiation and propagation models need to account for a plethora of uncertainties associated with material properties, stress factors, environmental conditions, etc. They are expressed, therefore, as stochastic equations, that is, PDFs with all parameters considered as random quantities. Parameter distributions are estimated via Monte Carlo simulations, probabilistic NNs, and others.

In such critical systems as aircraft, industrial, and manufacturing processes, etc., defect (e.g., cracks and anomalies) initiation and propagation must be estimated for fault prognosis to be effective. Dynamic models of fatigue and fault progression can be derived by using principles of physics. Recent studies in materials research focus on microstructural characterization of materials and modeling frameworks that describe robustly and reliably the anomaly initiation and propagation as a function of time (Friedman and Eidenoft, 1990). Such models addressing the constitutive material properties down to the single-particle level are generally not suitable for real-time treatment because of their immense computational complexity. Reduced-order models, called metamodels, are usually derived from the microstructural characterizations and used to estimate in almost real time the crack initiation and propagation with conventional computing platforms (Newman, 1992). A thorough description of the crack initiation constitutive equations can be found in Chestnut (1965) and Vachtsevanos et al. (2006).

Crack propagation models, clearly of interest from the prognosis viewpoint, are distinguished into two major categories: deterministic and stochastic models. Fatigue crack growth in such typical machinery components as bearings, gears, shafts, and aircraft wings is affected by a variety of factors, including stress states, material properties, temperature, lubrication, and other environmental effects. Variations of available empirical and deterministic fatigue crack propagation models are based on Paris' formula:

$$\frac{d\alpha}{dN} = C_o(\Delta K)^n \tag{5.6}$$

where, $\alpha =$ instantaneous length of dominant crack; $N =$ running cycles; C_o, $n =$ material-dependent constants; and $\Delta K =$ range of stress intensity factor over one loading cycle.

Model parameters are typically estimated via a nonlinear recursive least squares scheme with a forgetting factor (Ljung, 1999).

Stochastic crack growth models consider all parameters as random quantities. Thus, the resulting growth equation is a stochastic differential equation (Vachtsevanos et al., 2006).

5.2.4.2 Utility of Finite-Element Analysis in Model-Based Diagnosis

5.2.4.2.1 Finite-Element Method

The finite-element technique is a methodology used to study the behavior of systems in many engineering disciplines, including thermal engineering, fluid mechanics, electromagnetism, and mechanics of materials. The method derives its name from the idea that a complex system or structure can be analyzed by dividing it into small parts called elements, each of which is fully defined in its geometry and/or properties, that is, is discrete (Moaveni, 2003) and finite, as represented in Fig. 5.23. In the figure, the design has been divided

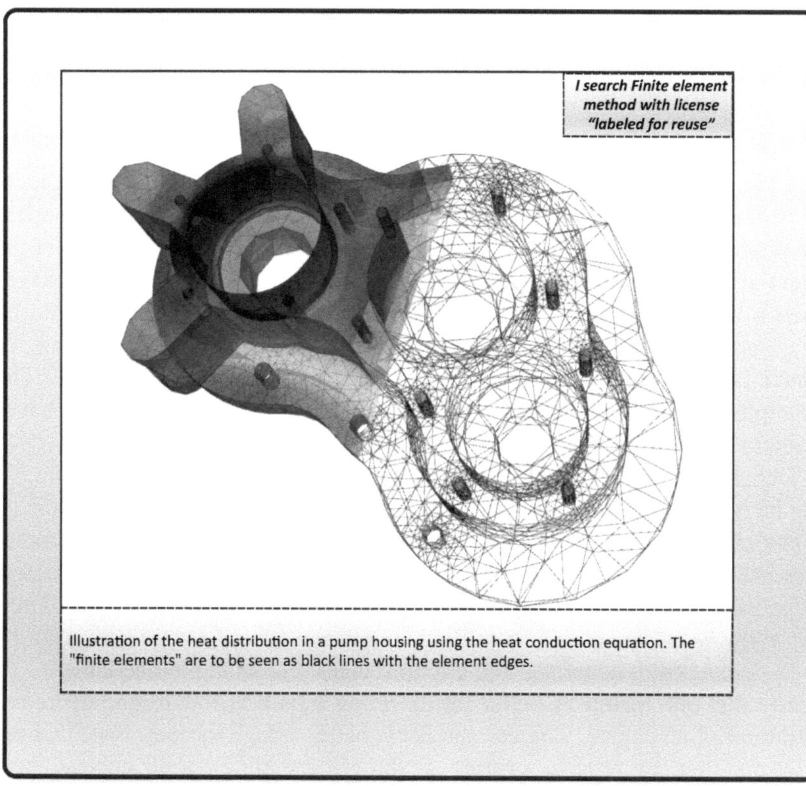

FIGURE 5.23 Computer Model of a mechanical component. *Adapted from Vachtsevanos, G., Lewis, F., Roemer, M., Hess, A., Wu, B., 2006. Intelligent Fault Diagnosis and Prognosis for Engineering Systems, John Wiley & Sons, Inc., Hoboken, New Jersey.*

into finite elements through a process known as meshing (Vachtsevanos et al., 2006). Elements can be studied more easily than the complete structure they are a part of, and each can be considered as a continuous structural member (Rao, 1990). The finite-element method calls for a search of simultaneous solutions to the reaction problem of the elements to applied loads or disturbances (i.e., forces, fields, heat, etc.), constraints (i.e., motion, position, temperature, etc.), or the interaction of adjacent elements (i.e., those connected by means of nodes). The problem is set up so that the solution converges toward the behavior of the total structure (Vachtsevanos et al., 2006).

Finite-element models allow a designer to analyze the static or dynamic response of a system to the application of loads or the presence of disturbances. In structural studies, a model for finite-element analysis (FEA) can be as simple as a single truss experiencing a point load or as complex as a complete engine with moving parts. Today's software tools based on FEA are usually available for a very specific engineering discipline; alternatively, they may be very large and complex program packages useful in a wide range of applications. The first case includes programs such as FASTRAN (Newman, 1992) and FRANC-3D (Cornell Fracture Group, 2003). These use finite elements or results from finite-element studies to simulate mechanical fracture growth. The second features powerful packages such as ANSYS (Swanson and DeSalvo, 1989) and ABAQUS (Vachtsevanos et al., 2006).

5.2.4.2.2 Finite-Element Analysis for Fault Diagnosis

FEA simulations can be applied to many different situations to approximate the behavior of systems when operating conditions vary or when faults appear and progress. Thus, the utility of FEAs for fault diagnosis is that they offer the potential to determine the way in which a condition of the system changes when a fault or anomaly is present. This condition of the system should be measured in the form of a variable value, formally known as state estimate, or should be extracted from a system's signal in the form of a feature. The assumption is, of course, that the state or feature under consideration behaves similarly in both the FEA model and the actual working system.

Depending on the behavior of one or several of these states or features, it may be possible not just to diagnose the presence of a fault but also to assess its magnitude and development stage. In addition, the results from FEA simulations under different operating conditions can be used to compare the relative performance of different states or features under these varying situations. Thus, they allow us to create a list or table that indicates what features best describe the system's health under specific circumstances. For example, the table might specify that one feature is better for detecting a fault at low torque operating conditions of an engine, whereas another is better at high torques. Note that the better qualifier can refer to a variety of things depending on our interest, such as response sensitivity, noise immunity, statistical variance, etc. This process of choosing features depending on the operating conditions is called *feature selection* (Vachtsevanos et al., 2006).

5.3 SUPERVISED, SEMISUPERVISED, AND UNSUPERVISED LEARNING: ISSUES AND CHALLENGES

To understand the nature of semi-supervised learning (SSL), it is useful to take a quick look at supervised and unsupervised learning (Chapelle et al., 2006).

5.3.1 Supervised and Unsupervised Learning

Traditionally, there have been two fundamentally different types of tasks in machine learning (ML): supervised and unsupervised learning.

5.3.1.1 Unsupervised Learning

Unsupervised learning refers to the process of finding structure in unlabeled data resulting in new data representations (including feature representations) and/or clustering data into categories of similar examples, based on such representations (Hinton and Sejnowski, 1999). The presence of unlabeled data distinguishes unsupervised learning from supervised learning and reinforcement learning. Important recent progress has been made in purely unsupervised learning, but these advances tend to be ignored by practitioners who continue using a handful of popular algorithms such as principal component analysis (PCA) and (independent component analysis) ICA (for feature extraction and dimensionality reduction), and K-means, and various hierarchical clustering methods for clustering (Jain et al., 1999).

Unsupervised learning can be expressed as the following: let $X = (x_1,\ldots, x_n)$ be a set of n examples (or points), where $x_i \in X$ for all $i \in [n]: = \{1,\ldots, n\}$. It is typically assumed that the points are independently and identically distributed (i.i.d.) from a common distribution on X. It is often convenient to define the $(n \times d)$-matrix $X = \left(x_i^{\mathrm{T}}\right)_{i \in [n]}^{\mathrm{T}}$ that contains the data points as rows. The goal of unsupervised learning is to find interesting structures in the data X.

It has been argued that the problem of unsupervised learning is fundamentally that of estimating a density, which is likely to have generated X. However, there are also weaker forms of unsupervised learning, such as quantile estimation, clustering, outlier detection, and dimensionality reduction (Chapelle et al., 2006).

Unsupervised learning can be approached from very different perspectives, from psychology and cognitive science to engineering. It is often called "learning without a teacher." This implies that a learning human, animal, or artificial system observes its surroundings and, based on these observations, adapts its behavior without being told to associate given observations to given desired responses (supervised learning) or without even given any hints about the goodness of a given response (reinforcement learning). Usually, the result of unsupervised learning is a new explanation or representation of the observation data, which will then lead to improved future responses or decisions.

In ML and AI, such a representation is a set of concepts and rules between these concepts, which give a symbolic explanation for the data. In ANNs, the representation may be a clustering of the data, a discrete map, or a continuous lower-dimensional manifold in the vector space of observations, which explains their structure and may reveal their underlying causes (Oja, 2002).

Unsupervised learning seems to be the basic mechanism for sensory adaptation, e.g., in the visual pathway (Barlow, 1989). On the engineering side, it is a highly powerful and promising approach to some practical data processing problems such as data mining (DM) and knowledge discovery from very large databases, or new modes of human–computer interactions in which the software adapts to the requirements and habits of the human user by observing his/her behavior.

In neural computation, there are two classical categories for unsupervised learning methods and models: first, extensions of PCA and factor analysis (FA), and second, learning vector coding or clustering methods that are based on competitive learning.

The more recent trend in unsupervised learning is to consider it in the framework of probabilistic generative models. If it is possible to build and estimate a model that explains the data in terms of some latent variables, key insights may be obtained into the true nature and structure of the data. Operations such as prediction and compression may become easier and rigorously justifiable (Oja, 2002).

5.3.1.1.1 Pluses and Minuses of Unsupervised Learning

- Unsupervised learning is more subjective than supervised learning, as there is no simple goal for the analysis, such as prediction of a response.
- Techniques for unsupervised learning are of growing importance in a number of fields:
  - Subgroups of breast cancer patients grouped by their gene expression measurements,
  - Groups of shoppers characterized by their browsing and purchase histories, and
  - Movies grouped by the ratings assigned by movie viewers (James et al., 2013).

5.3.1.1.2 Principal Components Analysis

Principal components analysis (PCA) produces a low-dimensional representation of a data set. It finds a sequence of linear combinations of the variables that have maximal variance and are mutually uncorrelated. Apart from producing derived variables for use in supervised learning problems, PCA also serves as a tool for data visualization.

The first principal component of a set of features $X_1, X_2, \ldots, X_p$ is the normalized linear combination of the features

$$Z_1 = \phi_{11}X_1 + \phi_{21}X_2 + \ldots + \phi_{p1}X_p$$

with the largest variance. By normalized, we mean that $\sum_{j=1}^{p} \phi_{j1}^2 = 1$; the elements $\phi_{11}, \ldots, \phi_{p1}$ are the loadings of the first principal component; together, the loadings make up the principal component loading vector, $\phi_1 = (\phi_{11}, \phi_{21}, \ldots \phi_{p1})^T$.

The loadings are constrained so that the sum of their squares is equal to one; otherwise, setting these elements to be arbitrarily large in absolute value could result in an arbitrarily large variance (James et al., 2013).

5.3.1.1.3 Clustering

Clustering refers to a very broad set of techniques for finding subgroups, or clusters, in a data set. It seeks a partition of the data into distinct groups so that the observations within each group are quite similar to each other. To do this, it is necessary to define what it means for two or more observations to be similar or different. This is often a domain-specific decision based on knowledge of the data being studied (James et al., 2013).

PCA versus clustering:

- PCA looks for a low-dimensional representation of the observations that explains a good fraction of the variance.
- Clustering looks for homogeneous subgroups among the observations.

Two clustering methods:

- *K-means clustering* seeks to partition the observations into a prespecified number of clusters.
- *Hierarchical clustering* does not know in advance how many clusters are required; we end up with a treelike visual representation of the observations, called a *dendrogram*, which allows us to view all clusterings obtained for each possible number of clusters, from 1 to n (James et al., 2013).

5.3.1.2 Supervised Learning

In *supervised learning*, the goal is to learn a mapping from x to y, given a training set made of pairs (x_i, y_i). Here, the $y_i \in Y$ are called the labels or targets of the examples x_i. If the labels are numbers, $y = \left(y_i^T\right)_{i \in [n]}^T$ denotes the column vector of labels. Again, a standard requirement is that the pairs (x_i, y_i) are sampled i.i.d. from some distribution, which, in this case, ranges over $X \times Y$. The task is well defined, as a mapping can be evaluated through its predictive performance on test examples. When $Y = R$ or $Y = R^d$ (or more generally, when the labels are continuous), the task is called regression. Most of this section focuses on classification, i.e., the case where y takes values in a finite set (discrete labels).

There are two families of algorithms for supervised learning. *Generative* algorithms try to model the class-conditional density $p(x|y)$ by some unsupervised learning procedure. A predictive density can then be inferred by applying Bayes theorem:

$$p(y|x) = \frac{p(x|y)p(y)}{\int y\, p(x|y)\, p(y)dy} \tag{5.7}$$

In fact, $p(x|y)p(y) = p(x, y)$ is the joint density of the data, from which pairs (x_i, y_i) can be generated. *Discriminative* algorithms do not try to estimate how the x_i have been generated but concentrate on estimating $p(y|x)$. Some discriminative methods even limit themselves to modeling whether $p(y|x)$ is greater than or less than 0.5; an example is the SVM. It has been argued that discriminative models are more directly aligned with the goal of supervised learning and therefore tend to be more efficient in practice (Chapelle et al., 2006).

Supervised learning is the ML task of inferring a function from supervised training data. The training data consist of a set of training examples. In supervised learning, each example is a pair consisting of an input object (typically a vector) and a desired output value (also called the supervisory signal). A supervised learning algorithm analyzes the training data and produces an inferred function, called a classifier (if the output is discrete, see classification) or a regression function (if the output is continuous, see regression). The inferred function should predict the correct output value for any valid input object. This requires the learning algorithm to generalize from the training data to unseen situations in a "reasonable" way (Wang et al., 2011).

5.3.1.2.1 How Supervised Learning Algorithms Work

Given a set of training examples of the form$\{(x_1, y_1), \ldots, (x_N, y_N)\}$, a learning algorithm seeks a function g: $X \rightarrow Y$, where X is the input space and Y is the output space. The function is an element of some space. The function g is an element of some space of possible functions G, usually called the *hypothesis space*. It is sometimes convenient to represent g using a scoring function f: $X \times Y \rightarrow \mathrm{R}$, such that g is defined as returning the y value that gives the highest score: $g(x) = \operatorname{argmax} f(x, y)$. Let F denote the space of scoring functions. Although G and F can be any space of functions, many learning algorithms are probabilistic models, where g takes the form of a conditional probability model $g(x) = P(x|y)$, or f takes the form of a joint probability model $f(x) = P(x, y)$. For example, naive Bayes and linear discriminant analysis are joint probability models, whereas logistic regression is a conditional probability model.

There are two basic approaches to choosing f or g: empirical risk minimization and structural risk minimization (James et al., 2003). Empirical risk minimization seeks the function that best fits the training data. Structural risk minimization includes a penalty function that controls the bias/variance trade-off.

In both cases, it is assumed that the training set consists of a sample of independent and identically distributed pairs, (x_i, y_i). A loss function, $L: Y \times Y \rightarrow \mathbb{R}^{\geq 0}$, is defined to measure how well a function fits the training data. For training example (x_i, y_i), the loss of predicting the value $\widehat{y}$ is $L(y_i, \widehat{y})$.

The risk $R(g)$ of function g is defined as the expected loss of g. This can be estimated from the training data as $R_{\text{emp}}(g) = \frac{1}{N} \sum_i L(y_i,\ g(x_i))$ (Wang et al., 2011).

5.3.1.2.2 Generalizations of Supervised Learning

The standard supervised learning problem can be generalized in several ways:

1. Semisupervised learning: The desired output values are provided only for a subset of the training data; the remaining data are unlabeled.
2. Active learning: Instead of assuming all training examples are given at the start, active learning algorithms interactively collect new examples, typically by making queries to a human user. Queries are often based on unlabeled data, creating a scenario that combines SSL with active learning.
3. Structured prediction: When the desired output value is a complex object, such as a parse tree or a labeled graph, the standard methods must be extended.
4. Learning to rank: When the input is a set of objects and the desired output is a ranking of those objects, the standard methods must again be extended (Wang et al., 2011).

5.3.1.2.3 Approaches and Algorithms

- Analytical learning
- Artificial neural networks (ANNs)
- Backpropagation
- Boosting
- Bayesian statistics
- Case-based reasoning
- Decision tree learning
- Inductive logic programming
- Gaussian process regression
- Kernel estimators
- Learning automata
- Minimum message length (decision trees, decision graphs, etc.)
- Naive Bayes classifier
- Nearest neighbor algorithm
- Probably approximately correct learning
- Ripple down rules
- Symbolic machine learning (ML) algorithms

- Subsymbolic ML algorithms
- SVMs
- Random forests
- Ensembles of classifiers
- Ordinal classification
- Data preprocessing
- Handling imbalanced data sets
- Statistical relational learning
- Proaftn, a multicriteria classification algorithm (Wang et al., 2011).

5.3.2 Semisupervised Learning

Semisupervised learning is halfway between supervised and unsupervised learning. In addition to unlabeled data, the algorithm is provided with some supervision information—but not necessarily for all examples. This information will frequently be the targets associated with some of the examples. For example, the data set $X = (x_i)_{i \in [n]}$ can be divided into two parts: the points $X_l = (x_1, \ldots, x_l)$, for which labels $Y_l = (y_1, \ldots, y_l)$ are provided, and the points $Xu = (x_{l+1}, \ldots, x_{l+u})$, the labels of which are not known. This is "standard" SSL.

Other forms of partial supervision are possible. For example, there may be constraints, such as "these points have (or do not have) the same target" (Abu-Mostafa, 1995). Other authors see SSL as unsupervised learning guided by constraints. Still others see SSL as supervised learning with additional information on the distribution of the examples x. The latter interpretation seems to be more in line with most applications, when the goal is the same as in supervised learning: to predict a target value for a given x_i. However, this interpretation does not readily apply if the number and nature of the classes are not known in advance but have to be inferred from the data. SSL as unsupervised learning with constraints may be more applicable in such situations.

A problem related to SSL, *transductive learning*, was introduced by Vapnik several decades ago. In this setting, we are given a (labeled) training set and an (unlabeled) test set. The idea of transduction is to perform predictions only for the test points. In contrast, in inductive learning, the goal is to output a prediction function defined on the entire space X. Most methods are transductive, a natural method for inference based on graph representations of data. This issue is addressed again in Section 5.3.2.1.4 (Chapelle et al., 2006).

5.3.2.1 When Can Semisupervised Learning Work?

Is SSL meaningful? More precisely, compared to a supervised algorithm that uses only labeled data, can we hope to have a more accurate prediction by taking into account the unlabeled points? In principle, the answer is "yes." However, there is an important prerequisite: the distribution of examples, which the unlabeled data will help elucidate, must be relevant to the classification problem.

In a more mathematical formulation, we could say that the knowledge on $p(x)$ that we gain through the unlabeled data has to carry information that is useful in the inference of $p(y|x)$. If this is not the case, SSL will not yield an improvement over supervised learning. Using the unlabeled data may even degrade the prediction accuracy by misguiding the inference.

For SSL to work, certain assumptions have to hold. In this context, plain supervised learning also has to rely on assumptions. One of the most popular such assumptions can be formulated as follows:

Smoothness assumption of supervised learning: If two points, x_1, x_2, are close, the corresponding outputs, y_1, y_2, should also be close.

Clearly, without such assumptions, it would not be possible to generalize from a finite training set to a set of possibly infinite unseen test cases (Chapelle et al., 2006).

5.3.2.1.1 Semisupervised Smoothness Assumption

A generalization of the smoothness assumption that is useful for SSL is called, naturally enough, the semisupervised smoothness assumption. While, in the supervised case, the output varies smoothly with the distance, we now also take into account the density of the inputs. The assumption is that the label function is smoother in high-density regions than in low-density regions:

Semisupervised smoothness assumption: If two points, x_1, x_2, in a high-density region are close, then so should be the corresponding outputs, y_1, y_2.

Note that by transitivity, this assumption implies that if two points are linked by a path of high density (e.g., if they belong to the same cluster), their outputs are likely to be close. If, on the other hand, they are separated by a low-density region, their outputs need not be close.

Note also that the semisupervised smoothness assumption applies to both regression and classification. In the next section, we show that in classification, it reduces to assumptions commonly used in SSL. At present, it is less clear how useful the assumption is for regression problems (Chapelle et al., 2006).

5.3.2.1.2 Cluster Assumption

Suppose we know that the points of each class tend to form a cluster. In this case, the unlabeled data can help us find the boundary of each cluster more accurately: we could run a clustering algorithm and use the labeled points to assign a class to each cluster. That is, in fact, one of the earliest forms of SSL. The underlying, now classical, assumption may be stated as follows:

Cluster assumption: If points are in the same cluster, they are likely to be of the same class.

This assumption may be considered reasonable on the basis of the sheer existence of classes: if there is a densely populated continuum of objects, it may seem unlikely that they were ever distinguished into different classes.

Note that the cluster assumption does not imply that each class forms a single, compact cluster: it only means we usually cannot observe objects of two distinct classes in the same cluster.

The cluster assumption can easily be seen as a special case of the above-proposed semisupervised smoothness assumption, considering that clusters are frequently defined as sets of points connected by short curves, which traverse only high-density regions. Therefore the cluster assumption can be formulated in an equivalent way as:

Low density separation: The decision boundary should lie in a low-density region.

The equivalence is easy to see: a decision boundary in a high-density region would cut a cluster into two different classes, and many objects of different classes in the same cluster would require the decision boundary to cut the cluster, i.e., to go through a high-density region.

Although the two formulations are conceptually equivalent, they can inspire different algorithms. The low-density version also gives additional insight into why the assumption is sensible in many real-world problems. Consider digit recognition, for instance, and suppose we want to learn how to distinguish a handwritten digit "0" from digit "1." A sample point taken from the decision boundary will be between a 0 and a 1, most likely a digit looking like a very elongated zero. But the probability that someone has written this "weird" digit is very small (Chapelle et al., 2006).

5.3.2.1.3 Manifold Assumption

A different but related assumption that forms the basis of several SSL methods is the manifold assumption:

Manifold assumption: The (high-dimensional) data lie (roughly) on a low-dimensional manifold.

How can this be useful? A well-known problem of many statistical methods and learning algorithms is the curse of dimensionality. Volume grows exponentially with the number of dimensions, and an exponentially growing number of examples is required for statistical tasks such as the reliable estimation of densities. This problem directly affects generative approaches based on density estimates in input space. A related problem of high dimensions, which may be more severe for discriminative methods, is that pair-wise distances tend to become more similar and thus less expressive. If the data happen to lie on a low-dimensional manifold, however, the learning algorithm can essentially operate in a space of corresponding dimension, thus avoiding the curse of dimensionality.

As above, we can argue that algorithms working with manifolds may be seen as approximately implementing the semisupervised smoothness assumption: such algorithms use the metric of the manifold to compute geodesic distances. If the manifold is viewed as an approximation of the high-density regions, it becomes clear that, in this case, the semisupervised smoothness assumption reduces to the standard smoothness assumption of supervised learning, applied on the manifold.

Note that if the manifold is embedded into the high-dimensional input space in a curved fashion (i.e., it is not just a subspace), geodesic distances differ from those in the input space. By ensuring more accurate density estimates and more appropriate distances, the manifold assumption may be useful for classification, as well as for regression (Chapelle et al., 2006).

5.3.2.1.4 Transduction

As mentioned, some algorithms naturally operate in a transductive setting. According to Vapnik, high-dimensional estimation problems should attempt to follow the following principle:

Vapnik's principle: When trying to solve a problem, we should not solve a more difficult problem as an intermediate step.

As an example, consider supervised learning, where predictions of labels y corresponding to some objects x are desired. Generative models estimate the density of x as an intermediate step, while discriminative methods directly estimate the labels.

In a similar way, if label predictions are only required for a given test set, transduction can be argued to be more direct than induction: while an inductive method infers a function f: $X \rightarrow Y$ on the entire space X and afterward returns the evaluations $f(x_i)$ at the test points, transduction consists of directly estimating the finite set of test labels, i.e., a function f: $X_u \rightarrow Y$ only defined on the test set. Note that transduction (as defined in this book) is not the same as SSL: some semisupervised algorithms are transductive, but others are inductive.

Now suppose we are given a transductive algorithm that produces a solution superior to an inductive algorithm trained on the same labeled data (but discarding the unlabeled data). In this case, the performance difference may be due to one of the following two points (or a combination thereof):

1. Transduction follows Vapnik's principle more closely than induction does, or
2. The transductive algorithm takes advantage of the unlabeled data in a way similar to SSL algorithms.

There is ample evidence of improvements due to the second of these points, but at present, we also find empirical results that selectively support the first point. In particular, the evaluation of the related benchmark suggests the systematic advantage of transductive methods. However, the properties of transduction remain a topic of debate (Chapelle et al., 2006).

5.4 NO FAULT FOUND (NFF) AND ISSUES OF COMPLEX SYSTEMS

No-Fault-Found (NFF) events occur when a system level test, such as Built—In Test (BIT), indicates a failure but no such failure is found during repair. With more electronics continuously monitored by BIT, it is more likely that an intermittent glitch will trigger a call for a maintenance action resulting in NFF. NFFs are often confused with false alarm, cannot duplicate (CND), or retest OK (RTOK) events, but as we go on to show, they are actually caused by FAs, cannot duplicate events (CNDs), Retest OK events (RTOKs), and a number of other complications. Attempting to repair NFFs wastes precious resources, compromises confidence in the product, and creates customer dissatisfaction; in addition, the repair quality remains a mystery. The problem is compounded by previous work showing that most failure indications calling for repair action are invalid. NFFs can be caused by real failures or may be a result of false alarms. Understanding the cause of the problem may help distinguish between units under test (UUTs) that can be repaired and those that cannot. In calculating the true cost of repair, we must take into account the effort wasted in attempting to repair unrepairable UUTs (Ungar, 2015).

A typical maintenance process within an organization can be observed in Fig. 5.24. It is important to understand the concept of how NFF instances can manifest themselves at various levels. When an operator records a system error, maintenance personnel are notified; they will investigate the reason for the system malfunction. If no causes are discovered, the failure is tagged "NFF." There are various reasons for NFFs. Perhaps the operator (or maintainer) lacks knowledge of the system or has received insufficient support to carry out the fault diagnosis. He or she may have minimal understanding of the manuals, or there may be a lack of equipment or other operational pressures. NFFs are predominantly associated with electronic equipment (Soderholm, 2007; Pecht, 2006) but can also affect mechanical systems.

The common causes of NFF events can be categorized as: technical, organizational, procedural, and behavioral. Technical causes involve things such as undefined or limited performance measures, inadequate information on operating environment, designs unsuitable for robust testing, or a lack of robust fault models. Organizational and procedural issues are driven by the business and commercial environment and encompass a lack of training and training tools, time pressures on maintenance operations, and organizational cultures with no cross-functionality; other problems include a lack of commitment or a lack of shared information and knowledge between designers, manufacturers, service providers, and operators. Behavioral causes, commonly called "human factors," arise directly from the maintenance personnel themselves; staff may show discrepancies in test procedures, report faults incorrectly or incoherently, or apply the wrong processes. There may

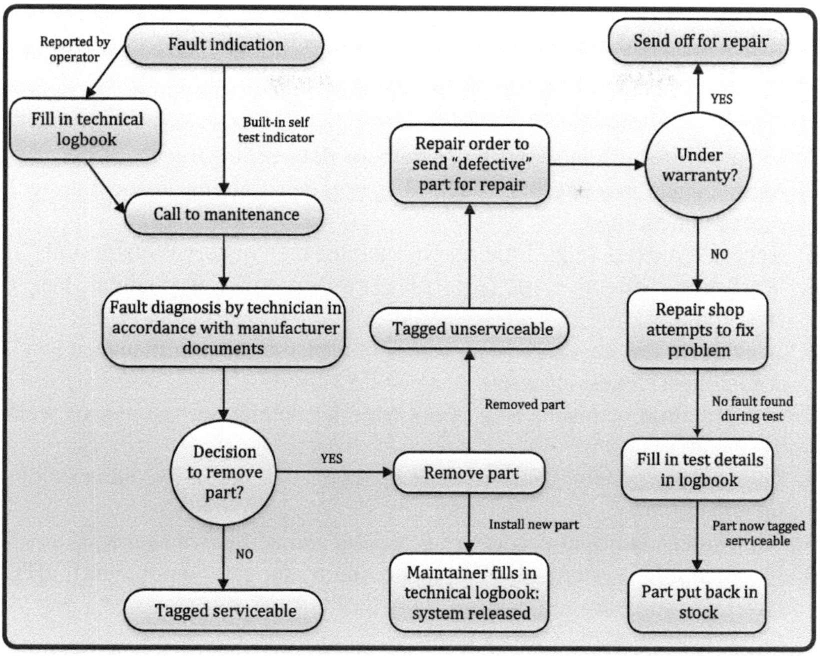

FIGURE 5.24 Simplified repair process during maintenance action. *Adapted from Khan, S., Phillips, P., Jennions, I., Hockley, C., 2014. No fault found events in maintenance engineering part 1: current trends, implications and organizational practices. Reliability Engineering and System Safety 123, 183–195.*

also be a lack of communication between maintenance personnel and other experts.

Khan et al. (2012) attempt to address the scale of the problem as a sequence of events. The sequence begins during operational service when a BIT fails (or the operator reports the possibility of an error); independent functionality tests are then triggered to verify the fault/failure. If it cannot be repeated, a failure to diagnose the problem is recorded. If the functionality tests fail, further off-line tests in the maintenance shop are used to slot the system fault into a group of shop replaceable units (SRUs) suspected of being the source of the line replaceable unit failure. Accurate diagnosis at this level will identify one SRU, but less precise diagnostics might call out two, three, or more SRUs. The components are then sent back to the depot/workshop for functional testing using automatic test equipment (ATE); these tests will conclude if the component is healthy and falsely reported, or definitely faulty and the diagnostic testing is inadequate (Ungar and Kirkland, 2008).

James (2003) investigates various NFF situations, focusing on developing practical guidance for designers and project managers to facilitate understanding of unknown failures and procedural improvements. The research gives a comprehensive breakdown of the potential reasons for NFF events:

1. Operator policies (e.g., short turn round times, availability of spares, different priorities).
2. Failure in recording/reporting (e.g., poor data coding).
3. Maintenance practice (e.g., lack of maintainer training, inaccurate technical publications).
4. Repeat removals (e.g., little use of maintenance history, "rogue units").
5. Workshop ineffectiveness (e.g., pressure to produce throughput, lack of staff training).
6. Inadequate test coverage (e.g., test philosophy across maintenance levels, comprehensiveness of tests).
7. Interpretation of results (e.g., fault code interpretation, training of workshop staff).
8. Intermittent system connection (e.g., connector integrity, harness/loom integrity).
9. Intermittent faults in product (e.g., solder joints, PCB weakness).
10. Incompatible system design (e.g., (built in test equipment) BITE coverage, software tolerances).

A great many diagnostic failure events of interest occur in electrical and electromechanical systems, but similar events occur in mechanical systems. In both systems, the causes of failures include aging, poor maintenance, and incorrect installation or usage. In mechanical systems, because it is easier to predict the effects of mechanical errors on systems operation, inspection criteria can be developed during the design phase. It should be noted that, like electrical failures, mechanical failures can be intermittent and occur under specific operating conditions (Khan et al., 2012).

In electronic systems, loose connections cover most of the fault cases published as NFF. Some of the more common and well known include poor solder joints, faulty electrical components, damaged PCBs, and problems with internal wiring. Electronic failures are not considered static (or random) events but a process of mechanical and material changes. These changes will not always lead to a loss of operation for the system, even though their components are no longer within specified parameters. Electronics have an inherent self-compensating aspect, which makes the task of diagnostics difficult. In addition, the degradation of failure modes will manifest differently depending on the operating environment and may offset one another depending on circuit configuration (Line and Krishnan, 2008).

5.4.1 Introduction to NFF

System level tests come in various forms and are run for many different reasons. In production, a system test is run to gain confidence that the product is ready for the end user. It is also used to ensure continued operation and is implemented in the form of a BIT. By implication, a system test can also be considered to be running when the end user is performing normal system operation. He/she may observe anomalies and inconsistencies, leading to a call for a repair action. System level tests encompass all these forms. When a system level test fails, one or more subsystem UUT is suspected of being at the root of the system failure. The repair at the system level consists of replacing the suspected UUT or UUTs and sending replaced UUTs to a depot-level repair facility, often the factory (Ungar, 2015).

Fig. 5.25 shows the outcome of failures found by system level tests; they include persistent faults (PFs) and NFFs. PFs, sometimes called confirmed faults, cause the system level to fail and will cause the UUT at the depot to also fail. We call them PFs to indicate they persist into the depot. In contrast, NFFs will pass the UUT test at the depot. There are two categories of NFFs, intermittent failures and false alarms (Ungar, 2015). As explained by Ungar (2014) the majority of system test failures result from system-level false alarms.

A common example of a failure that escapes a system level test, creating NFFs at the system level, is a computer hang-up. It is "repaired" by rebooting the software. No maintenance action is taken, and none of the subsystems is returned to the depot or factory, so the NFF does not penetrate to the depot. Unless the problem repeats, it is treated as a normal anomaly, and what would probably result in NFF at the depot is avoided.

To avoid confusion, it is important to understand the terminology. No-fault-found is a situation where the UUT passes the first test at the depot test station. Intermittent failures (IFs) are real failures that are exposed only under certain conditions. When they are not exposed, they result in NFFs (Ungar, 2015).

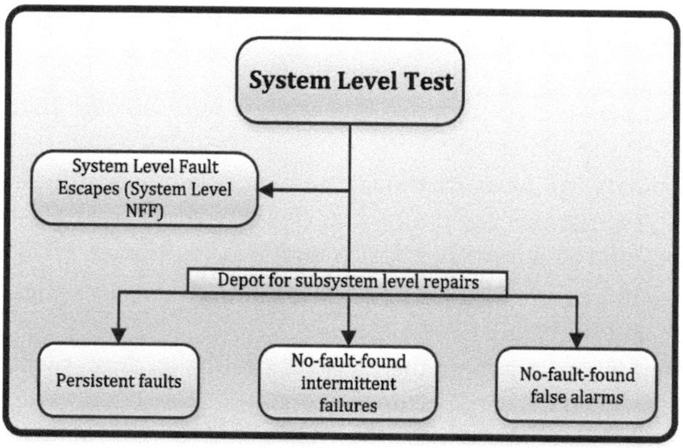

FIGURE 5.25 Outcome of failures found at system level. *Adapted from Ungar, L.Y., February 2015. Causes and Costs of No Fault Found Events, Advanced Test Engineering (A.T.E.) Solutions, Inc., El Segundo, CA.*

False alarms are indications at the system level of a failure where no fault exists (US Navy and MIL-HDBK-2165, 1995). Alternatively, an false alarm can be defined as a call for maintenance action when none is needed (Ungar, 2014). System-level false alarms may send some subassemblies to the depot for repair, or if the result is questioned, the same system level test is run again to gain confidence in the result. When the system level test is run several times, it increases the likelihood of distinguishing FAs from IFs, making it more likely that the UUT returned to the depot is a result of IF (Ungar, 2015).

The repeated tests can result in CNDs or inconsistent results of two or more tests. In response to CNDs, either no repair action is taken, or some subassembly UUTs are sent to repair. When the UUTs are tested at the depot, they may pass the test, a condition called RTOK. A UUT that malfunctions in a specific manner during system level testing, but performs that same function satisfactorily at the next level maintenance facility, is called an RTOK (Ungar and Kirkland, 2008). All RTOKs are NFFs, but not all NFFs are alike. Some are the result of FAs that cannot be repaired (because they are not really faulty), while others may be the result of IFs, which are real and repairable. It is important to know the difference (Ungar, 2015).

Another contribution to NFFs may come from false rejects (FRs). In some instances, tolerances, accuracies, calibration, and other uncertainties within the system level tester cause the test to fail. In essence, perfectly good systems are judged to be faulty and, consequently, UUTs are sent to the depot. These UUTs are not faulty and will likely pass the depot test as NFF. Since no fault exists, these are also FAs. FRs, however, can also be created by system-level interaction failures that do not have an analogous cause within a single UUT. Say, for example, due to some environmentally induced condition, a system failure is detected, causing one or more UUT to be sent to the depot for repair. The failure may require specific interaction of UUTs or it may require a specific environment, neither of which can be created in the depot test. The result is NFF, but it is possible that different UUTs in the system would not have failed the same environmental condition. This implies the UUT is flawed and, in a strict sense, repairable. In this case, it will fall into the IF category rather than the false alarm one.

It is necessary to have methods to distinguish between IFs and FAs. Repairing UUTs that are not faulty (i.e., FAs) is futile, and the only logical corrective action is to return such UUTs to the system level. UUTs from FRs, however, may not be clearly IFs of FAs; returning them to system level may pose damage and perhaps safety issues.

In this section, we examine the causes of NFFs to draw such distinctions when possible. We try to predict how many NFFs are IFs and how many are FAs. We evaluate the cost of repairing NFFs, recognizing that many, indeed most, are unrepairable, and our time and resources are being wasted when we attempt to do so. Finally, we consider various corrective strategies to deal with NFFs at the depot and calculate their economic performances (Ungar, 2015).

5.4.2 What Causes NFFs?

UUTs at a depot repair facility experiencing NFF are caused by one or more of the following:

- CNDs, including both real failures and false alarms
- RTOKs
- Measurement uncertainty and test equipment problems
- System level failures not detectable at the subsystem UUT level
- Intermittent failures

In the following, we explore each of these causes to determine whether the resulting UUT experiencing NFF is repairable at the depot (IF) or not (FA).

5.4.2.1 Cannot Duplicate

A CND can occur only when a test or a condition is repeatedly executed. Even a single disagreement in the results is considered a CND. The following failures can cause CND events (Simpson, 1986):

1. Operator error
2. Latent BIT design error
3. Environmentally induced BIT error
4. BIT transient failure
5. BIT hard failure
6. Latent design error manifestation
7. Transient (functional) failure
8. Environmentally induced functional error
9. Test equipment failure
10. Human error at shop level

Situations 1, 2,and 6 are FAs and are not repairable at the depot. Situation 1—operator error—may actually be the greatest contributor to FAs. A complex system requiring constant interaction with a human operator can conceivably find that operator pushing the wrong buttons or at the wrong time creating responses the operator does not expect. The operator can misinterpret such unexpected responses and (mistakenly) call for a repair action. Situation 6—latent design error manifestation—may be a subtle design flaw that only exhibits itself in certain situations, often intermittently. This, however, is not an intermittent fault because there really is no faulty circuit element. The product does what it was designed to do—though not necessarily what we want it to do.

Another manifestation of this CND situation can be caused by a sneak circuit, an unexpected path or logic flow within a system, which can sometimes initiate an undesired function or inhibit a desired function. It can be caused by software, hardware, end users, or a combination of these elements, but because it is the way the circuit was designed, it is not a fault that can be repaired. Sneak circuits are design errors—situations that are difficult to uncover. The more functions performed by the system, the more likely that a sneak circuit will sneak up and create an FA. Here is an illustrative example. Suppose we have an electronic circuit whose job is to tell the time when the user provides the position of the small and large hands on a classic analog clock. So when the user enters "small hand on 5, large hand on 9," the circuit states "the time is 5:45." But due to a design error, when the user enters "small hand on 12, large hand on 12," the circuit responds "the time is 12:00 noon." This will be correct whenever the circuit is used during the day but incorrect if it is used at midnight! It is important to recognize, however, that the uncovered error cannot be corrected by sending the circuit to the repair depot, and it is therefore an FA.

Environmentally and transient-induced failures in situations 3, 4, and 8 may be caused by circuit failure behavior but are usually temporary and no UUT repair is warranted. NFFs caused by these conditions are clearly FAs.

Test equipment and support personnel errors, situations 9 and 10, send nonrepairable UUTs to depot and should be considered FAs.

CND failures in BIT and in circuit function, situations 5 and 7, are most likely repairable (Ungar et al., 2015).

5.4.2.2 Retest OK–Produced False Alarms

Typically, RTOK is a result of a failure of BIT at the system level and subsequent passing of ATE at the depot level. Since RTOK compares the results of two distinctly different tests, RTOK NFFs result from different sources. The following RTOK events produce NFFs:

- *Misdiagnosis or false diagnosis* is a situation where the test incorrectly identifies the replaceable unit containing the fault. Such events are FAs.
- A more common source of RTOK NFF is *diagnostic ambiguity (DA)*. This situation arises when the test is able to detect a fault, but the smallest fault group to which that fault can be isolated contains more than one repair item. In this case, in addition to faulty UUTs, some good units will be delivered to the next level of assembly and when these good UUTs pass the depot ATE test, they will be rendered NFFs. If all the UUTs in the same system experience NFF, it cannot be determined which one is due to an intermittent failure, so all will be considered to have intermittent failures (Ungar, 2015).

5.4.2.3 Measurement Uncertainties and Test Equipment Problems

Measurement uncertainties and test equipment problems can create misleading test results. NFF could be caused by instruments on the tester rendering a good system as faulty, a concept referred to as FR. An FR may be due to incorrect selection or calibration of tolerances and test accuracy ratios, or it may be due to other test equipment problems. The FR will produce an NFF at the depot, and this will be an FA (Dobbert, 2012).

Similarly, the test equipment at the depot could experience a false accept (FX) condition. Like the FR, an FX is created by incorrect selection or calibration of tolerances and test accuracy ratios or it may be caused by other test equipment problems. An FX is an NFF that contains a failure, but one that the depot tester cannot confirm. Returning the UUT to the system will continue to show it as faulty, but it does not fit either the FA or IF category exactly (Ungar, 2015).

5.4.2.4 System Level Failures That are not Detectable at the Subsystem Unit Under Test Level

Environmental conditions can play havoc with the UUT circuit, with the tester and with interpretation of BIT or other monitoring techniques. A good example is when avionics fail at high altitudes, but those failures cannot be repeated at the depot. In these cases, it is not clear whether the UUT is experiencing IF or FA in the system, but since no such determination can be made at the depot, the UUT must be treated as FA. Attempts are sometimes made to simulate the environment by environmentally stressing the UUT, but there is no assurance that such actions result in replicating the failure.

Another common example is when a cable is temporarily or permanently disconnected within the system. The failure is not related to a single UUT, and all UUTs returned to the depot, as a result of this condition, will be FAs (Ungar, 2015).

5.4.3 Classifying Depot Level Repair Causes

As you will recall, Fig. 5.25 (above) identifies three categories of UUTs arriving at the depot, persistent faults, NNF IFs, and NFF FAs. Table 5.1 shows the causes that would place a UUT in one of the categories. It also classifies UUTs as "indeterminate" when it is not clear which category the UUT fits best.

To differentiate between IFs and FAs, we will explore each one in detail in the following sections.

5.4.3.1 Intermittent Failures

Fig. 5.26 is a wishbone diagram showing the cause and effect of intermittent failures in electronic assemblies (Qi et al., 2008). It details many of the factors that could create NFF IFs (Ungar, 2015).

A number of reports deal with NFFs caused by IFs. For example, Sorensen (2003) suggests that only 50% of all failures fall into the persistent failure category and IFs causing NFFs represent the other 50%. Since not all NFFs are caused by IFs, it is not clear from this analysis, where FAs should be placed. Other studies are equally ambiguous, often using the terms "IF" and "NFF" interchangeably.

TABLE 5.1 Classification of NFF Sources

Persistent Faults	No-Fault-Found IF	No-Fault-Found False Alarms	Indeterminate
Assembly faults within each UUT—shorts, opens, missing components, wrong component, wrong component insertion	CND BIT Hard failure	CND operator error	False accept conditions that occur when the depot tester inaccuracy renders a UUT as faulty. The UUT may in fact be faulty or not.
Parts faults—bad component, out of spec component	CND transient (functional) failure	CND Latent BIT design error	System level failures not detectable at subsystem UUT level. Usually FA, but could be IF.
Functional failures from hard faults	RTOK DA of all UUTs from same system if all UUTs are NFF	CND Latent design error manifestation	Environmentally induced failures at operating conditions that are not available at the depot. e.g., failures at high altitudes only.
	Connectors—corrosion and wiring failures	CND environmentally Induced BIT error	Some system level cable becomes disconnected temporarily, perhaps due to corrosion.
	Components—IC crosstalk, electromigration, dielectric breakdown, metallization, ionic contamination, tin whiskers, wire bond lifts, creep corrosion, etc.	CND BIT transient failure	
	PCB—electrochemical migration via cracking, warpage, and whiskers.	CND environmentally Induced functional error	

	Component–PCB interactions—solder problems, loads, manufacturing defects, socket problems	CND flight line test equipment error	
		CND flight line human error	
		CND test equipment failure	
		CND human error at shop level	
		RTOK misdiagnoses or false diagnosis	
		RTOK DA of 2nd, 3rd, etc. UUTs sent to depot if 1st UUT is not NFF	
		False reject—measurement tolerances and uncertainties causing a good system to fail system test.	

CND, cannot duplicate; *DA*, diagnostic ambiguity; *IC*, integrated circuit; *IF*, intermittent failures; *PCB*, printed circuit board; *RTOK*, retest OK; *UUT*, unit under test.
Adapted from Ungar, L.Y., February 2015. Causes and Costs of No Fault Found Events, Advanced Test Engineering (A.T.E.) Solutions, Inc., El Segundo, CA.

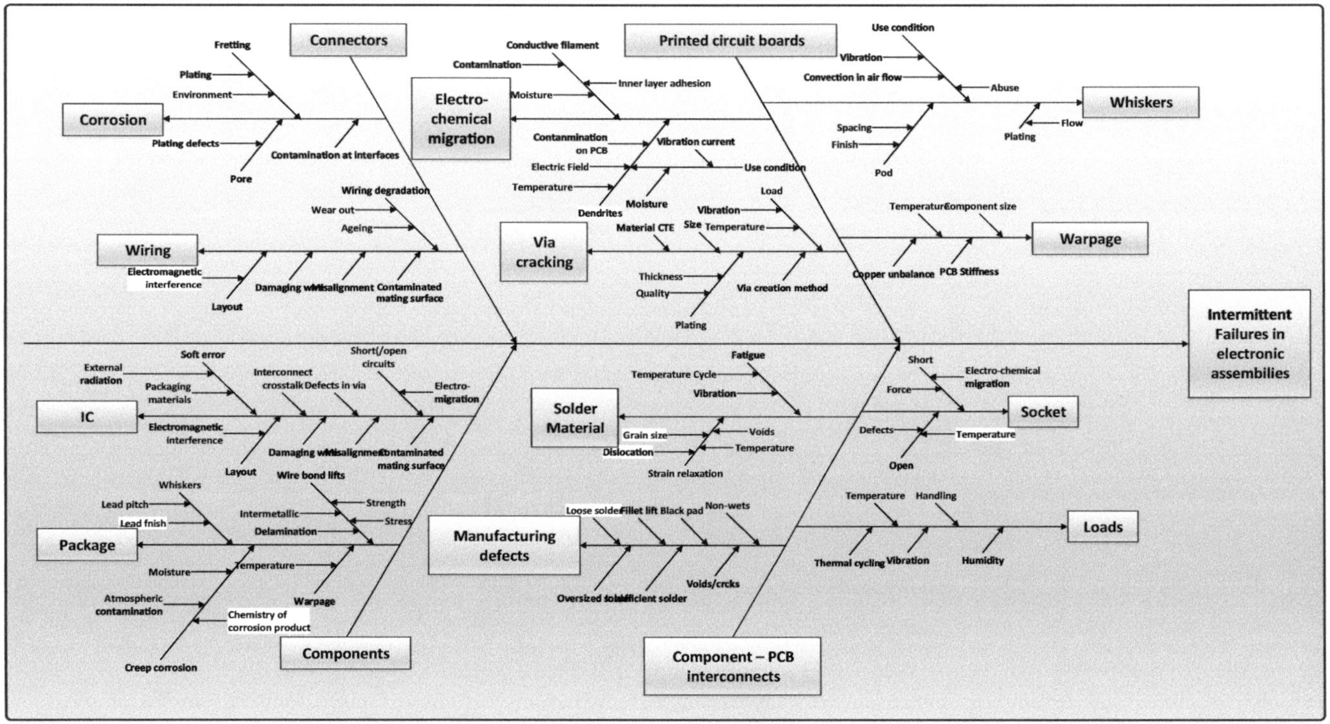

FIGURE 5.26 Cause-and-effect diagram of intermittent failures in electronic assemblies. *CTE*, coefficient of thermal expansion; *IC*, integrated circuit; *PCB*, printed circuit board. *Adapted from Qi, H., Ganesan, S., Pecht, M., 2008. No-fault-found and intermittent failures in electronic products. Microelectronics Reliability 48, 663–674 (Elsevier Ltd).*

Perhaps a better understanding of FAs will yield some distinguishing characteristics (Ungar, 2015).

5.4.3.2 False Alarms

FAs are caused by statistical occurrences of false positives (FPs) (Ungar, 2014). Four possible outcomes to a test are shown in Table 5.2. A graphical representation of the Gaussian distribution is shown in Fig. 5.27.

FP or type I error is an indication of a failure when, in fact, none exists.

TABLE 5.2 Possible Test Outcomes

Test Outcome	Label	Typical Percentage (%) of Cases
True negative—Good UUT passes all tests.	TN	95
True positive—Faulty UUT fails one or more tests.	TP	95
False negative—A faulty UUT results in UUT passing the test. Also called "missed fault" or type II error.	FN	5
False positive—A good UUT indicates that the test failed. Also called "false alarm" or type I error.	FP	5

UUT, unit under test
Ungar, L.Y., 2014. The true nature of false alarms. AUTOTEST '14 (IEEE Press).

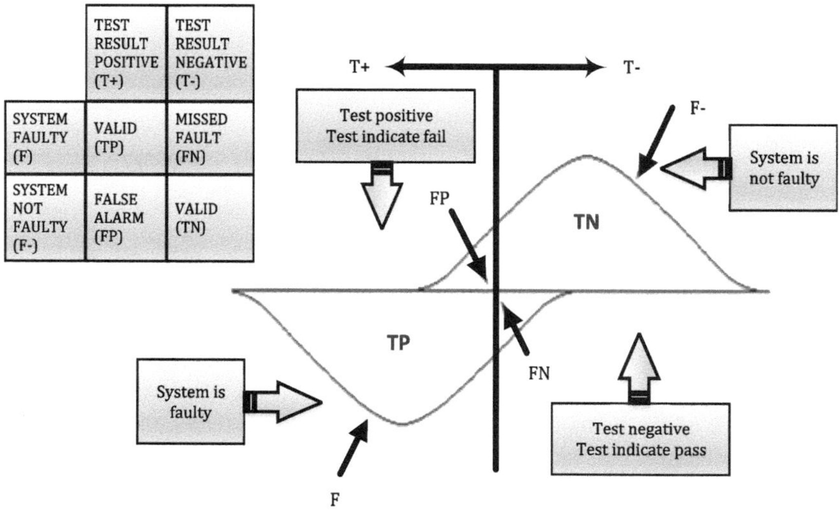

FIGURE 5.27 Graphical representation of possible test outcomes. *FN*, false negative; *FP*, false positive; *TN*, true negative; *TP*, true positive. *Adapted from Ungar, L.Y., 2014. The true nature of false alarms. AUTOTEST '14 (IEEE Press).*

While FN, or type II errors are equal to FP in the Table 5.2 example and Fig. 5.27 illustration, this relationship is not necessary. A coupling relationship between FP and FN, however, does exist. If we were to increase the fault coverage, true positive, the origin of the T+ and T− axes would shift to the right in the figure, reducing the area under FN. This makes sense, as increasing fault coverage should lower the number of missed faults. Because of the coupling between FP and FN, however, reducing the area under FN will also increase FP. The implication that a more comprehensive test creates more FPs appears nonsensical at first, but the implication is confirmed when we consider that tighter measurements result in higher numbers of anomalies.

While FPs may constitute only a small percentage of possible test outcomes (5% in Table 5.2), they can be responsible for a high percentage of the faulty results being invalid. Ungar (2014) shows that even with 5% FPs, the invalid faulty result (i.e., false alarms) can be as high as 77%. This implies that a system sent back to repair is actually fault free in 77% of the cases, and they will all result in NFFs (Ungar, 2015).

Using Bayes formula, Ungar (2014) reveals the following impact of FA. The probability of a valid faulty result, P(F/TP), is calculated as

$$P(F/TP) = \frac{P(TP/F) * P(F)}{P(TP/F) * P(F) + P(TP/F-) * P(F-)} \tag{5.8}$$

where, $P(F)$ is the probability that the system is faulty; $P(F-)$ is the probability that the system is not faulty; clearly, $P(F-) = 1 - P(F)$; $P(TP/F)$ is the probability that a "system faulty" result is caused by the existence of a fault; $P(TP/F-)$ is the probability that a system which is not faulty is still called faulty.

It also follows that if P(F/TP) is the probability of the valid faulty result, then $P(F-/TP)$ is the probability of an invalid faulty result, another way of saying false alarm. This can be quantified as

$$\text{Probability of False Alarm} = P(F - /TP) = 1 - P(F/TP) \tag{5.9}$$

As an example, Malcolm (1983) uses the following assumptions to calculate FA:

- Ninety-five percent fault detection will result in having 95% of the faulty system fail the test. Thus, $P(TP/F) = 0.95$.
- Ten percent of the systems diagnosed as faulty are, in fact, good. Therefore fraction of false alarms (FFA) = $P(TP/F-) = 0.10$.
- Only 3% of the items tested are faulty. $P(F) = 0.03$. Thus, the probability that a system is not faulty, $P(F-) = 0.97$.

Then the probability of a truly faulty system resulting from a faulty test, $P(F/TP)$ is

$$P(F/TP) = \frac{(0.95)(0.03)}{[(0.95)(0.03) + (0.1)(0.97)]} = 0.23 \tag{5.10}$$

The probability of invalid faulty result (FA) is

$$P(F-/TP) = 1 - P(F/TP) = 1 - 0.23 = 77\% \tag{5.11}$$

This implies that in 77% of the cases when a system test fails, the UUTs sent to the depot are FAs. The number of UUTs with FAs is actually greater because of diagnostic ambiguity and, in some cases, more than one UUT is sent to repair. Reducing the FFA in the above example to 2% (a typical specification, but hard to achieve for many systems), the probability of a valid faulty result in Eq. (5.10) would increase to about 0.60 and FA in Eq. (5.11) would decrease to 40% (Ungar, 2015).

5.4.3.3 All NFFs

Say we know the quantitative distribution of the categories of faults depicted in Fig. 5.25 and Table 5.1. We estimate from Eq. (5.11) and from the discussion above that FA can range from about 40% to 77%. We approximate FA to be generated by 60% of the systems that fail system tests (Ungar, 2015).

Empirical data support such high FA results. For example, for the military aircraft *F/A−18 A/B/C/D* shown in the table, 75% of the CNDs result from false alarms. In some cases, the units returned due to false alarms exceed 83%. This is not unusual; in the United States alone, O-Level and I-Level wasted maintenance labor resulting from BIT false alarms during 1999 resulted in unnecessary aircraft downtime of 2.96 years (Office of the Assistant Secretary of the Navy, 2001). Other sources show similar results, with one reporting 100% false alarm for a particular system (Hecht, 2007).

Using the diagnostic resolution of 90% isolation to a single module, 95% to two or less modules, and 100% to three or less modules, we can express 100 system failures as

$$\begin{aligned}\text{All modules removed} &= 100 + 2*(100-95) + 3(100-10)\\ &= 140\text{UUTs}/100 \text{ system failures}\end{aligned} \tag{5.12}$$

Assuming 60% of system failures are FAs and 100% of all ambiguous replacements are also FAs, we can also state the following

$$\begin{aligned}\text{FA modules removed} &= (60\% \text{ of } 100) + 100\%\ (2*5) + 100\%(3*10)\\ &= 100 \text{ FA UUTs}/100 \text{ system failures}\end{aligned} \tag{5.13}$$

$$\text{Percentage of FA UUTs in the depot} = 100/140 = 71.4\% \text{ of UUTs} \tag{5.14}$$

$$\text{Valid faulty modules} = 140 - 100 = 40 \text{ UUTs have persistent faults or IFs} \tag{5.15}$$

$$\text{Percentage of non} - \text{FA UUTs in the depot} = 40/140 = 28.6\% \text{ of UUTs} \tag{5.16}$$

From the above, we can predict the probable distribution of UUTs arriving at the depot repair as shown in Table 5.3. By implication, while 28.6% of the

TABLE 5.3 Distribution of Units Under Test Arriving at Depot Repair

Fault Type	Derivation	Probable UUT Distribution (%)
No-fault-found false Alarms	Eq. (5.8–5.14)	71.4
Not false Alarm faults (not FAs)	Eq. (5.16)	28.6
Persistent faults	50% of non-FAs per (Sorensen, 2003)	14.3
No fault found intermittent faults (NFF IFs)	50% of non-FAs per (Sorensen, 2003)	14.3

Adapted From Ungar, L.Y., February 2015. Causes and Costs of No Fault Found Events, Advanced Test Engineering (A.T.E.) Solutions, Inc., El Segundo, CA.

UUTs are faulty (not FA), only half of those, or 14.8% of the UUTs, will fail the first time depot repair (Ungar, 2015).

Kimseng (1999) studies intermittent failure in cruise control modules used in automobiles. The empirical data show 96% of the modules returned to the manufacturer because of customer complaints actually operated properly and passed the bench test. The study concludes the depot test is "inappropriate" because it does not test the UUTs in a real automotive environment. It is possible that the conclusions misinterpreted what the data suggest. When we include FAs, the 71.4% from Eq. (5.14) is not far from the 96% result. Considering that cruise control modules can be easily misused or misinterpreted by the driver (operator error), FA rates would likely be higher, and if we also consider IFs, the NFF of 96% may well represent the actual NFF rate. FA already takes into account the "inappropriateness" of the test environment and is a real and expected cause of NFF.

A survey of mobile phone NFFs in 2005 (Overton, 2005) concludes that 63% of devices returned had no faults. They were categorized as "struggling with functionality/usability" in 24% of the cases, as "device miss old/does not meet expectations" in 8% of the FA cases and as "handset configuration/setting issue" in the remaining 31%. Adams (2014) comes to a similar conclusion. Both results compare well with our prediction of 71.4%. Of the faults found, in 24% of the cases, the fault was from hardware, and in 12% of the cases, a software fault was found. This is also close to our estimate of 14.8% persistent hardware faults.

We should also consider an inherent bias technicians may have when documenting NFFs at the depot. When a technician decides to "shot gun" (repair based on a guess) a UUT that experiences NFF in the initial depot test, he/she is more likely to record this situation as a "repair" than an NFF. Since a correct repair and an NFF will both pass the test, the technician, having expended time and parts for the repair, will unlikely decide that the repair was unnecessary. Documenting this action as a correct repair will skew the NFF rate (Ungar, 2015).

5.4.4 Standards for NFF

Standards can be used to guide industries to promote efficient operation and improve reliability, thus reducing the NFF phenomenon. Best practices can be tailored to different application areas; in maintenance, best practices advocate the development of policies to meet process objectives. Specific standards have been introduced over the years to improve system maintenance and to reduce overall scheduling costs, e.g., RCM standards, i.e., IEC 60300-3-11 and SAE JA1012, and IEC 60706-5 for improved equipment maintainability through better testability features.

A top-down approach introduced for aircraft maintenance to provide significant improvements in availability and operational safety, while optimizing the costs of ownership, has gone through a number of revisions. The Maintenance Steering Group-3 (MSG-3) (MSG-3, 1993) program can be tailored around reliability by any operator depending on historical data. Under MSG-3, maintenance tasks are broken down into zones, where all potential components are serviced/replaced in one shot, instead of having to go back multiple times to rectify the issue. The maintenance tasks and component replacements are no longer based on hard time limits but on reliability trending. This greatly reduces repetitive tasks and provides time and costs savings for airlines/operators. It should be noted that NFF events (called hidden failures in MSG-3) are analyzed as parts of multiple failures, and such failures on their own have no consequence. Here, the goal of preventive maintenance is to guarantee the availability of those system components that are essential to avoid the effects of multiple failures on safety, operation, or economy. The question is whether there are any NFF failures that can affect components with a direct impact on system operation (or, if we extend the example beyond aviation, to production) (Khan et al., 2014).

A set of procedures introduced in the ARINC 672 Report (2008) are directly aimed at providing the basis for a structured process to address the NFF problem in the aviation industry. As illustrated in Fig. 5.28, the report provides criteria for decision-making with respect to root causes and describes the importance of taking maintenance actions at an early stage of the component repair cycle. It highlights how to reduce costs by avoiding unwanted removing units from the aircraft. It should be noted that the guidelines can be customized for specific operations environments (Khan et al., 2014).

It may be possible to develop a more fault-tolerant system with reduced NFF rates, one that is applicable in areas beyond aviation. An essential part is establishing cross-discipline features and design solutions that can be applied

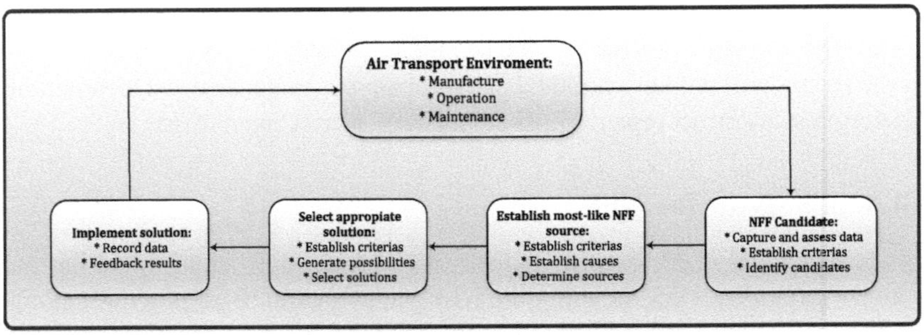

FIGURE 5.28 ARINC 672 NFF reduction interactive framework. *Adapted from Khan, S., Phillips, P., Jennions, I., Hockley, C., 2014. No fault found events in maintenance engineering part 1: current trends, implications and organizational practices. Reliability Engineering and System Safety 123, 183–195.*

to any engineering design. This will involve multidisciplinary modeling of the root causes of NFF achieved by modeling electronic, mechanical, and software interactions. Such an approach may lead to the development of design guides (or handbooks) covering the processes to avoid the root causes of NFF at the design stage. Of course, these will include accurate fault models, fault trees, system understanding (to aid in recognizing false BIT alarm caused by, e.g., sensors), and system synchronization problems (allowing root causes of BIT deficiencies). To complement any design guidance and rule sets, solutions are required to link service experience with design knowledge to generate an official standard to reduce NFF occurrences throughout a system's life cycle. These must be evaluated in a series of practical cases by collaborating companies and with the help of expert judgment.

Some system designers say that even establishing specific standards on NFF may not guarantee a solution to tackle NFF events. Their view is that a generic NFF standard will not be applicable for some specific (specialized) requirements (e.g., in safety critical applications). In fact, customers do not always specify NFF resolution as a requirement at the early stages of system design. This creates a gray area—who will absorb the implications when NFF events take place? Companies are not always familiar with such instances, and asking them to adapt or apply such standards may be deemed difficult to implement within their custom designs.

In any case, there is a lack of comprehensive international standards on NFFs. The issue is further clouded by inconsistent terminology. It would be useful to work toward harmonizing a set of generic maintenance-related standards that use a common terminology and framework; the ARINC 672 is a positive development moving toward the right direction (Khan et al., 2014).

5.4.5 Organizational Procedures and Administration

It is commonly accepted that NFFs often appear on two test levels, or more (Soderholm, 2007). At any test level, a fault may be recognized and localized as belonging to an individual piece of equipment; when the equipment is retested at a subsequent level, the fault may not be recognized/localized. The test at the first level may have been correct, but the attempts to replicate it at a subsequent level have failed, possibly because of an inability to recreate the conditions under which the original test was carried out, discrepancies in test procedure, or human error. Alternatively, the error may have been incorrectly recognized in the first place (Block et al., 2009).

There is a clear distinction between CND and RTOK. CND occurs at the same level of maintenance as the reported fault, while RTOK occurs at subsequent levels of maintenance (Ungar and Kirkland, 2008). This leads to the conclusion that RTOK is a "true" term for NFF. Even though it appears in much of the literature, CND should not be technically described as NFF, as it will only occur at second-line maintenance. This maintenance process is illustrated in Fig. 5.29 (Khan et al., 2014).

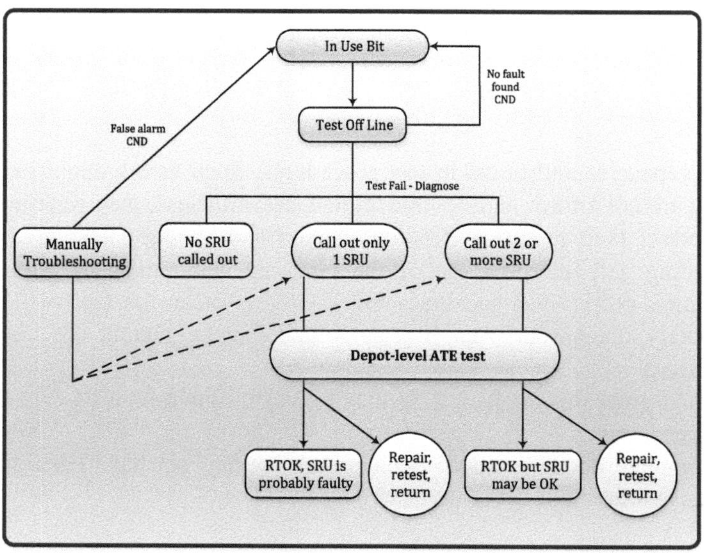

FIGURE 5.29 Tests and diagnostic processes in electronics. *ATE*, automatic test equipment; *CND*, cannot duplicate; *RTOK*, retest OK; *SRU*, shop replaceable unit. *Adapted from Khan, S., Phillips, P., Jennions, I., Hockley, C., 2014. No fault found events in maintenance engineering part 1: current trends, implications and organizational practices. Reliability Engineering and System Safety 123, 183–195.*

Jones and Hayes (2001) say NFF events occur on three identifiable levels in electronics:

1. Equipment level: Operators find the equipment does not function correctly, and, as a result, a maintenance action is scheduled. The engineer finds no problems during maintenance testing. This is carried out in second line maintenance.
2. Board level: Operators find the equipment does not function correctly; during maintenance testing, the engineer finds a problem does exist and identifies a board to be removed and returned for repair. When subsequently tested, however, the board functions correctly. This usually represents third line maintenance.
3. Component level: The board that has been removed is tested and a faulty component is identified. This component is replaced, and the removed component is found on subsequent failure analysis to be functioning correctly. After the component has been replaced, the board operates normally. This maintenance activity is usually carried out by the original equipment manufacturer.

It is necessary to remember that, at some stage, an event has occurred, whether a false alarm or component degradation, and this has resulted in maintenance actions as described above (ending with an NFF decision). This event is known as the root cause of NFF. The literature mentions several distinct types of root causes. Some typical root causes include (Soderholm, 2007; Khan et al., 2012; ARINC Working Group 672, 2008) the following:

1. Discrepancies and faults in test procedures: Such errors might exist, but they are not known and lie unidentified. Nevertheless, they generate NFF.
2. Incorrect fault reporting: Communication and having a common understanding and interpretation of the fault are sometimes not achievable. People's perceptions and interpretations often mean the fault will not be properly diagnosed, even if it is found. Human factors are obviously involved.
3. Wrong processes applied: This may reflect misinterpretation or a lack of proper training.
4. Incomplete documentation: Such processes may not have been comprehensive when first devised.

5.4.6 Implications of NFF

5.4.6.1 Financial Implications

Williams et al. (1998) claims NFF failures can make up more than 85% of all observed field failures in avionics components and account for more than 90% of all maintenance costs. This can be attributed to the following:

1. A limited understanding of root cause failure characteristics of a complex system,

2. Inappropriate means of diagnosing the condition of the system,
3. The inability to duplicate the field conditions in the laboratory.

NFF events pose problems to almost everyone involved with the operational service, from customers to manufacturers and their suppliers. There are other major impacts on business costs that are not so easily quantifiable, such as the supply chain, maintenance performance, and capacity, or indirect effects, such as customer perception (Cockram and Huby, 2009).

Janamanchi and Jin (2010) recently proposed a financial model to analyze the trade-off between the financial benefits of reliability improvement and the costs associated with the implementation of highly accelerated stress screening in the context of manufacturing ATE. The direct economic issues of NFF events are caused by putting units through the maintenance chain (replacement, logging, packaging, shipping, teasing, and documenting) for no apparent reason. The impact of NFF could range from mere nuisance to increased financial costs and safety risks. Without high levels of confidence that a reported fault is not fixed correctly the first time, along with a high probability of reoccurrence, there will be a measurable impact on the business output (Khan et al., 2014).

Direct maintenance costs of any system can make a significant contribution to its overall cost of ownership. Wu et al. (2004) says design and fault diagnosis are the key factors behind such costs, and a large proportion of cost spent on NFF events can be attributed to warranty claims (Wu, 2013; Pecht, 2006). Depending on how the maintenance contract is set up, claims can be made to include human factors or intermittent failures (a large proportion of the entire claim population) among the causal factors. Prakash and Ceglarek (2009) presents a methodology to determine optimal process adjustments to eliminate warranty-related NFF product failures. Another method to minimize product warranty costs is to embed reliability in the early development phase of the system; that is, by designing out potential failure modes related to hardware, software, process, or customer usage issues (Jin and Ozalp, 2009). Either way, there will be a need for adequate data and evidence on the cost of NFFs to make management recognize the need for change. The costs may not be so easy to establish, however, and some evidence suggests that under many circumstances, there is not even a willingness to establish what the NFF-related costs are.

As far as procedures and rules are concerned, defense organizations will point out that military aircraft are built to safe-life criteria with no redundancy, unlike commercial aircraft built to fail-safe standards; this would preclude the military from adopting many of the new practices used by commercial operators, such as the subject to aircraft check approach. Consider the UK military for a moment. There used to be no real incentive to solve the problem because their own maintenance and supply organizations would cope with the extra repairs required and, in fact, actually justified their existence. The organization is also too big and cumbersome to make the necessary changes easily. Although there are plenty of people, there is little interest in incurring the rising costs associated with such wasteful efforts (Khan et al., 2014).

5.4.6.2 Safety Considerations of NFFs

It can be argued that unless an NFF failure (with numerous potential causes) is a repetitive fault, influencing the system performance, an isolated incident cannot be considered a safety issue. However, the root cause might well be a safety concern. Soderholm (2007) classifies NFF events into two distinct categories, those that affect safety and those that do not.

An NFF occurrence with an impact on safety would be the case where, at any level of test, there is a failure to recognize and localize an actual fault. Conversely, tests resulting in false alarms, that is, recognizing faults that do not exist, will cause inconvenience but will not usually affect safety. In the case of inadequate diagnostic tools, the cause of a failure may be indeterminate—making repair impossible. This leads to fully functioning units being replaced; as a result of poor maintenance practice and inadequate testing methods, faulty units often pass subsequent tests and are returned to the field even though they are potential safety hazards.

After studying NFF occurrences related to a faulty ignition device on vehicles manufactured by the Ford Motor Company, Thomas et al. (2002) concludes that manufacturers should assume more responsibility for ascertaining root causes, including design/manufacturing flaws, inadequate testing, or even operator/customer error, of the failures in returned safety-related products. Manufacturers should stop assuming that a safety-related returned product is fit for service if it has passed the required routine checklist tests; instead, all safety-related products that are returned should be treated as field failures.

In the context of safety, the most crucial NFF failure type is intermittency. A high profile organization that has suffered from safety-related intermittent problems is NASA. In 2005, not for the first time, a space shuttle launch was delayed by 3 weeks. An intermittent fault appeared in a fuel-level sensor during tanking tests, but the equipment was returned to operational status without NASA ever discovering the root problem and fixing it. The circuitry with the intermittent fault was one of four that would cut off the shuttle's three main engines if at least two showed hydrogen fuel was running low. Should environmental launch conditions cause intermittent conditions to reappear inadvertently, the shuttle's safety would have been compromised. A multitude of back-to-back incidents within systems operated by NASA indicate that increased intermittency is a significant issue in aging electronics; the inability to detect intermittence and certify testing practices as adequate to ensure safe and reliable launch operations is, therefore, a major concern.

The danger of not dealing adequately with NFF events related to intermittent faults is further demonstrated by an incident on board a BMI A321 aircraft at 36,000 ft. The aircraft was carrying 43 passengers on route from Khartoum to Beirut. An intermittent failure in the electrical power-generator systems presented numerous symptoms, including an uncontrollable rudder trim. This caused the left wing to dip by 10 ft and the aircraft to deviate from its intended course by 37 km. In addition, the pilot and copilot's instruments were affected when the primary and navigational flight displays (among other instruments) flickered or went entirely blank (Kaminski-Morrow, 2010). In this case, the aircraft landed safely, but the near-accident shows the need for intermittent faults to be detected and localized during maintenance testing (Khan et al., 2014).

REFERENCES

Abarbanel, H.D.I., 1996. Analysis of Observed Chaotic Data. Springer-Verlag, New York.

Abebe, A.J., Solomatine, D.P., Venneker, R., 2000a. Application of adaptive fuzzy rule-based models for reconstruction of missing precipitation events. Hydrological Sciences Journal 45 (3), 425–436.

Abebe, A.J., Guinot, V., Solomatine, D.P., 2000b. Fuzzy Alpha-Cut vs. Monte Carlo techniques in assessing uncertainty in model parameters. In: Proceedings of 4th International Conference on Hydroinformatics, Cedar Rapids.

Abrahart, R.J., See, L.M., Solomatine, D.P., 2008. Practical hydroinformatics. Computational intelligence and technological developments in water applications. In: Singh, V.P. (Ed.), Water Science and Technology Library, vol. 68. Springer. Texas A&M University, College Station, U.S.A. ISBN: 978-3-540-79880-4.

Abu-Mostafa, Y., 1995. Financial market applications of learning from hints. In: Refenes, A. (Ed.), Neural Networks in the Capital Markets. Wiley, London, UK, pp. 221–232.

Adams, C., February/March 2014. Getting serious about no fault found. Aviation Maintenance Magazine 26–32.

An, D., Kim, N.H., Choi, J.-H., 2015. Practical options for selecting data-driven or physics-based prognostics algorithms with reviews. Reliability Engineering and System Safety 133, 223–236. Journal homepage: www.elsevier.com/locate/ress.

ARINC Working Group 672, 2008. Guidelines for the reduction of No Fault Found (NFF). ARINC.

Babovic, V., Keijzer, M., Stefansson, M., 2000. Optimal embedding using evolutionary algorithms. In: Proceedings of 4th International Conference on Hydroinformatics, Cedar Rapids.

Bárdossy, A., Duckstein, L., 1995. Fuzzy Rule-based Modeling with Applications to Geophysical, Biological and Engineering Systems. CRC Press Inc., Boca Raton, Florida, USA.

Barlow, H., 1989. Unsupervised learning. Neural Computation 1, 295–311.

Block, J., Tyrberg, T., Soderholm, P., 2009. No Fault Found events during the operational life of military aircraft items. In: Proceedings of the 8th International Conference on IEEE Reliability, Maintainability and Safety ICRMS, pp. 920–924.

Bodén, M., 2002. A guide to recurrent neural networks and backpropagation. In: Holst, A. (Ed.), The DALLAS Project. Report from the NUTEK-Supported Project AIS-8: Application of Data Analysis with Learning Systems, 1999–2001, SICS Technical Report T2002:03. SICS, Kista, Sweden.

Bunks, C., McCarthy, D., Ani, T., 2000. Condition-based maintenance of machines using hidden Markov models. Mechanical Systems and Signal Processing 14 (4), 597–612. Oakland, CA.

Chapelle, O., Schölkopf, B., Zien, A., 2006. In: Dietterich, T. (Ed.), Semi-Supervised Learning. ©2006 Massachusetts Institute of Technology.

Chen, J., Patton, R.J., 1999. Robust Model-based fault diagnosis for dynamic systems. In: The International Series on Asian Studies in Computer and Information Science, vol. 3. ISBN: 978-1-4613-7344-5 (Print) 978-1-4615-5149-2.

Chestnut, H., 1965. Systems Engineering Tools. Wiley, New York.

Chow, E.Y., Willsky, A.S., 1980. Issue in the development of a general algorithm for reliable failure detection. In: Proceedings of the 19th Conference on Decision & Control, Albuquerque, NM.

Cockram, J., Huby, G., 2009. No Fault Found (NFF) occurrences and intermittent faults: improving availability of aerospace platforms/systems by refining maintenance practices, systems of work and testing regimes to effectively identify their root causes. In: Proceedings of CEAS European Air and Space Conference.

Coppe, A., Haftka, R.T., Kim, N.H., 2011. Uncertainty identification of damage growth parameters using nonlinear regression. AIAA Journal 49 (12), 2818–2821.

Coppe, A., Pais, M.J., Haftka, R.T., Kim, N.H., 2012. Using a simple crack growth model in predicting remaining useful life. Journal of Aircraft 49 (6), 1965–1973.

Cordón, O., Herrara, F., 1995. A general study on genetic fuzzy systems. In: Winter, G., Périaux, J., Gálan, M., Cuesta, P. (Eds.), Genetic Algorithms in Engineering and Computer Science. JohnWiley & Sons, Chichester, pp. 33–57.

Cornell Fracture Group, 2003. FRANC3D & OSM 3D Tutorial Version 2.6. https://sites.coecis.cornell.edu/fracture/files/2016/05/F3D_OSM_Tutorial_CFG_V2.6-2ajplzi.pdf.

Dalianis, P.J., Tzafestas, S.G., Anthopoulos, G., October 1993. A study of the generalization capability versus training in backpropagation neural networks. In: IEEE/SMC.93 Conference, (Le Toquet, France).

Dibike, Y.B., Velickov, S., Solomatine, D.P., Abbott, M.B., 2001. Model induction with support vector machines: introduction and applications. ASCE Journal of Computing in Civil Engineering 15 (3), 208–216.

Dobbert, M., 2012. Understanding measurement risk. First Presented at the 2007 NCSL International Workshop and Symposium. http://cp.literature.agilent.com/litweb/pdf/5991-1265EN.pdf.

Duch, W., Jankowski, N., 1999. Survey of neural transfer functions. Neural Computing Surveys 2, 163–212.

Filippetti, F., Franceschini, G., Tassoni, C., Vas, P., October 6–10, 1996. AI techniques in induction machines diagnosis including the speed ripple effect. In: IEEE-IAS Annual Meeting Conference, San Diego, QCI, 655–662.

Firth, A.E., Lahav, O., Somerville, R.S., 2003. Estimating photometric redshifts with artificial neural networks. Monthly Notices of the Royal Astronomical Society 339 (4), 1195–1202.

Frank, P.M., 1991. Enhancement of robustness in observer based fault detection. Proceedings of the IFAC Symposium on SAFEPROCESS'91, Baden-Baden 275–287.

Frapporti, G., Vriend, S.P., Van Gaans, P.F.M., 1993. Hydrogeochemistry of the shallow Dutch groundwater: classification of the national groundwater quality monitoring network. Water Resources Research 29 (9), 2993–3004.

Friedman, R., Eidenoft, H.L., 1990. A two-stage fatigue life prediction method. Grumman Advanced Development Report.

Galeati, G., 1990. A comparison of parametric and non-parametric methods for runoff forecasting. Hydrology Sciences Journal 35 (1), 79–94.

Goldberg, D.E., 1989. Genetic Algorithms in Search Optimisation and Machine Learning. Addison-Wesley, USA.

Giustolisi, O., Savic, D.A., 2006. A symbolic data-driven technique based on evolutionary polynomial regression. Journal of Hydroinformatics 8 (3), 202–207.

Hadden, G., Bergstrom, P., Bennett, B., Vachtsevanos, G., Van Dyke, J., May 9–12, 1999a. Shipboard machinery diagnostics and prognostics/condition based maintenance: a progress report. In: MARCON 99, Maintenance and Reliability Conference, Gatlinburg, TN, 73.01–73.16.

Hadden, G., Bennett, B., Vachtsevanos, G., Van Dyke, J., April 1999b. Machinery diagnostics and prognostics/condition based maintenance: a progress report, failure analysis: a foundation for diagnostics and prognostics development. Proceedings of the 53rd Meeting of the Society for Machinery Failure Prevention Technology, Virginia Beach, VA.

Hadden, G., Nelson, K., Edwards, T., 1994. State-based feature recognition. In: 17th Annual AAS Guidance and Navigation Conference, Scottsdale, AZ, 1994.

Hadden, G., Nelson, K., 1991. A state-based approach to trend recognition and failure prediction for the space station freedom. Proceedings of the Space Operations, Applications, and Research Symposium (SOAR-91).

Hall, M.J., Minns, A.W., 1999. The classification of hydrologically homogeneous regions. Hydrological Sciences Journal 44, 693–704.

Hannah, D.M., Smith, B.P.G., Gurnell, A.M., McGregor, G.R., 2000. An approach to hydrograph classification. Hydrological Processes 14, 317–338.

Hardman, W., Hess, A., Sheaffer, J., 1999. SH-60 helicopter integrated diagnostic system (HIDS) program: diagnostic and prognostic development experiences. In: Proceedings of Aerospace Conference, Big Sky, MT, vol. 1, pp. 473–491.

Harris, N.M., Gurnell, A.M., Hannah, D.M., Petts, G.E., 2000. Classification of river regimes: a context for hydrogeology. Hydrological Processes 14, 2831–2848.

Hecht, D., 2007. Boeing Aircraft Company, Diagnostic – Lessons Learned, Chart Presented at AutoTestCon 2007.

Hinton, G.E., Sejnowski, T.J., 1999. Unsupervised Learning: Foundations of Neural Computation. Computational Neuroscience. MIT Press. ISBN: 9780262581684.

Isermann, R., Ballé, P., 1997. Trends in the application of model-based fault detection and diagnosis of technical processes. Control Engineering Practice 5 (5), 709–719.

Isermann, R., 1997. Supervision, fault-detection and fault-diagnosis methods: an introduction. Control Engineering Practice 5 (5), 639–652.

Jain, A.K., Murty, M.N., Flynn, P.J., 1999. Data Clustering: A Review.

James, G., 2003. Variance and bias for general loss functions. Machine Learning 51, 115–135. http://www-bcf.usc.edu/~gareth/research/bv.pdf.

James, G., Witten, D., Hastie, T., Tibshirani, R., 2013. An introduction to statistical learning: with applications in R. In: Springer Texts in Statistics. Springer New York Heidelberg Dordrecht London.

James, I., Lumbard, D., Willis, I., Goble, J., 2003. Investigating no fault found in the aerospace industry. In: Proceedings of IEEE Annual Reliability and Maintainability Symposium, pp. 441–446 (iD:1).

Janamanchi, B., Jin, T., 2010. Reliability growth vs. hass cost for product manufacturing with fast-to-market requirement. The International Journal of Productivity and Quality Management 5 (2), 152–170.

Jang, J.-S., Sun, C.-T., Mizutani, E., 1997. Neuro-Fuzzy and Soft Computing. Prentice Hall.

Jardine, A.K.S., Lin, D., Banjevic, D., 2006. A review on machinery diagnostics and prognostics implementing condition-based maintenance. Mechanical Systems and Signal Processing 20 (7), 1483–1510.

Jensen, F.V., 2001. Bayesian Networks and Decision Graphs. Springer-Verlag, New York.
Jin, T., Ozalp, Y., 2009. Minimising warranty cost by design for reliability in product development phase. International Journal of Six Sigma and Competitive Advantage 5 (1), 42–58.
Jones, J., Hayes, J., 2001. Investigation of the occurrence of: No-Faults-Found in electronic equipment. IEEE Transactions on Reliability 50 (3), 289–292.
Kaminski-Morrow, D., 2010. BMI A321 strayed off course as pilots battled electrical failure. In: Proceedings of Flight Global.
Karlsson, M., Yakowitz, S., 1987. Nearest neighbour methods for non-parametric rainfall runoff forecasting. Water Resources Research 23 (7), 1300–1308.
Karr, C.L., 1991. Genetic algorithms for fuzzy logic controllers. AI Expert 6, 26–33.
Khan, S., Phillips, P., Jennions, I., Hockley, C., 2014. No fault found events in maintenance engineering part 1: current trends, implications and organizational practices. Reliability Engineering and System Safety 123, 183–195.
Khan, S., Phillips, P., Hockley, C., Jennions, I., 2012. Towards standardisation of no-fault found taxonomy. In: Proceedings of the 1st International Through-Life Engineering Services Conference, 2012, pp. 246–253.
Khu, S.-T., Liong, S.-Y., Babovic, V., Madsen, H., Muttil, N., 2001. Genetic programming and its application in real-time runoff forecasting. Journal of the American Water Resources Association 37 (2), 439–451.
Kimseng, K., Holt, M., Precht, M., 1999. Physics-of-failure assessment of a cruise control module. Microelectronics Reliability 39, 1423–1444.
Konrad, H., Isermann, R., 1996. Diagnosis of different faults in milling using drive signals and process models. Proceedings of the 13th World Congress, International Federation of Automatic Control, Vol. B., Manufacturing, San Francisco, CA, June 30–July 5, 1996 91–96.
Kosko, B., 1997. Fuzzy Engineering. Prentice-Hall, Upper Saddle River.
Koza, J.R., 1992. Genetic Programming: on the Programming of Computers by Means of Natural Selection. MIT Press, Cambridge, MA.
Liao, L., Köttig, F., 2014. Review of hybrid prognostics approaches for remaining useful life prediction of engineered systems, and an application to battery life prediction. IEEE Transactions on Reliability 63 (1), 191–207.
Line, J.K., Krishnan, G., 2008. Managing and predicting intermittent failures within long life electronics. In: Proceedings of the IEEE Aerospace Conference, pp. 1–6 (iD: 1).
Ling, Y., Mahadevan, S., 2013. Quantitative model validation techniques: new insights. Reliability Engineering and System Safety 2013 (111), 217–231.
Liong, S.Y., Sivapragasam, C., 2002. Flood stage forecasting with SVM. Journal of American Water Resources Association 38 (1), 173–186.
Liu, J., Saxena, A., Goebel, K., Saha, B., Wang, W., 2010. An adaptive recurrent neural network for remaining useful life prediction of lithium-ion batteries. In: Annual Conference of the Prognostics and Health Management Society, Portland, Oregon; October 10–16, 2010.
Ljung, L., 1999. System Identification. John Wiley & Sons, Inc.
Lobbrecht, A.H., Solomatine, D.P., 1999. Control of water levels in polder areas using neural networks and fuzzy adaptive systems. In: Savic, D., Walters, G. (Eds.), Water Industry Systems: Modelling and Optimization Applications. Research Studies Press Ltd., Baldock, pp. 509–518.
Malcolm, J.G., 1983. Practical application of Bayes' formulas. Proceedings of the Annual Reliability and Maintainability Symposium.
Mitchell, T.M., 1997. Machine Learning. McGraw-Hill, New York.

Moaveni, S., 2003. Finite Element Analysis: Theory and Application with ANSYS, second ed. Prentice-Hall, Upper Saddle River, NJ.

Moody, T.J., Darken, C.J., 1989. Fast leaning in networks of locally tuned processing unit. Neural Computation 1, 281–294.

MSG-3, 1993. Task Force: Maintenance Program Development Document. Air Transport Association of America.

Mylaraswamy, D., Venkatasubramanian, V., 1997. A hybrid framework for large scale process fault diagnosis. Computers & Chemical Engineering 21, S935–S940.

Nandi, S., Toliyat, H.A., 1999. Condition Monitoring and Fault Diagnosis for Electrical Machines – A Review. Electric Machines & Power Electronics Laboratory. Department of Electrical Engineering. Texas A&M University, College Station, TX, pp. 77843–83128.

Newman, J.C., February 1992. FASTRAN-II: A Fatigue Crack Growth Structural Analysis Program, Revised Copy. NASA Technical Memorandum 104159, Langley, VA.

Oden, J.T., Prudencio, E.E., Bauman, P.T., 2013. Virtual model validation of complex multiscale systems: applications to nonlinear elastostatics. Computer Methods in Applied Mechanics and Engineering 266, 162–184.

Office of the Assistant Secretary of the Navy, October 2001. Built-in Test, Design and Optimization Guidelines. TB # ABM 1001-01.

Oja, E., 2002. Unsupervised learning in neural computation. In: Theoretical Computer Science, vol. 287. Helsinki University of Technology, Neural Networks Research Centre, P.O. Box 5400, 02015 HUT, Finland, pp. 187–207.

Overton, D., 2005. Survey of Mobile Phones No Fault Found. http://www.wds.co/no-fault-found-returns-cost-the-mobile-industry-4-5-billion-per-year/.

Palade, V., Bocaniala, C.D., 2006. Computational Intelligence in Fault Diagnosis. ©Springer-Verlag London Limited.

Parker, B.E., Ware Jr., H.A., Wipf, D.P., Tompkins, W.R., Clark, B.R., Larson, E.C., Poor, H.V., 2000. Fault diagnostics using statistical change detection in the bispectral domain. Mechanical Systems and Signal Processing 14 (4), 561–570.

Patton, R.J., Chen, J., 1996. Robust Fault Detection and Isolation (FDI) Systems. In: Leondes, C. (Ed.), Con. and Dyn. System, vol. 74. Mita Press, pp. 171–224.

Pecht, M., 2006. Establishing a relationship between warranty and reliability. IEEE Transactions on Electronics Packaging Manufacturing 29 (3), 184–190.

Pesti, G., Shrestha, B.P., Duckstein, L., Bogárdi, I., 1996. A fuzzy rule-based approach to drought assessment. Water Resources Research 32 (6), 1741–1747.

Phoon, K.K., Islam, M.N., Liaw, C.Y., Liong, S.Y., 2002. A practical inverse approach for forecasting nonlinear hydrological time series. ASCE Journal of Hydrologic Engineering 7 (2), 116–128.

Prakash, I.L., Ceglarek, D., 2009. Functional process adjustments to reduce no-fault-found product failures in service caused by in-tolerance faults. CIRP Annals—Manufacturing Technology 58 (1), 37–40.

Qi, H., Ganesan, S., Pecht, M., 2008. No-fault-found and intermittent failures in electronic products. Microelectronics Reliability 48, 663–674. Elsevier Ltd.

Rabiner, L.R., 1989. A tutorial on hidden Markov models and selected applications in speech recognition. Proceedings of the IEEE 77 (2), 257–286.

Rao, S.S., 1990. Mechanical Vibrations, second ed. Addison-Wesley, Reading, MA.

Rebba, R., Mahadevan, S., Huang, S., 2006. Validation and error estimation of computational models. Reliability Engineering System Safety 91, 1390–1397.

Reeves, C.R., Rowe, J.E., 2003. Genetic Algorithms – Principles and Perspectives. A Guide to GA Theory. Kluwer Academic Publishers Group.

Rosenberg, B.J., 1989. The Navy IDSS program: adaptive diagnostics and feedback analysis: precursors to a fault prognostics capability. In: Proceedings of the IEEE 1989 National Aerospace and Electronics Conference NAECON 1989, Dayton, OH, Vol. 3, May 1989, pp. 1334–1338.

Rumelhart, D.E., Hinton, G.E., Williams, R.J., 1986. Learning internal representations by error propagation. In: Parallel Distributed Processing: Explorations in the Microstructure of Cognition, Vol. 1. Foundations. MIT Press, pp. 318–362.

Sargent, R.G., 2013. Verification and validation of simulation models. Journal of Simulation 7, 12–24.

Schoen, R.R., Lin, B.K., Habetler, T.G., Schlag, J.H., Farag, S., November–December 1995. An unsupervised on-line system for induction motor fault detection using stator current monitoring. IEEE Transactions on Industry Applications 31 (6), 1280–1286.

Sehitoglu, H., 1997. Thermal and thermal–mechanical fatigue of structural alloys. ASM Handbook. In: Fatigue and Fracture, vol. 19. ASM International, New York, pp. 527–556.

Shamseldin, A.Y., O'Connor, K.M., 1996. A nearest neighbour linear perturbation model for river flow forecasting. Journal of Hydrology 179, 353–375.

Simpson, W.R., 1986. Prediction and Analysis of Testability Attributes: Organizational-Level Testability Prediction, RADC-TR-85-268. Rome Air Development Center, Griffis AFB, NY.

Soderholm, P.A., 2007. System view of the No Fault Found (NFF) phenomenon. Reliability Engineering and System Safety 92 (1), 1–14.

Solomatine, D.P., 1999. Two strategies of adaptive cluster covering with descent and their comparison to other algorithms. Journal of Global Optimization 14 (1), 55–78.

Solomatine, D.P., Rojas, C., Velickov, S., Wust, H., 2000. Chaos theory in predicting surge water levels in the North Sea. In: Proceedings of the 4th International Conference on Hydroinformatics, Cedar-Rapids.

Solomatine, D.P., Maskey, M., Shrestha, D.L., 2007. Instance-based learning compared to other data-driven methods in hydrologic forecasting. Hydrological Processes 21. http://dx.doi.org/10.1002/hyp.6592.

Sorensen, B., February 24, 2003. Digital averaging – the smoking gun behind 'No-Fault-Found'. Air Safety Week.

Sugeno, M., Kang, G.T., 1988. Structure identification of fuzzy model. Fuzzy Sets and Systems 28 (1), 15–33.

Svozil, D., Kvasnička, V., Pospíchal, J., 1997. Introduction to multi-layer feed-forward neural networks. Chemometrics and Intelligent Laboratory Systems 39, 43–62.

Swanson, J.A., DeSalvo, G.J., 1989. ANSYS-engineering analysis system user's manual. Swanson Analysis Inc., Elizabeth, Pennsylvania, 15037.

Takagi, T., Sugeno, M., 1985. Fuzzy identification of systems and its applications to modelling and control. IEEE Transactions on Systems, Man, and Cybernetic SMC-15 116–132.

Thomas, D., Ayers, K., Pecht, M., 2002. The "trouble not identified" phenomenon in automotive electronics. Microelectronics Reliability 42 (4), 641–651.

Toth, E., Brath, A., Montanari, A., 2000. Comparison of short-term rainfall prediction models for real-time flood forecasting. Journal of Hydrology 239, 132–147.

Tzafestas, S.G., Dalianis, P.J., 1994. Fault diagnosis in complex systems using artificial neural networks. In: 3rd IEEE CCA. Intelligent Robotics and Control Unit (IRCU). Computer Science Division, Electrical & Computer Engineering Department. National Technical University of Athens, Zografou 15773, Athens, Greece.

Ungar, L.Y., Kirkland, L.V., 2008. Unravelling the cannot duplicate and retest OK problems by utilizing physics in testing and diagnosis. In: Proceedings of the AUTO-TESTCON, pp. 550–555.

Ungar, L.Y., 2014. The true nature of false alarms. AUTOTEST '14. IEEE Press.

Ungar, L.Y., February 2015. Causes and Costs of No Fault Found Events. Advanced Test Engineering (A.T.E.) Solutions, Inc., El Segundo, CA.

US Navy, MIL-HDBK-2165, July 1995. Testability handbook for systems and equipment. In: Superseding MIL-STD-2165A, 01, vol. 31.

Vachtsevanos, G., Lewis, F., Roemer, M., Hess, A., Wu, B., 2006. Intelligent Fault Diagnosis and Prognosis for Engineering Systems. John Wiley & Sons, Inc., Hoboken, New Jersey.

Vapnik, V.N., 1998. Statistical Learning Theory. Wiley & Sons, New York.

Vas, P., 1999. Artificial-Intelligence-Based Electrical Machines and Drives: Applications of Fuzzy, Neural, Fuzzy-Neural and Genetic Algorithm Based Techniques. Oxford University Press, New York.

Velickov, S., Solomatine, D.P., Yu, X., Price, R.K., 2000. Application of data mining techniques for remote sensing image analysis. In: Proceedings of the 4th International Conference on Hydroinformatics, USA.

Velickov, S., Solomatine, D.P., Price, R.K., 2003. Prediction of nonlinear dynamical systems based on time series analysis: issues of entropy, complexity and predictability. In: Proceedings of the XXX IAHR Congress, Thessaloniki, Greece.

Vernieuwe, H., Georgieva, O., De Baets, B., Pauwels, V.R.N., Verhoest, N.E.C., De Troch, F.P., 2005. Comparison of data-driven Takagi–Sugeno models of rainfall–discharge dynamics. Journal of Hydrology 302 (1–4), 173–186.

Vijayaraghavan, L., 2010. Machine learning approach to fault diagnosis in engineering systems. In: IJEST (Ed.), International Journal of Engineering, Science and Technology (IJEST).

Wang, L.X., 1994. Adaptive Fuzzy Systems and Control: Design and Stability Analysis. PTR Prentice Hall Inc., Englewood Cliffs, NJ.

Wang, B., Zou, D., Ding, R., 2011. Support vector regression based video quality prediction. Multimedia (ISM), 2011 IEEE International Symposium on. IEEE.

Wilamowski, B.M., Iplikci, S., Kaynak, O., Efe, M.Ö., 2001. An algorithm for fast convergence in training neural networks. In: Proceedings of the International Joint Conference on Neural Networks, vol. 3, pp. 1778–1782.

Williams, R., Banner, J., Knowles, I., Dube, M., Natishan, M., Pecht, M., 1998. An investigation of "cannot duplicate" failures. Quality and Reliability Engineering International 14 (5), 331–337.

Witten, I.H., Frank, E., 2000. Data Mining. Morgan Kaufmann, San Francisco.

Wu, H., Liu, Y., Ding, Y., Liu, J., 2004. Methods to reduce direct maintenance costs for commercial aircraft. Aircraft Engineering and Aerospace Technology 76 (1), 15–18.

Wu, S., 2013. A review on coarse warranty data and analysis. Reliability Engineering and System Safety 114, 1–11.

Xing, Y., Miao, Q., Tsui, K.-L., Pecht, M., 2011. Prognostics and health monitoring for lithium-ion battery. In: IEEE International Conference, pp. 242–247.

Xu, J., Xu, L., 2011. Health management based on fusion prognostics for avionics systems. Journal of Systems Engineering and Electronics 22 (3), 428–436.

Yao, X., Liu, Y., 1997. A new evolutionary system for evolving artificial neural networks. IEEE Transactions on Neural Networks 8 (3), 694–713.

Zadeh, L.A., 1965. Fuzzy sets. Information and Control 8, 338–353.

Zhang, X., Song, X., September 2006. Spatial pattern identification of soil moisture based on self-organizing neural networks. In: Proceedings of the 7th International Conference on Hydroinformatics, Nice.

Zou, M., Green, I., September 4–6, 1997. Real-time condition monitoring of mechanical face seal. In: Proceedings of the 24th Leeds-Lyon Symposium on Tribology. Imperial College, London, pp. 423–430.

Zou, M., Dayan, J., Green, I., 2000. Dynamic simulation and monitoring of a non contacting flexibly mounted rotor mechanical face seal. IMechE, Proceedings of the Institution of Mechanical Engineers Part C: Journal of Mechanical Engineering Science 214 (C9), 1195–1206.

Chapter 6

Prognosis

6.1 INTRODUCTION

The maintenance of all types of assets is increasingly important in both the industrial and the scientific sectors. Industries want to improve their maintenance techniques to increase the lifetime of their equipment. All equipment deteriorates over time, as it operates under a certain voltage or load in the real environment, thus generating maintenance activities more often as it ages.

Maintenance combines various methods, tools, and techniques in a bid to reduce maintenance costs while increasing reliability, availability, and security of equipment (Otilia E. Dragomir et al., 2009). The most common type of maintenance is corrective maintenance (also called unplanned maintenance, or run-to-failure maintenance), which takes place only at breakdowns. Time-based preventive maintenance (also called planned maintenance) sets a periodic interval to perform preventive maintenance regardless of the health status of a physical asset.

With the rapid development of modern technology, products have become increasingly complex while better quality and higher reliability are required. This raises the cost of preventive maintenance. Preventive maintenance has become a major expense of many industrial companies (Jardine et al., 2006). Therefore, many companies are now implementing condition-based maintenance (CBM).

CBM is designed to avoid unnecessary maintenance tasks. Prognostics form a key element of a CBM program. Prognostics are of great interest to both industry and research communities as they can significantly improve the efficiency of a maintenance program. Prognostics basically try to predict how much time remains before a fault or failure will occur, given the current state of the asset and the operation. In other words, prognostics are based on predicting the remaining useful life (RUL) of a system based on condition monitoring of the asset and subsequent analysis of the information and data collected for a given task occurs.

With prognostics, industries can potentially reduce the costs of both preventive and corrective maintenance. If they can predict when a machine or asset may fail, they will be able to prevent or limit maintenance activities. Prognostics are mainly based on mathematical models for predicting the RUL.

eMaintenance. http://dx.doi.org/10.1016/B978-0-12-811153-6.00006-3

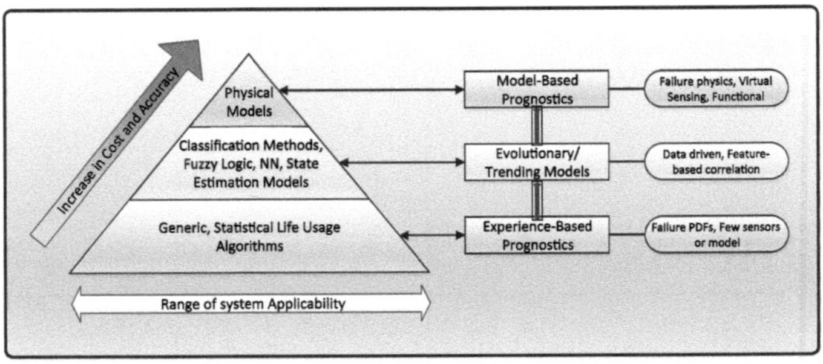

FIGURE 6.1 Hierarchy of prognostics approaches (Luo et al., 2003a). *NN*, neural network.

These models are built using one of four approaches: experience based, data based, model based, and hybrid methodology (discussed later).

Fig. 6.1 illustrates the hierarchy of possible approaches in relation to their applicability and relative costs.

A typical example in the application of prognostics is the use of models for modeling fatigue initiation and propagation of cracks in structural components. Other examples include the study of the estimation of RUL for bridge structures using a system based on experience or a study of vibrations in rotating machines using a data-driven approach. The objective of this chapter is to compare the techniques and models used to estimate the RUL of different assets. It conducts a review of the relevant literature on prognostics techniques and their use in the industrial sphere.

6.1.1 Maintenance and Prognosis

Since the industrial revolution, maintenance has gone through a number of different stages. At the beginning of the industrial revolution, workers were responsible for the repair of equipment. As machines became more complex and the amount of repair work increased, the first maintenance departments, distinct from production departments, were established. Tasks in both eras were basically remedial, with efforts devoted to addressing failures.

The concept of reliability appeared in the 20th century after the end of World War II. Maintenance departments now sought not only to fix the faults but to prevent them. This involved creating a new figure in maintenance departments: staff whose purpose was to study what maintenance should be performed to avoid failures. With the advent of the computer age, maintenance changed again, with the introduction of new technologies.

Clearly, maintenance techniques have evolved. Not all maintenance models apply to all companies. Some models interact and others do not. But all have been adapted to new uses in industry. Today, the specific needs of each team and each industry will determine the most appropriate maintenance model to optimize resources and needs.

6.1.2 Types of Maintenance

6.1.2.1 Corrective Maintenance

Corrective maintenance is used to repair damage that has already occurred. Usually, when this type of maintenance is performed, the manufacturing process is stopped, decreasing production and increasing costs. Repair time cannot be predicted, nor can the expenses resulting from the breakdown and consequent disturbances on the production line. Therefore, corrective maintenance is applied on assets with low criticality, whose faults do not involve large temporal or economic problems. It is often used for specific equipment where other techniques would be more costly.

6.1.2.2 Preventive Maintenance

Preventive maintenance is planned in a time horizon and aims to prevent breakdowns. Unlike corrective maintenance, because it is planned, it is not done during production time.

The intention of this type of maintenance is to reduce the number of corrective interventions, performing periodic reviews and replacing worn components.

It is a demanding type of maintenance, as it requires strict supervision and development of a plan to be carried out by qualified personnel. In addition, as it involves routine tasks, personnel may not be motivated. Furthermore, if it is not done correctly, there will be a cost overrun with no significant improvements in productivity.

6.1.2.3 Condition-Based Maintenance

CBM aims to determine the condition of equipment, so that operation remains safe, efficient, and economic. Monitoring techniques are aimed at measuring physical variables that indicate the condition of the machine and comparing these with normal values to determine if the machine is in good condition or deteriorating. CBM assumes there are measurable and observable characteristics that are indicators of the condition of the machine.

Condition monitoring studies the evolution of selected time-dependent parameters; it identifies trends indicating the existence of a fault, its severity, and the likely time to failure (TTF). Timely decision-making avoids the occurrence of faults and eliminates the possibility of catastrophic failure. CBM can be performed while the machine is running (Gerardo Trujillo and América, 2003).

CBM consists of three key steps (see Fig. 6.2):

1. Data acquisition (information collecting), to obtain data relevant to system health.
2. Data processing (information handling), to handle and analyze the data or signals collected in step 1 for better understanding and interpretation of the data.

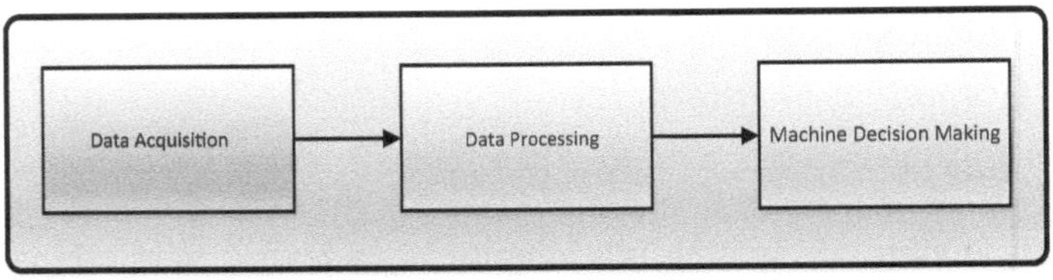

FIGURE 6.2 Three steps in condition-based maintenance.

3. Maintenance decision-making (decision-making), to recommend efficient maintenance policies.

   Steps to Implement CBM:

   Data Acquisition

Data acquisition is a process of collecting and storing useful data (information) from targeted physical assets for the purpose of CBM. This is an essential step in implementing a CBM program for machinery faults (or failure, usually caused by one or more faults). Data collected in a CBM program can be categorized into two main types: event data and condition-monitoring data. Event data include information on what happened (e.g., installation, breakdown, overhaul, etc., along with the causes) and/or what was done (e.g., minor repair, preventive maintenance, oil change, etc.) to the targeted physical asset. Condition-monitoring data are measurements related to the health condition/state of the physical asset. These are very versatile and can include vibration, acoustic, oil analysis, temperature, pressure, moisture, humidity, weather or environment data, etc.

Data Processing

Data processing consists of two stages. The first is data cleansing; this step is important because usually the data are entered manually. This leads to frequent errors, requiring data cleaning to increase the probability that the data are clean (no errors). The second step is the analysis of the data. There are a variety of models, algorithms, and tools for analysis; selection depends on the types of data collected.

Maintenance Decision Support

The last step of a CBM program is maintenance decision-making. Sufficient and efficient decision support is crucial for determining maintenance actions. Techniques for maintenance decision support in a CBM program can be divided into two main categories: diagnostics and prognostics. Fault diagnostics focus on detection, isolation, and identification of faults when they occur. Prognostics attempt to predict faults or failures before they occur.

Obviously, prognostics are superior to diagnostics in the sense that prognostics can either prevent faults or failures, or be ready (with spare parts and human resources) for the problems, thus reducing the costs of unplanned maintenance.

Nevertheless, prognostics cannot completely replace diagnostics since, in practice, there are always some faults and failures, which are not predictable. In addition, prognostics, like any other prediction technique, cannot be 100% accurate. In the case of unsuccessful prediction, diagnostics can be a complementary tool for maintenance decision-making. Diagnostics are also helpful for improving prognostics; diagnostic information can result in more accurate event data, and a better CBM model can be built for prognostics. Furthermore, diagnostic information can be used as feedback information for system redesign.

6.2 PROGNOSIS TECHNIQUES

Maintenance activity combines various methods, tools, and techniques to reduce maintenance costs while increasing reliability, availability, and security of assets. We usually speak about fault detection, failure diagnostics, response development (choice of management actions, i.e., preventive and/or corrective), and scheduling actions.

Briefly, these steps correspond to the need, first, to "perceive" phenomena, second, to "understand" them, and third, to "act" correctly. However, rather than understanding a phenomenon that appears as a failure (a posteriori), it may be convenient to "anticipate" its manifestation to quickly resort to protective actions. This could be defined as the "prognostic process" (Otilia E. Dragomir et al., 2009).

Prognosis is a very promising maintenance activity as it should permit plants to improve safety, plan, and schedule successful maintenance, and reduce maintenance cost and down time (Brotherton et al., 2000).

6.2.1 Concept of Prognostics

Prognostics refers to an engineering discipline focused on predicting the time at which a system or a component will no longer perform its intended function. This lack of performance is most often a failure beyond which the system can no longer be used to meet its desired goals. Thus, the predicted time becomes the RUL of the system or component, an important concept in decision-making for contingency mitigation. Prognostics predict future performance by assessing the extent of deviation or degradation of a system or component from its expected normal operating conditions. The science of prognostics is based on the analysis of failure modes, detection of early signs of wear and aging, and fault conditions. These signs are correlated in a damage propagation model. Prognostics can potentially be used in CBM (Pecht, 2008).

Prognostics should be based on assessment criteria; limits depend on the system itself and on performance objectives. Prognosis can be split into two subactivities: first, to predict the evolution of a situation at a given time; second, to assess this predicted situation using an evaluation referential (Dragomir et al., 2007) (see Fig. 6.3).

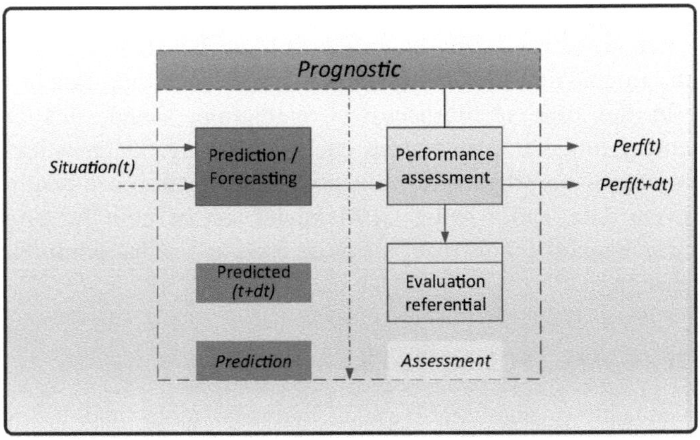

FIGURE 6.3 Prognostics as a prediction and assessment process (Perf(t) = performance assessment at t) (Dragomir et al., 2007).

6.2.2 Remaining Useful Life

A main concept of prognostics is the RUL, the output generated by a prognostic algorithm describing the distribution in time of likely equipment failure times. Fig. 6.4 illustrates the key concepts of RUL. At time t_P, a prediction is made and an estimate of the RUL is generated. Once the RUL has been generated, the next question is when to carry out corrective maintenance. Ideally, the time chosen for maintenance action will avoid equipment failure and maximize the useful life of the equipment. However, these are connecting requirements and, as a consequence, selecting when to perform maintenance is typically an exercise in risk management.

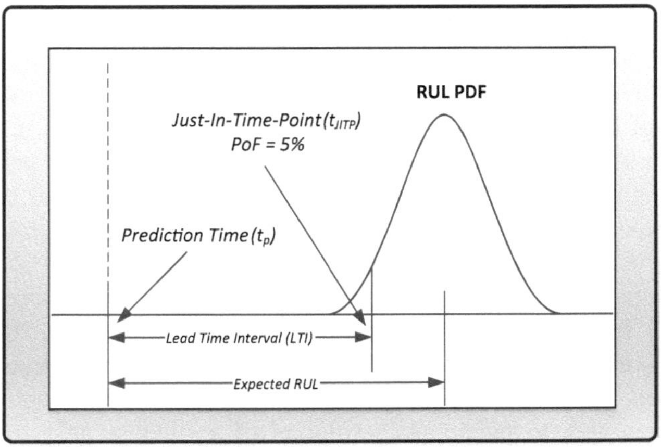

FIGURE 6.4 Remaining useful life (Byington et al., 2002). *PoF*, probability of failure; *RUL*, remaining useful life.

To develop the requirements for a prognostic algorithm, the maximum allowable probability of failure (PoF) must be considered. This value defines the maximum acceptable level of risk of equipment failure, beyond which equipment can no longer be operated, as the risk of equipment failure is deemed excessive. Using the defined maximum allowable PoF and the estimated RUL, an important value known as the just-in time-point (JITP) can be identified. The JITP defines the latest point in time before which corrective maintenance actions must be carried out to avoid operating equipment beyond the maximum allowable PoF.

In real-life applications, selecting the maximum allowable PoF usually requires the consideration of a number of factors, including safety, criticality, and economic concerns.

In certain scenarios, where safety considerations are primary, the requirement might be to avoid as many in-service failures as possible; thus, a conservative value for the maximum allowable PoF might be chosen. Alternatively, a plant operator may accept a higher maximum allowable PoF value in situations where maximizing the useful life of expensive equipment/components might be more economical than avoiding an occasional failure. An example of such a scenario might be the use of a diamond-headed cutting tool.

In Fig. 6.5, a maximum allowable PoF value of 5% is assumed for illustrative purposes. Once the JITP has been identified, the lead-time interval (LTI) can be computed. The LTI is defined as the time interval between the time the prediction is generated t_P and the JITP t_{JITP} (Shane Butler, 2012) and is expressed as:

$$t_{\text{LTI}} = t_{\text{JITP}} - t_P$$

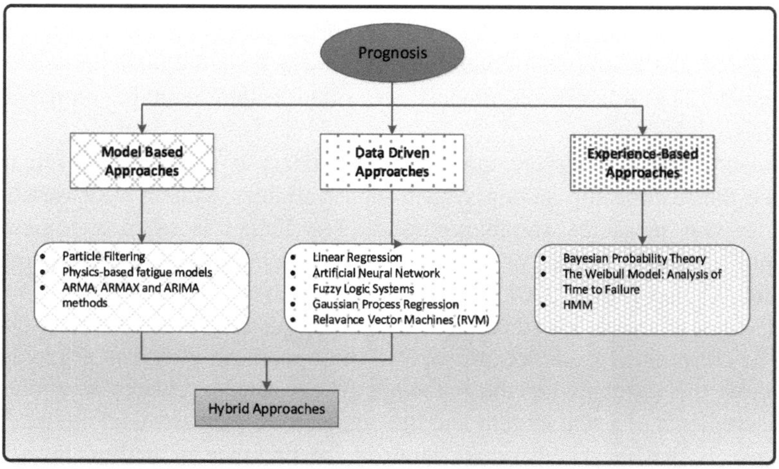

FIGURE 6.5 Approaches to and techniques of prognostics. *ARIMA*, autoregressive integrated moving average; *ARMA*, autoregressive moving average; *ARMAX*, autoregressive moving model with exogenous input; *HMM*, hidden Markov model.

Another way to estimate RUL is as follows:

$$T - \frac{t}{T} > t, Z(t),$$

where T denotes the random variable of TTF, t is the current age, and $Z(t)$ is the past condition profile up to the current time. Since RUL is a random variable, the distribution of RUL would be of interest to fully understand the RUL. In the literature, "remaining useful life estimate" has two possible meanings. In some cases, it means finding the distribution of RUL. In other cases, it simply means the expectation of RUL (Jardine et al., 2006):

$$E\left[T - \frac{t}{T} > t, Z(t),\right]$$

6.2.3 Technical Approaches

Technical approaches to building forecasting models can be classified broadly into data-based approaches, model-based approaches, and experience-based or symbolic approaches.

The following Fig. 6.5 illustrates the approaches and their respective techniques used in prognostics.

6.2.3.1 Model-Based Approaches

Model-based prognostics attempt to incorporate physical understanding (physical models) of the system into the estimation of the RUL. Modeling can be accomplished at different levels, for example, micro and macro levels. At the micro level (also called the material level), physical models are based on a series of dynamic equations that define relationships at a given time or load cycle, between damage (or degradation) of a system/component and environmental and operational conditions under which the system/component is operated. The micro-level models are often called damage propagation models.

Macro-level models are mathematical models at the system level; they define the relationship among system input variables, system state variables, and system measures variables/outputs. The model is often a somewhat simplified representation of the system, for example, a lumped parameter model (Yu and Harris, 2001).

Generally, model-based methods assume an accurate mathematical model can be constructed from first principles. These methods often use residuals as features; the residuals are the outcomes of consistency checks between the measurements of a real system and the outputs of a mathematical model. The premise is that the residuals are large in the presence of malfunctions and small in the presence of normal disturbances, noise, and modeling errors (Otilia E. Dragomir et al., 2009).

The main advantage of model-based approaches is their ability to incorporate a physical understanding of the monitored system. In addition, in many situations, the changes in feature vectors are closely related to model parameters, and a functional mapping between the drifting parameters and the selected prognostic features can be established. Moreover, if the understanding of the system degradation improves, the models can be adapted to increase accuracy and address subtle performance problems.

Consequently, they can significantly outperform data-driven approaches (see next section). But this closed relation with a mathematical model may also be a weakness: it can be difficult, even impossible, to catch the system's behavior.

Further, some authors think the monitoring and prognostic tools must evolve as the system does (Byington et al., 2002). Fig. 6.6 schematically represents physical-based prognostics (model-based approach).

6.2.3.2 Data-Driven Approaches

In many situations, the complexity of the systems under observation makes it impossible to derive robust and accurate models for prognosis.

However, historical data that capture the signal behavior of measured signals or extracted features from the incipient fault stage to equipment failure are often available. In such cases, data-driven methods that model how such signals and features evolve can be utilized to generate predictions of RUL.

Data-driven prognostic approaches typically adopt one of two strategies. The first is a two-stage process. Appropriate dimensionality reduction, feature extraction, or pattern matching techniques are employed to map system signals or features onto a single dimension damage, degradation, or health index. Technically, this first step falls under the realm of fault diagnostics since it is concerned with posterior event analysis. Once the current level of degradation is identified, it is extrapolated into the future until a predefined critical threshold limit is exceeded. A range of techniques can be applied in both steps. The second strategy is to directly model the relationship between monitored signals or features and the remaining life of the system. In this situation, the remaining life of the system is the output generated by the models.

The next section presents a brief overview of some data-based techniques, which have been applied to prognostic problems.

Time Series Approaches: The simplest data-driven approaches to prognostics rely on projection methods, which project the current level of degradation into the future. This task is essentially a time series prediction problem and within the realm of prognostics has been addressed by a variety of approaches, including autoregressive models and exponential smoothing techniques.

The autoregressive integrated moving average model (ARIMA) is part of a general class of linear models that have historically seen wide use in modeling and forecasting time series. Unsurprisingly, such approaches have also been

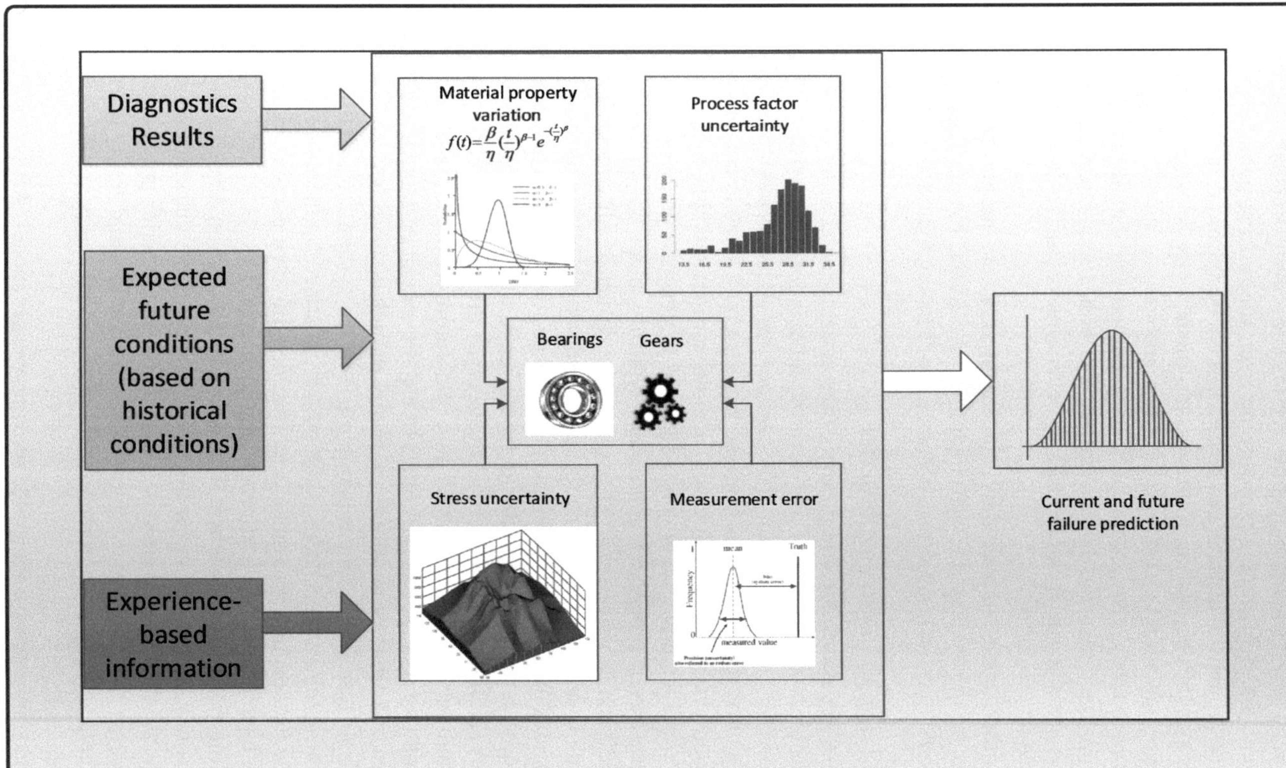

FIGURE 6.6 Physics-based modeling approach (Byington et al., 2002).

applied to prognostic problems, which are similar, in many respects, to forecasting problems. ARIMA models are derived from the more common autoregressive moving average (ARMA) model, which models a time series using two parts, an autoregressive (AR) part and a moving average (MA) part. However, since ARMA models can only be used to model stationary processes, ARIMA models are employed to model nonstationary time series signals. Examples of ARIMA models applied to prognostic problems can be found in Shane Butler (2012).

Artificial Neural Networks: Perhaps the most common data-driven technique applied to prognostic problems is an artificial neural network (ANN). ANNs model relationships between input and output variables with a model structure inspired by the neural structure of the brain. The network weights and biases, which define the interconnections between the neurons, are adapted during a training process to maximize the fit between the input and output data on which the models are trained. ANNs have been applied in a number of ways in prognostics. Their most common use is in time series prediction, where the current degradation state is predicted into the future until it exceeds a threshold value.

Typically, in a feed-forward ANN, previous values of the degradation index are used as the inputs to generate a one-step-ahead prediction. The generated output is then fed back as an input to the next iteration to generate long-term predictions. ANNs can also be employed to estimate the current degradation index, using system features as inputs. The degradation index can be predicted into the future by once again using ANNs, or via alternative prediction methods. A dynamic wavelet neural network (DWNN) is used to predict the fault propagation process into the future and estimate RUL. The first ANN used, the WNN, is a static feed-forward neural network (NN) used to derive a static relationship between inputs and outputs. The second ANN used, the DWNN, is a recurrent neural network (RNN), which incorporates feedback within the network structure to predict the time series evolution. ANNs have also been used to directly model the relationship between system features and RUL. ANNs learn by example and, as such, require sufficient instances of historical failure examples for training and are typically data hungry. As a result, ANNs can generate poor prediction performance when future failure examples on which they have not been trained do not exhibit behavior similar to the examples in the training set.

While the intrinsic ability of Gaussian processes for regression (GPR) and relevance vector machine (RVM) to generate confidence limits on generated predictions makes such approaches attractive, ANN approaches, in general, do not provide confidence limits associated with predictions. The availability of confidence limits associated with RUL predictions are highly desirable and facilitate uncertainty management. Other data-based techniques applied to prognostic problems include hidden Markov models (HMM) and neuro-fuzzy networks (Wang et al., 2004a), but in many situations, the complexity of the systems under observation makes it impossible to derive robust and accurate models for prognostic purposes.

In summary, data-driven approaches are relatively simple to implement and faster, with a huge variety of generic data mining and machine learning available. They also help our understanding of physical behaviors when data are large. They do not model physical cause—effect relationships, however.

6.2.3.3 Experience-Based Prognostics

If a physical model of a subsystem or component is absent and there is an insufficient sensor network to assess the condition, an experienced-based prognostic model may be the only alternative. This model is the least complex and requires the failure history or "by-design" recommendations of the component under similar operation. Typically, failure and/or inspection data are compiled from legacy systems, and a Weibull distribution or other statistical distribution is fitted to the data. An example of this type of distribution is given in Fig. 6.7. Although simplistic, an experience-based prognostic distribution can be used to drive interval-based maintenance practices that can be updated at regular intervals. An example is maintenance scheduling for a low criticality component with few or no sensed parameters. In this case, the prognosis of when the component will fail or degrade to an unacceptable condition must be based solely on analysis of past experience or original equipment manufacturer (OEM) recommendations. Depending on the maintenance complexity and criticality associated with the component, the prognostic system may be set up for a maintenance interval (i.e., replace every 1000 ± 20 effective operating hours) and updated as more data become available. Having an automated maintenance database is important for the application of experience-based prognostics (Byington et al., 2002).

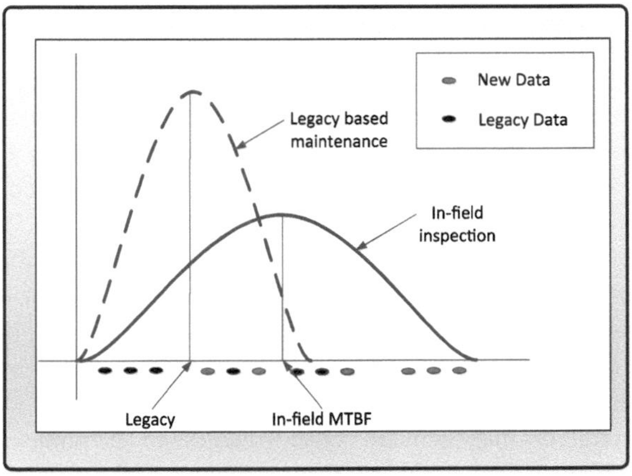

FIGURE 6.7 Experience-based approach. *MTBF*, mean time between failures.

The simplest prognostic approaches rely on the collection of statistical information, which examines historical failure rates of systems or components. Such data can be used to develop life-usage models showing distributions of failure rates over time. These approaches are helpful in the development of preventive maintenance schedules in which maintenance is performed on the basis of mean time between failures, as derived from life-usage models. However, they lack predictive capability and, as such, cannot be described as truly predictive prognostic techniques. At the same time, they have wide applicability in systems or components with low criticality or cost or in situations where sensor data, which can be used to infer condition, are not available (Shane Butler, 2012).

6.2.3.4 Hybrid Approaches

The hybrid approach attempts to harness the power of both data-based approaches and model-based approaches. In fact, we rarely see a case where only one model is used. Hybrid approaches can be classified into two types: preestimate and postestimate fusion.

Preestimate fusion is usually applied when there are no data on the ground. This can occur in situations where the diagnosis does a good job detecting faults before a system failure occurs; as a result, there is almost no information on failure. But for maintenance, it is always useful to know when a system will make better use of the RUL while avoiding unscheduled maintenance (usually more costly). Preestimate fusion can also be done by a process outside the combined online process. This requires the use of a simulation model based on physics to understand the relationships of sensor responses to error states in the online mode. First, data can be used to identify the current status of the damage. The data can then be tracked to determine the spread of damage. Finally, a propagation model based on individual data for remaining life prediction can be applied.

Postestimate fusion can be used to reduce the uncertainty ranges of approaches based on data or models while improving accuracy. The underlying notion is that multiple sources of information can improve the performance of an estimator. This principle has been successfully applied in the context of the fusion classifier; in this case, the output of multiple classifiers is used to achieve a better result than any single classifier. In the context of forecasts, fusion can be performed by assigning quality assessments to individual estimates based on a variety of inputs, e.g., heuristics, performance known a priori, horizon prediction, and robustness of the prediction (Wikipedia, The Free Encyclopedia). In summary, the advantage of a hybrid approach is that it does not necessarily require high-fidelity models or large volumes of data. In addition, it works in a complementary fashion, retains the intuitiveness of a model but explains observed data, and helps in uncertainty management. Yet because it needs both data and models, there is a risk of incorrect models or noisy data that may bias the approach.

6.3 REMAINING USEFUL LIFE AND PROGNOSTICS

This section introduces the procedure used to implement prognostics in a particular asset to estimate the RUL and explains the prognostics techniques used for the estimation. Fig. 6.8 shows the steps used to estimate the RUL of an asset. The steps are explained in more detail in the following subsections.

Once the machine or asset has been selected for analysis, depending on the type of asset, the most appropriate monitoring technique must be chosen.

Possible techniques include vibration analysis, thermography, tribology, oil analysis, etc. This will enable the collection of data on the physical variables (temperature, vibration, viscosity, voltage, acoustic waves, etc.) to be analyzed to estimate useful life.

Data are collected using sensors placed on specific points on the assets; the types of sensors represent the dependent variable and can be analyzed.

When the data are already collected from the monitoring techniques, we must filter the data, because most of the time, they are contaminated by the environment where the asset is used. This contamination can be, for example, noise signals. These alter the collected data, causing erroneous results.

In addition to filtering, the following must be done:

Feature extraction:

1. Extract patterns from raw signals.
2. Compute numeric or symbolic information from the observations and build features.
3. Use techniques from signal/image processing, data mining, statistics, etc.

Feature selection:

1. Reduce dimensionality.
2. Use techniques from data mining, statistics, machine learning, genetic algorithms, etc. (Gouriveau, 2011).

Fig. 6.9 shows an example of feature extraction from a vibration signal.

After the data are monitored, analyzed, and processed, we proceed to make a diagnosis of the asset under analysis.

The goal of diagnosis is to determine the state of the asset's health to identify whether it is failing or not operating correctly. Algorithm fault diagnosis is designed to detect system performance, control levels of degradation, and identify failures (or faults) based on changes in physical properties, through detectable phenomena. It also identifies the subsystem or component, which is failing and the specific mechanism of failure that has occurred (Shane Butler, 2012). After making the diagnosis and determining the health of the asset, most maintainers want to know when to stop using it; this requires an estimation of RUL, something always done after diagnosis. As noted above, prognostics can be used to estimate the RUL. The approach can be model-based, data-driven,

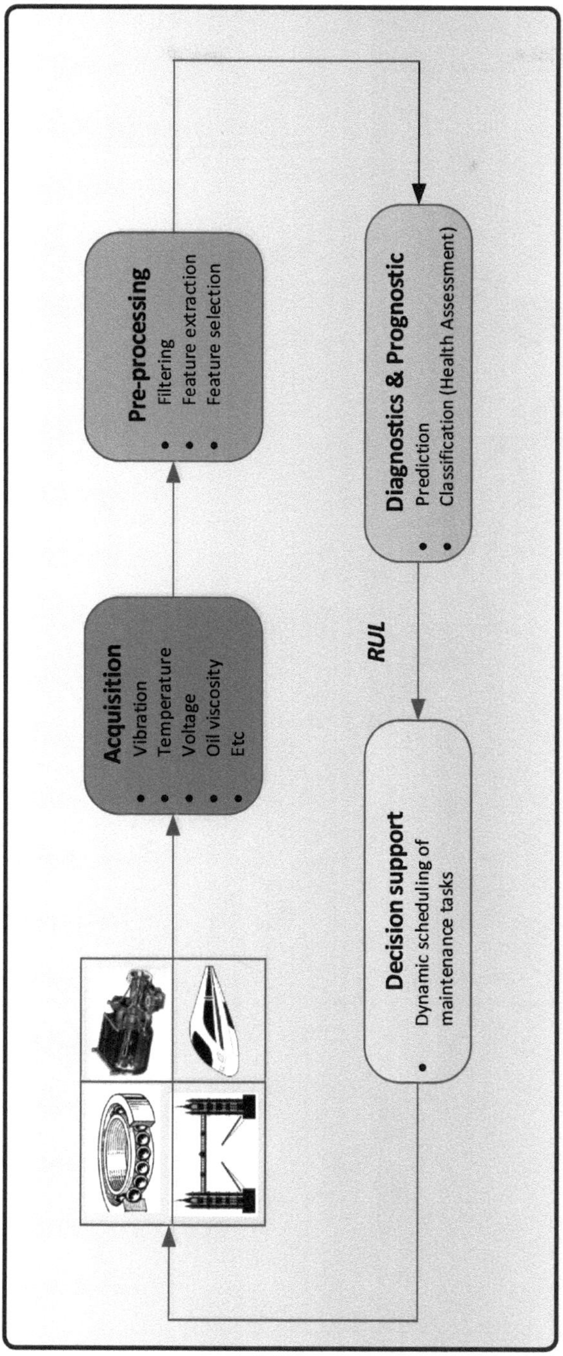

FIGURE 6.8 Implementing prognostics. *RUL*, remaining useful life.

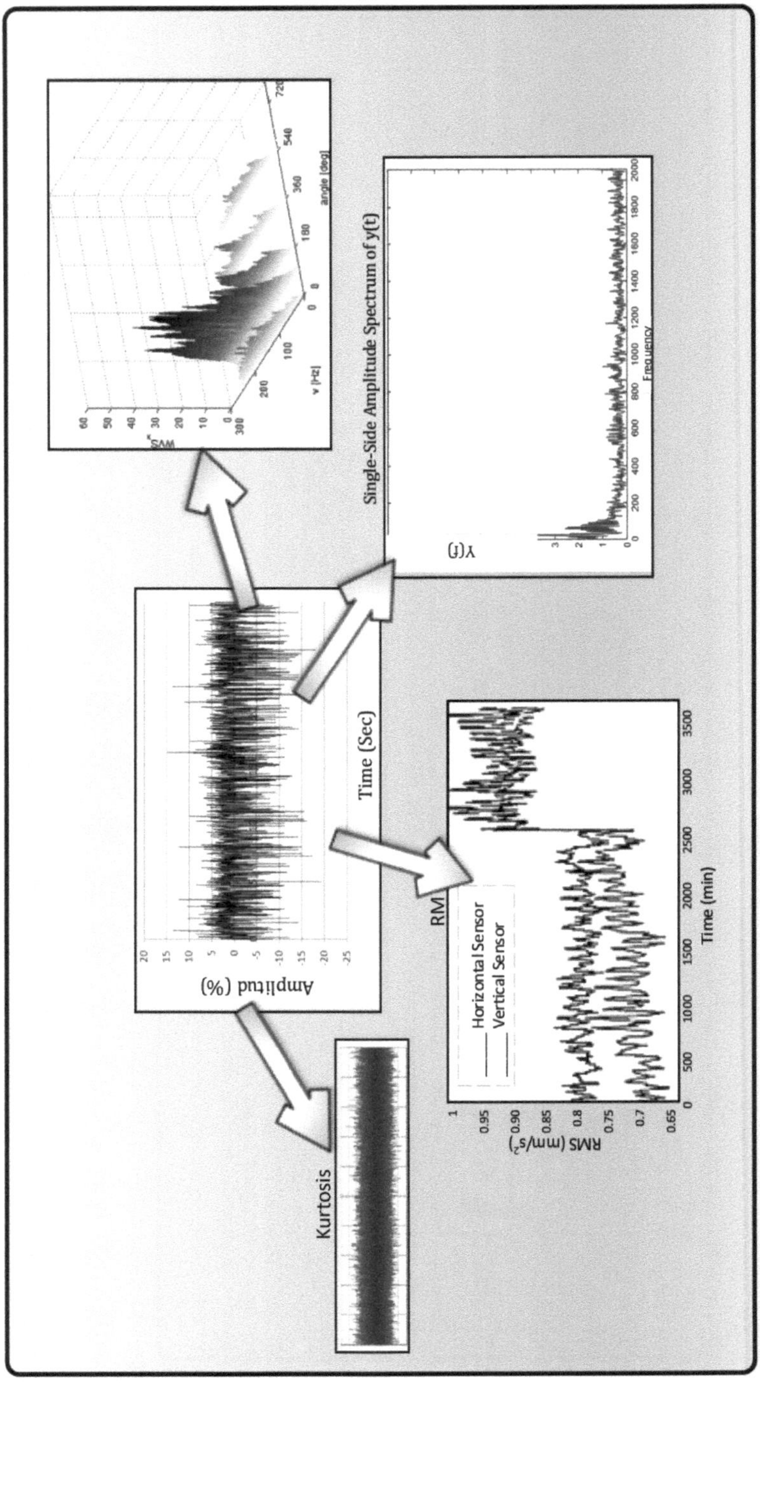

FIGURE 6.9 Feature extraction from a vibration signal (Gouriveau, 2011).

or experience-based. A hybrid approach fuses elements of the others. Each can be implemented using the techniques shown in the diagram.

The last step, performed after estimating the RUL of the asset or system, is a series of operation and maintenance decisions based on the type of asset or system. Clearly, such decisions are very important to save costs. Hence, a good estimate of RUL is crucial.

6.3.1 Prognostic Techniques

As explained above, prognostics are based on three types of approaches, along with an approach fusing two of the three. The techniques they use are explained below.

6.3.1.1 Techniques of Model-Based Approaches

6.3.1.1.1 Particle Filtering for Prognostics

One way to estimate the RUL of a component or system is to use a Bayesian technique. This technique uses a dynamic state model and a measurement model to predict the density function posterior probability of the state, i.e., to predict the time evolution of a fault.

A particulate filter is a recursive Bayesian estimation technique used to avoid the assumption of linearity and Gaussian noise. Kalman filtering provides a solid framework for long-term prognosis. A methodology for accurate and precise prediction, it presents a failing component on the basis of particle-filtering strategies. We use fatigue failure as an example here.

Prediction of the evolution of a fault or fault indicator entails large-grain uncertainty. Accurate and precise prognosis of the TTF of a failing component/subsystem must consider critical-state variables such as crack length, corrosion pitting, etc. as random variables with associated probability distribution vectors. Once the probability distribution of the failure is estimated, other important prognosis attributes such as confidence intervals can be computed.

A possible solution to the prognosis problem is the use of recursive Bayesian estimation techniques that combine both the information from fault-growth models and online data obtained from sensors monitoring key fault parameters (observations). Prognosis or long-term prediction of failure evolution is based on both an accurate estimation of the current state and a model describing the fault progression. If the incipient failure is detected and isolated at the early stages of the fault initiation, it is reasonable to assume sensor data will be available for a certain time window, allowing corrective measures to be taken. Thus, there will be improvements in model parameter estimates so that prognosis will provide accurate and precise prediction of the TTF. At the end of the observation window, the prediction outcome is passed on to the user (operator, maintainer);

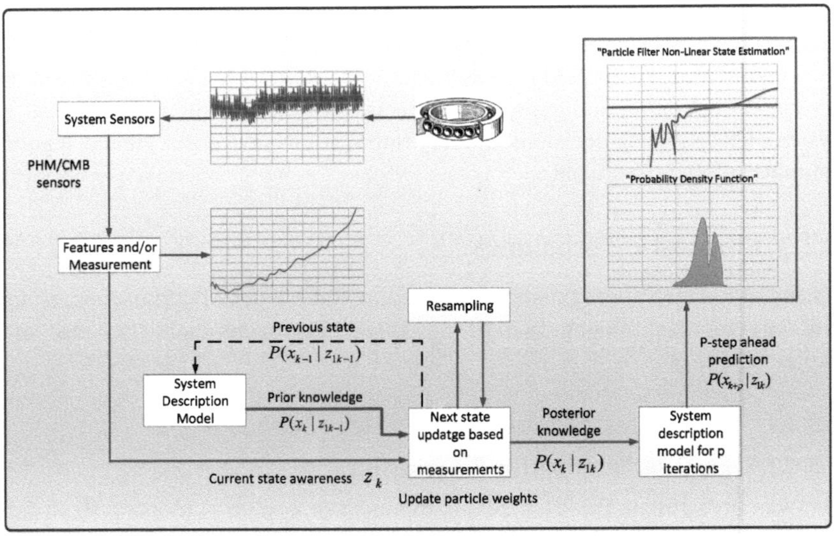

FIGURE 6.10 Fault prognosis based on particle filtering.

additional adjustments are not feasible because corrective action must be taken to avoid a catastrophic event.

Fig. 6.10 depicts a conceptual schematic of a particle-filtering framework aimed at addressing the fault prognosis problem. CBM sensors and the feature-extraction module provide the sequential observation (or measurement) data of the fault growth process Z_k at time instant k. We assume the fault progression can be explained through the state evolution model and the measurement model:

$$X_k = f_k(X_{k-1}, \omega_k) \leftrightarrow p(X_k | X_{k-1}) \tag{6.1}$$

$$Z_k = h_k(X_k, \nu_k) \leftrightarrow p(Z_k | X_k) \tag{6.2}$$

where, X_k is the state of the fault dimension (such as the crack size); $f_k : \mathbb{R}^{n_x} \times \mathbb{R}^{n_\omega} \rightarrow \mathbb{R}^{n_x}$ is the state transition function (possibly nonlinear), which describes the evolution of the system state; $h_k : \mathbb{R}^{nz} \times \mathbb{R}^{n_\nu} \rightarrow \mathbb{R}^{n_z}$ is the measurement/observation function, which describes the sequence of measurements Z_k collected at successive time steps t_k; $\omega_k \in \mathbb{R}^{n_\omega}$ is an independent identically distributed (i.i.d.) process noise sequence of known distribution; $\nu_k \in \mathbb{R}^{n_\nu}$ is an i.i.d. measurement noise sequence of known distribution.

The first part is state estimation, that is, estimating the current fault dimension as well as other important changing parameters in the environment. The second part is long-term prediction based on the current estimate of the fault dimension and the fault-growth model with parameters refined in the posteriori state estimation. A novel recursive integration process based on both importance sampling and the probability density function (pdf) approximation using Kernel functions is then applied to generate state predictions from $(k+1)$ to $(k+p)$ as the following:

$$P(X_{k+p}|Z_{1k}) = \int P(X_{k+p}|Z_{0k}) \prod_{j=k+1}^{k+p} P(x_j|x_{j-1}) dx_{kk+p-1} \tag{6.3}$$

$$= \sum_{i=1\$}^{N} w_k \int \dots \int P\left(x_{k+1} \middle| x_k^{(i)}\right) \prod_{j=k+2}^{k+p} P(x_j|x_{j-1}) dx_{k+1k+p-1} \tag{6.4}$$

Long-term predictions can be used to estimate the PoF in a process, given a hazard zone defined by its lower and upper bounds (H_{lb} and H_{up}, respectively). The prognosis confidence interval, as well as the expected TTF, can be deduced from the TTF pdf as

$$P_{TTF}(TTF) = \sum_{i=1}^{N} \Pr\left\{ H_{lb} \leq x \overset{(i)}{TTF} \leq H_{up} \right\} w_{TTF} \tag{6.5}$$

The procedure described forms the basis for the determination of the optimal Bayesian solution. However, the recursive computation of the posterior state pdf is more conceptual than practical and, in general, cannot be determined analytically. In a restricted set of cases, such as when using linear Gaussian state space models, the optimal solution leads to the well-known Kalman filter (mentioned above). However, in the presence of a nonlinear process model and/or non-Gaussian noise processes, an alternative approach must be considered. A common choice is particle filtering, which approximates the optimal Bayesian solution.

6.3.1.1.2 Physics-Based Fatigue Models

Physical models quantitatively characterize the behavior of a failure mode using physical laws. This means we must have deep knowledge about the behavior of the system being studied. Physical models perform an estimate of the RUL of an asset or system by solving equations or a deterministic set of equations derived from empirical data collected in the data acquisition. Some of these data are obtained using common scientific and engineering knowledge, while others are collected through specific laboratory or field experimentation.

Physical models for the estimation of the RUL of a particular system must identify one or more specific parameters of the system (e.g., exact physical properties, corrosion rates, equation constants). These models are generally described in two ways: first, using dynamic Lagrangian or Hamiltonian dynamic ordinary differential or partial equations (approximation methods applied to partial differential equations, distributed models, and other techniques); second, using methods of state space (i.e., no differential equations), resolved accordingly (Vachtsevanos et al., 2006). Once a physical model is available, the sensor measurements are compared to actual process model outputs.

The difference between reality and the model is called waste; large differences (residuals) are assumed to indicate a fault, while small ones occur in normal conditions or as a result of noise or modeling errors. The residuals are calculated using parameter estimation, state-space methods, or parity equations (Luo et al., 2003b).

RUL estimates are based on projecting degradation behaviors into expected future operating conditions. To construct a physics of failure model, the following features and associated levels of certainty must be characterized:

- Set of likely initiating failure modes for which behavioral models are required.
- Process of behavior across possible/typical operating ranges.
- Degradation behavior under aforementioned process conditions.
- Relationship between process measurements and degradation behavior(s).
- Process and measurement of noise (Sikorska, 2010).

A few failure mechanisms by which physical models have been developed include fatigue, overload failure, corrosion, and ductile to brittle transitions. When discussing effects, failure mechanisms can be divided into two categories.

The first can be considered failures in effort. These occur when the load exceeds the material strength; there is no long-term effect once the load has been removed. Examples include brittle fracture, ductile fracture, and buckling performance.

The second category is failures; these are characterized by irreversible, cumulative damage that does not disappear when the load is removed, such as fatigue, corrosion, stress-corrosion cracking, and wear and creep. Once the mechanism allowed for the particular damage tolerance is exhausted, the normal operating loads will exceed the strength the asset has left and a failure will occur in effort (Blischke and Murthy, 2000).

In a physical forecasting model, if we want to implement the failure modes, we must be able to track the damage and its rate of progression under any or all conditions of operation.

However, when talking about failures in effort, we only need to monitor the current state to determine the immediate risk of failure and carry out a stochastic analysis of force and distributed load (Todinov, 2005). In practice, when discussing behavior, forecasting applies only when we have identified the relevant failures. The main advantage of this group of models is the ability to incorporate the current understanding of the physical mechanisms of failure; this has been the subject of extensive and exhaustive empirical testing. Crack propagation models, of clear interest to prognosis, include deterministic and stochastic models. Growth of fatigue cracks in the components of typical equipment, such as bearings, gears, shafts, and aircraft wings, is affected by a variety of factors, including the states of stress, material properties, temperature, lubrication, and other environmental effects. A variation of empirical and deterministic models can model fatigue crack propagation based on Paris' formula (Vachtsevanos et al., 2006).

Fatigue crack growth can be expressed as

$$\frac{d\alpha}{dn} = C_0(\Delta K)^n \tag{6.6}$$

where, α is the instantaneous length of the dominant crack; N is the running cycles; C_0 are the material dependent constants; and ΔK is the range of the stress-intensity factor over one loading cycle.

$\Delta K = Y(\alpha)(\Delta_S)\sqrt{\pi\alpha}$ where, K is a stress-intensity factor, $Y(\alpha)$ is related to crack geometry, and Δ_s is the stress range.

Growth models of stochastic cracks consider all parameters as random quantities. Thus, the resulting equation is a growth stochastic differential equation. Here, to estimate the distribution parameters, we use Monte Carlo simulations, probabilistic NNs, and others (Vachtsevanos et al., 2006).

6.3.1.1.3 ARMA and ARIMA Methods

Parameter estimation prognosis is another model possibility when the physics-based process model is unavailable or too complicated for implementation. In such cases, some of the system identification procedures must be used. Based on the knowledge of the system, a model structure and complexity can be proposed. System identification reduces the estimation of the unknown model parameters vector using the observed input and output sequences.

As explained at greater length below, ARMA models (p, q) capture the entire process when a simple model AR (p) or MA (q) cannot.

ARIMA models (p, d, q) solve the problem where the test range is not stationary due to the existence of a trend.

Finally, multiplicative seasonal ARIMA models (p, d, q) × (p, d, Q)s solve the problem where the test range is not stationary due to the existence of a seasonality and/or a trend.

***6.3.1.1.3.1 Autoregressive Moving Average Models* (p, q)** An ARMA model combines the two basic models, AR (p) and MA (q), to model partially and partially autoregressive series of moving averages.

These combined models are abbreviated by the acronym ARMA (p, q), where p is the order of the autoregressive part and q is the order of the moving average.

The general expression for an ARMA model (p, q) is

$$\begin{aligned} Y_t = \phi_1 Y_{t-1} + \phi_2 Y_{t-2} + \ldots + \phi_p Y_{t-p} + a_t + \theta_1 a_{t-1} + \theta_2 a_{t-2} \\ + \ldots + \theta_q a_{t-q} + \mu \end{aligned} \tag{6.7}$$

As the model is a combination of the two basic models, the value of the series Y_t is a linear combination of the most recent disturbances of q and p.

6.3.1.1.3.2 *Autoregressive Integrated Moving Average Models* (p, d, q)

A time series (Y_t) follows an ARIMA model if the 10th differentiation (where d is an integer) $W_t = \nabla^d Y_t$ is a stationary ARMA process. If (W_t) follows an ARMA (p, q), we can say (Y_t) is a process ARIMA (p, d, q). For practical purposes, we generally use a d of 1 or 2.

The differential equation of the form of an integrated moving average ARMA (p, d, q) autoregressive model is

$$\begin{aligned} Y_t = (1+\phi_1)Y_{t-1} + (\phi_2 - \phi_1)Y_{t-2} + \ldots + (\phi_p - \phi_{p-1})Y_{t-p} + a_t + \theta_1 a_{t-1} \\ + \ldots + \theta_2 a_{t-2} + \ldots + \theta_q a_{t-q} + \mu \end{aligned} \tag{6.8}$$

6.3.1.1.3.3 *Multiplicative Seasonal Autoregressive Integrated Moving Average Models* (p, d, q) × (p, d, q)*s*

A time series (Y_t) is a multiplicative seasonal ARIMA model with no seasonal orders (regular) p, d, and q, with seasonal orders P, D, and Q, and with a seasonal period s if the series differentiated $W_t = \nabla^d \nabla_s{}^D Y_t$ satisfies an ARMA model (p, q) × (P, Q)s with seasonal period s. Thus, the (Y_t) process is an ARIMA (p, d, q) × (P, D, Q)s with seasonal period s.

In its most general form, multiplicative seasonal ARIMA model (p, d, q) × (P, D, Q)s can be written as

$$\begin{aligned} Y_t = \phi_1 Y_{t-1} + \phi_2 Y_{t-2} + \ldots + \phi_{ps+p+Ds+d} Y_{t-ps-p-Ds-d} + \ldots + \\ a_t + \theta_1 a_{t-1} + \theta_2 a_{t-2} + \ldots + \theta_{qs+q} a_{t-qs-q} + \mu \end{aligned} \tag{6.9}$$

This expression defines which is the most generic of the ARIMAs model; all others are special cases of it. When we want to do prognostics using ARIMA models, we must take into account the need to understand the equation defining the delay model to a set in $t + 1$, where l is the moment we want to predict. As can be seen, no delay is actually calculated, and if time t is considered "current", the predicted time is at a distance l in the future.

$$\begin{aligned} \widehat{Y}(l) = \phi_1 \widehat{Y}_t(l-1) + \phi_2 Y_t + \phi_{l-1} Y_t(l) + \ldots + \phi_{l+1} Y_{t-1} + \ldots + \\ \phi_{PS+p+Ds+d} Y_{t+l-ps-p-Ds-d} + \theta_l a_t + \theta_{l+1} a_{t-1} + \theta_{Qs+q} a_{t+l-Qs-q} + \mu \end{aligned} \tag{6.10}$$

The formula to predict the value of the variable for moment l for a general ARIMA (p, d, q) × (P, D, Q)s is shown above.

The following aspects of prognostics are important:

- Prognostics values are calculated sequentially. To enable the implementation of the autoregressive part of a model in periods other than the first prediction, the predicted value is taken in the immediately preceding $\widehat{Y}_t(l-1)$.
- All known random perturbations are considered to be in the sample period and to have the character of white noise in prognostics periods.

- The optimal predictor (the expected value of the series in the period of prognostics is equal to that predicted if optimal) must show that random perturbations used in the prediction are those of previous periods, i.e.,

$$a_{t+j} \quad \forall j \leq 0 \tag{6.11}$$

Since we assume that in the prognostics period, there are no random perturbations, to make subsequent predictions, $Y_t\ (l)$, we must have values for a_t, a_{t-1},…$t + lq$. This is expressed as

$$a_{t+j} = Y_{t+j} - \widehat{Y}_{t+j-l}(l) \quad \forall j \leq 0 \tag{6.12}$$

ARIMA prognostics are adaptive and the results obtained for $(t + l)$, with the information available at time t, are the same as those obtained for the same period, taking $t-1$ as baseline information and adding a term error (Vachtsevanos et al., 2006).

6.3.1.1.4 Data-Driven Approaches

6.3.1.1.4.1 Linear Regression Regression analysis uses existing data and determines the relationships, if any, between the measurable outcome and the variables contributing to that outcome (e.g., life expectancy is the outcome and exercise and diet are the variables contributing to that outcome.) Neter et al. (1996) offer a framework to use a statistical relationship to predict remaining machine life. Their general linear regression model is given by

$$Y_i = \beta_0 + \beta_1 X_{i,1} + \beta_2 X_{i,2} + \ldots + \beta_{p-1} X_{i,p-1} + \varepsilon_j, \quad i = 1, 2, \ldots n \tag{6.13}$$

where Y_i is a random variable denoting the value of the ith trial's response, b_0, b_1, b_2 are estimated parameters, $X_{i,1}$, $X_{i,2}$,…$X_{i,p-1}$ are the values of the prediction or contributing variables, and ε_i is the random error with mean = 0, variance = σ^2, and covariance = 0. Regression analysis estimates the parameters of the regression function, β_0, β_1, β_2, β_{p-1}, to find a representative model using the method of least squares. The method of least squares defines a variable Q where

$$Q = \sum_{i=1}^{n} \left(Y_i - \beta_0 - \beta_1 X_{i,1} - \beta_2 X_{i,2} - \ldots \beta_{p-1} X_{i,p-1} \right)^2 \tag{6.14}$$

and attempts to find estimates for β_0, β_1, β_2, β_{p-1} denoted by b_0, b_1, b_2…, b_{p-1}, which minimize Q for the observations (X_1, Y_1), (X_2, Y_2), … (X_n, Y_n). The simultaneous solution to the equations formed by taking the derivative of Q with respect to β_0, β_1, β_2, β_{p-1} provides the least squares estimates, b_0, b_1, b_2…, b_{p-1}. The advantage of least squares estimates is that they are unbiased and have minimum variance, resulting in

$$\widehat{Y} = b_0 + b_1 X_1 + b_2 X_2 + b_{p-1} X_{p-1} \tag{6.15}$$

The method of maximum likelihood can also be used to estimate b_0, b_1, b_2..., b_{p-1}, if the probability distribution of the error terms is known (Neter et al., 1996).

6.3.1.1.4.2 Neural Network Model An NN is basically a computer program designed to simulate the way the human brain processes information. By using the concept of learning through experience, an NN pools knowledge and identifies patterns and relationships in data. The NN structure constitutes a computational model containing hundreds of artificial neurons and connects with known coefficients as weights. Fig. 6.11 shows the model structure of an NN. Input signals X_1, X_2,...X_n are propagated through the network with weights. W_1, W_2,...W_n represent the connection between neuron input and hidden neurons. The combined input signals and weights pass through an activation procedure to produce the output value of the neuron y_k (Agatonovic-Kustrin and Beresford, 2000).

NNs have been widely applied in time series prediction. The time series prediction model uses a feedforward neural network (FNN) and employs a sliding window over the input sequence. To construct the multistep prediction, the model employs previous predicted values to forecast future values iteratively until the expected future values are obtained. This multistep prediction is illustrated in Fig. 6.12 (Tan and Van Cauwenberghe, 1999).

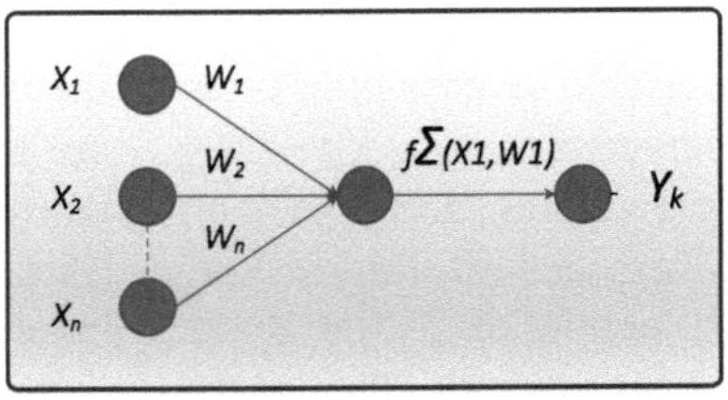

FIGURE 6.11 Neural network structure.

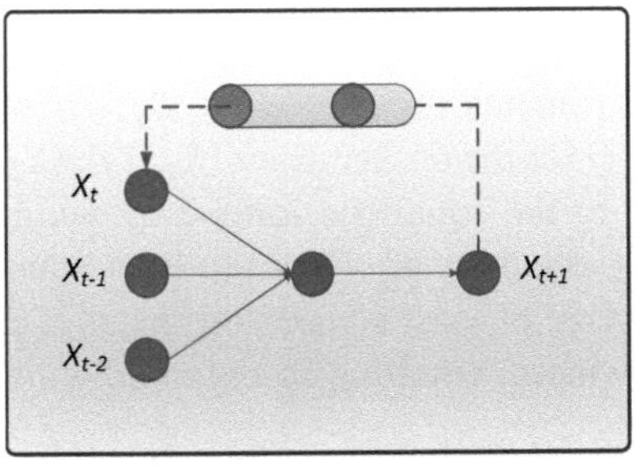

FIGURE 6.12 Multistep prediction.

This multistep technique can predict a series of future values based on the required number of time step values d. The prediction technique uses the previous values to forecast the future d time step values iteratively. Given the observation Yt = [X_{t-r+1}, X_{t-r-2},...X_t], the first future value can be predicted by using

$$\widehat{y}_{t+1} = f(y_t) = f(x_{t-r+1}, x_{t-r+2} \ldots x_t) \tag{6.16}$$

where r denotes the number of inputs or the size of sliding window dimension. To predict the next value, the same prediction model can be given as:

$$\widehat{y}_{t+2} = f\left(x_{t-r+2}, x_{t-r+3} \ldots \widehat{y}_{t+1}\right) \tag{6.17}$$

Then, the procedure is repeated recursively depending on the required number of time series d (Siti Azirah Asmai et al., 2011).

$$\widehat{y}_{t+d} = f\left(x_{t-r+d}, x_{t-r+d+1} \ldots \widehat{y}_{t+d-1}\right) \tag{6.18}$$

The prognostic process is accomplished by predicting and extrapolating the dynamic failure probabilities (FPs) over time from the performance degradation model using FNN. To use FNN, three other key parameters need to be considered: the number of hidden layers, the choice of activation functions, and the number of neurons. As a single hidden layer can compute a uniform approximation of any continuous function (Sun et al., 2010), the proposed FNN architecture is composed of an input layer, a hidden layer, and an output layer with one output neuron. To introduce nonlinearity into the network model, nonlinear activation functions are needed. Logistic and tanh functions can be used to create a combination of activation functions from the input layer to the output layer.

In applying NNs, determining the number of hidden input neurons has always been an issue. Having a smaller number of hidden neurons tends to result in ineffective performance, while having too many may increase the risk of overfitting the data and impeding generalization. Ultimately, the selection of the architecture of an NN comes down to trial and error (Heaton, 2008). There is a selection method for the process of trial and error to determine the number of input neurons and hidden neurons, with the final combination resting on the activation functions. In this method, the training data are iteratively trained, with the number of input and output activation functions changed until the error produced by the network is minimal based on the root mean square error (RMSE) calculated as follows:

$$\text{RMSE} = \sqrt{\frac{1}{n} \sum_{t=1}^{n} (x_t - \overline{x}_t)^2} \tag{6.19}$$

where n is length of time series data, x_t represents the target values, and $\overline{X}_t$ represents actual values. After the FNN architecture is identified, the training data set is used to train the network in adjusting the synaptic weights. Once the network is completely trained, the weights are frozen, and the network is ready to predict and extrapolate FP. To validate the performance of the network, a set of validation data is used to compare the network output with the predicted output (Siti Azirah Asmai et al., 2011).

It is important to note that the technique of NN is frequently used in the medical realm, especially to determine the estimated risk of relapse in patients with cancer. Many institutions and medical laboratories are using this prognostic technique to determine the health of patients. In other words, prognostics techniques used in the field of engineering can be applied in other fields such as health.

6.3.1.1.4.3 Fuzzy Logic Systems Neuro-fuzzy system (NFS) is a data-driven method employed in the prediction of machine-condition degradation. NFS has been shown to outperform conventional neural network–based predictors such as FNNs, radial basis functions, and RNN-based models. Through off-line training with available data sets, NFS is used to model machine dynamics to enable accurate predictions of machine health. However, since the machine dynamics in real applications change with time, the trained NFS cannot carry out accurate predictions if the new dynamics/states are not taken into account during the prediction process. As Bayesian algorithms can update system states in real time via new data, NFS is integrated with Bayesian algorithms so online data can be used to improve the prediction accuracy (Wang et al., 2004b).

The NFS predictor is, in essence, a fuzzy logic system, where the system parameters are optimized via NN training. The NFS architecture is schematically shown in Fig. 6.13. It consists of five layers: input layer, membership

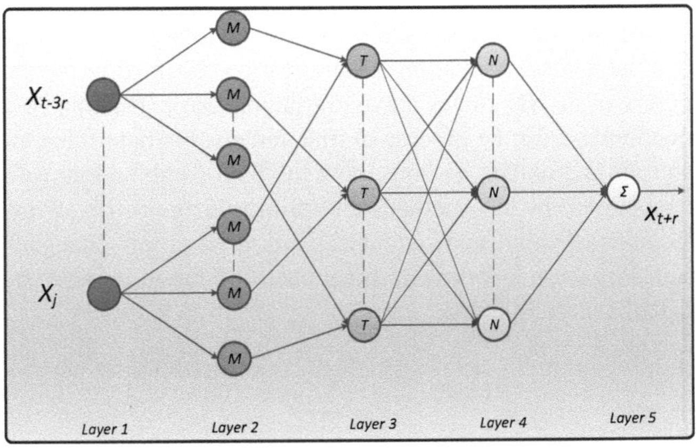

FIGURE 6.13 An example of an architecture of a fuzzy system login (neuro-fuzzy system).

function (MF) layer, rule layer, normalized layer, and output layer. There are l input nodes in the input layer; each is related to m term nodes in the MF layer. Thus, the number of nodes in the MF layer is $l \times m$, where m denotes the rule number (Lin and Lee, 1996). The NFS signal propagation proceeds as follows:

Layer 1 (input layer): The input values are transmitted directly to the next layer without any computation. The outputs of this layer can be expressed by

$$O_i^{(1)} = X_i^{(1)} \tag{6.20}$$

Layer 2 (MF layer): Each node in this layer performs an MF calculation. Sigmoid MFs are used here, as shown below:

$$u_{ij}^{(2)} = \frac{1}{1 + \exp\left(-b_{ij}^{(2)}\left(O_i^{(1)} - m_{ij}^{(2)}\right)\right)} \tag{6.21}$$

where $u_{ij}^{(2)}$ is the MF layer's output and is associated with the jth term of the ith input and $O_i^{(1)}$, $b_{ij}^{(2)}$ and $m_{ij}^{(2)}$ are the parameters of the sigmoid function.

Layer 3 (rule layer): The following max-product operation is carried out in this layer and is expressed as

$$O_i^{(3)} = \prod_i u_{ij}^{(2)} \tag{6.22}$$

where the output $O_i^{(3)}$ represents the firing strength of the jth fuzzy rule.

Layer 4 (normalized layer): This layer performs the normalization operation for all the rule firing strengths. The resulting output is given by

$$O_i^{-(4)} = \frac{o_i^{(3)}}{\sum_i o_j^{(3)}} \tag{6.23}$$

Layer 5 (output layer): The output of the NFS is calculated by using the centroid defuzzification procedure.

$$X_{t+r} = O_k^{(5)} = \sum_j o_j^{-(4)} w_{JK}^{(5)} \tag{6.24}$$

where $W_{jk}^{(5)}$ is the kth estimated output associated with the jth rule (Chen, 2012).

Another way to apply fuzzy data-driven prognostics is to use recurrent neuro-fuzzy systems (RNFS). The RNFS predictor (Fig. 6.14) possesses additional feedback links added in Layer 2 (MF layer). Each node in Layer 2 functions as a memory unit that performs the following operation:

$$u_{ij}^{(2)} = \frac{1}{1 + \exp\left(-b_{ij}^{(2)}\left(x_{ij}^{(2)} - m_{ij}^{(2)}\right)\right)} \tag{6.25}$$

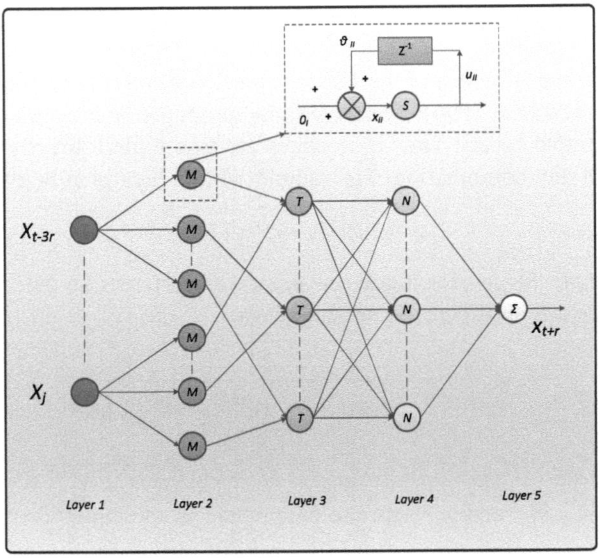

FIGURE 6.14 Architecture of recurrent neuro-fuzzy systems predictor; $Z-1$ is a unit delay operator; S is a sigmoid function; T is max-product operation; N is normalized operation.

where $u_{ij}^{(2)}$ is the MF layer's output associated with the jth term of the ith input and $x_{ij}^{(2)}$, $b_{ij}^{(2)}$ and $m_{ij}^{(2)}$ are the parameters of the sigmoid function. Note that the inputs of this layer involve the feedback components

$$X_{ij}^{(2)}(t) = O_i^{(1)}(t) + \theta_{ij}^{(2)} u_{ij}^{(2)}(t-1) \tag{6.26}$$

where $\theta_{ij}^{(2)}$ is the feedback link weight. The activation responses $u_{ij}^{(2)}(t-1)$ at the previous time step are used as part of the current input values; this allows the RNFS predictor to memorize past information to deal with temporal issues. More information on recurrent NFSs can be found in Lee and Teng (2000).

6.3.1.1.4.4 Gaussian Process Regression A Gaussian process (GP) is a collection of random variables, any finite number of which has a joint Gaussian distribution. A real GP $f(x)$ is completely specified by its mean function $m(x)$ and covariance function $k(x,x')$ defined as

$$\begin{aligned} &m(x) = E[f(x)], \\ &k(x,x') = E[(f(x) - m(x))(f(x') - m(x'))], \\ &f(x) \approx GP(m(x), k(x,x')). \end{aligned} \tag{6.27}$$

The index set $X \in \Re$ is the set of possible inputs, which need not necessarily be a time vector. Given prior information about the GP and a set of training points $\{(x_i, f_i)\ i = l, \ldots, n\}$, the posterior distribution over functions is derived by imposing a restriction on prior joint distribution to contain only those functions that agree with the observed data points (Rasmussen and Williams, 2006). These functions can be assumed to be noisy, as in real world situations, we have access to only noisy observations not exact function values; i.e., $y_i = f(x) + \varepsilon$, where ε is additive i.i.d. $N\left(0, \sigma_n^2\right)$. Once we have a posterior distribution, it can be used to assess predictive values for the test data points. The following equations describe the predictive distribution for GPR (Williams et al., 1996):

$$\text{PRIOR} \quad \begin{bmatrix} y \\ f_{\text{test}} \end{bmatrix} \approx N\left(0, \begin{bmatrix} K(X,X) + \sigma_n^2 & K(X, X_{\text{test}}) \\ K(X_{test}, X) & K(X_{\text{test}}, X_{\text{test}}) \end{bmatrix}\right) \tag{6.28}$$

$$f_{\text{test}} \Big| X, y, X_{\text{test}} \approx N\left(\bar{f}_{\text{test}}, \text{cov}(f_{\text{test}})\right)$$

$$\text{POSTERIOR} \quad \bar{f}_{\text{test}} \equiv E[f_{\text{test}} | X, y, X_{\text{test}}] = K(X, X_{\text{test}})\left[K(X,X) + \sigma_n^2 I\right]^{-1}$$
$$\text{cov}(f_{\text{test}}) = K(X_{\text{test}}, X_{\text{test}}) - \left[K(X_{\text{test}}, X) + \sigma_n^2 I\right]\right]^{-1} K(X, X_{\text{test}}) \tag{6.29}$$

where X is the input, y is the target, K is the covariance function, σ_n^2 is the noise level, and X_{test} are test inputs.

A crucial ingredient in a GP predictor is the covariance function that encodes the assumptions about the functions to be learned by defining the relationship between data points. The covariance structure also incorporates prior beliefs of the underlying system noise. GPR requires prior knowledge of the form of the covariance function, which must be derived from the context if possible. Furthermore, covariance functions consist of various hyperparameters that define their properties. Setting right values of hyperparameters is another challenge in learning the desired functions. Although the choice of covariance function must be specified by the user, corresponding hyperparameters can be learned from the training data using a gradient-based optimizer, such as maximizing the marginal likelihood of the observed data with respect to hyperparameters (Mardia and Marshall, 1984).

6.3.1.1.4.5 Relevance Vector Machine The RVM is a Bayesian form representing a generalized linear model identical in its functional form to the support vector machine (SVM). Although SVM is a state-of-the-art technique for classification and regression, it suffers from a number of disadvantages, one of which is the lack of probabilistic outputs that make more sense in health-monitoring applications. RVM addresses these issues in a Bayesian framework and uses fewer kernel functions for comparable generalization of performance (Tipping, 2000).

This type of supervised machine learning starts with a set of input vectors $\{x_n\}$, $n = 1, \ldots, N$, and their corresponding targets $\{t_n\}$. The aim is to learn a model of the dependency of the targets on the inputs to make accurate predictions of t for unseen values of x. Typically, the predictions are based on some function $F(x)$ defined over the input space, and learning is the process of inferring the parameters of this function. The targets are assumed to be samples from the model with additive noise:

$$t_n = F(x_n; w) + \varepsilon_n \tag{6.30}$$

where, ε_n represents independent samples from some noise process (Gaussian with mean 0 and variance σ^2). Assuming the independence of t_n, the likelihood of the complete data set can be written as

$$p(t|w, \sigma^2) = (2\pi\sigma^2)^{-N/2} \exp\left\{ -\frac{1}{2\sigma^2} \left\| t - \Phi W \right\|^2 \right\} \tag{6.31}$$

where, $w = (w_1, w_2, \ldots, w_M)^T$ is a weight vector and Φ is the $N \times (N+1)$ design matrix with $\Phi = [\phi(t_1), \phi(t_2), \ldots, \phi(T_n)]^T$; in which $\phi(t_N) = [1, K(x_n, x_1), K(x_n, x_2), \ldots, K(x_n, x_N)]^T$, and $K(x, x_i)$ is a kernel function.

To prevent overfitting, a preference for smoother functions is encoded by choosing a zero-mean Gaussian prior distribution over w parameterized by the hyperparameter vector η. To complete the specification of this hierarchical prior, the hyperpriors over η and the noise variance σ^2 are approximated as delta functions at their most probable values η_{MP} and $\sigma^2{}_{\mathrm{MP}}$. Predictions for new data are then made according to

$$p(t^*|t) = \int p(t^*|w, \sigma^2_{\mathrm{MP}}) p(w|\eta_{\mathrm{MP}}, \sigma^2_{\mathrm{MP}}) dw \tag{6.32}$$

6.3.1.1.5 Experience-Based Approaches

6.3.1.1.5.1 Bayesian Probability Theory Methods of confronting uncertainty include Bayesian methods, the Dempster–Shafer theory, and fuzzy logic. Probability methods are generally based on Bayes' theorem (Lewis, 1986), which says

$$f_{X,Y}(x,y) = f_{X/Y}(x/y) f_Y(y) = f_{Y/X}(y/x) f_x(x) \tag{6.33}$$

where $f_{X,Y}(x,y)$ is the joint pdf, the marginal pdfs are $f_X(x)$ and $f_Y(y)$, and the conditional pdfs are $f_{X/Y}(x/y)$ and $f_{Y/X}(y/x)$. We compute the marginal pdfs from the joint pdf by using the following:

$$f_X(x) = \int f_{X,Y}(x,y) dy \quad f_Y(y) = \int f_{X,Y}(x,y) dx \tag{6.34}$$

where the integrals are over the entire region of definition of the pdf. Therefore, we can write (Vachtsevanos et al., 2006)

$$f_{Y/X}(y/x) = \frac{f_{X,Y}(x,y)}{\int f_{X,Y}(x,y) dy} \tag{6.35}$$

Similarly, for $f_{X/Y}(x/y)$, the mean or expected value of a random variable is given in terms of the pdf by

$$\bar{x} = E(x) = \int x f_x(x)dx \tag{6.36}$$

and the conditional mean is given by

$$E(Y/X) = \int y f_{Y/X}(y/x)dy = \frac{\int y f_{X,Y}(x,y)dy}{\int f_{X,Y}(x,y)dy} \tag{6.37}$$

All these quantities may be computed from the joint pdf. Therefore, in applications, the computation of the joint pdf, given the available data, is of high priority. The joint probability function is shown in Fig. 6.15.

6.3.1.1.5.2 Weibull Model: Analysis of Time to Failure In general, a typical Weibull probability distribution function (PDF) is defined by

$$f(t) = \frac{\beta}{\eta}\left(\frac{t}{\eta}\right)^{\beta-1} e^{-\left(\frac{t}{\eta}\right)^{\beta}} \tag{6.38}$$

where $t \geq 0$ represents time, $\beta > 0$ is the shape or slope parameter, and $\eta > 0$ is the scale parameter of the distribution. Eq. (6.38) is usually referred to as the two-parameter Weibull distribution. The slope of the Weibull distribution, β, is very important, as it determines which member of the family of Weibull failure distributions best fits or describes the data. It also indicates the class of failures in the "bathtub curve" failure modes as shown in Fig. 6.16. The Weibull shape parameter β indicates whether the failure rate is increasing, constant, or decreasing. If $\beta < 1$, the system or component has a decreasing failure rate. This scenario is typical of "infant mortality" and indicates the system or component is failing during its "burn-in" period. If $\beta = 1$, there is a constant

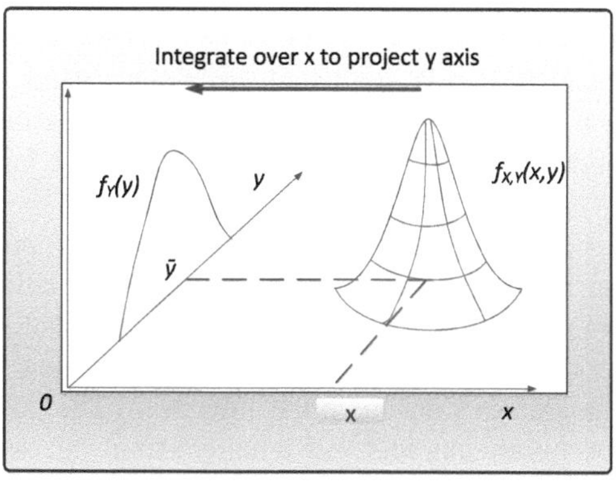

FIGURE 6.15 Joint probability function of $f(X,Y)$.

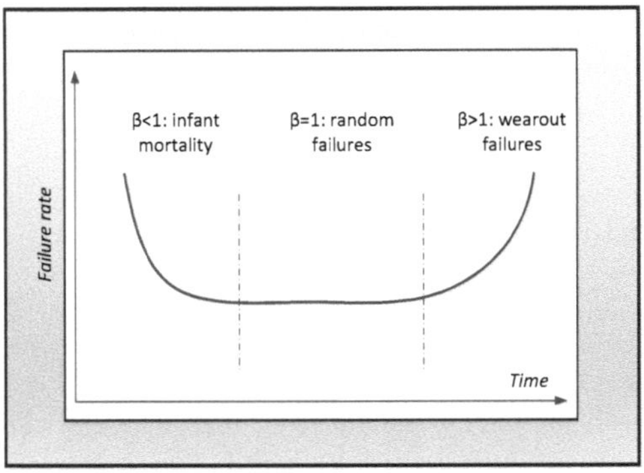

FIGURE 6.16 Bathtub curve failure modes.

failure rate. Components that have survived burn-in will frequently exhibit a constant failure rate. If $\beta > 1$, there is an increasing failure rate. This is typical for items that are wearing out. To summarize:

- $\beta < 1$ indicates infant mortality;
- $\beta = 1$ means random failures (i.e., independent of time);
- $\beta > 1$ indicates wear-out failures.

The information about the β value is extremely useful for reliability-centered maintenance planning and product lifecycle management because it can provide a clue to the physics of the failures and tell the analyst whether scheduled inspections and overhauls are needed. For instance, if β is less than or equal to one, overhauls are not cost-effective. With β greater than one, the overhaul period or scheduled inspection interval can be read directly from the plot at an acceptable or allowable PoF. For wear-out failure modes, if the cost of an unplanned failure is much greater than the cost of a planned replacement, there will be an optimum replacement interval for minimum cost (Zhai et al., 2013).

The scale parameter or spread, η, sometimes called the characteristic life, represents the typical TTF in Weibull analysis and is related to the mean-time-to-failure (MTTF). In Weibull analysis, η is defined as the time at which 63.2% of systems or components under analysis will have failed (Pasha et al., 2006).

There are basically two fitting methods for parameter estimation in widespread use in reliability analysis, namely the maximum likelihood estimation (MLE) and regression methods. MLE involves developing a likelihood function based on the available data and finding the values of the parameter estimates that maximize the likelihood function. The regression method generally works best with data sets with smaller sample sizes that contain only complete data (i.e., data in which all of the units under consideration have been run or tested to failure). These failure-only data are best analyzed with rank regression on time, as it is preferable to regress in the direction of uncertainty. In Weibull analysis, the median rank regression method is often deployed to find out the shape and scale parameters for complete life data (Abernethy, 2006).

The PoF at time t, also referred to as the Weibull distribution or the cumulative distribution function, can be derived from Eq. (6.38) and expressed as

$$F(t) = 1 - e^{-\left(\frac{t}{\eta}\right)^{\beta}} \tag{6.39}$$

Thus, the Weibull reliability at time t, which is $1 - F(t) = R(t)$, is defined as

$$R(t) = 1 - F(t) = e^{-\left(\frac{t}{\eta}\right)^{\beta}} \tag{6.40}$$

This can be written as

$$\frac{1}{1 - F(t)} = e^{\left(\frac{t}{\eta}\right)^{\beta}} \tag{6.41}$$

Taking two times the natural logarithms of both sides gives an equation of a straight line:

$$\ln \ln\left(\frac{1}{1 - F(t)}\right) = \beta \ln t - \beta \ln \eta \tag{6.42}$$

The above equation represents a straight line in the form of "$y = ax + b$" on *log*/*log*(Y) versus *log*(X), where the slope of the straight line in the plot is β, namely the shape parameter of Weibull distribution. Through the above transformation, the life data samples can be fitted in the Weibull model and the two Weibull parameters can be estimated.

The mean of the Weibull PDF, T, which is the MTTF in Weibull analysis, is given by:

$$\overline{T} = \eta \cdot \Gamma\left(\frac{1}{\beta} + 1\right) \tag{6.43}$$

where Γ is the gamma function.

When $\beta = 1$, MTTF is equal to η. In fact, as a rough approximation, in practices of Weibull analysis where β is equal to or slightly larger than 1, the characteristic life can be approximated as MTTF. However, for β much larger than *1*, MTTF should be calculated using an equation mean of T (Zhai et al., 2013).

6.3.1.1.5.3 Hidden Markov Models An HMM is a statistical model used to represent stochastic processes defined by the following parameters:

- N: number of states in the model.
- M: the number of distinct observations for each state.
- A: the state transition probability distribution.
- B: the observation probability distribution of each state.
- π: the initial state distribution π.

For simplicity and clarity of presentation, a compact notation ($\lambda = \pi$, A, B) is used for each HMM. In practice, HMMs are used to solve three typical problems: detection, decoding, and learning.

Discrete HMMs usually consider observations as discrete symbols and use discrete probability densities to model the transition and the observation probabilities. The problem with this approach is that in condition monitoring, the observations are typically continuous signals. To use a continuous observation density, some restrictions are set to insure the parameters of the pdf can be reestimated as the following:

$$b_i(o) = \sum_{m=1}^{M} C_{jm}\xi\left(o, \mu_{jm}, U_{jm}\right), \ 1 \leq j \leq N \tag{6.44}$$

In the equation, O is the observation vector, C_{jm} is the mixture coefficient for the mth mixture in state i and ξ is any log concave or elliptically symmetric density with mean vector μ_{jm} and covariance matrix U_{jm} for the mth mixture component in state j. A Gaussian density is generally used for ξ, and the corresponding model is called an MoG-HMM (mixture of Gaussians hidden Markov model), completely defined by the A matrix, the B matrix, and the initial probability π. For an MoG-HMM, the observation matrix B is modeled by a Gaussian density with a mean μ, a standard deviation σ, and a mixture matrix M (Tobon-Mejia et al., 2011). Fig. 6.17 below gives the states of HMM.

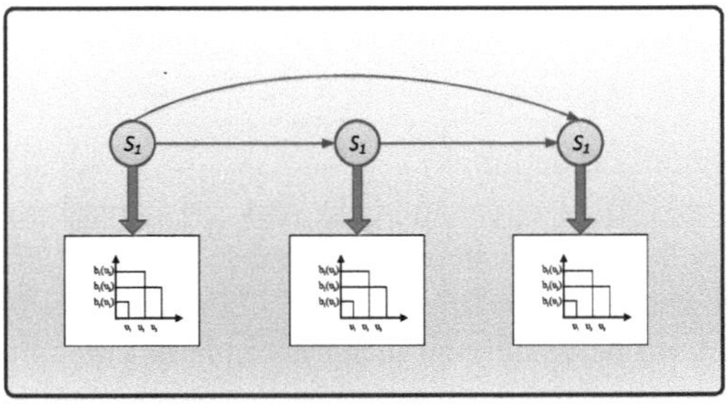

FIGURE 6.17 Three state left-to-right hidden Markov models.

There are three types of HMM-based prognostics: the first is traditional, the second is based on hidden semi-Markov models (HSMM), and the third uses MoG-HMM.

6.3.1.1.5.3.1 The HMMs Case. To perform diagnostics and prognostics, traditional HMMs have a learning and an exploitation phase:

- *Learning phase:* features extracted from complete monitoring histories (from the normal condition to the failure state) are transformed into HMMs using the well-known Baum–Welch algorithm. An HMM model is created for each failure; the model is stored in a model base containing all the HMMs, with a diagnostic label associated with each learned history.
- *Exploitation phase*: online data are used as inputs to the learned models to prognosticate the health state. The online extracted features are used to find the model that best fits the actual observation sequence by computing the probability $P(O|\lambda)$. Then, the parameters of the selected model are used to assess the current health state and estimate the RUL.

When the actual health state is known, the Chapman–Kolmogorov Eq. (6.45) can be used to reestimate the health state after n iterations. When the predicted probability of being in the last state reaches a predefined limit ε, RUL can be calculated (6.46) (Ross, 2001) as

$$\widehat{P}(n) = \pi_t A^n \tag{6.45}$$

$$\text{RUL} = n \Leftrightarrow \widehat{P}(s = s_N|n) = \varepsilon \tag{6.46}$$

6.3.1.1.5.3.2 The HSMMs Case. The problem with traditional HMMs is that the durational behavior is usually characterized by a geometrically decaying function. This assumption can cause inaccurate duration modeling because most real-life applications do not obey this function. Thus, to solve this problem, the use of a model with explicit time durations, the HSMM, has been proposed. The learning and exploitation phases are the following:

- *Learning phase*: like a traditional HMM, the parameters of the HSMM are defined by using the history data and the Baum–Welch algorithm. For an HSMM, the shape is constrained to a left-to-right model. In addition, for each state, the stay durations $D(s_i)$ are learned by using the Viterbi algorithm. The idea is to use the learned parameters (π, A, B) and the history data to obtain the whole observation sequence. Then, by taking into account the transition instant between the states, the duration D can be defined.

  Finally, by assuming the sojourn time follows a Gaussian distribution, the mean time duration $\mu(D(s_i))$ and the standard deviation $\sigma(D(s_i))$ of the same state can be estimated for different histories of the same fault:

$$\mu(D(s_i)) = \frac{1}{H}\sum_{h=1}^{H} D(s_i)_h \tag{6.47}$$

$$\sigma(D(s_i)) = \sqrt{\frac{1}{H}\sum_{h=1}^{H}\left[D(s_i)_h - \mu(D(s_i))\right]^2} \tag{6.48}$$

In the equations above, $D(s_i)$ stands for the stay duration, i is the state index, h is the history index, and H is the total number of histories from a particular fault state (Dong and He, 2007).

- *Exploitation phase*: the learned models and their associated stay duration are used in this phase. First, a competitive model selection is performed by using the raw data and the forward-backward algorithm to compute the probability $P(O|\lambda)$. The label of the winning model is used to prognosticate the monitored system; the actual health state is defined in the same way as in traditional HMMs. The selected model is used to estimate RUL.

  The HSMM permits us to estimate RUL in two ways:

- First, by using the expression where the stay durations are merged with the state probability transitions:

$$\text{RUL}_l = t_{lc} + a_{l,l+1}(\text{RUL}_{l+1}) \tag{6.49}$$

  where l is the actual state index, t_{lc} is the state changing point, and $a_{l,l+1}$ is the probability of transition to the next state.

- Second, by using a more intuitive expression adding the state duration from the current state until the last state and subtracting the time spent in the actual state:

$$\text{RUL}_l = \sum_{i=l}^{N}\mu(D(s_i)) - t_l \tag{6.50}$$

A confidence interval with different recovering values can be easily estimated using the Bonferroni method:

$$\text{RUL}_{\text{limits}} = \left[\sum_{i=l}^{N}\mu(D(s_i)) \pm cf\cdot\sigma(D(s_i))\right] - t_l \tag{6.51}$$

In the equations, l is the actual state, i is the state index, N is the total number of states, $\mu(D(s_i))$ is the mean time duration in the state i, $\sigma(D(s_i))$ is the standard deviation, t_l is the time spent in the actual state, α is the confidence interval between [0,1], and cf is the confidence factor defined by using Φ, i.e., the cumulative distribution function of a Gaussian probability distribution (Dempster et al., 1977).

6.3.1.1.5.3.3 The MoG-HMM–Based Method. The originality of the MoG-HMM–based method is that raw signals are processed using the wavelet packet decomposition (WPD) to extract the relevant information to learn behavior. In the generated MoG-HMM, the states' stay durations are not assumed to be a geometrically decaying function as in the HSMMs case but are learned from the monitoring data (note that multiple continuous signals are considered in both learning and simulation phases, instead of the traditional mono observations). Moreover, in the proposed method, there is no limitation on the type of the generated MoG-HMM (it can be ergodic, left-right, or parallel left-right).

- *Learning phase*: this first phase is executed off-line; the raw data recorded by the sensors are processed to determine the energy of each node at the last decomposition level using the WPD technique. These features are used to learn several behavior models (in the form of MoG-HMMs) corresponding to different histories related to several initial states and/or operating conditions of the component or system. One global left-to-right MoG-HMM is learned for each type of fault as well. There are X states, i.e., the different asset health states. Each raw data history corresponding to a given condition of the component or system is transformed into a feature matrix F, by using the WPD. In this matrix, each column vector (of C features at time t) corresponds to a snapshot on the raw signal, and each cell fact represents the node c of the last WPD level at time t.

$$\text{Raw signal} \xrightarrow{\text{WPD}} F = \begin{pmatrix} f_{11} & \cdots & f_{lt} \\ . & .. & .. \\ f_{cl} & \cdots & f_{ct} \end{pmatrix} \tag{6.52}$$

with $1 \leq t \leq T$ and $1 \leq c \leq C$.

The nodal energy (features) is used to estimate the parameters (π, A, B) and the temporal parameters (stay duration in each state) of the MoG-HMMs. This algorithm permits us to obtain the state sequence and compute the time the component has been in each state of the corresponding MoG-HMM. Thus, by assuming the duration in each state follows a normal law, it is possible to estimate the mean duration (6.53) and the corresponding standard deviation (6.54) by computing the duration and the number of visits in each state. The Viterbi algorithm permits us to identify the final state representing the physical component's failure state.

$$\mu(D(s_i)) = \frac{1}{\Omega}\sum_{\omega=1}^{\Omega} D(s_{i\omega}) \tag{6.53}$$

$$\sigma(D(s_i)) = \sqrt{\frac{1}{\Omega}\sum_{\omega=1}^{\Omega}\left[D(s_{i\omega})_h - \mu(D(s_i))\right]^2} \tag{6.54}$$

In the equations, $D(s_{i\omega})$ stands for the visit duration, i is the state index, ω is the visit index, and Ω corresponds to the total visits. A compact representation of each learned MoG-HMM used to perform diagnostics and prognostics is given by the following expression:

$$\lambda = (\pi, A, B, \mu(D(s_i)),\ \sigma(D(s_i)), S_{\text{final}}) \tag{6.55}$$

where λ is the fully defined model and S_{final} is the final state (corresponding to the end of the considered condition-monitoring history).

- *Exploitation phase*: this phase is performed online; it consists of exploiting the learned models to detect the component's current condition (using the Viterbi algorithm) and computing the corresponding RUL. The processed data and the extracted nodal energy using the Wavelet toolbox from Matlab are continuously fed to the learned models to determine the global MoG-HMM, which best fits the observed sequence. Diagnostics are then performed, and the current health state is defined. The selection process is based on the calculation of a likelihood $P(O|\lambda)$ of the model over the observations (HMM problem 1). The current model state and the stay durations learned in the off-line phase can be used to estimate the component or system's RUL and the associated confidence value.

The generated MoG-HMMs are used during the online phase to estimate RUL and the associated confidence value of the physical component using the following procedure:

a. Detecting the appropriate global left-to-right general MoG-HMM that best fits and represents the online observed sequence of nodal energy. The diagnostic label of the selected model is used to establish the current condition.
b. Selecting the nearest RUL model knowing the active failure mode.
c. Identifying the current state of the selected RUL model.
d. Identifying the critical path from the current state to the end state. The idea is to identify all the nonzero probabilities in the transition matrix as potential transitions and to choose the minimal path among all possible ones (Fig. 6.18) with only one visit per state.
e. Estimating RUL by using the temporal parameters of the stay duration in each state. In addition, a confidence value is calculated based on the

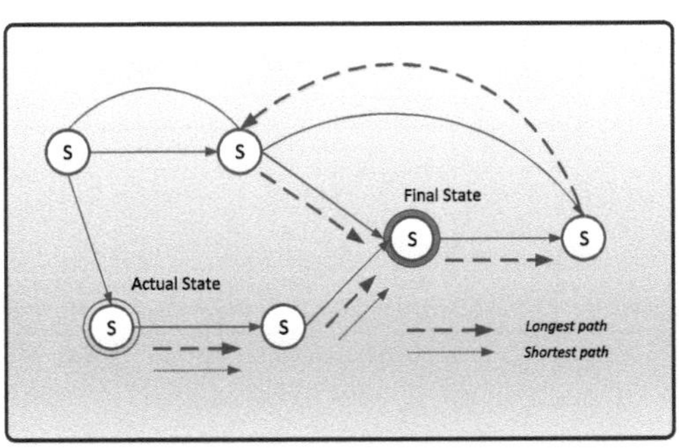

FIGURE 6.18 Path estimation.

standard deviation values of the stay durations and the Bonferroni confidence interval (Zarei and Poshtan, 2007):

$$\mathrm{RUL}_{\mathrm{lower}} = \left[\sum_{i=\text{current state}}^{N} \mu(D(s_i)) \pm cf \cdot \sigma(D(s_i)) \right] \tag{6.56}$$

6.4 SELECTION OF PROGNOSIS TECHNIQUES FOR DIFFERENT TYPES OF ASSETS

In this section, we discuss the calculation of the RUL of three types of assets: rotating machines (compressors, pumps, turbines, etc.), structures (bridges, buildings, etc.), and complex systems (cars, planes, etc.). We then compare the various techniques used to estimate each asset's RUL.

6.4.1 Rotating Machines

Rotating machinery is a major and critical component of many mechanical systems in industrial plants, air and ground transportation vehicles, and in many other applications. Rotating elements have unique characteristics both in performance and acoustics that can be predicted. For example, the rotating elements exhibit high harmonic oscillations when on the verge of failure. Thus, fault diagnosis is commonly employed as a maintenance methodology for rotating machinery.

Recently, maintenance methodology has shifted toward CBM. CBM employs a continuous maintenance methodology to maximize the availability of equipment. It is accomplished by continuously monitoring equipment condition and performing maintenance actions only when faults exist. At the operational level, CBM alerts operators to evolving faults and allows maintenance actions to be taken before catastrophic failure occurs.

A schematic of a typical CBM system is shown in Fig. 6.19. The key components are machinery, sensors, fault detectors, fault classifiers, predictive models, model/data fusion, and outputs.

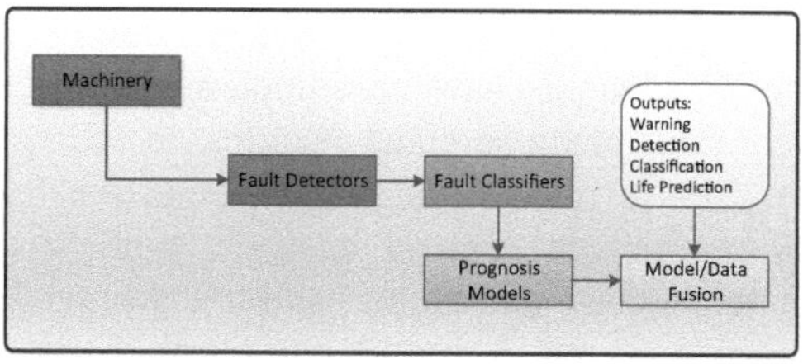

FIGURE 6.19 Key elements of condition-based maintenance system.

6.4.1.1 Sensing Techniques and Sensors

6.4.1.1.1 Monitoring Parameters

The monitoring of the performance and operating condition of machinery can be accomplished by carefully selecting the parameters to measure. The most common parameters are vibration and acoustic signatures, temperature, pressure, motor current, wear debris, and electrostatic exhaust measurements. To detect failure inception, deterioration mechanisms are continuously monitored to identify the initiation of cracks, fractures, and other failures, such as shaft cracks, seal leakage, and corrosions.

6.4.1.1.2 Sensing Techniques

Sensing techniques have undergone continuous development. State-of-the-art techniques include active interrogation approaches, nondestructive evaluations such as acoustic emission and stress wave monitoring, holographic imaging, oil debris analysis, chemical composition and analysis, use of in situ and embedded microsensors, electrostatic exhaust measurements, remote sensing, and electrical resistance measurements. Also important is the optimal placement of sensors, along with a sensing strategy that monitors small changes in state variables that could capture fault precursors and features. The virtual sensor concept, a simulation and modeling of signals based on limited data and measurements, has good potential to reduce the number of sensors and, thus, cost.

6.4.1.1.3 Sensors

Recent advances in materials and sensor development, combined with new manufacturing techniques (e.g., nanofabrication) have allowed manufacturers to build rugged sensors that are both affordable and small. The new materials and processing include piezoelectric elements, piezofilm, piezocomposites, optical fibers, and microelectromechanical systems. In addition, biosensors are being developed to sense and identify chemical species; this has good potential for use in fault diagnostics (Natke and Cempel, 1996).

6.4.1.2 Feature Extraction

6.4.1.2.1 Detection and Identification

With the increasing need for identification and location of incipient failures, new signal processing techniques have been implemented to enhance detection and identification of fault precursors and features, including signal enhancement or background noise removal and signal detection techniques. Signal processing techniques include cepstral analysis, time—frequency analysis, nonlinear dynamic techniques, and higher- and lower-order spectral analyses. Time—frequency analyses such as the Wigner distribution and wavelet transforms have been successfully used to detect and identify fault features.

Sources of higher harmonic oscillations in rotating machinery are caused by imbalance, misalignment, and nonlinear excitation; nonlinear dynamic techniques and lower- and higher-order spectral (polyspectral) analysis techniques can be used to detect and identify higher harmonic oscillations, as well as fault features. In particular, alpha-stable distributions, a lower-order statistics method, can detect impulsive-type features embedded in non-Gaussian signals.

6.4.1.2.2 Feature Classification

Once fault features are detected, identified, and extracted using signal processing techniques, the fault features are transformed to a fault vector space where classification is performed. Classification techniques include conventional statistical methods, NNs, fuzzy logic, and pattern recognition. NNs have commonly been applied to fault vectors to train and classify the various faults. Fuzzy logic, a rule-based reasoning approach, has been used successfully to classify faults. Efforts are ongoing to combine NNs and fuzzy logic, the so-called neuro-fuzzy network approach, to classify fault precursors and features. Pattern recognition is still used to classify faults.

6.4.1.3 Prognosis Model

Failure analysis plays an important role in understanding how components fail and forms the basis of the failure predictive models. As mentioned previously, failure mode identification and classification is a technical issue. Though failure mechanisms have been studied for a long time, understanding how failures occur and theoretical developments in predicting mechanical failures are far from satisfactory. In particular, there is a need for physics-based models relating fundamental theory with experimental data to predict the RUL (safe operating) of structures and components. The challenge is to model fault initiation and propagation as a unified approach and then predict state variables related to mechanical response or vibration. A statistical approach has been used to predict the fatigue life expectancy of structural components subject to cyclic loading. This approach deals with the statistical variability in such things as material properties and initial flaw sizes, as well as the statistical nature of the fatigue process. RUL can be estimated using either data or experience-based approaches. Data-based approaches use ANNs, fuzzy logic, and regression; experience-based approaches frequently use the Weibull.

6.4.1.4 Data/Model Fusion

Data monitoring and predictive models must be fused to enhance decision-making and provide reliable information, including warnings, to operators. Occasionally, signals from multiple sensor inputs and models can be contradictory; therefore, selection of the proper data and integration or fusion of this information are critical in the reasoning and decision-making processes (Natke and Cempel, 1968–1997).

In summary, to estimate the RUL of rotating machines, a number of different approaches can be used. Approaches based on experience generally opt for the Weibull, while model-based approaches use physics-based models (Paris' law crack growth modeling, fatigue spall initiation, and progression modeling) and particle filtering. Time series prediction uses NNs, fuzzy logic, and regression analysis.

6.4.2 Infrastructures

Many existing structures are aging and a large installed base of civil infrastructure requires maintenance, rehabilitation, or replacement. The main method used to identify structural deterioration is visual inspection. Inspections provide the engineer with data on the number of cracks and the rate of crack growth, but quantitative data are often necessary to distinguish between structures that can be kept in service and those requiring replacement or modernization. However, there are few studies on patterns of maintenance replacement for bridges and buildings. Most are based on ranking factors against failure (evaluation of capacity). One of the most critical types of structural deterioration is induced fatigue fracture; hence, a maintenance strategy based on the remaining useful fatigue life is increasingly used by maintainers. This is usually based on a combination of load voltage specifications of Miner's rule and the fatigue curve.

6.4.2.1 Implementation of Condition-Based Maintenance on Bridges

CBM for a bridge can be initiated based on the degradation monitored by various sensors. Once a degradation characteristic crosses a specified threshold, maintenance actions are triggered. In CBM, the characteristic degradation measures of the structure are obtained and updated using real-time data from both structural and durability health-monitoring systems. These systems detect degradation by continuously updating information on the performance of the structure with respect to its load-carrying capacity and durability as these are affected by aging and environment. Determining the deterioration rates of bridges and buildings is essential to evaluate the effectiveness of maintenance/repair options. As the deterioration rate is a function of time, the conventional inspection approach is inadequate because of its ad hoc nature and lack of continuous time-history supporting data. Long-term real-time monitoring systems and analytical tools are required to estimate the deterioration rates at key locations/components of the bridges under in-service conditions.

Fig. 6.20 illustrates various potential degradation mechanisms for bridges. Maintenance decision-making for a reinforced concrete structure depends on two factors: the deterioration status of its embedded steel tendons or steel reinforcements and its load-carrying capacity. It therefore requires both durability and structural health monitoring. For durability health monitoring,

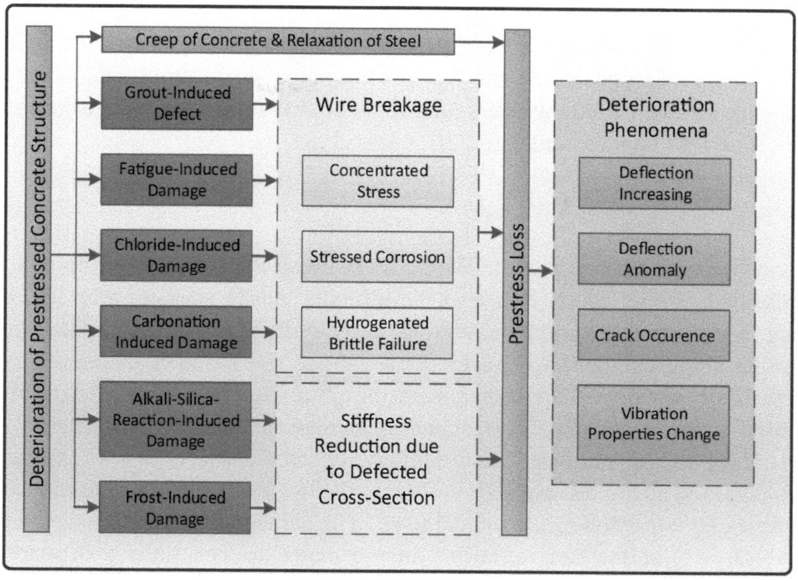

FIGURE 6.20 Degradation mechanisms for reinforced concrete structures.

analytical tools must be developed to evaluate the formation and propagation of cracking and the time for the spalling of concrete cover based on the monitoring data from corrosion sensors (corrosion potential, corrosion currents, concrete resistivity, linear polarization resistance, concrete temperature, concrete relative humidity, and gas concentration in decks and piers).

After developing such analytical tools, a linkage must be established to transfer the monitoring/analysis results into an updatable and evolving rating system to facilitate CBM decision-making and the prioritization of maintenance activities. The bridge rating system is generally developed in terms of the criticality and vulnerability indices that can be formulated/updated using the design information, inspection results, and monitoring data. As the structural condition is obtained from CBM, a new synthetic rating system (Rating Cubic) is formulated as a rating matrix in terms of the criticality rating, vulnerability rating, and condition rating of the structural components. The action and prioritization of condition-based inspection/maintenance for structural components is based on the results (dynamically updated over time) of the synthetic rating (rating results within Cube 1 correspond to the highest priority; the rating results within Cube 8 correspond to the lowest priority) (Ni and Wong, 2012).

6.4.2.2 *Remaining Fatigue Life Estimation of Structural Components*

Suppose a component is subjected to a certain stress amplitude or stress range σ_i for n_i number of cycles at load level i and N_i is the fatigue life (failure number of cycles) corresponding to σ_i. Hence, the residual life at load level i can be obtained as $(N_i - n_i)$. The stress $\sigma_{eq(i)}$, which corresponds to the failure life $(N_i - n_i)$, is considered the ith level damage stress amplitude or stress range (otherwise called the stress amplitude or stress range relevant to residual life). The new damage indicator, D_i is stated as

$$D_i = \frac{\sigma_{(i)eq} - \sigma_i}{\sigma_u - \sigma_i} \tag{6.57}$$

where σ_u is the intercept of the Wöhler curve with the ordinate at one-quarter of the first fatigue cycle. It is also the ultimate tensile strength amplitude or range for rotating bending test-based S–N curves and the ultimate shear strength amplitude or range for torsional fatigue test-based S–N curves.

In the first cycle, the damage stress amplitude or range $\sigma_{(i)eq}$ is equal to applied stress σ_i, and the corresponding damage indicator becomes $D_i = 0$. According to the proposed methodology, current damage has to be transformed to the next load level. Therefore, in the last cycle, the damage indicator becomes $D_i = 1$ when $\sigma_{(i)eq}$ is equal to σ_u. The damage indicator is normalized to 1 ($D_i = 1$) at the fatigue failure of the material. The same procedure is followed until $D_i = 1$. Here, the defined fatigue failure is the time taken for the occurrence of the first through-thickness crack at the location of maximum stress of the structural component. In the case of railway bridge components, it can probably be understood as the time taken for the initiation of a crack near a connection (rivet or bolt).

6.4.2.3 *Identification of Critical Components/Connections*

Components or structural members of a bridge with the lowest remaining fatigue life of each member set (set of members with the same load capacity) are called "critical members" here. Generally, these members are subjected to more attention in member replacement–based maintenance. From the previously obtained remaining fatigue lives, we can easily identify the critical members. The connections joining these critical members are called "critical connections."

6.4.2.4 *Remaining Fatigue Life Estimation of Critical Connections*

Taking bridges as an example, the stress concentration effect in connections between the primary members of bridges is a main reason for fatigue damage. Most such connections are subjected to multiaxial fatigue. To capture this effect at riveted connections or discontinuities, the detail class (BS 5400, 1980) of riveted connection-based Wöhler curves has been considered in previous life estimation. However, the variation of real rotational fixity, clamping force, and geometry at the connection causes the real stress distribution to change at the connections. Such changes may result in over- or underprediction of the estimated fatigue life. As a result, replacement of members based on previously determined remaining lives may not be an appropriate maintenance procedure. Replacement of members based on fatigue lives of critical connections is a more appropriate strategy.

Initially, all critical connections should be investigated nondestructively to determine the current condition by conducting such tests as X-ray, ultrasonic, magnetic particle, or radiographic examinations. The connections that do not show significant change from the initial state or condition are not subject to any unexpected stress concentration. For connections showing significant change, we may need to assess the remaining fatigue life to determine the degree of criticality.

The remaining fatigue life of the connections not subject to significant deviation from the initial condition is finalized as the lowest remaining fatigue life of the members joined to those particular connections. The remaining fatigue lives of connections where the conditions have significantly changed must be evaluated based on the current geometric condition, secondary stress distribution, etc.

This procedure is based on a recently proposed multiaxial fatigue model. Initially, the accumulated plastic strain per each stabilized cycle must be obtained from a finite-element secondary stress analysis of the connection or part of the member (submodel):

$$\varepsilon_S^{pc} = \frac{4}{\sqrt{3}} \frac{2k^* - k_{max} - k_{min}}{c}, \tag{6.58}$$

where $c = b + 2\eta$; b and η are the mesoscopic linear hardening modulus and the shear modulus respectively; k^* is the radius of the smallest hypersphere, which contains the entire history of the macroscopic deviatoric stress amplitude of the stabilized cycle; k_{max} and k_{min} are the maximum and the minimum values of mesoscopic yield stresses that can be reached during the loading cycle. The fatigue life is now calculated from the new damage indicator as follows:

$$D_i = \frac{\left(\varepsilon_s^{pc}\right)_{(i)eq} - \left(\varepsilon_s^{pc}\right)_i}{\left(\varepsilon_s^{pc}\right)_u - \left(\varepsilon_s^{pc}\right)_i} \tag{6.59}$$

where $\left(\varepsilon_s^{pc}\right)_i$ is the accumulated plastic mesostrain per stabilized cycle of n_i number of cycles at load level i. N_i is the fatigue life (number of cycles to crack nucleation) corresponding to $\left(\varepsilon_s^{pc}\right)$ and can be estimated from

$$N = A\left(\varepsilon_s^{pc}\right)^{-\xi} \tag{6.60}$$

where ξ and A are material parameters determined from fatigue tests. The accumulated plastic mesostrain $\left(\varepsilon_s^{pc}\right)_{(i)eq}$ corresponds to the failure life $(N_i - n_i)$ and is considered the ith level damage of the accumulated plastic mesostrain; $\left(\varepsilon_s^{pc}\right)_u$ is the accumulated plastic mesostrain corresponding to one-quarter of the first fatigue cycle. According to the proposed methodology, current damage is now transformed to the next load level in a manner similar to 0 in the sequential law discussed in Section 6.3.1.1. The damage indicator is again normalized to 1 ($D_i = 1$) on the fatigue failure of the material, and same procedure is followed until $D_i = 1$.

6.4.2.5 Member Replacement/Strengthening Scheme

The lowest remaining fatigue life of the connections refers to the remaining fatigue life of the bridge. Once the age of the bridge reaches this value (when bridge life becomes zero), it is advisable to replace the corresponding critical member with a new member having a longer fatigue life. At the same time, the associated connection should be strengthened. After this essential repair, creating a new sequence for future member replacement by following a similar procedure from the beginning is recommended. This type of maintenance strategy extends the service life (fatigue capacity) of the bridge in the safest manner.

In summary, the most widely used approach for estimating the remaining life of structures, such as bridges (buildings use the same approach), is the model-based technique for estimating the lifetime of fatigue (Kohout, 2001).

6.4.3 Complex Systems

System maintenance is important to manufacturers of complex systems, such aircraft or cars, such as it can improve the reliability, security, safety, and final cost of their products. To improve the maintenance capabilities of a complex system, an onboard health-monitoring system must be deployed; it must provide enough informative pieces of data to determine which type of maintenance operation is required. Complex systems are an assemblage of heterogeneous components (continuous, discrete, hybrid components); each may require a different technique to monitor its health. A global health-monitoring system would be useful, but there are many difficulties to deal with.

First, in classical maintenance, the objective is to replace faulty components (also called line replacement units). Components are faulty when they do not perform the set of functions they are supposed to perform. To identify these components, the monitoring system requires online diagnostic capabilities (Vachtsevanos et al., 2006).

Second, the optimality of the maintenance is crucial. To decrease the maintenance cost, it is necessary to perform preventive maintenance, and this requires embedding prognostic capabilities in the health-monitoring system to determine the aging state of the system.

Third, without a good level of abstraction to describe the system knowledge to be used in the health-monitoring system, it is impossible to guarantee the global consistency of the health-monitoring architecture. The objective is to get an abstracted description that is also homogeneous (Engel et al., 2000).

Complex systems must be continuously monitored (CBM) using sensors able to capture all information related to the health status of the entire system. These data are then used to make a prognosis and estimate the RUL.

6.4.3.1 Prognosis Methods

As noted above, prognostics basically consist of estimating at time $t_{\text{prog}} \geq t_j$ the time $t_{j+1} \geq t_{\text{prog}}$ of the future mode change, relying on the aging models available in the system mode m_j^{Σ}. A set of aging models (or life models) is available for each private parameter $\text{pp}^{i,k}$ ε PP^i. In the following, $\text{pp}^{i,k}$ will be generically noted pp for the sake of simplicity. Aging models of a private parameter describe the evolution of the parameter value pp with environmental constraints (temperature, humidity, vibration, stress conditions, etc.). An aging model $\text{ag}_x^{i,k}$ can be represented as an algebraic relation depending on the operational conditions of the system mode m_x^{Σ}. According to the current system mode m_i^{Σ} defining the range of the private parameters of the system, well-suited aging models are selected for each private parameter pp ε PP. The remaining time until the parameter pp becomes faulty is noted as $\text{RTF}_{(\text{pp})}$. A fault probability for each private parameter in PP is established from the selected aging models.

Basically, let f_{pp} denote a pdf representing the fault probability of a private parameter pp in the mode m_x^{Σ}, and $P_{\max}$ be the maximal fault probability value acceptable for the parameter pp. The remaining time to fault (RTF) of pp is the time t_p when the fault probability has reached the threshold $P_{\max}$:

$$\text{RTF}_{(\text{pp})} = t_p \quad \text{such that} \int_0^{t_p} f_{\text{pp}}(t)dt = P_{\max} \tag{6.61}$$

It follows that $t_{j+1} = t_{\text{prog}} + \min(\text{RTF}_{(\text{pp}),\ \text{pp}\ \varepsilon\ \text{PP}})$ and the system mode m_{j+1}^{Σ} is such that pp is out of range (Ribot et al., 2009).

6.4.3.2 Generic Modeling for Prognostics

Private parameters usually represent physical attributes of the component and are totally heterogeneous. The difficulty is determining a common representation of the prognosis for any private parameter; the parameter must be as flexible as possible to represent any type of pdfs. For these reasons, the Weibull distribution is often used. The pdf is

$$W(t, \beta, \eta, \theta) = \frac{\beta}{\eta}\left(\frac{t-\theta}{\eta}\right)^{(\beta-1)} e^{-\left(\frac{t-\theta}{\eta}\right)\beta} \tag{6.62}$$

where β characterizes the shape of the distribution, η characterizes the scale, and θ characterizes the location of the distribution. For a given private parameter pp ε PP^i and a given probability threshold $P_{\max}$, we get

$$\mathrm{RTF}_{(\mathrm{pp})} = t_p \quad \text{such that} \quad \int_0^{t_p} W\left(t, \beta_{\mathrm{pp}}, \eta_{\mathrm{pp}}, \theta_{\mathrm{pp}}\right) dt = P_{\max} \tag{6.63}$$

β_{pp}, η_{pp}, and θ_{pp} fully characterize the FP distribution and model the way the parameter pp is aging. The description of the characteristics β_{pp}, η_{pp}, θ_{pp} relies on the aging models associated with pp. Whatever the technique used to obtain such models, the available knowledge is characterized by algebraic relations that define β_{pp}, η_{pp}, θ_{pp}: $\beta_{pp} = ar_\beta(ip^{i,1},\ldots,ip^{i,n}),\theta_{pp} = ar_\theta(ip^{i,1},\ldots, ip^{i,n})$, $n_{pp} = ar_\eta(ip^{i,1},\ldots, ip^{i,n})$.

6.4.3.3 Functional Prognosis

RUL of a system is defined as the remaining time until the system cannot successfully perform its full set of complex functions resulting from the composition of the basic functions FU^i of any component C^i ε Comps. In systems such as aircraft, for instance, the system functions usually rely on redundant implementations of the basic functions (the functions of one component can be performed by another). Therefore, prognostics rely on the functional view of the system. To get the prognosis of the availability of any system function in the future, we need to get the prognosis of the availability of every basic function $F \varepsilon FU^i$ implemented in every component C^i of the system. The aging of F naturally depends on the aging of the private parameters of the component that implements F. The remaining time until a basic function F becomes failed is noted as ettf(F), which stands for estimated TTF (Voisin et al., 2009), and is expressed as:

$$\mathrm{ettf}(F) = t_f \text{ such that } \int_0^{t_p} W(t, \beta_F, \eta_F, \theta_F) dt = P_F \tag{6.64}$$

where P_F is a probability threshold, and β_F, η_F, and θ_F are fully determined by a combination of the characteristics $\{\beta_{pp}, \eta_{pp}, \theta_{pp}\}_{pp2PP(F)}$. By extension, the ettf(F_{sys}) of any system function F_{sys} is defined as ettf(F) where F is the basic function whose availability is necessary for F_{sys} to be available and whose ettf(F) is minimal. The RUL of a component is then RUL(C^i) = min(ettf($Fu^{i,j}$), $Fu^{i,j}$ ε FU^i), and the RUL of the system is estimated to the minimal ettf(F_{sys}). Finally, maintenance can be scheduled based on the ettf(F_{sys}) and consists at least of replacing a component C^i implementing the basic function $F \varepsilon FU^i$ with a new one so that ettf(F_{sys}) is increased.

In summary, prognosis to implement complex systems should take into account that a complex system consists of several subsystems and these subsystems are, in turn, composed of components. Each component may require a different type of prognosis. The most common way to implement prognosis is by using the Weibull distribution, but depending on the component, other methods may be suitable (fuzzy NNs, etc.). Each complex system has a different mathematical model; when we want to estimate RUL, we must first make a mathematical model of the system in question.

6.5 CONTEXT-BASED PROGNOSIS: THE INFLUENCE OF INFORMATION AND COMMUNICATION TECHNOLOGY IN REMAINING USEFUL LIFE ESTIMATION

The concept of context-aware decision-making is now popular in IT, web services, and medical fields; it is pervasive and ubiquitous, in many computing areas (Baldauf et al., 2007; Hong et al., 2009). The idea is to predict the possible scenarios based on different levels of context using data aggregation by machine-learning methods to provide output to the user that is necessary, efficient, and effective (Derntl and Hummel, 2005). Applying context-aware decision-making to maintenance activities may lead to more efficient diagnosis and prognosis.

A context-aware system actively and autonomously adapts and provides the most appropriate services or information to users, taking advantage of people's contextual information but requiring little interaction. Context-aware computing is defined by Schilit and Theimer (1994) and Schilit et al. (1994) as "the ability of a mobile user's applications to discover and react to changes in the environment they are situation".

Context-aware systems are usually complicated and are responsible for many jobs, such as representation, modeling (Strang and Linnhoff-Popien, 2004), management, reasoning (Wang et al., 2004c), and analysis of context information. They require the collaboration of many different components in the various relevant systems. There are several types of context-aware systems, making it hard to generalize a context-aware system process; however, a context-aware system usually follows four steps. The typical process for context-aware systems is shown in Fig. 6.21.

The first step is acquiring context information from sensors. Sensors convert real world context information into computable context data. Using physical and virtual sensors, the system can capture various types of context-aware information. The system then stores the data into its repository. When storing context data, the kind of data model used to represent the context information is very important; context models are diverse and each has its own unique characteristics. To easily use the stored context data, the system controls the abstraction level of the data by interpreting or aggregating them. Finally, the system uses the abstracted context data for context-aware applications. One such representation of context-aware system was developed by Chen et al. (2003) for intelligent broker systems. They suggest creating a

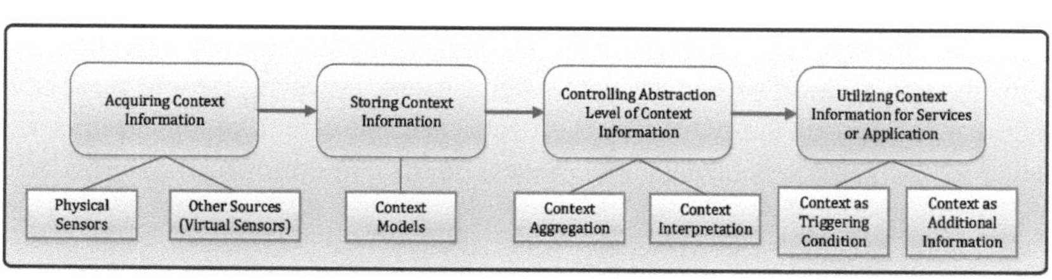

FIGURE 6.21 General process in context-aware systems.

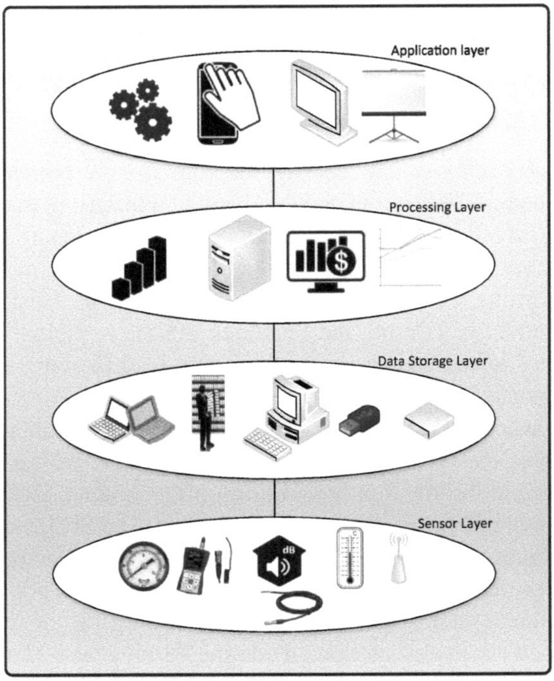

FIGURE 6.22 Layered architecture of context-aware systems.

context broker architecture for context acquisition using different sensors among users, machines, and agents to provide adequate support for context modeling. The features can then be extracted from the data and brokerage activities applied. The typical context-aware system consists of four layers: sensor layer for data acquisition, data storage layer for maintaining data, preprocessing layer for all analysis and modeling, and an application layer for final representation as shown in Fig. 6.22.

6.5.1 Context-Aware Condition Monitoring

In the past, the different functional areas, e.g., process monitoring, equipment monitoring, and performance monitoring, all performed independently; each tried to "optimize" its associated functional area without regard to the effect that any actions might have on the other functional areas (Belvins et al., 2001). A low-priority equipment problem might not be corrected; yet the failure to reach a desired or critical level of performance might cause a large problem. With the asset cloud providing data to end users, however, people now have access to a view of the plant based on two or more of equipment-monitoring data, process-performance data, and process control—monitoring data. Similarly, diagnostics performed for a plant may take into account data associated with process operation and the equipment operation and provide a better overall diagnostic analysis.

With the advancement in computing and data acquisition capabilities, the competences of condition monitoring have also improved (Remboski et al., 2002). In dynamic and adaptive environments, context-aware condition monitoring reduces risks and improves safety by effectively predicting RUL (Bottazzi et al., 2006).

6.5.2 Diagnosis With Anomaly Detection

The anomaly detection task is to recognize the presence of an unusual (and potentially hazardous) state within the behaviors or activities of a system, with respect to some model of "normal" behavior, which may be either hard-coded or learned from observation (Chandola et al., 2009). We focus here on learning models of normalcy at the user behavioral level, as observed. An anomaly detection agent faces many learning problems, including learning from streams of temporal data, learning from instances of a single class, and adapting to a dynamically changing concept (Lane, 2000). The domain is complicated by considerations of the trusted insider problem (recognizing the difference between innocuous and malicious behavior changes by a trusted user) and the hostile training problem (avoiding learning the behavioral patterns of a hostile user who is attempting to deceive the agent).

Anomaly detection refers to the problem of finding patterns in data that do not conform to expected behavior. These nonconforming patterns are called anomalies, outliers, discordant observations, exceptions, aberrations, surprises, peculiarities, or contaminants in different application domains. Noise removal is driven by the need to remove unwanted objects before data analysis is performed (Ye and Chen, 2001).

Given the various challenges, the anomaly detection problem, in its most general form, is not easy to solve (Garcia-Teodoro et al., 2009). In fact, most of the existing anomaly detection techniques solve a specific formulation of the problem. The formulation is induced by various factors such as the nature of the data, availability of labeled data, type of anomalies to be detected, etc. Often these factors are determined by the application domain in which the anomalies need to be detected. Accordingly, researchers have adopted concepts from diverse disciplines, such as statistics, machine learning, data mining, information theory, and spectral theory and have applied them to specific problem formulations.

If a data instance is anomalous in a specific context (but not otherwise), it is termed a contextual anomaly (also called a conditional anomaly) and is particularly applied in video recognition (Zhu et al., 2013). The notion of a context is induced by the structure in the data set and must be specified as a part of the problem formulation. Each data instance is defined using either contextual or behavioral attributes (Chandola et al., 2009).

a. Contextual attributes are used to determine the context (or neighborhood) of a data instance. For example, in spatial data sets, the longitude and latitude of a location are the contextual attributes. In time-series data, time is a contextual attribute, which determines the position of an instance.

b. Behavioral attributes define the noncontextual characteristics of a data instance. For example, in a spatial data set describing the average rainfall of the entire world, the amount of rainfall at any location is a behavioral attribute.

Anomalous behavior is determined using the values for the behavioral attributes within a specific context. A data instance might be a contextual anomaly in a given context, but an identical data instance (in terms of behavioral attributes) could be considered normal in a different context. This property is key in identifying contextual and behavioral attributes for a contextual anomaly detection technique.

Contextual anomalies have been most commonly explored in time-series data and spatial data. Fig. 6.23 displays one such example for a temperature time series, which shows the monthly temperature of an area over the last few years (Chandola et al., 2009). A temperature of 35°F might be normal during the winter (at time t1) at that place, but the same value during summer (at time t2) would be an anomaly. The choice of applying a contextual anomaly detection technique is determined by the meaningfulness of the contextual anomalies in the target application domain. Another factor is the availability of contextual attributes. In some cases, defining a context is straightforward and applying a contextual anomaly detection technique makes sense (George et al., 2011). In other cases, defining a context is not easy, making it difficult to apply such techniques.

Note: the temperature at time t1 is the same as at time t2 but occurs in a different context and, hence, is not considered an anomaly.

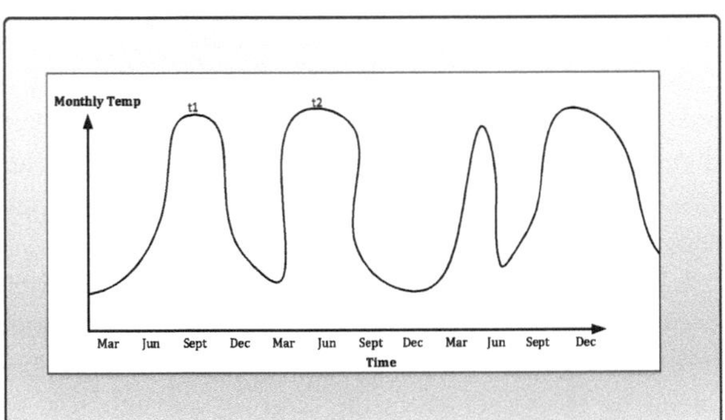

FIGURE 6.23 Contextual anomaly t2 in a temperature time series.

6.5.3 Context-Driven e-Maintenance

Once connectivity is sorted out, sense making becomes the real challenge for data sets. It is time to take concepts from e(lectronic)-Maintenance into i(ntelligent)-Maintenance (Galar et al., 2012). Maintainers must deal with many different sources of information. Therefore, we use a system framework supporting the integration of various data sources, which could have different formats and natures. To handle those differences, the system framework should provide facilities for data wrapping and mediation between different data formats, along with interfaces for external data wrappers and mediators. The system should also be able to add new sources and mediation procedures and handle the necessary data validation and consistency checking. From the operation point of view, different data spaces must be managed at different levels of the system. At the data management space, the following agents and databases must be managed and merged:

- Synthetic database, containing derived calculations from the database or from external sources not included in the database;
- Information on managing the databases;
- Information on managing wrappers and mediators; and
- Archived data.

An example of typical data and their relations, taken from transportation facilities for a Swedish case study, can be seen in Fig. 6.24.

6.6 CONCLUSIONS AND DISCUSSION

The chart in Fig. 6.25 sums up the previous discussion, showing the different prognosis techniques used to estimate the RUL of certain types of assets.

Note: A refers to rotating machines (pumps, turbines, compressors, motors); B refers to structures (bridges, buildings); and C refers to complex systems (planes, cars, etc.).

As the chart suggests, prognosis can be applied to many different industrial assets.

Rotating machines only require particle filter and physics-based fatigue analysis; other techniques are rarely used for this type of asset. ANNs and fuzzy logic may be used in data-based models; on occasion, linear regression is used. For experience-based models, the most common choice is Weibull because it is the most conventional.

Models for structures such as bridges and buildings use fatigue; the monitoring applied to these assets is used to calculate the RUL of fatigue, as this is usually the main problem with such assets. But depending on the type of structure, Weibull can be used.

Finally, the calculation of the RUL of more complex systems uses mathematical models; most complex systems require the development of a model

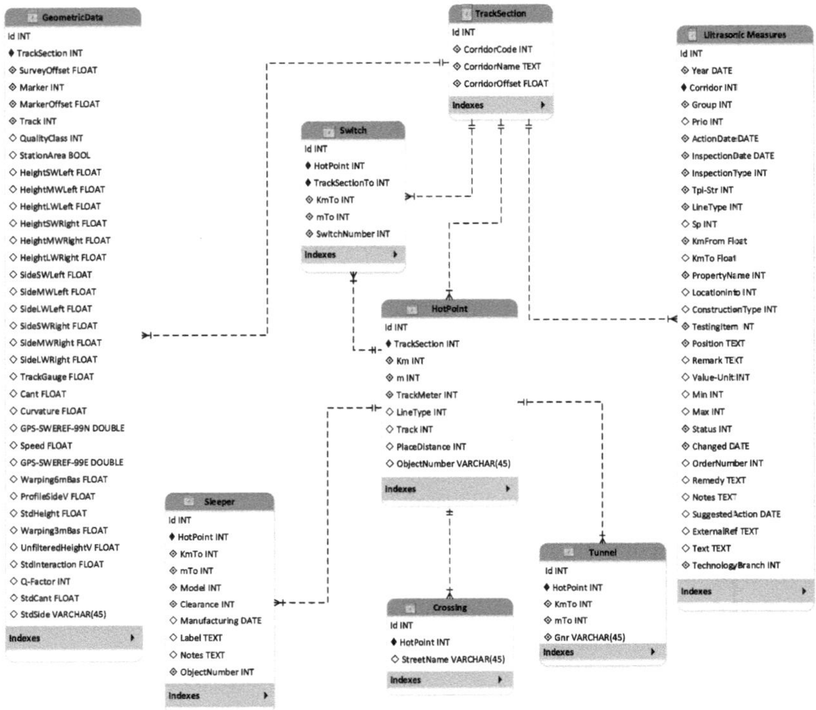

FIGURE 6.24 Database design for Swedish case study (OPTIRAIL project).

Approaches Prognosis in Diffierent Industrial Assets			
Assets			***Approaches***
A	B	C	
			Model Based
✱	✱		Particle Filtering
✱	✱		Physics-based fatigue models
			ARMA, ARMAX and ARMA Methods
			Data Driven
✱			Linear Regression
✱		✱	Artificial Neural Networks
✱		✱	Fuzzy Logic Systems
			Gaussian Process Regression
			Relevance Vector Machines (RVM)
			Experience Based
			Bayesian Probability Theory
✱	✱	✱	The Weibull Model: Analysis of Time to Failure
			HMM
			Models
		✱	Mathematical

FIGURE 6.25 Different prognostic techniques. *ARMA*, autoregressive moving average; *ARIMA*, autoregressive integrated moving average models; *HMM*, hidden Markov model.

because these systems are not alike. Similar systems often use techniques such as NNs, fuzzy networks, and occasionally Weibull.

As the chapter makes clear, CBM program can be applied to estimate RUL. It has reviewed various techniques, models, and algorithms following the three main steps of a CBM program: data acquisition, data processing, and maintenance decision-making. Although advanced maintenance techniques are described in the literature, there are two common extremes in industry. One is to adopt a run-to-failure (breakdown) policy. Another is to apply an as-frequent-as-possible maintenance policy. The two can be applied to some special cases with satisfactory results, but in many situations, especially when both maintenance and failure are very costly, CBM is a better choice. Expert knowledge in both application and theory is required to choose the best maintenance policies.

Advanced maintenance technologies have not been well implemented in industry for the following possible reasons:

- lack of data because of incorrect data collecting approach, or no data collection and/or data storage;
- lack of efficient communication between theory developers and practitioners in the area of reliability and maintenance;
- lack of efficient validation approaches; and
- difficulty of implementation with frequent change of design, technologies, business policies, and management executives.

Next generation prognostic systems will likely focus more on various aspects of continuous monitoring and automate the prognostics.

The chapter also discusses concepts associated with prognostics in different industrial assets. It reviews prognostic approaches and implementation issues, including current CBM developments, and gives examples. It discusses variations in data, modeling, and reasoning for the various prognostic approaches and illustrates these in rotating machines (pumps, turbines, compressors, motors), structures (bridges and buildings), and complex systems (planes, cars, etc.).

Data availability, dominant failure or degradation mode of interest, modeling and system knowledge, accuracy, and criticality are some of the variables that determine the choice of prognostic approach.

The ability to predict the time to conditional or mechanical failure (on a real-time basis) is of enormous benefit; health management systems that can effectively implement the capabilities presented here offer the opportunity to reduce the overall life cycle costs of operating systems and decrease maintenance tasks.

The measurement data from all models must be consistent, valid, and reliable to guarantee the development of a successful diagnostic/prognostic strategy and to assure an adequate safety level for the whole system. However, the different nature of the data requires data wrapping into several levels and categories to find meaningful links. This calls for the use of context awareness.

In future work, it would be interesting to study how to apply prognosis in other types of industrial assets. In addition, it would be useful to conduct an experimental study using a real case with the approaches outlined here to compare the results. Studies could also compare two approaches to the same data. Finally, the hybrid approaches could be explored in greater detail, especially their applicability in industry.

For CBM, it would be interesting to study the progress and trends of this type of program at present (data cleaning, data quality, issues of context awareness) and to explore the use of data when estimating RUL. Another future work could study the advances in sensors and data acquisition systems, along with the computer programs used for this purpose.

REFERENCES

Abernethy, R.B., 2006. The New Weibull Handbook, fifth ed.

Agatonovic-Kustrin, S., Beresford, R., 2000. Basic concepts of artificial neural network (ANN) modelling and its application in pharmaceutical research. Journal of Pharmaceutical and Biomedical Analysis 22, 717–727.

Baldauf, M., Dustdar, S., Rosenberg, F., 2007. A survey on context-aware systems. International Journal of Ad Hoc and Ubiquitous Computing 2 (4), 263–277.

Belvins, T., Gudaz, J., Keyes, M., Nixon, M., Schleiss, T., 2001. Fusion of process performance monitoring with process equipment monitoring and control. http://www.google.com/patents/US20020077711.

Blischke, W.R., Murthy, D.N., 2000. Reliability: Modelling, Prediction and Optimization. John Wiley & Sons Ltd.

Bottazzi, D., Corradi, A., Montanari, R., 2006. Context-aware middleware solutions for anytime and anywhere emergency assistance to elderly people. Communications Magazine, IEEE 44 (4), 82–90.

Brotherton, T., Jahns, G., Jacobs, J., Wroblewski, D., 2000. Prognosis of faults in gas turbine engines. Proceedings of IEEE International Conference on Aerospace 6, 163–171.

BS 5400, 1980. Part 10. Code of practice for fatigue. In: Steel Concrete and Composite Bridges. British Standard Institutions, UK.

Byington, C.S., Roemer, M.J., Thomas, G., 2002. Prognostics enhancements to diagnostic systems for improved condition based maintenance. IEEE.

Chandola, V., Banerjee, A., Kumar, V., 2009. Anomaly detection: a survey. ACM Computing Surveys (CSUR) 41 (3), 15.

Chen, C., Zhang, B., Vachtsevanos, G., 2012. Prediction of machine health condition using Neuro-Fuzzy and Bayesian algorithms (School of Electrical and Computer Engineering. Georgia Institute of Technology, Atlanta, Georgia 30332 USA. Impact Technologies LLC, Rochester, NY 14623).

Chen, H., Finin, T., Joshi, A., October 2003. An intelligent broker for context-aware systems. Adjunct Proceedings of Ubicomp 3, 183–184.

Dempster, A., Laird, N., Rubin, D., 1977. Maximum likelihood from incomplete data via the EM algorithm. Journal of the Royal Statistical Society 39, 1–38.

Derntl, M., Hummel, K.A., March 2005. Modeling context-aware e-learning scenarios. In: Third IEEE International Conference on Pervasive Computing and Communications Workshops (PerCom 2005 Workshops). IEEE, pp. 337–342.

Dong, M., He, D., 2007. A segmental hidden semi-markov model (HSMM)-based diagnostics and prognostics framework and methodology. Mechanical Systems and Signal Processing 21, 2248–2266.

Dragomir, O., Gouriveau, R., Zerhouni, N., Dragomir, F., 2007. Framework for a distributed and hybrid prognostic system. 4th IFAC Conference on Management and Control of Production and Logistics.

Dragomir, O.E., Gouriveau, R., Dragomir, F., Minca, E., Zerhouni, N., September 2009. Review of prognostic problem in condition-based maintenance (hal-00418761, version 1–21).

Engel, S., Gilmartin, B., Bongort, K., Hess, A., 2000. Prognostics, the real issues involved with predicting life remaining. IEEE Aerospace Conference, USA 6, 457–469.

Galar, D., Wandt, K., Karim, R., Berges, L., August 2012. The evolution from e(lectronic) Maintenance to i(ntelligent)Maintenance. Insight – Non-Destructive Testing and Condition Monitoring 54 (8), 446–455(10).

Garcia-Teodoro, P., Diaz-Verdejo, J., Maciá-Fernández, G., Vázquez, E., 2009. Anomaly-based network intrusion detection: techniques, systems and challenges. Computers & Security 28 (1), 18–28.

George, J., Crassidis, J., Singh, T., Fosbury, A.M., 2011. Anomaly detection using context-aided target tracking. Journal of Advances in Information Fusion 6 (1), 39–56.

Gerardo Trujillo, C., América, N.L., October 2003. Condition monitoring – a strategy integration technologies. 1st Mexican Congress of Reliability and Maintenance, Mexico.

Gouriveau, R., 2011. An Introduction to Prognostics. FEMTO-ST Institute. UMR CNRS 6174 FCLAB Research Federation, FR CNRS 3539.

Heaton, J., 2008. Introduction to Neural Networks for Java. Heaton Research Inc.

Hong, J.Y., Suh, E.H., Kim, S.J., 2009. Context-aware systems: a literature review and classification. Expert Systems With Applications 36 (4), 8509–8522.

Jardine, A.K.S., Lin, D., Banjevic, D., 2006. A review on machinery diagnostics and prognostics implementing condition – based maintenance. Mechanical Systems & Signal Processing 20, 1483–1510.

Kohout, J., Vechet, S., 2001. A new function for fatigue curves characterization and its multiple merits. International Journal of Fatigue 23 (2), 175–183.

Lane, T.D., 2000. Machine learning techniques for the computer security domain of anomaly detection.

Lee, C.-H., Teng, C.-C., August 2000. Identification and control of dynamic systems using recurrent fuzzy neural networks. IEEE Transactions on Fuzzy Systems 8 (4).

Lewis, D., 1986. Philosophical Papers, vol. II. Oxford University Press, Oxford.

Lin, C., Lee, C., 1996. Neural Fuzzy System: A Neuro-Fuzzy Synergism to Intelligent Systems. Prentice-Hall, Upper Saddle River, NJ.

Luo, J., Namburu, M., Pattipati, K., Qiao, L., Kawamoto, M., Chigusa, S., 2003a. Model-based prognostic techniques. In: Proceedings of IEEE Autotestcon, pp. 330–340.

Luo, J., Namburu, M., Pattipati, K., Qiao, L., Kawamoto, M., Chigusa, S., 2003b. Model-Based Prognostic Techniques (Anaheim, CA, United States: Institute of Electrical and Electronics Engineers Inc., Piscataway, NJ, United States), pp. 330–340.

Mardia, K.V., Marshall, R.J., 1984. Maximum likelihood estimation for models of residual covariance in spatial regression. Biometrika 71 (1), 135–146.

Natke, H.G., Cempel, C., 1996. Model-Aided Diagnosis of Mechanical Systems Fundamentals, Detection, Localization, and Assessment. Springer-Verlag.
Natke, H.G., Cempel, C., 1968–1997. Proceedings of the society for machinery failure prevention technology. MFPT, Formerly MFPG.
Neter, J., Kutner, M.H., J Nachtsheim, C., Wasserman, W., 1996. Applied Linear Statistical Models. Irwin.
Ni, Y., Wong, K.Y., 2012. Integrating Bridge Structural Health Monitoring and Condition-Based Maintenance Management. Department of Civil and Environmental Engineering. The Hong Kong Polytechnic University, Hong Kong.
Pasha, G.R., Shuaib Khan, M., Ahmed Hesham Pasha, 2006. Empirical analysis of the Weibull distribution for failure data. Journal of Statistics 13 (1), 33–45.
Pecht, M.G., 2008. Prognostics and Health Management of Electronics. Wiley. ISBN: 978-0-470-27802-4.
Rasmussen, C.E., Williams, C.K.I., 2006. Gaussian Processes for Machine Learning. The MIT Press.
Remboski, D., Brooks, K.M., Canavan, P.J., Douros, K., Gardner, J.L., Gardner, R.M., et al., 2002. U.S. Patent No. 20020151297. U.S. Patent and Trademark Office, Washington, DC.
Ribot, P., Pencolé, Y., Combacau, M., 2009. Functional prognostic architecture for maintenance of distributed systems. In: 7th IFAC Symposium on Fault Detection, Supervision and Safety of Technical Processes.
Ross, S.M., 2001. Probability Models for Computer Science. Academic Press.
Schilit, B.N., Theimer, M.M., 1994. Disseminating active map information to mobile hosts. IEEE Network 8 (5), 22–32.
Schilit, B., Adams, N., Want, R., December 1994. Context-aware computing applications. In: Mobile Computing Systems and Applications, 1994. WMCSA 1994. First Workshop on. IEEE, pp. 85–90.
Shane Butler, September 2012. Prognostic Algorithms for Condition Monitoring and Remaining Useful Life Estimation. A thesis submitted in partial fulfillment for the degree of Doctor of Philosophy.
Sikorska, J.Z., Hodkiewicz, M., Ma, L., 2010. Prognostic modelling options for remaining useful life estimation by industry. Mechanical Systems, Signal Processing.
Siti Azirah Asmai, Abd. Samad hasan Basari, Abdul Samad Shibghatullah, Nuzulha Khilwani Ibrahim, Burairah Hussin, 2011. Neural Network Prognostics Model for Industrial Equipment Maintenance. Department of Industrial Computing, Faculty of Information and Communication Technology. Universiti Teknikal Malaysia Melaka, Malaysia.
Strang, T., Linnhoff-Popien, C., September 2004. A context modeling survey. In: Workshop Proceedings.
Sun, Y., Babovic, V., Chan, E., 2010. Multi-step-ahead model error prediction using time-delay neural networks combined with chaos theory. Journal of Hydrology.
Tan, Y., Van Cauwenberghe, A., 1999. Neural-network-based d-stepahead predictors for nonlinear systems with time delay. Engineering Applications of Artificial Intelligence 12, 21–35.
Tipping, M.E., 2000. The relevance vector machine. In: Advances in Neural Information Processing Systems, vol. 12. Cambridge MIT Press, pp. 652–658.
Tobon-Mejia, D.A., Medjaher, K., Zerhouni, N., June 2011. Hidden markov models for failure diagnostic and prognostic (version 1–21).
Todinov, M., 2005. Reliability and Risk, "Models Setting Reliability Requirements". John Wiley & Sons Ltd., Chichester, England.

Vachtsevanos, G., Lewis, F., Roemer, M., Hess, A., Wu, B., 2006. Intelligent Fault Diagnosis and Prognosis for Engineering Systems. John Wiley and Sons Inc., Hoboken, New Jersey.

Voisin, A., Levrat, E., Cocheteux, P., Iung, B., 2009. Generic prognosis model for proactive maintenance decision support: application to preindustrial e-maintenance test bed. Journal of Intelligent Manufacturing. Accepted article.

Wang, W.Q., Golnaraghi, M.F., Ismail, F., 2004a. Prognosis of machine health condition using neuro-fuzzy systems. Mechanical Systems and Signal Processing 18 (4), 813–831.

Wang, W., Golnaraghi, F., Ismail, F., 2004b. Prognosis of machine health condition using neuro-fuzzy systems. Mechanical Systems and Signal Processing 18, 813–831.

Wang, X.H., Zhang, D.Q., Gu, T., Pung, H.K., March 2004c. Ontology based context modeling and reasoning using OWL. In: Pervasive Computing and Communications Workshops, 2004. Proceedings of the Second IEEE Annual Conference on. IEEE, pp. 18–22.

Wikipedia the Free Encyclopedia. Prognostics, Hybrid Approaches. http://en.wikipedia.org/wiki/Prognostics#Hybrid_approaches.

Williams, C.K.I., Rasmussen, C.E., 1996. Gaussian processes for regression. In: Touretzky, D.S., Mozer, M.C., Hasselmo, M.E. (Eds.), Advances in Neural Information Processing Systems, vol. 8. The MIT Press, Cambridge, MA, pp. 514–520.

Ye, N., Chen, Q., 2001. An anomaly detection technique based on a chi-square statistic for detecting intrusions into information systems. Quality and Reliability Engineering International 17 (2), 105–112.

Yu, W.K., Harris, 2001. A new model based on stress fatigue life of ball bearings. Tribology Transactions 11–18. http://dx.doi.org/10.1080/10402000108982420.

Zarei, J., Poshtan, J., 2007. Bearing fault detection using wavelet packet transform of induction motor stator current. Tribology International 40 (5), 763–769.

Zhai, L.-Y., Lu, W.-F., Liu, Y., Li, X., Vachtsevanos, G., 2013. Analysis of time-to-failure data with Weibull model in product life cycle management. In: 20th CIRP International Conference on Life Cycle Engineering, Singapore.

Zhu, Y., Nayak, N.M., Roy-Chowdhury, A.K., 2013. Context-aware activity recognition and anomaly detection in video. IEEE Journal of Selected Topics in Signal Processing 7 (1), 91–101.

FURTHER READING

Adolfsson, E., Dahlström, T., 2011. Efficiency in Corrective Maintenance—A Case Study at SKF Gothenburg (Master thesis). Report No. E2011:096. Chalmers University of Technology, Gothenburg, Sweden.

Knezevic, J., 1996. In: Maintenance. Isdefe.

Lee, J., Abujamra, R., Jardine, A.K.S., Banjevic, D., 2004. An integrated platform for diagnostics, prognostics and maintenance optimization. In: The IMS, 2004 International Conference on Advances in Maintenance and in Modelling, Simulation and Intelligent Monitoring of Degradations, Arles, France.

Mishra, R.C., Pathak, K., August 1, 2004. Maintenance Engineering and Management. PHI Learning Pvt. Ltd., p. 228

Orchard, M.E., Vachtsevanos, G.J., 2009. A particle filtering approach for on-line fault diagnosis and failure prognosis. Transactions of the Institute of Measurement and Control 31, 221–246.

Orchard, M., Kacprzynski, G., Goebel, K., Saha, B., Vachtsevanos, G., October 2008. Advances in uncertainty representation and management for particle filtering applied to prognostics. In: International Conference on Prognostics and Health Management, PHM, p. 16.

Roemer, M.J., Byington, C.S., Kacprzynski, G.J., Vachtsevanos, G., Goebel, K., 2011. Prognostics. In: Johnson, S.B., Gormley, T., Kessler, S.S., Mott, C., Ann, P.-H., Reichard, K., Scandura, P.A. (Eds.), System Health Management: With Aerospace Applications. John Wiley & Sons, West Sussex, UK.

Sondalini, M. How to Use Condition Based Maintenance Strategy for Equipment Failure Prevention, Lifetime Reliability Solutions HQ. http://www.lifetime-reliability.com.

Tian, Z., 2009. An artificial neural network method for remaining useful life prediction of equipment subject to condition monitoring. Journal of Intelligent Manufacturing. N/A:1{11.

Tsang, A.H.C., 1995. Condition-based maintenance: tools and decision making, 1 (3), 3–17.

Chapter 7

Maintenance Decision Support Systems

7.1 A NEW ERA IN INDUSTRY 4.0: MAINTENANCE 4.0

Asset management (AM) enables the realization of value from assets throughout their full life cycle. It involves the coordinated and optimized planning, selection, acquisition/development, utilization, care (maintenance), and ultimate disposal or renewal of assets and asset systems. The layered structure of AM is presented in Fig. 7.1.

Assets are complex mixes of complex systems. Each system is built from components which, over time, may fail. Problems occurring in systems used to be diagnosed by experienced personnel with in-depth training and experience (Chiang et al., 2001). Computer-based systems are now being used to automatically diagnose problems and overcome some of the disadvantages of relying on experienced personnel (Price and Price, 1999).

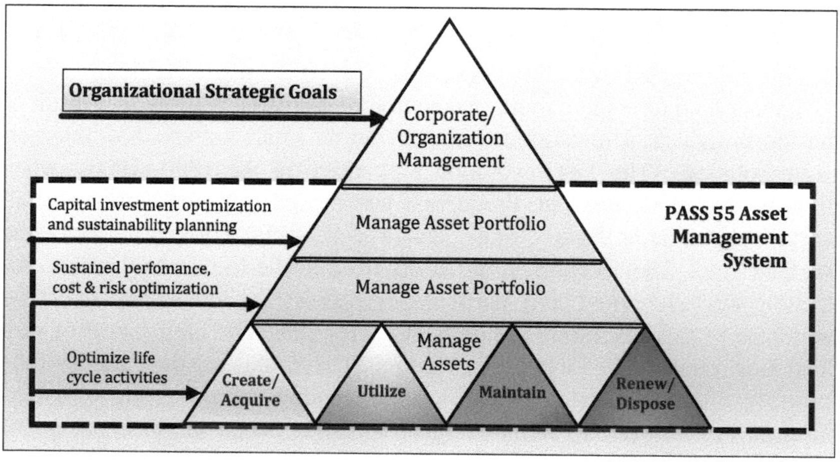

FIGURE 7.1 Layered structure of asset management based on PAS 55. *Adapted from Woodhouse, J., 2013. Setting a Good Standard in Asset Management. Chair of Experts Panel, Institute of Asset Management and Managing Director, TWPL.*

eMaintenance. http://dx.doi.org/10.1016/B978-0-12-811153-6.00007-5

ISO 55000 is an international standard for AM launched in 2014. It is based on PAS 55 (Publicly Available Specification) published by British Standard Institution in 2004; PAS 55 gives guidance for good practices in physical AM. ISO 55000 extends the definition of assets, as it covers any asset that creates value for an organization, not just physical production assets. The standard provides a framework for developing an AM system and enables organizations to achieve business objectives through the effective and efficient management of its assets (ISO, 2014).

The challenge is to provide intelligent tools to monitor and manage assets (machines, plants, products, etc.) proactively through information and communications technology (ICT), focusing on health degradation monitoring and prognosis instead of fault detection and diagnostics. Maintenance effectiveness depends on the quality, timeliness, accuracy, and completeness of information on machine degradation. This translates into the following key requirements: the ability to prevent data overload, the ability to differentiate and prioritize data (during collection and reporting), and the ability to prevent, as far as possible, the occurrence of information islands.

Data relevant to AM are gathered, produced, and processed on different levels by different information technology (IT) systems, e.g., enterprise resource planning (ERP) for business functions, supervisory control and data acquisition (SCADA) for process functions, computerized maintenance management system (CMMS) and condition monitoring (CM) for maintenance functions, and safety instrumented systems for safety-related functions. Combining the different data sources into one seamless system can exponentially increase the positive effects on operations and maintenance (O&M) performance. Only a few years ago, linking CMMS, SCADA, and CM technologies was only an extremely expensive vision. With the technology available today, it is relatively easy and inexpensive to combine the information supplied by these systems for maintenance knowledge discovery, i.e., the process targeted by Industry 4.0.

Industry 4.0 refers to a fourth generation of industrial activity as a result of the fourth industrial revolution characterized by smart systems and Internet-based solutions. The first revolution took place in the 1800–1900s, when production was mechanized. Production was moved from the home or small workshops to large factory units, and a new social class was born—the working class. The second revolution occurred in the last century when production was electrified and parts and processes were standardized. The archetype of this revolution is Ford's assembly line. The digitization of production is usually called the third revolution. Two characteristic features of the fourth revolution, Industry 4.0, are computerization with the help of cyber-physical systems (CPS) and intelligent factories based on the concept of "Internet of things" (IoT).

CPS are integrated computer-based or digital components that monitor and control physical devices, also called embedded systems. These systems communicate over a network usually based on Internet technology, creating, in other words, an IoT (as opposed to social media that could be described as an

"Internet of persons"). Combining these two concepts, we get a distributed network of embedded systems communicating with each other in an ad hoc and dynamic way. One of the application areas of Industry 4.0 is maintenance in the form of self-learning and smart systems that predict failure, make diagnoses, and trigger maintenance actions. These systems have high demands with respect to data access and data quality and use multiple data sources to extract relevant information. Several research projects have focused on the cyber-physical approach to developing intelligent maintenance management systems for failure detection, diagnostics, and prognostics. So far, the main application area has been process and manufacturing industries, but what has been called Maintenance 4.0 has huge potential in other areas as well, for example, the railway transportation industry (Koskinen et al., 2016).

7.1.1 What is Industry 4.0?

7.1.1.1 Definition and Development

Industry 4.0 refers to a further developmental stage in the organization and management of the entire value chain process of manufacturing. Another term for this process, as noted above, is the fourth industrial revolution.

The concept of Industry 4.0 is widely used across Europe, particularly in Germany's manufacturing sector. In the United States and the English-speaking world more generally, some commentators also use the terms the "Internet of things", the "Internet of everything," or the "industrial Internet."

What all these terms and concepts have in common is the recognition that traditional manufacturing and production methods are in the throes of a digital transformation. For some time now, industrial processes have increasingly embraced modern IT, but the most recent trends go beyond simply the automation of production that has, since the early 1970s, been driven by developments in electronics and IT (see Fig. 7.2) (Deloitte, 2014).

The widespread adoption of ICT is increasingly blurring the boundaries between the real world and the virtual world in what are known as cyber-physical production systems (CPPSs).

CPPSs are online networks of social machines organized much like social networks. Simply put, they link IT with mechanical and electronic components that then communicate with each other via a network. Radio-frequency identification (RFID) technology, in use since 1999, was a very early form of this technology.

Smart machines continually share information about current stock levels, problems or faults, and changes in orders or demand levels. Processes and deadlines are coordinated with the aim of boosting efficiency and optimizing throughput times, capacity utilization, and quality in development, production, marketing, and purchasing.

CPPSs not only network machines with each other but they also create a smart network of machines, properties, ICT systems, smart products, and individuals across the entire value chain and the full product life cycle (PLC). Sensors and control elements enable machines to be linked to plants, fleets,

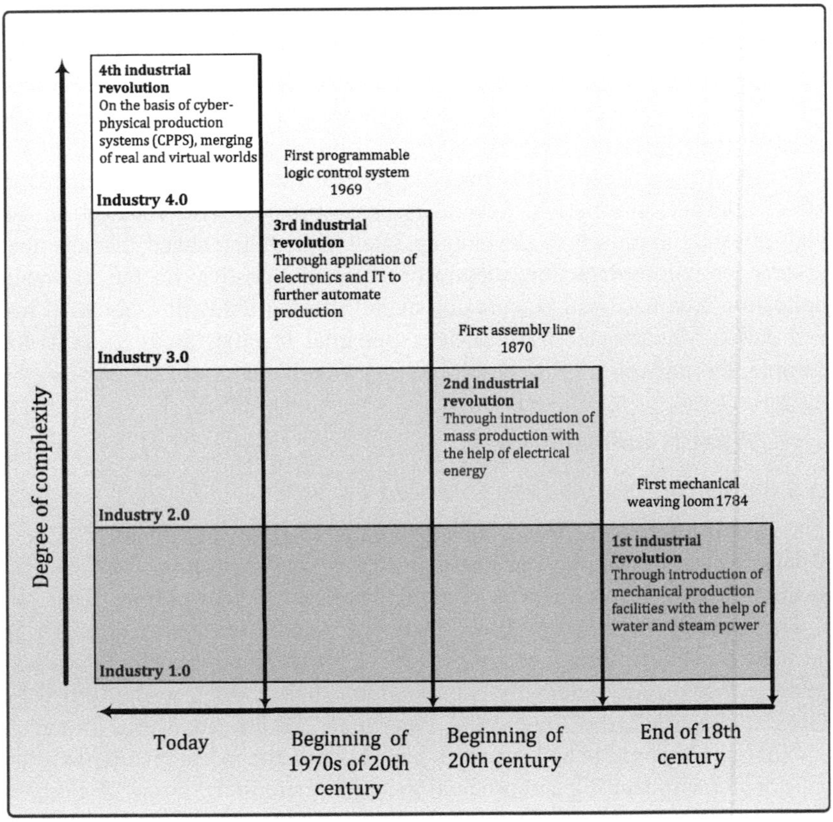

FIGURE 7.2 Definition of Industry 4.0. *Adapted from Deloitte, 2014. Industry 4.0. Challenges and Solutions for the Digital Transformation and Use of Exponential Technologies. Audit. Tax. Consulting. Corporate Finance. March–August, 2014.*

networks, and human beings. Smart networks of this kind are the bedrock of smart factories, which themselves underpin Industry 4.0. An overview is given in Fig. 7.3 (Deloitte, 2014).

Of central importance for Industry 4.0 is its interface with other smart infrastructures, such as those for smart mobility, the smart grid, smart logistics, and smart homes and buildings. Links to both business and social networks—the business web and the social web—also play an increasingly important role in the digital transformation to Industry 4.0. All these new networks and interfaces offered within an IoT, services, data, and people mean manufacturing is set to undergo enormous changes. This trend is still in its infancy in some manufacturing companies and industrial sectors, but in others, the transformation to Industry 4.0 is well under way.

Traditional industrial economies, such as Germany and the United States, expect this fourth industrial revolution to bring many advantages, ranging from enhanced global competitiveness to a reversal of the trend to relocate production in low-wage countries and the opening of more domestic production locations in Europe and North America.

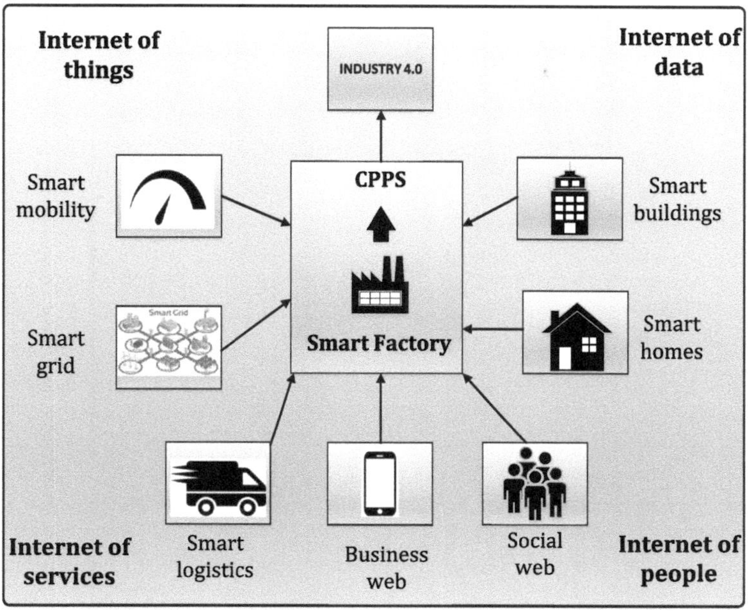

FIGURE 7.3 Industry 4.0 environment. *Adapted from Deloitte, 2014. Industry 4.0. Challenges and Solutions for the Digital Transformation and Use of Exponential Technologies. Audit. Tax. Consulting. Corporate Finance. March–August, 2014.*

Exponentially growing technologies will be the key to the transformation to Industry 4.0, a concept illustrated in Fig. 7.4. Research shows that Moore's law—which states that the capacity of microchips, bandwidth, and computers doubles every 18 months, representing exponential growth—also applies to other technological developments. Three-dimensional (3D) printing, sensor technology, artificial intelligence (AI), robotics, drones, and nanotechnology are a few examples of exponentially growing technologies that are radically changing industrial processes, accelerating them, and making them more flexible. Many of these technologies are not new and appeared some 20 or 30 years ago, but the recent massive boost in computing power (Moore's law) and the reduction in cost, along with miniaturization, now make them suitable for industrial use (Deloitte, 2014).

7.1.1.2 Main Characteristics

The following four main characteristics of Industry 4.0 demonstrate the huge capacity for change in industry and traditional manufacturing:

1. The vertical networking of smart production systems, such as smart factories and smart products, and the networking of smart logistics, production and marketing, and smart services, with a strong needs-oriented, individualized, and customer-specific production operation;

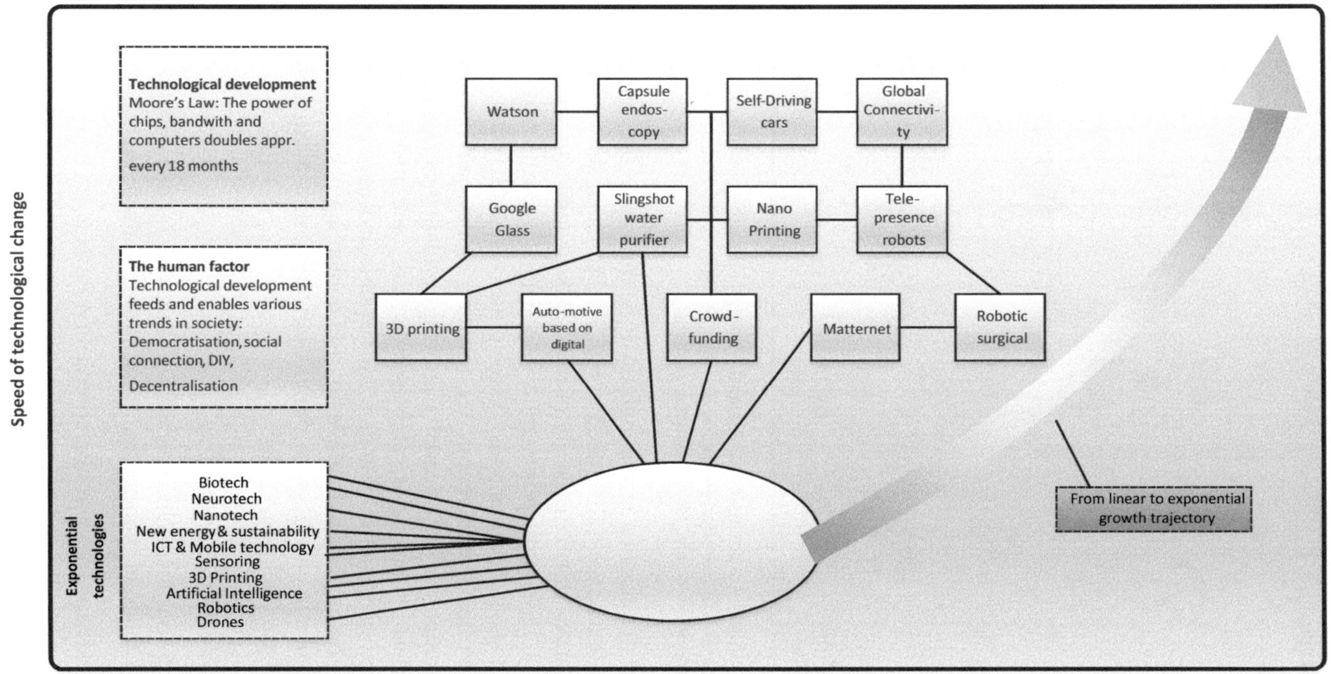

FIGURE 7.4 Exponential technologies. *Adapted from Deloitte, 2014. Industry 4.0. Challenges and Solutions for the Digital Transformation and Use of Exponential Technologies. Audit. Tax. Consulting. Corporate Finance. March–August, 2014.*

2. Horizontal integration by means of a new generation of global value creation networks, including integration of business partners and customers, and new business and cooperation models across countries and continents;
3. Through-engineering throughout the entire value chain, taking in not only the production process but also the end product—that is, the entire PLC;
4. Acceleration through exponential technologies that, while not new in terms of their development history, are only now capable of mass-market application as their cost and size have come down (e.g., sensor technology) and their computing power has risen immensely (Deloitte, 2014).

These four characteristics are shown in Fig. 7.5.

7.1.1.2.1 Vertical Networking of Smart Production Systems

The first main characteristic of Industry 4.0 is the vertical networking of smart production systems in the factories of the future. Vertical networking uses CPPSs to enable plants to react rapidly to changes in demand or stock levels and to faults. Smart factories organize themselves and enable customer-specific and individualized production. This requires data to be extensively integrated. Smart sensor technology is also needed to help with monitoring and autonomous organization.

CPPSs enable the autonomous organization of both production and maintenance management. Resources and products are networked, and materials and parts can be located anywhere and at any time. All processing stages in the production process are logged, with discrepancies registered automatically. Amendments to orders, fluctuations in quality, or machinery breakdowns can be

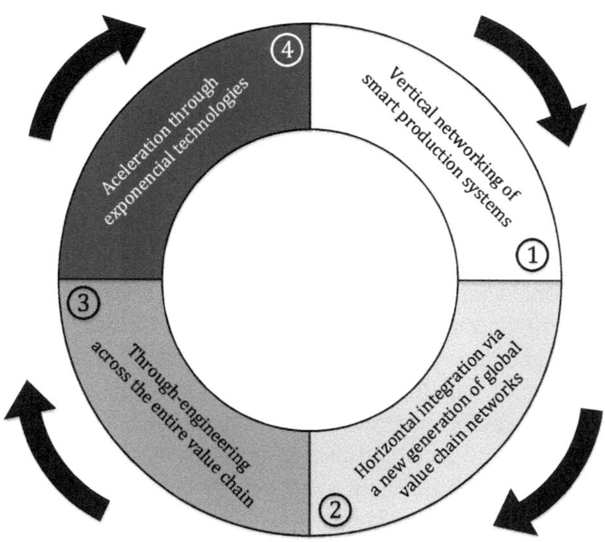

FIGURE 7.5 Four characteristics of Industry 4.0. *Adapted from Deloitte, 2014. Industry 4.0. Challenges and Solutions for the Digital Transformation and Use of Exponential Technologies. Audit. Tax. Consulting. Corporate Finance. March–August, 2014.*

dealt with more rapidly. Such processes also enable wear and tear on materials to be monitored more effectively or preempted. All in all, waste is reduced.

Significant emphasis is attached to resource efficiency, especially the efficient use of materials, energy, and human resources. The demands on workers engaged in operational tasks such as production, warehousing, logistics, and maintenance are changing as well, and new skills are required to work efficiently with CPPSs (Deloitte, 2014).

7.1.1.2.2 Horizontal Integration via a New Generation of Global Value Chain Networks

The second main characteristic of Industry 4.0 is horizontal integration via a new generation of global value chain networks. These new value-creation networks are real-time optimized networks that enable integrated transparency, offer a high level of flexibility to respond more rapidly to problems and faults, and facilitate better global optimization.

Like networked production systems, these (local and global) networks provide networking via CPPSs, from inbound logistics through warehousing, production, marketing, and sales to outbound logistics and downstream services. The history of any part or product is logged and can be accessed at any time, ensuring constant traceability (a concept known as "product memory").

This creates transparency and flexibility across entire process chains—from purchasing through production to sales, for example, or from the supplier through the company to the customer. Customer-specific adaptations can be made, not only in the production but also in the development, ordering, planning, composition, and distribution of products, enabling factors such as quality, time, risk, price, and environmental sustainability to be handled dynamically, in real time and at all stages of the value chain.

This kind of horizontal integration of both customers and business partners can generate completely new business models and new models for cooperation, representing a challenge for all those involved. Legal issues and questions of liability and protection of intellectual property are becoming increasingly important (Deloitte, 2014).

7.1.1.2.3 Through-Engineering Across the Entire Value Chain

The third main characteristic of Industry 4.0 is cross-disciplinary and is called through-engineering across the entire value chain and across the full life cycle of both products and customers. This engineering occurs seamlessly during the design, development, and manufacture of new products and services. New products need new and/or modified production systems. The development and manufacture of new products and production systems is integrated and coordinated with PLCs, enabling new synergies to be created between product development and production systems.

Characteristic of through-engineering is that data and information are available at all stages of a product's life cycle, enabling new, more flexible processes to be defined from data via modeling and applied to prototypes and the product stage (Deloitte, 2014).

7.1.1.2.4 Acceleration Through Exponential Technologies

The fourth main characteristic of Industry 4.0 is the impact of exponential technology accelerants or catalysts allowing individualized solutions, flexibility, and cost savings in industrial processes. Industry 4.0 already requires automation solutions to be highly cognitive and highly autonomous. AI, advanced robotics, and sensor technology have the potential to increase autonomy even more and to speed up individualization and flexibilization.

AI can help to plan driverless vehicle routes in factories and warehouses more flexibly, save time and cost in supply chain management (SCM), increase reliability in production, and analyze big data. It can also help to find new construction and design solutions or enhance the cooperation between humans and machines.

Functional nanomaterials and nanosensors can be used in production control functions to make quality management more efficient or allow the production of next generation robots that work hand in hand and safely with humans. Flying maintenance robots in production halls or using drones to make inventories of warehouse stock levels and deliver spare parts, at any time of day or night and in any terrain and weather, are further applications that will become simply routine in the autonomous and smart factories of the future (Deloitte, 2014).

7.1.2 Industry 4.0 Key Components

Industry 4.0 enables the manufacturing of individual and customized products at the same cost as mass production, in highly automated and efficient smart factories. For production companies, this specifically means they can overcome major interface issues between product development, production, and product usage, and orient all main value-adding processes toward the customer's requirements. New development processes, e.g., integrated product and production system development, intensify exchange between departments and companies. Furthermore, Industry 4.0 enables intelligent and flexible production control using IT-based intercommunicating and interacting machines, products, services, equipment, and tools. Machine-to-machine (M2M) communication and networking bridge department boundaries and promote a company's orientation toward the customer. In general, as depicted in Fig. 7.6 and described at more length below, Industry 4.0 has four components (Wang, 2016):

1. CPS;
2. IoT;
3. Data mining (DM);
4. Internet of services (IoS).

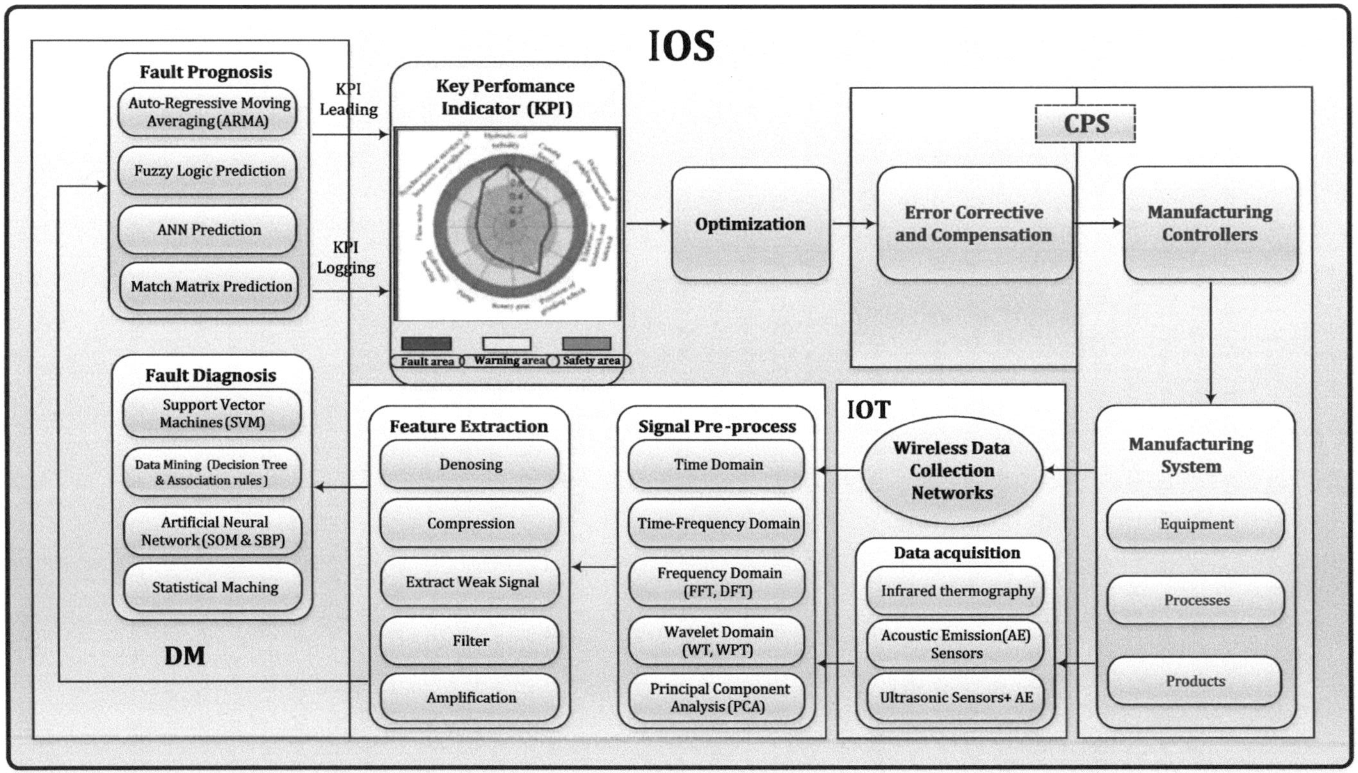

FIGURE 7.6 Framework of intelligent predictive maintenance systems in Industry 4.0. *Adapted from Wang, K., 2016. Intelligent predictive maintenance (IPdM) system – industry 4.0 scenario. WIT Transactions on Engineering Sciences 113, © 2016 WIT Press: www.witpress.com, ISSN: 1743-3533. http://dx.doi.org/10.2495/IWAMA150301.*

7.1.2.1 Cyber-Physical System

CPSs transfer the physical world into the virtual one. They can be understood as a basic unit in the system. The development of CPS can be divided into three stages. The first generation of CPS includes identification technologies such as RFID tags, which allow unique identification of objects; storage and analytics have to be provided as a centralized service. The second generation of CPS is equipped with sensors and actuators with a limited range of functions. CPS of the third generation can store and analyze data, are equipped with multiple sensors and actuators, and are network compatible.

7.1.2.2 Internet of Things

IoT allows things or objects to interact with each other and cooperate with their "smart" components to reach common aims. Therefore, IoT can be thought of as a network where CPSs cooperate in unique schemas (more details are provided in Section 7.3.4).

7.1.2.3 Data Mining

DM enables us to analyze and discover patterns, rules, and knowledge from big data collected from multiple sources so we can make the right decision at the right time and right place.

7.1.2.4 Internet of Services

The IoS enables service vendors to offer their services via the Internet. The IoS consists of business models, an infrastructure for services, the services themselves, and the participants. Services are offered and combined into value-added services by various suppliers. They are communicated to users and consumers and accessed by them via various channels.

Over IoT, CPS communicate and cooperate with each other and humans in real time. Via IoS, both internal and cross-organizational services are offered and used by participants in the value chain (Wang, 2016).

7.1.3 Principles of Industry 4.0

Companies interested in developing an Industry 4.0 strategy should consider the following seven advantages of implementation: interoperability, virtualization, decentralization, real-time capability, service orientation, modularity, and security (Hermann et al., 2015).

7.1.3.1 Interoperability

In Industry 4.0, companies, CPS, and humans are connected over the IoT and the IoS. Standards are a key factor in the communication between CPS of various manufacturers.

7.1.3.2 Virtualization

Virtualization means the CPSs are able to monitor physical processes. Sensor data are linked to virtual plant models and simulation models, creating a virtual copy of the physical world.

7.1.3.3 Decentralization

The rising demand for individual products makes it increasingly difficult to control systems centrally. Embedded computers enable CPS to make decisions on their own. Only in cases of failure, are tasks delegated to a higher level. Nevertheless, for quality assurance and traceability, it is necessary to keep track of the whole system at any time.

7.1.3.4 Real-Time Capability

In Industry 4.0, data are collected and analyzed in real time. The status of the plant is permanently tracked and analyzed, so the plant can react to the failure of a machine and reroute products to another machine.

7.1.3.5 Service Orientation

The services of companies, CPS, and humans are available over the IoS and can be used by other participants. They can be offered both internally and across company borders. All CPSs offer their functionalities as an encapsulated Web service. As a result, the product-specific process operation can be composed based on customer-specific requirements.

7.1.3.6 Modularity

Modular systems are able to adapt to changing requirements by replacing or expanding individual modules flexibly. Therefore, they can be easily adjusted in case of seasonal fluctuations or changed product characteristics.

7.1.3.7 Security

Industry 4.0 uses ICT technologies for data transmission and processing. The security and privacy of the information is emphasized in the data exchange processes using both hardware and software (Wang, 2016).

7.2 VIRTUALIZATION AND EMULATION: THE E-FACTORY FOR FAULT RATE REDUCTION

Emulation and virtualization are both digital preservation strategies but have some important differences.

7.2.1 Terminology

Terminology in this area is somewhat confusing, so before continuing the discussion, it may be useful to take stock. Data preservation is, in fact, not the main use of either emulation or virtualization, and terms have been reused with somewhat different meanings to those useful for preservation. The following definitions are slightly idiosyncratic, but should add some much-needed clarity (Rosenthal, 2015):

- *Virtual machine (VM)* is a computer that has no separate physical existence but is part of the behavior of a physical computer, called the VM's host computer. VMs mimic the instruction set and hardware configuration of some physical machine or an abstract machine such as the Java VM (Lindholm et al., 2014).
- *Virtualization* is a technique for implementing a VM on a host computer. It depends on the host computer's instruction set being the same as (mostly a superset of) the VM's instruction set and having certain specific hardware properties that enable virtualization. Almost all instructions executed by the VM are directly executed by the host computer's central processing unit (CPU); only a few instructions are intercepted using the specific properties and performed by host software. Virtualized systems run unmodified software binaries.
- *Paravirtualization* is a close relative of virtualization, often used to substitute for obsolete input/output (I/O) devices. Software binaries, typically the operating systems (OSs), are modified by replacing the drivers for the obsolete I/O devices by newly constructed ones for virtual I/O devices that translate I/O to the host system's devices.
- *Virtual machine monitor (VMM)* is the name of the host software that enables virtualization by creating and monitoring the VM and performing the intercepted operations. An alternate name is a hypervisor.
- *Emulation* is a technique for implementing a VM on a host computer whose instruction set is different from the host computer's. None of the instructions executed by the VM is directly executed by the host computer's CPU; all are translated by host computer software from the VM's instruction set to the host computer's instruction set by host software before being executed. Emulated systems run unmodified software binaries.
- *Emulator* refers to the host computer software that does this translation, mimicking the VM's instruction set and its virtual hardware configuration.
- *Preserved system image* is a set of stored data that are input to an emulator or a VMM and thereby executed. The set of data will normally include metadata describing the VM's configuration, the contents of its storage media, and, optionally, the state of the VM's memory. A preserved system image might be the result of imaging a physical computer's disks, as for example, when accessing a deceased faculty member's "papers." Or it might be created by installing from the vendor's install media for the OS and applications needed, together with preserved files requiring that environment.

As can be seen, the difference between emulation and virtualization is not absolute. In virtualization, some VM instructions are executed in host software. It is common to run an emulator to execute a VM whose instruction set is a subset of the host's. Emulating an early PC with an Intel 386 CPU on a modern Intel CPU is an example. Some emulators are implemented using just-in-time compilation, in which the first time a VM instruction sequence is encountered it is in effect replaced by its translation into the corresponding host machine sequence. This can greatly reduce the cost of emulation.

Thus, except where specifically indicated otherwise, we use "emulation" as shorthand for "emulation or virtualization depending on the overlap between the VM and host instruction sets" (Rosenthal, 2015).

7.2.2 History

Emulation and virtualization technologies have been a feature of the IT landscape for a long time, but their importance for preservation was first brought to public attention in Jeff Rothenberg's seminal 1995 *Scientific American* article "Ensuring the Longevity of Digital Documents" (Rothenberg, 1995). As he wrote, Apple was using emulation in the transition of the Macintosh from the Motorola 68000 to the Power PC. The experience he drew on was the rapid evolution of digital storage media, such as tapes and floppy disks, and of applications such as word processors, each with its own incompatible format. His vision can be summed up as follows: documents are stored in off-line media, which decay quickly, whose readers become obsolete quickly, as do the proprietary, closed formats in which they are stored. In addition, OSs and hardware change quickly in ways that break the applications that render the documents (Rosenthal, 2015).

Rothenberg identified two techniques by which digital documents could survive in this unstable environment, contrasting the inability of format migration to guarantee fidelity with emulation's ability to precisely mimic the behavior of obsolete hardware (Rothenberg, 1995).

Rothenberg's advocacy notwithstanding, most digital preservation efforts since then have used format migration as their preservation strategy. The isolated demonstrations of emulation's feasibility, such as the collaboration between the UK National Archives and Microsoft (BBC News, 2007), have had little effect. Emulation has been regarded as impractical because it is thought to require more skill and knowledge to both create and invoke emulations than scholars wanting access to preserved materials possess (Rosenthal, 2015).

Emulation advocates, including Rothenberg, Raymond Lorie of IBM, and the Dutch KB, were side tracked by the goal of building a "universal virtual computer" (UVC) (Van Der Hoeven et al., 2005). A UVC can be thought of as a metaemulator or a specification language for emulators. As the UVC was of interest only for preservation, it was unable to leverage the emulators that were concurrently being developed by hardware developers, game enthusiasts, and others. The utopian goal of UVC-diverted resources that could otherwise have been applied to making the existing emulations easily usable thus addressing the valid criticisms of their practicality (Rosenthal, 2015).

7.2.3 Virtualization for Manufacturing and Internet of Things

Integrated Industry brings even more IT and software to the shop floor. Virtualization offers numerous economic and technical benefits and has been a standard in server farms and the Cloud for over a decade. Yet VMs add significant overhead, which has so far precluded the technology from making serious inroads in embedded systems common to production environments.

In a new development, Linux containers (LXC) now offer a virtualization method that can be used on small computers and embedded systems. Contrary to full VMs, containers do not emulate hardware and share a single kernel with the base system using standard Linux technologies. Nevertheless, containers can sandbox and encapsulate applications. This way every application and even every sensor or actor can be virtualized in a container and given its own IP address (Walther and Regtmeier, 2016).

7.2.3.1 Virtualization

Virtualization is an accepted part of server and Cloud infrastructure. Its many technical and economic advantages include the following:

- Minimizing the number of servers to be purchased and maintained.
- Easy scalability.
- Straightforward sharing of resources.
- Sandboxing of systems and applications to provide additional reliability and security.

In most cases, virtualization is achieved by virtualizing entire OSs (see Fig. 7.7) by running a hypervisor on a base OS to manage the individual VMs. Every VM contains a complete OS, which is then used to run individual applications (Walther and Regtmeier, 2016).

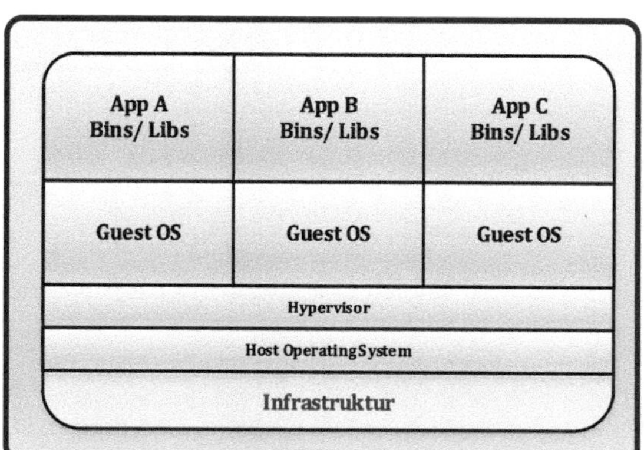

FIGURE 7.7 Classical virtualization using virtual machines. *Adapted from Walther, K., Regtmeier, J., 2016. Virtualization for manufacturing and IoT. HARTING IT Software Development GmbH & Co. KG. Marienwerder Str. 2, 32339 Espelkamp, Germany.*

This approach has some downsides, including the following:

- Every VM contains an OS, which has to be managed, patched, and serviced.
- There is a large CPU overhead and memory use by OSs.

These disadvantages have precluded the wide adoption of virtualization in embedded systems.

7.2.3.2 Integrated Industry

The goal of Integrated Industry is merging the IT world and manufacturing. Although virtualization has not been a consideration in manufacturing, the IoT and Integrated Industry are introducing more software into manufacturing environments. It would be natural to introduce virtualization to small computers and embedded devices to achieve the same benefits as in other IT systems, such as simplified management, more effective use of hardware resources, and improved security through sandboxing to manufacturing systems. Unfortunately, the overhead added by classical virtualization is inacceptable on the type of embedded systems used in manufacturing and machine environments (Walther and Regtmeier, 2016).

7.2.3.3 Lightweight Virtualization Using Linux Containers

As described in a previous section, LXC do not use VMs but containers. Rather than a complete OS, each container only contains what is needed to run the specific virtual environment: code, runtime engines, libraries, system tools, and applications. All containers share a single kernel—albeit in separate protected memory spaces. At the most basic level, a container is simply a file system containing the application and the libraries and files needed to run it. A visualization of a container appears in Fig. 7.8 (Walther and Regtmeier, 2016).

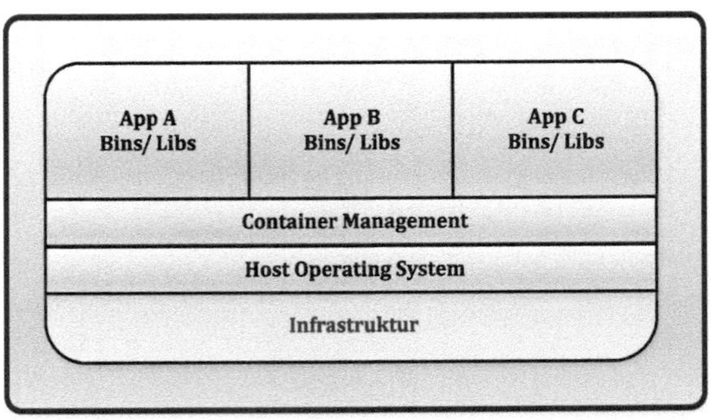

FIGURE 7.8 Lightweight virtualization using Linux containers. *Adapted from Walther, K., Regtmeier, J., 2016. Virtualization for manufacturing and IoT. HARTING IT Software Development GmbH & Co. KG. Marienwerder Str. 2, 32339 Espelkamp, Germany.*

Containers sandbox and isolate an application from the OS without the disadvantages of classic virtualization. They do not have a hardware emulation layer, but own their own processes. The processes run in separate kernel namespaces. Cgroups manage resources, so each container can be assigned necessary resources. In effect, the container architecture provides an extended chroot environment similar to BSD-jails and Linux VServer but is easier to use than chroot and more streamlined than a full VM.

This minimal virtualization lets containers start up almost instantly and keeps processing speeds close to "bare metal."

The fundamental technologies for containers have been around for a few years but were complicated and hard to use for quite some time. But starting in 2013, the proliferation of Cloud technology and the following exponential growth in the number of VMs has shifted containers into the mainstream of IT and significantly improved their usability.

7.2.3.4 Virtualization in Manufacturing Environments

HARTING IIC MICA (modular industry computing architecture) is the first system offering usable virtualization on embedded devices for manufacturing use and is depicted in Fig. 7.9. Using LXC on top of a streamlined base OS, it is capable of virtualizing field devices without the overhead of conventional virtualization.

Each application runs sandboxed in its own container, which also contains all the necessary libraries and device drivers. This completely eliminates package dependencies or incompatibilities. All communication between containers is IP-based, providing a simple and secure application program interface (API) between applications (Walther and Regtmeier, 2016).

FIGURE 7.9 Modular industry computing architecture—virtualization on a compact device. *Adapted from Walther, K., Regtmeier, J., 2016. Virtualization for manufacturing and IoT. HARTING IT Software Development GmbH & Co. KG. Marienwerder Str. 2, 32339 Espelkamp, Germany.*

7.2.3.5 *Reference Architectural Model Industry 4.0: Integrated Industry Layers on Modular Industry Computing Architecture*

The Reference Architectural Model Industry 4.0 (RAMI 4.0), developed jointly by the German standard bodies ZVEI and VDE, consists of a 3D coordinate system that describes all crucial aspects of Integrated Industry. It allows complex interrelations to be broken down into smaller and simpler clusters.

The six layers on the vertical axis describe the decomposition of a machine into its properties structured layer by layer, i.e., the virtual mapping of a machine. Such representations originate from ICT, where properties of complex systems are commonly broken down into layers.

Simple pumps, sensors, or linear motors are usually not capable of talking to IT systems or each other. By using bar codes, RFID, or other means of identification, it is possible to assign them a unique ID and a virtual representation. This virtual instance contains all the relevant properties and capabilities of the device (Walther and Regtmeier, 2016) (Fig. 7.10).

MICAs containers offer a natural way to encapsulate each instance and RAMI layer in a container. This means every instance and layer has a unique IP address making it trivial to access them in a way consistent with the principles of IoT or Integrated Industry. To put it another way, any sensor or actor connected to MICA can be readily identified and accessed through an IP network—no matter how old it is or which protocols it supports (Walther and Regtmeier, 2016).

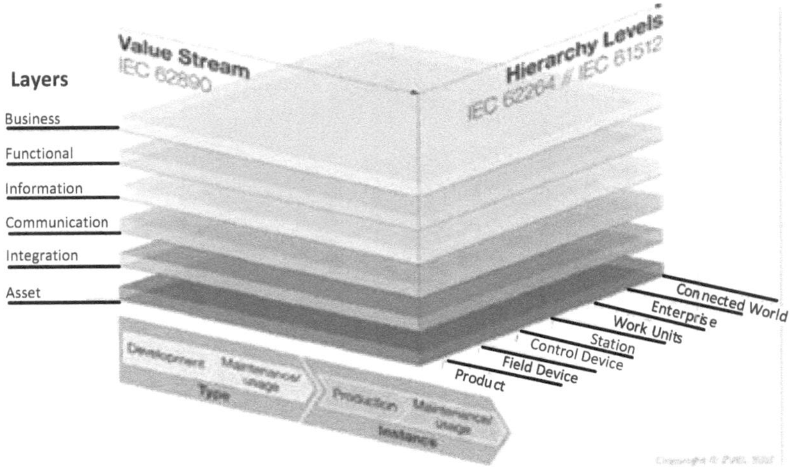

FIGURE 7.10 Reference Architectural Model Industry 4.0. *Adapted from Walther, K., Regtmeier, J., 2016. Virtualization for manufacturing and IoT. HARTING IT Software Development GmbH & Co. KG. Marienwerder Str. 2, 32339 Espelkamp, Germany.*

7.2.4 Embedded Virtualization

Embedded virtualization allows consolidation of multiple different workloads on a single piece of multicore silicon. An industrial process may use one system for data collection, one system for signal processing, one system for vision, one system for control, and one system for a human–machine interface. Each of these many different systems requires a piece of hardware that costs money, takes up space, and requires power, spare parts, and an update and replacement strategy—and in general has a big impact on capital expenses (CAPEX) and operating expenses (OPEX) (WIND, 2016). This is visualized in Fig. 7.11.

Embedded virtualization provides the capability to combine all these different functionalities in a single piece of hardware. While the new piece of hardware may be more powerful, and typically more expensive, than one of the original pieces of hardware, the consolidation of the multiple systems into one removes a lot of complexity from the end system, simplifies the maintenance and replacement process, and, in general, reduces CAPEX and OPEX.

Today's computing hardware allows efficient virtualization for embedded systems—so efficient that the impact on the performance of the end system is minimal. Indeed, embedded virtualization has minimal impact on memory size, memory performance, device access, interrupt latency, jitter, and so on (WIND, 2016).

7.2.4.1 Internet of Things and Embedded Virtualization

With the advent of IoT, industrial control devices are now connecting to the Cloud for benefits including optimizing production efficiency, minimizing environmental impact, and minimizing downtime. In IoT, the Cloud provides control for the devices, and the devices provide information back to business control logic in the Cloud.

One of the capabilities that vendors often associate with IoT is provisioning the industrial control devices. In other words, the devices are given a personality through the Cloud. Provisioning involves initial provisioning, updates, and upgrades.

When people think of provisioning, they tend to think of downloading applications, where the device has most of its personality already and is given final instructions through software download. Initial personality in such a case includes device drivers, OS, base management layer, and so on. But embedded virtualization allows an additional level of abstraction, making it possible to provision a device with a generic management layer only, and "empty slots" into which to provision functionality. These empty slots are called VMs and can be provisioned with logic consisting of OSs, as well as business logic. As produced in the factory, the device will have hardware, a virtualization layer, and a management layer. The management layer connects into the Cloud and can download its content from there; the downloaded content is the business logic that allows the device to perform its specific function (WIND, 2016) (Fig. 7.12).

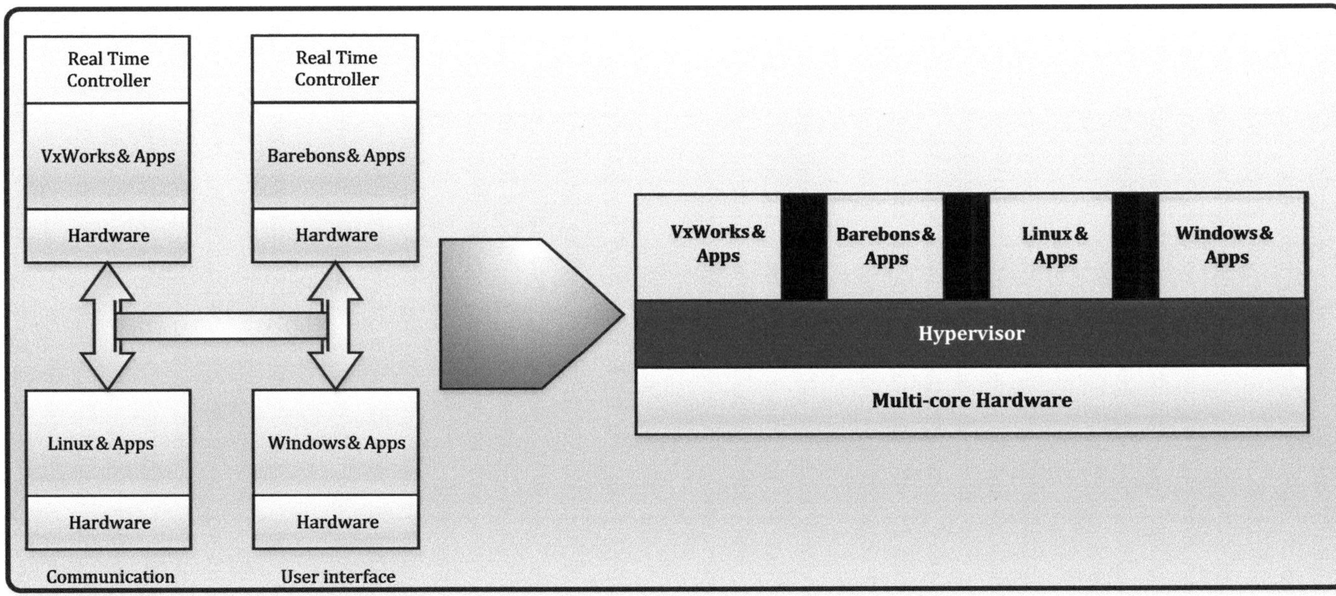

FIGURE 7.11 Consolidation of heterogeneous operating systems and functionality with virtualization profile for VxWorks. *Adapted from WIND, 2016. Virtualization and the Internet of Things. © 2016 Wind River Systems, Inc.*

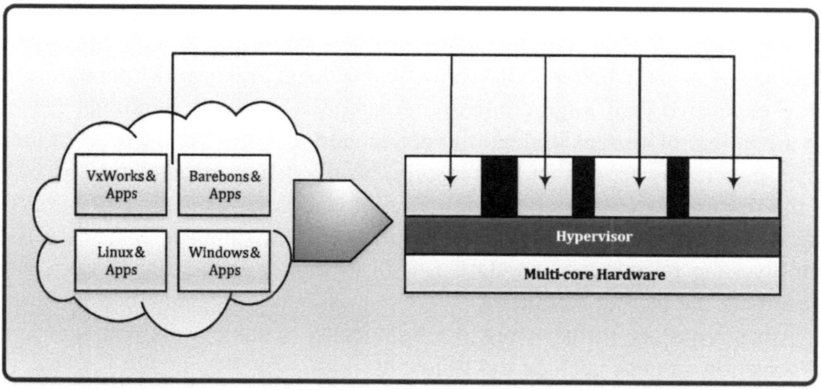

FIGURE 7.12 Provisioning functionality into virtual machines on a device with a generic management layer. *Adapted from WIND, 2016. Virtualization and the Internet of Things. © 2016 Wind River Systems, Inc.*

Because the device, as shipped, only has logic that ties it into the Cloud, device manufacturers can create a generic device that knows nothing about its eventual functionality. The device is connected to, for example, a power generator, and it obtains the generator's identity through a standard interface. It reports that identity into the Cloud, which has been prepared and is expecting the connection report; the Cloud then provides the device with the required software content. The device downloads the content, instantiates VMs, and is ready to function.

The management layer also makes it possible for the business content to safely communicate with the Cloud through the management interface (WIND, 2016).

7.2.5 Emulation Frameworks

Recent developments in emulation frameworks make it possible to deliver emulations to readers via the Web in ways that make them appear normal components of Web pages. This removes what was the major barrier to deployment of emulation as a preservation strategy. Barriers remain: for one thing, the tools for creating preserved system images are inadequate. For another, the legal basis for delivering emulations is unclear, and where it is clear, it is highly restrictive. Both raise the cost of building and providing access to a substantial, well-curated collection of emulated digital artifacts otherwise beyond reach.

If these barriers can be addressed, emulation will play a much greater role in digital preservation in coming years. It will provide access to artifacts that migration cannot, and even assist in migration where necessary by allowing the original software to perform it. The evolution of digital artifacts means that current artifacts are more difficult and expensive to collect and preserve than those from the past, and less suitable for migration. This trend is expected to continue.

Emulation is not a panacea. Technical, scale, and intellectual property difficulties make many current digital artifacts infeasible to emulate. Where feasible, even with better tools and a viable legal framework, emulation is more expensive than migration-based strategies. The most important reason for the failure of current strategies to collect and preserve the majority of their target material is economic; the resources available are inadequate. The bulk of the resources expended on both migration and emulation strategies are for ingest, especially metadata generation and quality assurance. There is a risk that diverting resources to emulation, with its higher artifact ingest cost, will exacerbate the lack of resources.

Areas requiring further work if emulation is to achieve its potential as a preservation strategy include the following:

- Standardization of the format of preserved system images, the way they are obtained by emulators, and the means by which emulations of them are exposed to readers. This would enable interoperability between emulation components, aiding contributions, and support from the open-source community.
- Improvements to the tools for associating technical metadata with preserved software to enable it to be emulated and the technical metadata databases on which they depend. This would reduce the cost of preserved system images.
- Clarification and, if possible, relaxation of the legal constraints on the creation and provision of access to collections of preserved system images. This would encourage institutions to collect software.

7.3 MULTIVARIATE MAINTENANCE DECISION SUPPORT: A CONSEQUENCE OF INTERNET OF THINGS

7.3.1 Decision Support Systems

A decision support system (DSS) is an interactive computer-based application that combines data and mathematical models to help decision makers solve complex problems involved in managing public and private enterprises and organizations. The analysis tools provided by business intelligence (BI) architecture are DSSs capable of transforming data into information and knowledge helpful to decision makers. In this respect, DSSs are basic components in the development of BI architecture (Vercellis, 2009).

7.3.1.1 Definition of System

The term system is often used in everyday language: for instance, we refer to the solar system, the nervous system, or the justice system. The entities that we intuitively denominate systems share a common characteristic: each is made up of a set of components in some way connected to each other so as to provide a single collective result and a common purpose.

Every system has boundaries that separate its internal components from the external environment. A system is said to be open if its boundaries can be crossed in both directions by flows of materials and/or information. When such flows are lacking, the system is said to be closed. In general terms, any given system receives specific input flows, carries out an internal transformation process, and generates observable output flows.

This abstract definition of system can be used to describe a broad class of real-world phenomena. For example, the logistic structure of an enterprise is a system whose input is a set of materials, services, and information and whose output is a set of products, services, and information. More generally, even an enterprise, as a whole or in part, may be represented as a system, provided the boundaries and the input and output flows are clearly defined.

Fig. 7.13 shows the structure we use here to describe the concept of system. In the figure, a system receives a set of input flows and returns a set of output flows through a transformation process regulated by internal conditions and external conditions. The effectiveness and efficiency of a system are assessed using measurable performance indicators (PIs) classified into different categories. The figure shows the main types of metrics used to evaluate systems embedded within enterprises (Vercellis, 2009).

A system will often incorporate a feedback mechanism. Feedback occurs when a system component generates an output flow that is fed back into the system as an input flow, possibly as a result of a further transformation. Systems able to modify their own output flows based on feedback are called

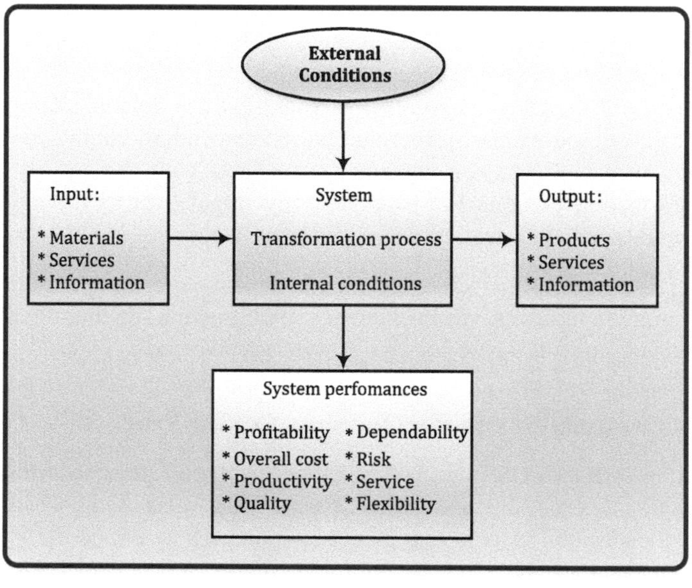

FIGURE 7.13 Abstract representation of a system. *Adapted from Vercellis, C., 2009. Business Intelligence: Data Mining and Optimization for Decision Making. © 2009 John Wiley & Sons, Ltd. ISBN:978-0-470-51138-1.*

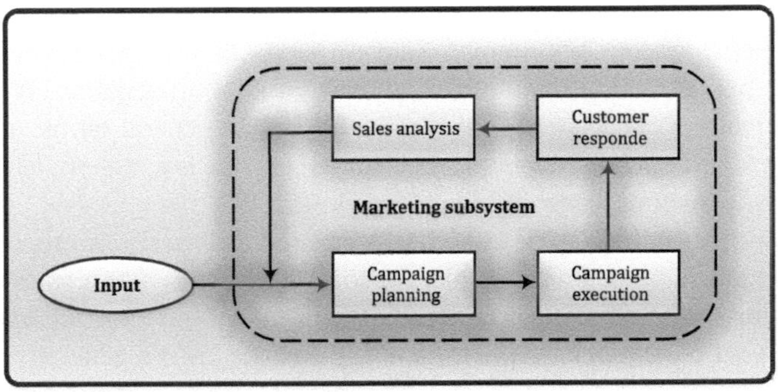

FIGURE 7.14 Closed-cycle marketing system with feedback effects. *Adapted from Vercellis, C., 2009. Business Intelligence: Data Mining and Optimization for Decision Making. © 2009 John Wiley & Sons, Ltd. ISBN:978-0-470-51138-1.*

closed-cycle systems. For example, the closed-cycle system outlined in Fig. 7.14 develops a sequence of marketing campaigns. The sales results for each campaign are gathered and become available as feedback input so as to design subsequent marketing promotions (Vercellis, 2009).

In connection with a decision-making process, whose structure is described in the next section, it is often necessary to assess the performance of a system. For this purpose, it is appropriate to categorize the evaluation metrics into two main classes: effectiveness and efficiency.

- *Effectiveness*. Effectiveness measurements express the level of conformity of a given system to the objectives for which it was designed. The associated PIs are therefore linked to the system output flows, such as production volume, weekly sales, and yield per share.
- *Efficiency*. Efficiency measurements highlight the relationship between input flows used by the system and the corresponding output flows. Efficiency measurements are therefore associated with the quality of the transformation process. For example, they might express the amount of resources needed to achieve a given sales volume.

Generally speaking, effectiveness metrics indicate whether the right action is being carried out or not, while efficiency metrics show whether the action is being carried out in the best possible way or not (Vercellis, 2009).

7.3.2 Representation of the Decision-Making Process

To build an effective DSS, we first need to describe in general terms how a decision-making process is articulated. In particular, we need to understand the steps that lead individuals to make decisions and the extent of the influence exerted on them by the subjective attitudes of the decision makers and the specific context within which decisions are made (Vercellis, 2009).

7.3.2.1 The Decision-Making Process

A compelling representation of the decision-making process was proposed in the early 1960s and still remains a major methodological reference. The model includes three phases: intelligence, design, and choice. Fig. 7.15 shows an extended version of the original scheme, which results from the inclusion of two additional phases: implementation and control (Vercellis, 2009).

- *Intelligence*. In the intelligence phase, the task of the decision maker is to identify, circumscribe, and explicitly define the problem that emerges in the system under study. Analysis of the context and all available information may allow decision makers to quickly grasp the signals and symptoms pointing to a corrective action to improve the system performance. For example, during the execution of a project, the intelligence phase may consist of a comparison of the current progress of the activities and the original development plan. In general, it is important to avoid confusing the problem with the symptoms. For example, suppose an e-commerce bookseller receives a complaint about late delivery of a book order placed online. Such inconvenience may be interpreted as the problem and be tackled by arranging a second delivery by priority shipping to circumvent the dissatisfaction of the customer. However, this may be the symptom of a broader problem, an understaffed shipping department where human errors are likely to occur under pressure.

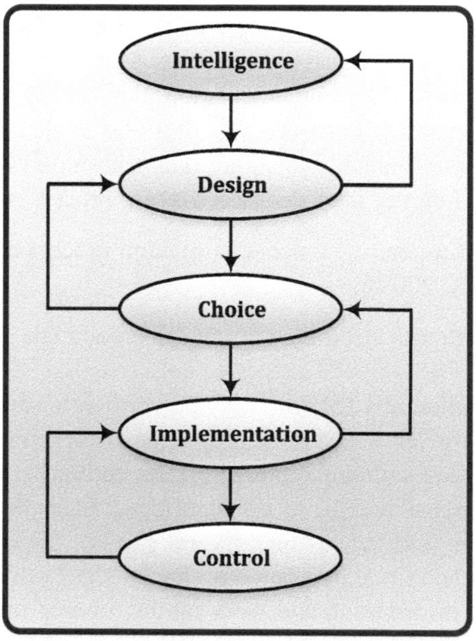

FIGURE 7.15 Phases of the decision-making process. *Adapted from Vercellis, C., 2009. Business Intelligence: Data Mining and Optimization for Decision Making. © 2009 John Wiley & Sons, Ltd. ISBN:978-0-470-51138-1.*

- *Design*. In the design phase, actions aimed at solving the identified problem should be developed and planned. At this level, the experience and creativity of the decision makers play a critical role, as they are asked to devise viable solutions that allow the intended purpose to be achieved. Where the number of available actions is small, decision makers can make an explicit enumeration of the alternatives to identify the best solution. Where the number of alternatives is very large, or even unlimited, their identification occurs in an implicit way, usually through a description of the rules that feasible actions should satisfy. For example, these rules may directly translate into the constraints of an optimization model.
- *Choice*. Once the alternative actions have been identified, it is necessary to evaluate them on the basis of the performance criteria deemed significant. Mathematical models and the corresponding solution methods usually play a valuable role during the choice phase. For example, optimization models and methods allow the best solution to be found in very complex situations involving countless or even infinite feasible solutions, while decision trees can be used to handle decision-making processes influenced by stochastic events.
- *Implementation*. When the best alternative has been selected by the decision maker, it is transformed into action by means of an implementation plan. This involves assigning responsibilities and roles to all those involved into the action plan.
- *Control*. Once the action has been implemented, it is necessary to verify that the original expectations have been satisfied and the effects of the action match the original intentions. In particular, the differences between the values of the PIs identified in the choice phase and the values actually observed at the end of the implementation plan should be measured. In an adequately planned DSS, the results of these evaluations translate into experience and information, which are then transferred into the data warehouse to be used during later decision-making processes (Vercellis, 2009).

The most salient aspects of a decision-making process can be summarized as follows (Vercellis, 2009):

- Decisions are often devised by a group of individuals instead of a single decision maker.
- The number of alternative actions may be very high, sometimes unlimited.
- The effects of a given decision usually appear later, not immediately.
- The decisions made within a public or private enterprise or organization are often interconnected and have broad effects. Each decision has consequences for many individuals and several parts of the organization.
- During the decision-making process, knowledge workers are asked to access data and information and work on them based on a conceptual and analytical framework.
- Feedback plays an important role in providing information and knowledge for future decision-making processes within an organization.

- In most instances, the decision-making process has multiple goals, with different PIs that may be in conflict with one another.
- Many decisions are made in a fuzzy context and entail risk factors. The level of propensity or aversion to risk varies significantly among individuals.
- Experiments carried out in a real-world system using a trial-and-error scheme are too costly and risky to be of practical use for decision-making.
- The dynamics in which an enterprise operates is strongly affected by the pressure of a competitive environment; this implies knowledge workers need to address situations and make decisions in a timely fashion.

7.3.2.2 Types of Decisions

Defining a taxonomy of decisions may prove useful during the design of a DSS, as decision-making processes with similar characteristics are likely to be supported by the same set of methodologies. Decisions can be classified in terms of two main dimensions, their nature and their scope. Each dimension can be subdivided into three classes, giving a total of nine possible combinations, as shown in Fig. 7.16 (Vercellis, 2009).

According to their nature, decisions can be classified as structured, unstructured, or semistructured.

- *Structured decisions.* A decision is structured if it is based on a well-defined and recurring decision-making procedure. In most cases, structured decisions can be traced back to an algorithm, which may be more or less explicit for decision makers, making them better suited for application.

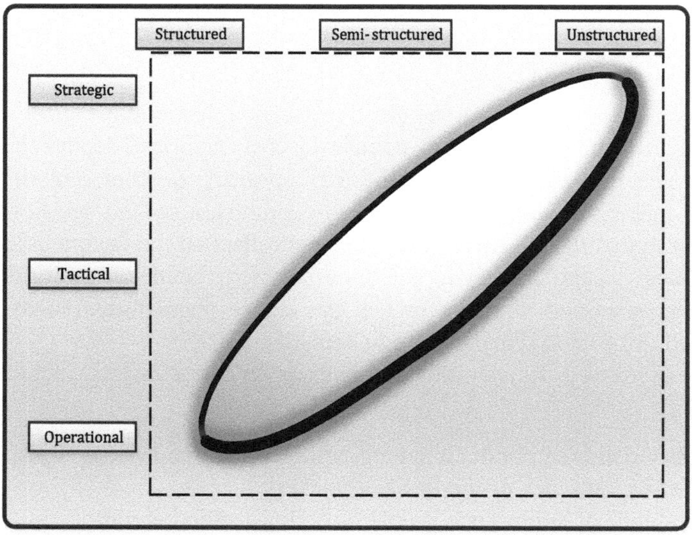

FIGURE 7.16 Taxonomy of decisions. *Adapted from Vercellis, C., 2009. Business Intelligence: Data Mining and Optimization for Decision Making. © 2009 John Wiley & Sons, Ltd. ISBN:978-0-470-51138-1.*

More specifically, a decision is structured if input flows, output flows, and the transformations performed by the system can be clearly described in the three phases of intelligence, design, and choice. In this case, we can say that each component phase is structured in its turn. Actually, even decisions that appear fully structured generally require the direct intervention of decision makers to cope with unexpected events, caused, for example, by unusual values of some input flows.

- *Unstructured decisions.* A decision is said to be unstructured if the three phases of intelligence, design, and choice are also unstructured. This means that for each phase, there is at least one element in the system (input flows, output flows, transformation processes) that cannot be described in detail and reduced to a predefined sequence of steps. Such an event may occur when a decision-making process is faced for the first time or if it happens very seldom. The role of knowledge workers is fundamental for this type of decision, and BI systems may provide support to decision makers through timely and versatile access to information.
- *Semistructured decisions.* A decision is semistructured when some phases are structured and others are not. Most decisions faced by knowledge workers in managing public or private enterprises or organizations are semistructured. This allows them to take advantage of DSSs and a business improvement (BI) environment in two ways. For the unstructured phases of the decision-making process, BI tools may offer passive support, which translates into timely and versatile access to information. For the structured phases, it is possible to provide an active form of support by using mathematical models and algorithms that allow significant parts of the decision-making process to be automated.

Sometimes the nature of a decision cannot be easily identified. When facing the same problem, such as establishing the sale price of a product, different decision makers operating in different organizations may come up with dissimilar choices. For example, one decision maker may believe that the best sale price can be obtained by comparing cost and price–demand elasticity curves. As a consequence, he or she may consider the choice phase of the decision-making process as structured. By contrast, a second decision maker may believe that the elasticity curve does not reflect all the factors influencing the response of the market to price variations, as some of these elements cannot be quantified. For him or her, the choice phase turns out to be unstructured or at most semistructured (Vercellis, 2009).

The nature of a decision process depends on many factors, including the following:

- Characteristics of the organization within which the system is placed;
- Subjective attitudes of the decision makers;
- Availability of appropriate problem-solving methodologies;
- Availability of effective decision support tools.

Depending on their scope, decisions can be classified as strategic, tactical, or operational.

- *Strategic decisions*. Decisions are strategic when they affect the entire organization or a substantial part of it for a long period. Strategic decisions strongly influence the general objectives and policies of an enterprise. As a consequence, strategic decisions are made at a higher organizational level, usually by top management.
- *Tactical decisions*. Tactical decisions affect only parts of an enterprise and are usually restricted to a single department. The time span is limited to a medium-term horizon, typically up to a year. Tactical decisions fall into the context of strategic decisions. In a company hierarchy, tactical decisions are made by middle managers, such as department heads.
- *Operational decisions*. Operational decisions refer to specific activities carried out within an organization, with a modest impact on the future. Operational decisions are framed by the elements and conditions determined by strategic and tactical decisions. Therefore, they are usually made at a lower organizational level by knowledge workers responsible for a single activity or task, such as subdepartment heads, workshop forepersons, back-office heads, etc.

The characteristics of the information required in a decision-making process will change depending on the scope of the decisions to be supported; consequently, the orientation of a DSS will vary accordingly. Fig. 7.17 shows variations in the characteristics of the information as the scope of the decisions changes. The scheme can be used as an assessment tool when designing a DSS: once the scope of the decisions for which the system is intended has been

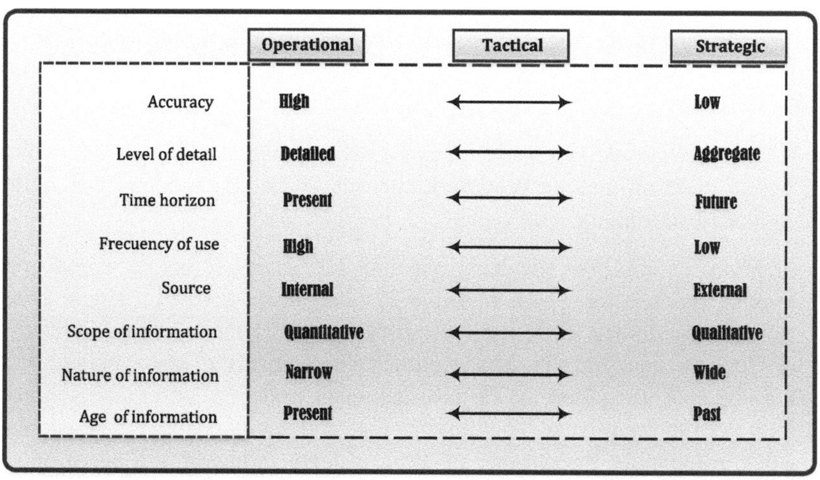

FIGURE 7.17 Characteristics of information in terms of scope of decisions. *Adapted from Vercellis, C., 2009. Business Intelligence: Data Mining and Optimization for Decision Making. © 2009 John Wiley & Sons, Ltd. ISBN:978-0-470-51138-1.*

set, the scheme can be used to establish whether the decision-making process is adequately supported by the right information (Vercellis, 2009).

Although nature and scope are not perfectly correlated, most real-world decisions fall within the ellipse shown in Fig. 7.16: most strategic decisions are unstructured, while most operational decisions are structured, and most tactical decisions are semistructured. This is useful when defining in advance the characteristics of a DSS to facilitate a decision-making process of a specific nature and scope (Vercellis, 2009).

7.3.2.3 Approaches to the Decision-Making Process

As observed above, the subjective orientation of decision makers across an organization strongly influences the structure of the decision-making process. This section reviews major approaches in the management of complex organizations and the implications for designing a DSS.

A preliminary distinction should be made between a rational approach and a political–organizational approach (Vercellis, 2009):

- *Rational approach.* When a rational approach is followed, a decision maker considers major factors, such as economic, technical, legal, ethical, procedural, and political factors; the decision maker establishes the criteria of evaluation to assess options and select the best decision. In this context, a DSS may help both in a passive way, through timely and versatile access to information, and in an active way, through the use of mathematical models for decision-making.
- *Political–organizational approach.* When a political–organizational approach is pursued, a decision maker proceeds in a more instinctual and less systematic way. Decisions are not based on clearly defined alternatives and selection criteria. As a consequence, a DSS can only help in a passive way, providing timely and versatile access to information. A DSS might also be useful during discussions and negotiations in decision-making processes involving multiple actors, such as managers of different departments.

Within the rational approach, we can further distinguish two ways in which the actual decision-making process influences decisions: absolute rationality and bounded rationality.

- *Absolute rationality.* The term absolute rationality refers to a decision-making process for which multiple PIs can be reduced to a single criterion; this naturally lends itself to an optimization model. For example, a production manager who has to put together a medium-term logistic plan may be able to convert all PIs into monetary units and derive the solution with the minimum cost. This implies that nonmonetary indicators, such as stock volumes or the number of days of delay in handling a given order, should be transformed into monetary measurement units. From a methodological perspective, a multiobjective optimization problem is transformed into a single-objective problem by expressing all the relevant factors in a common measurement unit that allows the heterogeneous objectives to be added together.

- *Bounded rationality.* Bounded rationality occurs whenever it is not possible to meaningfully reduce multiple criteria into a single objective; the decision maker considers an option to be satisfactory when the corresponding PIs fall above or below prefixed threshold values. For instance, a production plan is acceptable if its cost is sufficiently low, the stock quantities are within a given threshold, and the service time is below customers' expectations. Therefore, the concept of bounded rationality captures the rational choices that are constrained by the limits of knowledge and cognitive capability.

Most decision-making processes within enterprises are aimed at making a decision that appears acceptable with respect to multiple evaluation criteria; therefore, decision processes based on bounded rationality are more likely to occur in practice.

7.3.3 Condition-Based Maintenance Decision Support Systems

Condition-based maintenance is a maintenance technique that involves monitoring machine condition and predicting machine failure (Toolingu, 2013). Generally, the monitoring is controlled by computers. The device maintenance mode has two modes, break-down maintenance and planned maintenance. Break-down maintenance refers to maintenance conducted after a fault has occurred in a device. The fault has occurred, and loss has been induced. Planned maintenance refers to maintenance following a prescribed maintenance cycle.

"Excessive maintenance" and "insufficient maintenance" happen if devices under different conditions are maintained on a planned cycle. Because planned maintenance cannot give full emphasis to the unique traits and operating context of devices, it may waste financial and human resources. In addition, a fault may not be found in time, leading to a possibly disastrous accident.

Condition-based maintenance is conducted on devices before a fault occurs based on the evaluation of changes and the prediction of faults using information provided by CM, fault diagnosis technology, and production management information; maintenance is carried out in light of a device's health; this prevents excessive maintenance or disrepair.

Evaluating a device's condition in real time reduces the risk of hidden defects, improves the serviceability rate and health level, ensures safe operation, extends the maintenance cycle, reduces the waste of human resources and materials, and avoids excessive maintenance. Carrying out a reasonable condition-based maintenance strategy improves reliability, reduces maintenance cost, and extends asset life cycle (Haddad et al., 2011). Overall benefits include the following: reducing planned maintenance, avoiding break-down maintenance, developing condition-based maintenance, and promoting optimal maintenance. Fig. 7.18 shows an optimal maintenance model (Jamil et al., 2013).

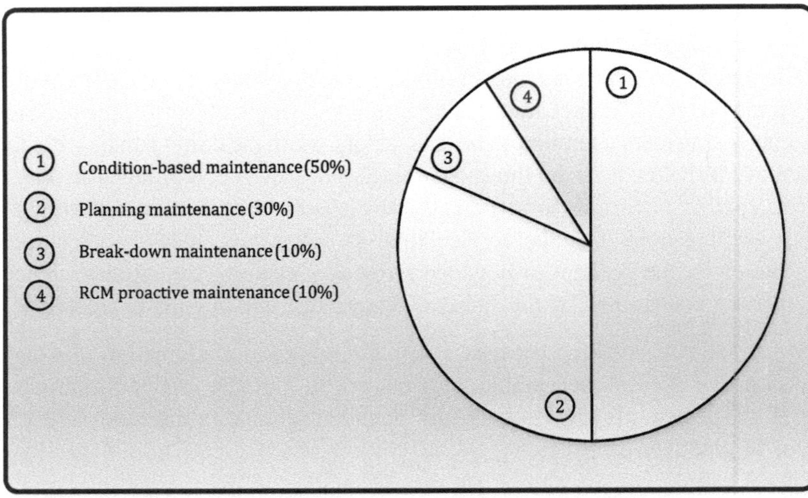

FIGURE 7.18 Composition of proportions of optimal maintenance model. *Adapted from Jamil, I., Jamil, R., Jinquan, Z., Ming, L., Dong, W.Y., Jamil, R., November 2013. Condition-based maintenance decision-making support system (DSS) of hydropower plant. Copyright © 2013 ISSR Journals 4 (3). ISSN:2028-9324.*

Development of condition-based maintenance: The evolution of a maintenance system reflects the development of both productivity and technology. The appropriate maintenance mode should be accompanied by both theoretical and technological knowledge.

Technological orientation of condition-based maintenance: For today's condition-based maintenance, we need a unified maintenance information platform to acquire condition data and perform feature calculation, real-time monitoring, fault recording, test recording, and technical diagnosis of primary devices. There should be a unified data access model and an analysis and diagnosis model for different devices. We will also require technical assurance to make the transition from planned maintenance to condition-based maintenance. For example, we can adopt the international network communication standard IEC61850-MMS to carry out data transmission and sharing and establish a unified maintenance information platform for device CM of an organization. Finally, we should implement a method that adopts a condition-based maintenance assistant decision-making mode, constructs a data center for device condition, applies diagnosis and analysis, and derives a reliability maintenance strategy. A number of advanced application algorithms and models are available for condition evaluation, fault diagnosis, and condition forecasting (Jamil et al., 2013).

7.3.3.1 Composition of Condition-Based Maintenance Systems

A condition-based maintenance technical system has a unit vibration/swing protection system comprised of the following: a circuit breaker and transformer monitoring system, a CM and tendency analysis system, a computer

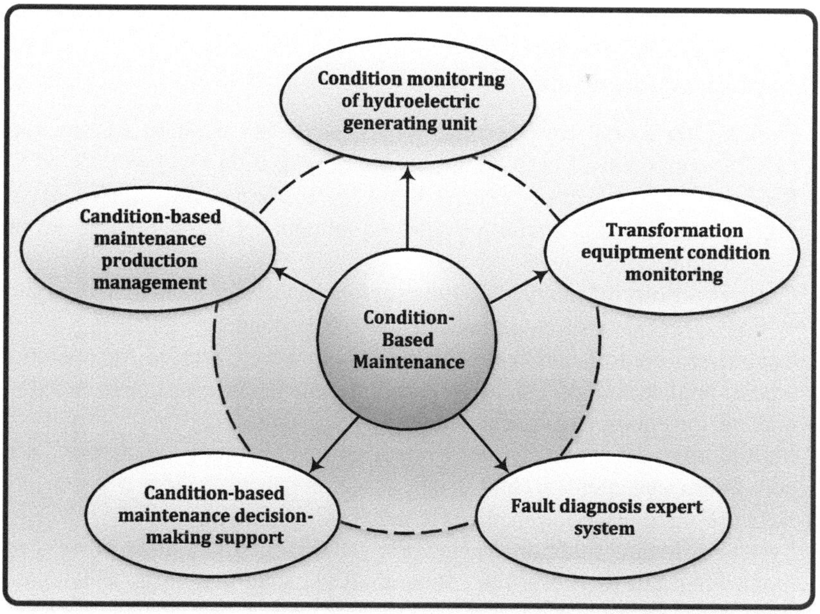

FIGURE 7.19 Maintenance information platform for device condition monitoring in a hydroelectric plant. *Adapted from Jamil, I., Jamil, R., Jinquan, Z., Ming, L., Dong, W.Y., Jamil, R., November 2013. Condition-based maintenance decision-making support system (DSS) of hydropower plant. Copyright © 2013 ISSR Journals 4 (3). ISSN:2028-9324.*

supervisory monitoring system, a production management system, and a condition-based maintenance DSS. Fig. 7.19 shows how this looks in a hydroelectric plant.

The following steps should be taken when setting up a technical system to perform CM on a device:

- Establish a unified maintenance information platform for CM of the organization;
- Acquire relevant basic data on primary devices, real-time/historical data of the devices, and other characteristic parameters that reflect device health condition and evaluate current health condition of the devices;
- Carry out effective risk evaluation;
- Implement comprehensive analysis, reasoning, and diagnosis through the optimal maintenance strategy model;
- Give maintenance suggestions; and
- Transmit the analysis, conclusions, and maintenance suggestions to the production information management system through a service bus to facilitate inquiry and reference, thus providing technical assurance of the transition from planned maintenance to condition-based maintenance (Xu et al., 2007).

7.3.3.2 Configuration of a Condition-Based Maintenance Decision Support System

7.3.3.2.1 Hardware Composition

- Plant station access server: service for access to data acquired and analyzed by CM equipment of each station in the organization.
- Real-time plant database server: service for storage and inquiry of the original "black box" data of CM equipment of each station in the organization.
- Centralized control monitoring data server: service for storage and inquiry of CM access data of equipment of the whole plant.
- Centralized control maintenance application server: service for operation access application of condition-based maintenance assistance decision-making for equipment of the whole plant.
- Mobile engineer work station: system equipment model definition, user permission management, and database maintenance service.
- Network auxiliary equipment: safe networking service.
- System software function framework: data acquisition, data processing, monitoring and early warning, state analysis, state diagnosis, state evaluation, forecast evaluation, risk assessment, and suggestions for decisions (Jamil et al., 2013).

For a schematic of the above elements, see Fig. 7.20.

7.3.3.3 Functions of a Condition-Based Maintenance Decision Support System

7.3.3.3.1 Data Acquisition Module

The function of a data acquisition module is to acquire from the external system or the relevant device various basic equipment data, real-time data, inspection and test data, and other data that reflect equipment health condition to provide data resources for further judgment and evaluation. This is shown in Fig. 7.21.

7.3.3.3.2 Data Processing Module

The function of the data processing module, as depicted in Fig. 7.22, is to process the acquired data in light of business demand and extract the data on equipment health state to reflect current operating performance for monitoring and early warning and state evaluation.

7.3.3.3.3 Monitoring and Early Warning Module

This module monitors the change of state index, gives early warning when the equipment state goes beyond the evaluation guidelines and regulations, and releases early warning information to equipment management personnel at all levels based on the various categories and levels. Its functions are depicted in Fig. 7.23.

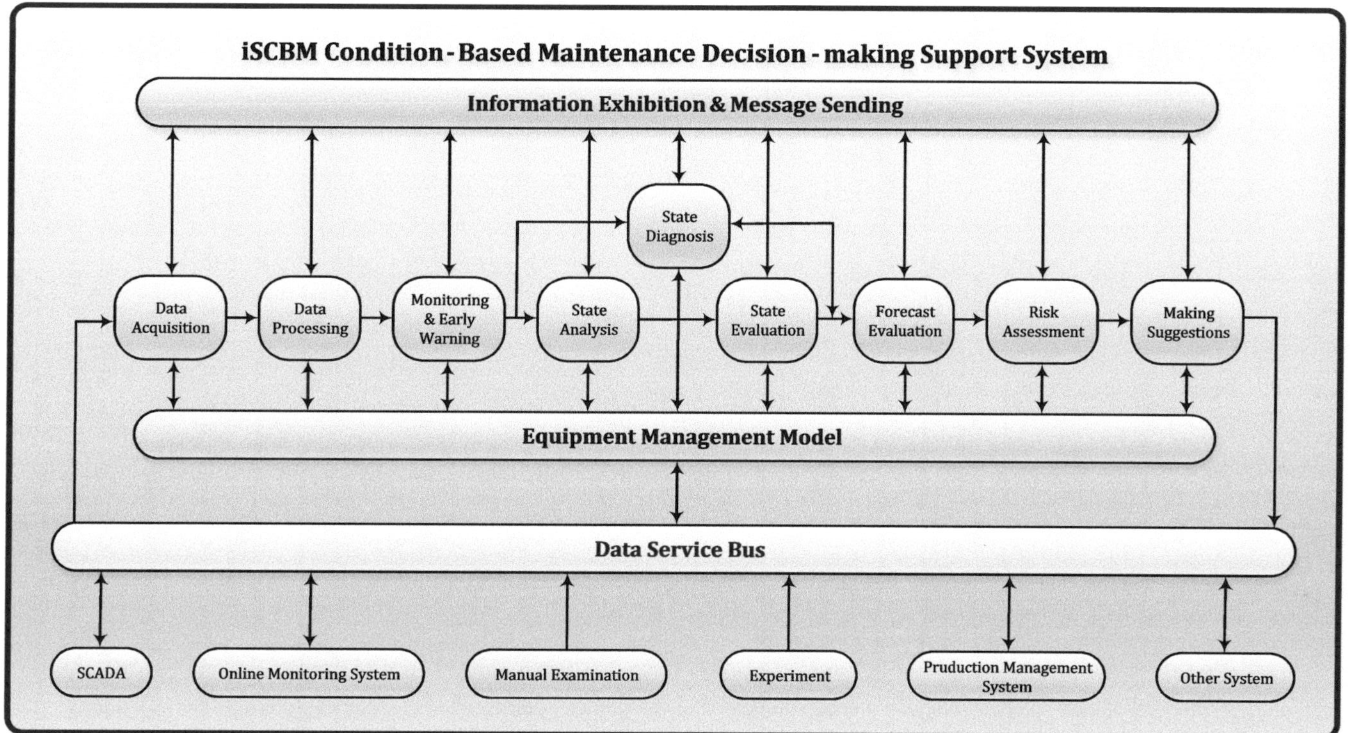

FIGURE 7.20 Diagram of software functions of condition-based maintenance decision support system. *Adapted from Jamil, I., Jamil, R., Jinquan, Z., Ming, L., Dong, W.Y., Jamil, R., November 2013. Condition-based maintenance decision-making support system (DSS) of hydropower plant. Copyright © 2013 ISSR Journals 4 (3). ISSN:2028-9324.*

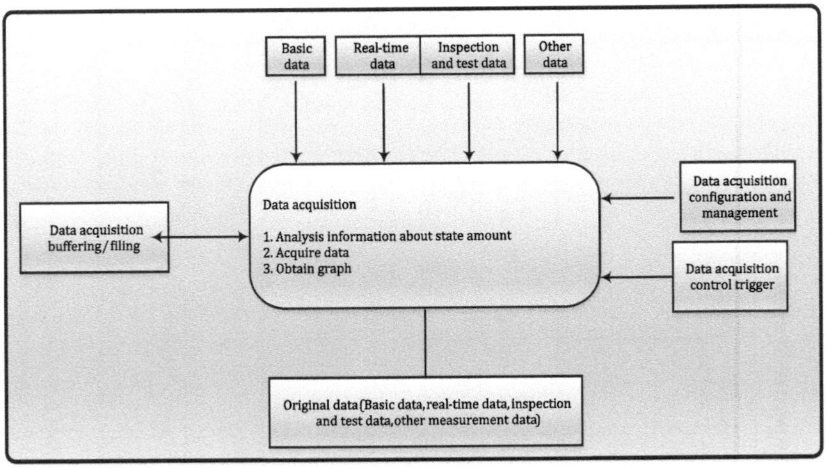

FIGURE 7.21 Data acquisition module. *Adapted from Jamil, I., Jamil, R., Jinquan, Z., Ming, L., Dong, W.Y., Jamil, R., November 2013. Condition-based maintenance decision-making support system (DSS) of hydropower plant. Copyright © 2013 ISSR Journals 4 (3). ISSN:2028-9324.*

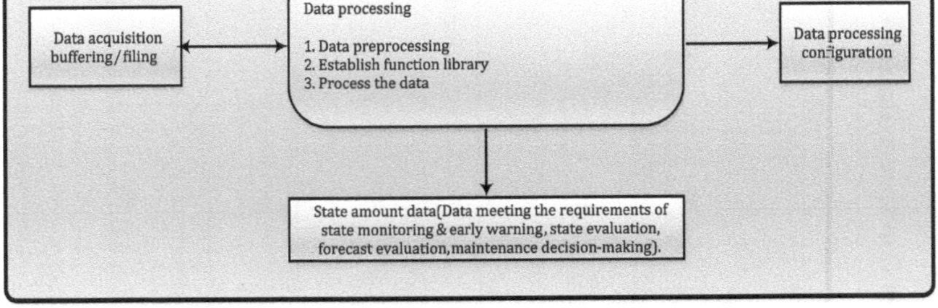

FIGURE 7.22 Data processing module. *Adapted from Jamil, I., Jamil, R., Jinquan, Z., Ming, L., Dong, W.Y., Jamil, R., November 2013. Condition-based maintenance decision-making support system (DSS) of hydropower plant. Copyright © 2013 ISSR Journals 4 (3). ISSN:2028-9324.*

7.3.3.3.4 State Analysis Module

The state analysis module adopts various state analysis methods (time-domain analysis, frequency-domain analysis, correlation analysis, wavelet analysis, tendency analysis, transient analysis, characteristic spectrogram analysis, etc.) to analyze the current operating condition and health level of the equipment/component. It may integrate the state analysis function of CM system and the historical tendency analysis function of the tendency analysis system. See Fig. 7.24 for a depiction.

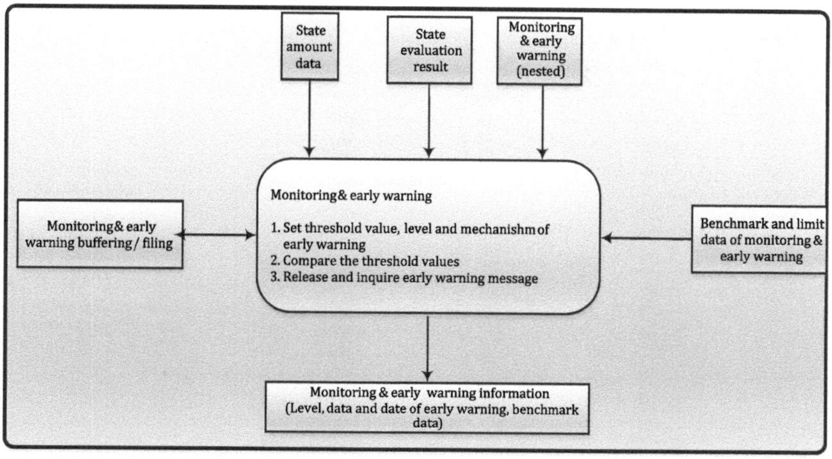

FIGURE 7.23 Monitoring and early warning module. *Adapted from Jamil, I., Jamil, R., Jinquan, Z., Ming, L., Dong, W.Y., Jamil, R., November 2013. Condition-based maintenance decision-making support system (DSS) of hydropower plant. Copyright © 2013 ISSR Journals 4 (3). ISSN:2028-9324.*

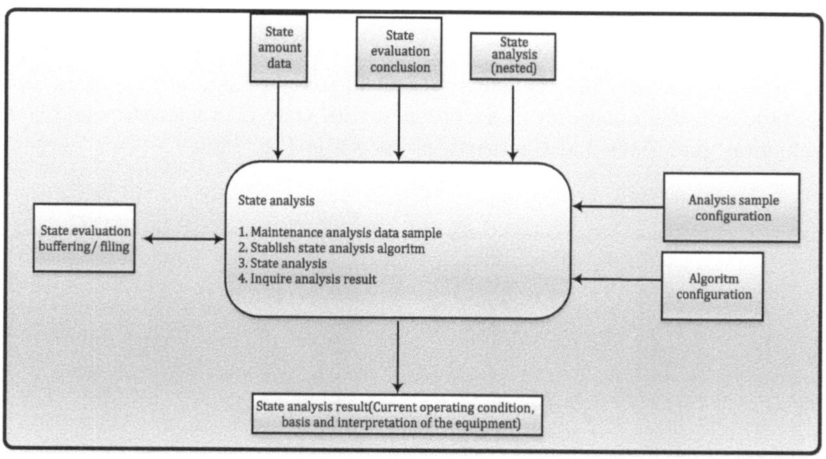

FIGURE 7.24 State analysis module. *Adapted from Jamil, I., Jamil, R., Jinquan, Z., Ming, L., Dong, W.Y., Jamil, R., November 2013. Condition-based maintenance decision-making support system (DSS) of hydropower plant. Copyright © 2013 ISSR Journals 4 (3). ISSN:2028-9324.*

7.3.3.3.5 State Diagnosis Module

When the state amount index goes beyond the early warning level or with descending health level, the state diagnosis module diagnoses the cause and position of possible faults and guides fault handling and state recovery, as indicated in Fig. 7.25.

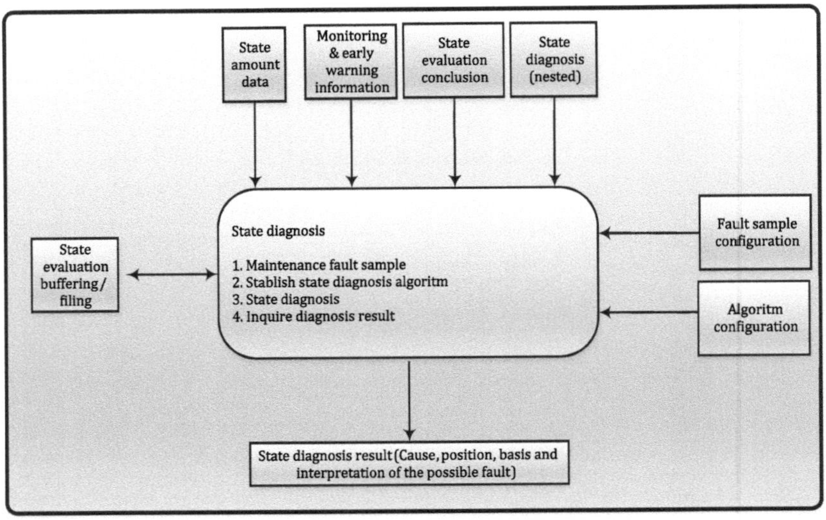

FIGURE 7.25 State diagnosis module. *Adapted from Jamil, I., Jamil, R., Jinquan, Z., Ming, L., Dong, W.Y., Jamil, R., November 2013. Condition-based maintenance decision-making support system (DSS) of hydropower plant. Copyright © 2013 ISSR Journals 4 (3). ISSN:2028-9324.*

7.3.3.3.6 State Evaluation Module

The module analyzes and evaluates data on all state amount indexes that reflect equipment health condition in accordance with state characteristics and state evaluation guidelines and standards, as shown in Fig. 7.26.

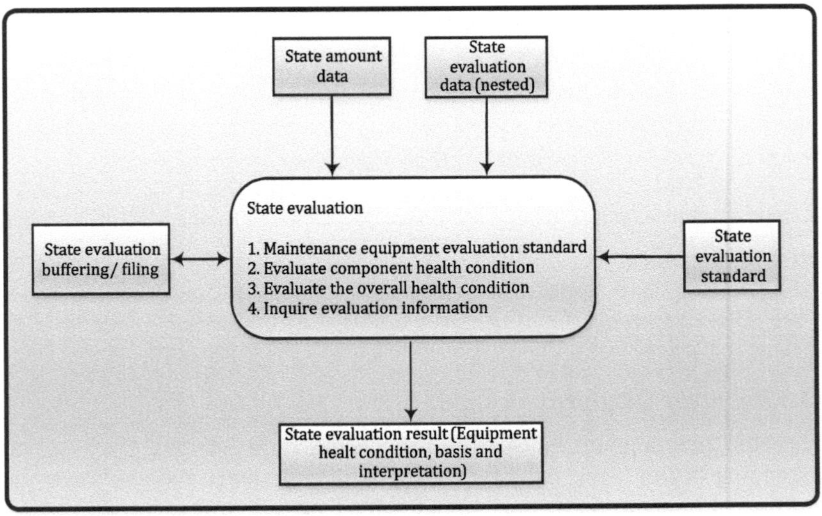

FIGURE 7.26 State evaluation module. *Adapted from Jamil, I., Jamil, R., Jinquan, Z., Ming, L., Dong, W.Y., Jamil, R., November 2013. Condition-based maintenance decision-making support system (DSS) of hydropower plant. Copyright © 2013 ISSR Journals 4 (3). ISSN:2028-9324.*

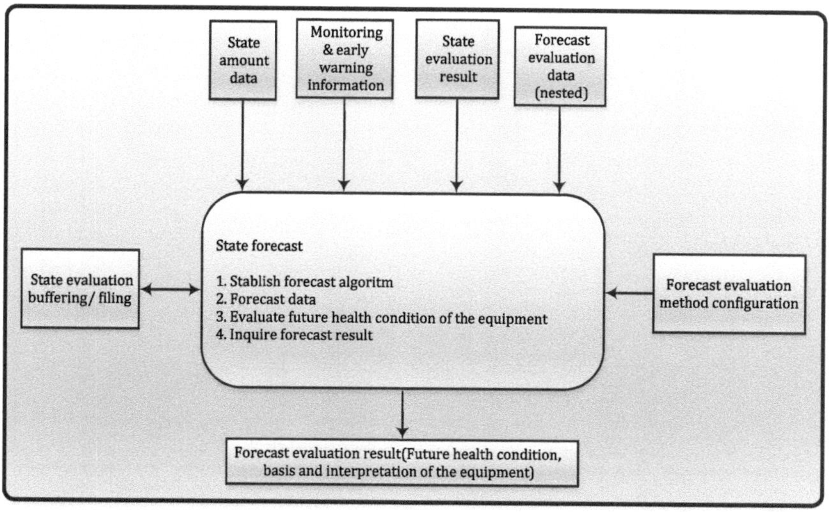

FIGURE 7.27 State forecast module. *Adapted from Jamil, I., Jamil, R., Jinquan, Z., Ming, L., Dong, W.Y., Jamil, R., November 2013. Condition-based maintenance decision-making support system (DSS) of hydropower plant. Copyright © 2013 ISSR Journals 4 (3). ISSN:2028-9324.*

7.3.3.3.7 State Forecast Module

The state forecast module uses current and historical state data of the equipment and adopts the appropriate forecast algorithm to diagnose and evaluate the state development tendency of the equipment in a future period. See Fig. 7.27.

7.3.3.3.8 Risk Assessment Module

The role of the risk assessment module is to analyze the degree of asset loss in cases of possible equipment failure and the probability of such threats by identifying potential internal defects and external threats; it uses a risk assessment algorithm to determine equipment risk level. See Fig. 7.28.

7.3.3.3.9 Decision Suggestion Module

Taking the equipment state evaluation result as its basis, this module considers risk assessments; optimizes the maintenance sequence, maintenance time, and maintenance grade arrangement of primary devices; and submits suggestions to equipment management personnel or delivers them to a related external production management system. The process is depicted in Fig. 7.29.

7.3.3.4 Key Technologies in the Fault Diagnosis System

There are three intelligent fault diagnosis methods: the expert system approach, fault tree analysis, and case-based diagnosis. These can be adopted simultaneously for system fault diagnosis; multiinformation fusion technology

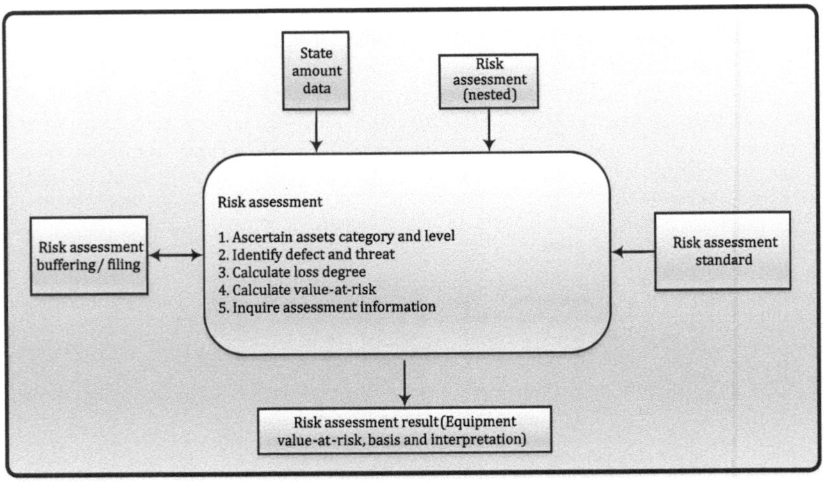

FIGURE 7.28 Risk assessment module. *Adapted from Jamil, I., Jamil, R., Jinquan, Z., Ming, L., Dong, W.Y., Jamil, R., November 2013. Condition-based maintenance decision-making support system (DSS) of hydropower plant. Copyright © 2013 ISSR Journals 4 (3). ISSN:2028-9324.*

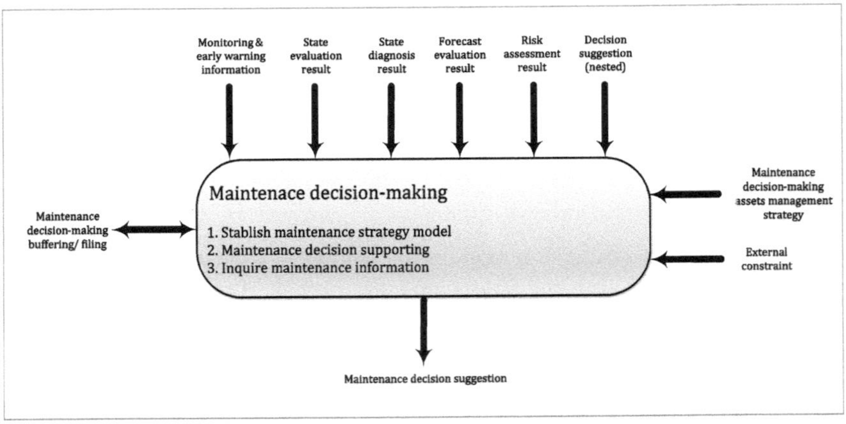

FIGURE 7.29 Decision suggestion module. *Adapted from Jamil, I., Jamil, R., Jinquan, Z., Ming, L., Dong, W.Y., Jamil, R., November 2013. Condition-based maintenance decision-making support system (DSS) of hydropower plant. Copyright © 2013 ISSR Journals 4 (3). ISSN:2028-9324.*

based on D–S evidence theory (Dempster/Shafer evidence theory) can be used to integrate the three to obtain a final diagnosis result.

A condition-based maintenance DSS is a primary device, which monitors machine condition and predicts machine failure. With a systematic strategy and the application of a condition-based maintenance management mode, a company can improve device reliability, reduce device maintenance cost, and extend the asset life cycle (Jamil et al., 2013).

7.3.4 Internet of Things

7.3.4.1 Introduction

In the IoT paradigm, many of the objects that surround us will be on the network in one form or another. Enormous amounts of data will be generated; these have to be stored, processed, and presented in a seamless, efficient, and easily interpretable form. Cloud computing can provide the virtual infrastructure required to integrate monitoring devices, storage devices, analytics tools, visualization platforms, and client delivery. It will enable end-to-end service provisioning for businesses and users to access applications on demand from anywhere (Gubbia et al., 2013).

Smart connectivity with existing networks and context-aware computation using network resources is an indispensable part of IoT. With the growing presence of Wi-Fi and 4G-LTE wireless Internet access, the evolution toward ubiquitous information and communication networks is already evident. However, for the IoT vision to succeed, the computing paradigm will need to go beyond traditional mobile computing scenarios with smartphones and portables and evolve into connecting everyday existing objects and embedding intelligence into our environment. For technology to disappear from the consciousness of the user, the IoT demands the following:

1. A shared understanding of the situation of its users and their appliances,
2. Software architectures and pervasive communication networks to process and convey contextual information to where it is relevant, and
3. Analytics tools that aim for autonomous and smart behavior.

With these three in place, smart connectivity and context-aware computation can be accomplished.

The term IoT was first coined by Kevin Ashton in 1999 in the context of SCM (Ashton, 2009). However, in the past decade, the definition has been more inclusive, covering a wide range of applications such as health care, utilities, transport, etc. (Sundmaeker et al., 2010). Although the definition of "things" has changed as technology has evolved, the main goal of making computers sense information without the aid of human intervention remains the same.

Basically stated, in IoT, we are witnessing a radical evolution of the current Internet into a network of interconnected objects that not only harvests information from the environment (sensing) and interacts with the physical world (actuation/command/control), but also uses existing Internet standards to provide services for information transfer, analytics, applications, and communications. Fueled by the prevalence of devices enabled by open wireless technology such as Bluetooth, RFID, Wi-Fi, and telephonic data services as well as embedded sensor and actuator nodes, IoT has stepped out of its infancy and is on the verge of transforming the current static Internet into a fully integrated future Internet (Buckley, 2006).

The Internet revolution led to the interconnection of people at an unprecedented scale and pace. The next revolution will be the interconnection of objects to create a smart environment. In 2011, the number of interconnected devices on the planet overtook the number of people. At the time of writing there are 9 billion interconnected devices, and this is expected to reach 24 billion by 2020. This amounts to $1.3 trillion revenue opportunities for mobile network operators alone, spanning segments such as health, automotive, utilities, and consumer electronics. A schematic of the interconnection of objects is depicted in Fig. 7.30; the application domains are chosen based on the scale of the impact of the data generated. The users go from individuals to national level organizations addressing a wide range of issues (Gubbia et al., 2013).

7.3.4.2 Definition, Trends, and Elements of Internet of Things

7.3.4.2.1 Definition

As identified by Atzori et al. (2010), IoT can be realized in three paradigms: Internet-oriented (middleware), thing-oriented (sensors), and semantic-oriented (knowledge). Although this type of delineation is required because of the interdisciplinary nature of the subject, the usefulness of IoT can be unleashed only in an application domain where the three paradigms intersect.

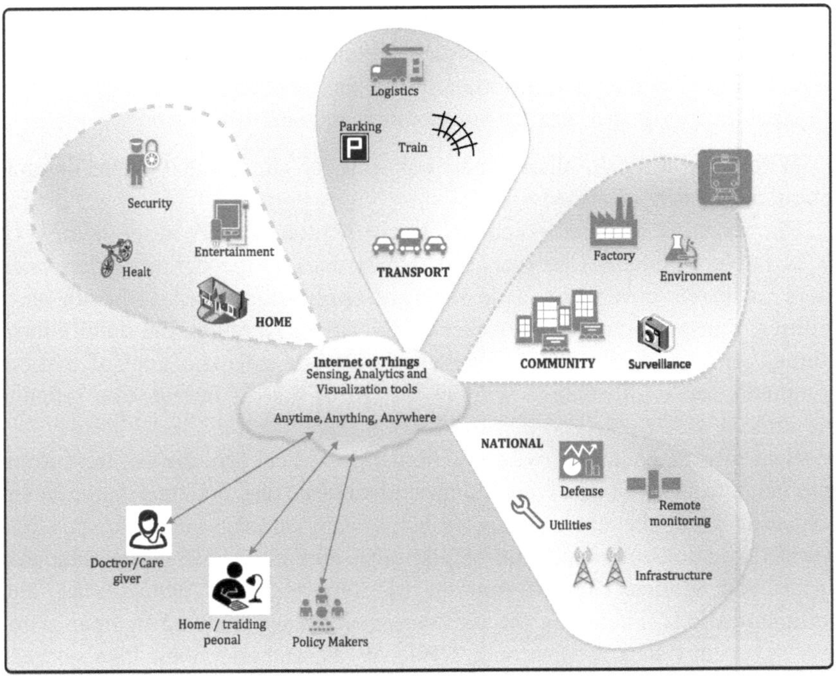

FIGURE 7.30 Internet of Things schematic. *Adapted from Gubbia, J., Buyyab, R., Marusic, S., Palaniswami, M., 2013. Internet of things (Iot): a vision, architectural elements, and future directions. Future Generation Computer Systems 29, 1645–1660.*

The RFID group defines IoT as follows:

- The worldwide network of interconnected objects uniquely addressable based on standard communication protocols.

A European research project on IoT defines things as the following (Sundmaeker et al., 2010):

- "Things" are active participants in business, information, and social processes where they are enabled to interact and communicate among themselves and with the environment by exchanging data and information sensed about the environment, while reacting autonomously to the real/physical world events and influencing it by running processes that trigger actions and create services with or without direct human intervention.

According to Forrester (Belissent, 2010), a smart environment

- Uses ICTs to make the critical infrastructure components and services of a city's administration, education, health care, public safety, real estate, transportation, and utilities more aware, interactive, and efficient.

This definition is more user centric and is not restricted to a standard communication protocol, thus allowing long-lasting applications to be developed and deployed using the available state-of-the-art protocols at any given point. A definition of IoT for smart environments is the following:

- Interconnection of sensing and actuating devices providing the ability to share information across platforms through a unified framework, developing a common operating picture for enabling innovative applications. This is achieved by seamless ubiquitous sensing, data analytics, and information representation with Cloud computing as the unifying framework (Gubbia et al., 2013).

7.3.4.2.2 Trends

IoT is one of the emerging technologies in IT as depicted diagrammatically in Gartner's IT Hype Cycle shown in Fig. 7.31. A Hype Cycle (Gartner et al., 2012) is a way to represent the emergence, adoption, maturity, and impact of IoT on applications of specific technologies. IoT will take 5–10 years for market adoption (Gubbia et al., 2013).

The popularity of paradigms changes with time. The Web search popularity of the terms IoT, wireless sensor networks (WSNs), and ubiquitous computing (ubicomp), as measured by Google search trends during the last 10 years, is shown in Fig. 7.32. As can be seen, since IoT came into existence, its search volume is consistently increasing, with a falling trend for WSNs. According to Google's search forecast (dotted line), this trend is likely to continue as other enabling technologies converge to form a genuine IoT (Gubbia et al., 2013).

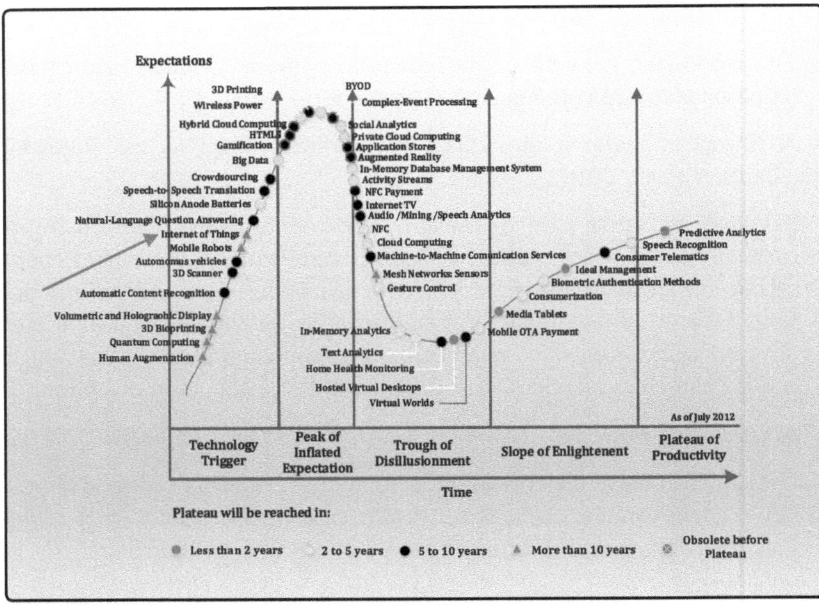

FIGURE 7.31 Hype Cycle of emerging technologies. *Adapted from Gartner Inc., 2012. Gartner's Hype Cycle Special Report for 2011. http://www.gartner.com/technology/research/hype-cycles/.*

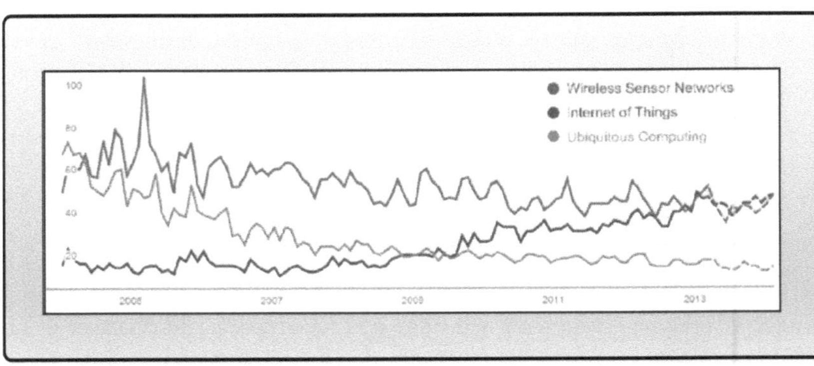

FIGURE 7.32 Google search trends since 2004 for Internet of Things, wireless sensor networks, and ubiquitous computing. *Adapted from Gubbia, J., Buyyab, R., Marusic, S., Palaniswami, M., 2013. Internet of things (Iot): a vision, architectural elements, and future directions. Future Generation Computer Systems 29, 1645–1660.*

7.3.4.2.3 Internet of Things Elements

This section provides a taxonomy that will aid in defining the components required for IoT from a high-level perspective; specific taxonomies of each component can be found elsewhere (Buyya et al., 2009; Tilak et al., 2002; Tory and Moller, 2004). Three IoT components that enable seamless ubicomp are as follows:

- Hardware: sensors, actuators, and embedded communication hardware;
- Middleware: on demand storage and computing tools for data analytics; and
- Presentation: novel easy to understand visualization and interpretation tools, which can be widely accessed on different platforms and designed for different applications.

A few technologies in these categories follow (Gubbia et al., 2013).

7.3.4.2.3.1 Radio-Frequency Identification RFID technology is a major breakthrough in the embedded communication paradigm, which enables the design of microchips for wireless data communication. These help in the automatic identification of anything they are attached to by acting as an electronic barcode (Welbourne et al., 2009; Juels, 2006). The passive RFID tags are not battery powered; they use the power of the reader's interrogation signal to communicate the ID to the RFID reader. This has resulted in many applications, particularly in retail and SCM. The applications can be found in transportation (replacement of tickets, registration stickers) and access control applications as well. The passive tags are currently being used in many bank cards and road toll tags. Active RFID readers have their own battery supply and can instantiate the communication. Of several applications, the main application of active RFID tags is in port containers for monitoring cargo (Juels, 2006). For more discussion of RFIDs, see Section 7.5.1.2 and the following.

7.3.4.2.3.2 Wireless Sensor Networks Recent technological advances in low-power integrated circuits and wireless communications have made available efficient, low cost, low-power miniature devices for use in remote sensing applications. The combination of these factors has improved the viability of using a sensor network consisting of a large number of intelligent sensors, enabling the collection, processing, analysis, and dissemination of valuable information, gathered in a variety of environments. Active RFID is nearly the same as the lower-end WSN nodes with limited processing capability and storage. The scientific challenges that must be overcome to realize the enormous potential of WSNs, however, are substantial and multidisciplinary in nature (Akyildiz et al., 2002).

In WSNs, sensor data are shared among sensor nodes and sent to a distributed or centralized system for analytics. The components that make up the WSN monitoring network include the following:

- WSN hardware: A node (WSN core hardware) typically contains sensor interfaces, processing units, transceiver units, and power supply. Nodes almost always have multiple A/D converters for sensor interfacing; more modern sensor nodes have the ability to communicate using one frequency band, making them more versatile (Akyildiz et al., 2002).

- WSN communication stack: The nodes are expected to be deployed in an ad hoc manner for most applications. Designing an appropriate topology, including a routing and media access control (MAC) layer is critical for the scalability and longevity of the deployed network. Nodes in a WSN need to communicate among themselves to transmit data in single or multihop to a base station. Node drop outs and consequent degraded network lifetimes are frequent. The communication stack at the sink node should be able to interact with the outside world through Internet to act as a gateway to the WSN subnet and Internet (Ghosh and Das, 2008).
- WSN middleware: This is a mechanism to combine cyber infrastructure with a service-oriented architecture and sensor networks to provide access to heterogeneous sensor resources in a deployment-independent manner (Ghosh and Das, 2008). It is based on the idea of isolating resources that can be used by several applications. It requires a platform-independent middleware for developing sensor applications, such as an open sensor web architecture (OSWA) (Sang et al., 2006). OSWA is built on a uniform set of operations and standard data representations, as defined in the Sensor Web Enablement Method by the Open Geospatial Consortium (Gubbia et al., 2013).
- Secure data aggregation: An efficient and secure data aggregation method is required to extend the lifetime of the network and ensure reliable data are collected from sensors (Sang et al., 2006). As node failures are a common characteristic of WSNs, the network topology should have the capability to heal itself. Ensuring security is critical, as the system is automatically linked to actuators, and protecting the systems from intruders is very important (Gubbia et al., 2013).

7.3.4.2.3.3 Addressing Schemes The ability to uniquely identify "Things" is critical for the success of IoT. This will allow us to uniquely identify billions of devices and also to control remote devices through the Internet. The most critical features of creating a unique address are uniqueness, reliability, persistence, and scalability.

Every element that is already connected and those that are going to be connected must be identified by their unique identification, location, and functionalities. The current Internet Protocol version 4 (IPv4) supports, to an extent, the identification of a group of cohabiting sensor devices geographically but not individually. The Internet mobility attributes in IPv6 may alleviate some of the device identification problems; however, the heterogeneous nature of wireless nodes, variable data types, concurrent operations, and confluence of data from devices exacerbates the problem (Zorzi et al., 2010).

Persistent network functioning to channel the data traffic ubiquitously and relentlessly is another aspect of IoT. Although the TCP/IP (transmission control protocol/Internet protocol) takes care of this mechanism by routing in a more reliable and efficient way from source to destination, IoT faces a bottleneck at the interface of gateway and wireless sensor devices. Furthermore, the scalability of the device address of the existing network must be sustainable. The addition of networks and devices must not hamper the performance of the network, the functioning of the devices, the reliability of the data over the network, or the effective use of the devices from the user interface. The Uniform Resource Name (URN) system is considered fundamental for the development of IoT as it can address these issues. URN creates replicas of the resources that can be accessed through the URL. With large amounts of spatial data being gathered, it is often important to take advantage of the benefits of metadata to transfer information from a database to the user via the Internet.

IPv6 also gives a very good option to access resources uniquely and remotely. Another critical development is the development of a lightweight IPv6 that will enable addressing home appliances uniquely.

WSNs (i.e., building blocks of IoT) run on a different stack than the Internet and cannot possess an IPv6 stack to address individually; hence, a subnet with a gateway having a URN will be required. With this in mind, we need a layer to address sensor devices by the relevant gateway. At the subnet level, the URN for the sensor devices could be the unique IDs, not human-friendly names as in the www, along with a lookup table at the gateway to address the device. Further, at the node level each sensor could have a URN (as numbers) for sensors to be addressed by the gateway. The entire network now forms a web of connectivity from users (high-level) to sensors (low-level) addressable (through URN), accessible (through URL), and controllable (through URC) (Gubbia et al., 2013).

7.3.4.2.3.4 Data Storage and Analytics One of the most important outcomes of this emerging field is the creation of an unprecedented amount of data. Storage, ownership, and expiry of the data are becoming critical issues. The Internet consumes up to 5% of the total energy generated today and, with these types of demands, it is sure to go up even further. Data centers that run on harvested energy and are centralized will ensure energy efficiency and reliability. The data have to be stored and used intelligently for smart monitoring and actuation. It is important to develop AI algorithms that could be centralized or distributed based on need. Novel fusion algorithms need to be developed to make sense of the data collected. State-of-the-art nonlinear, temporal machine learning (ML) methods based on evolutionary algorithms, genetic algorithms, neural networks (NNs), and other AI techniques are necessary to achieve automated decision-making. These systems show characteristics such as interoperability, integration, and adaptive communications. They also have a modular architecture both in terms of hardware system design and in terms of software development and are usually very well suited for IoT applications.

More importantly, a centralized infrastructure to support storage and analytics is required. This forms the IoT middleware layer; the numerous challenges involved are discussed in later sections. Cloud-based storage solutions are becoming increasingly popular and in the years ahead, Cloud-based analytics and visualization platforms are foreseen (Gubbia et al., 2013).

7.3.4.2.3.5 Visualization Visualization is critical for an IoT application as it allows the interaction of the user with the environment. With recent advances in touch screen technologies, the use of smart tablets and smartphones has become very intuitive. For a layperson to fully benefit from the IoT revolution, attractive and easy-to-understand visualization has to be created. As we move from 2D to 3D screens, more information can be provided in meaningful ways for consumers. This will also enable policy makers to convert data into knowledge, something critical for fast decision-making. Extraction of meaningful information from raw data is nontrivial; it encompasses both event detection and visualization of the associated raw and modeled data, with information represented according to the needs of the end user (Gubbia et al., 2013).

7.3.4.3 Applications

Several application domains will be impacted by the emerging IoT. The applications can be classified based on the type of network availability, coverage, scale, heterogeneity, repeatability, user involvement, and impact. Applications can be categorized into four domains:

1. Personal and home,
2. Enterprise,
3. Utilities,
4. Mobile.

Fig. 7.30 represents a personal and home IoT at the scale of an individual or home, an enterprise IoT at the scale of a community, a utility IoT at a national or regional scale, and a mobile IoT spread across other domains because of the nature of connectivity and scale. There is a huge crossover in applications and the use of data between domains. For instance, the personal and home IoT produces electricity usage data in the house and makes it available to the electricity (utility) company, which, in turn, can optimize the supply and demand in the utility IoT. The Internet enables sharing data between different service providers in a seamless manner, creating multiple business opportunities (Gubbia et al., 2013).

7.3.4.4 Cloud-Centric Internet of Things

The vision of IoT can be seen from two perspectives—"Internet" centric and "thing" centric. In the Internet-centric architecture, Internet services are the main focus, and data are contributed by the objects. In the thing-centric architecture (Lopez et al., 2012), smart objects take center stage. A conceptual framework integrating the ubiquitous sensing devices and the applications is shown in Fig. 7.33. To realize the full potential of Cloud computing and ubiquitous sensing, a combined framework with a Cloud at the center seems to be most viable. This not only gives the flexibility of dividing associated costs

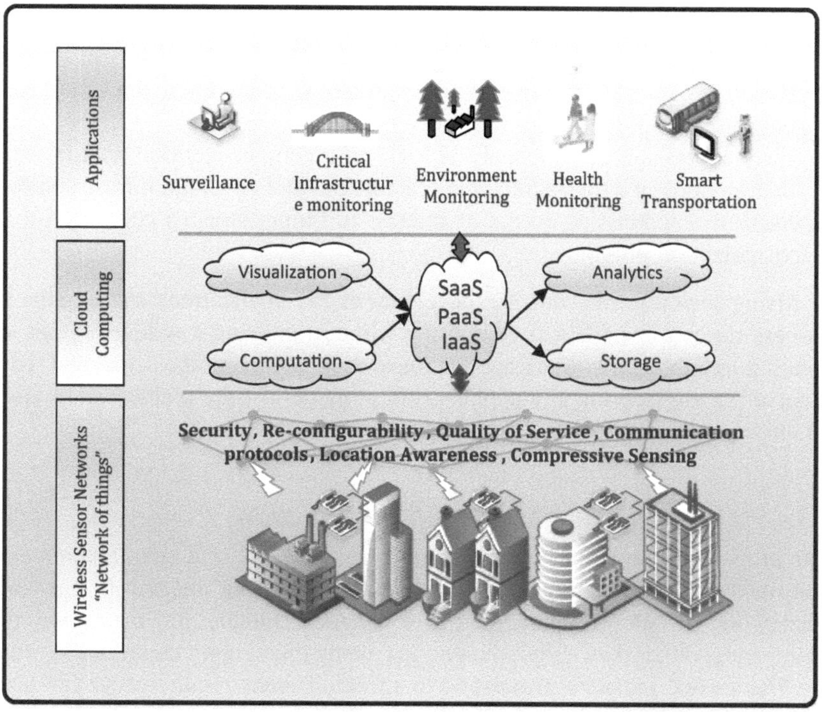

FIGURE 7.33 Conceptual Internet of Things framework with Cloud computing at the center. *Adapted from Gubbia, J., Buyyab, R., Marusic, S., Palaniswami, M., 2013. Internet of things (Iot): a vision, architectural elements, and future directions. Future Generation Computer Systems 29, 1645–1660.*

in the most logical manner but is also highly scalable. Sensing service providers can join the network and offer their data using a storage Cloud; analytic tool developers can provide their software tools; AI experts can provide their DM and ML tools to convert information to knowledge; and computer graphics designers can offer a variety of visualization tools. Cloud computing can offer these services as infrastructures, platforms, or software; the full potential of human creativity can be tapped using them as services. This, in some sense, agrees with the ubicomp vision of Weiser as well as Rogers' human-centric approach. The data generated, tools used, and the visualization created disappear into the background, tapping the full potential of IoT in various application domains. As seen in Fig. 7.33, the Cloud integrates all ends of ubicomp by providing scalable storage, computation time, and other tools to build new businesses (Gubbia et al., 2013).

However, developing IoT applications using low-level Cloud programming models and interfaces such as Thread and MapReduce models is complex. To overcome this, it is necessary to have an IoT application-specific framework for the rapid creation of applications and their deployment on Cloud infrastructures. This can be achieved by mapping the proposed framework to

Cloud APIs offered by platforms such as Aneka. The new IoT application-specific framework should be able to provide support for the following:

1. Reading data streams from sensors directly or to fetch the data from databases;
2. Easy expression of data analysis logic as functions/operators that process data streams in a transparent and scalable manner on Cloud infrastructures;
3. Passing outcomes of events of interest to output streams connected to a visualization program.

Using such a framework, the developer of IoT applications will be able to harness the power of Cloud computing without knowing low-level details of creating reliable and scale applications, thus reducing the time and cost involved in engineering IoT applications. A model for the realization of such an environment is shown in Fig. 7.34.

7.3.4.5 Open Challenges and Future Directions

The proposed Cloud-centric vision comprises a flexible and open architecture that is user centric and enables different players to interact in the IoT framework. It allows interaction in a manner suitable for their own requirements, rather than simply having IoT being thrust upon them. In this way, the framework includes provisions to meet different requirements for data ownership, security, privacy, and sharing of information.

The remaining challenges include IoT specific ones, such as privacy, participatory sensing, data analytics, Geographic Information System (GIS)–based visualization, and Cloud computing, as well as the standard WSN challenges, including architecture, energy efficiency, security, protocols, and quality of service (QoS). The end goal is to have plug-n-play smart objects, which can be deployed in any environment with an interoperable backbone allowing them to blend with other smart objects around them. The standardization of frequency bands and protocols plays a pivotal role in accomplishing this goal.

A road map of key developments in IoT research in the context of pervasive applications is shown in Fig. 7.35. It includes the technology drivers and key application outcomes expected in the next decade (Atzori et al., 2010).

7.3.4.5.1 Architecture

Most of the work on IoT architecture has been from the WSNs perspective (Castellani et al., 2010). The European Union projects of SENSEI and Internet of Things–Architecture (IoT-A) have also been addressing the challenges, particularly from the WSN perspective; they have been very successful in defining the architecture for different applications. In these applications, the IoT architecture has the user at the center. An architecture with Cloud computing at the center has been proposed here. However, this may not be the best option for every application domain, particularly for defense where

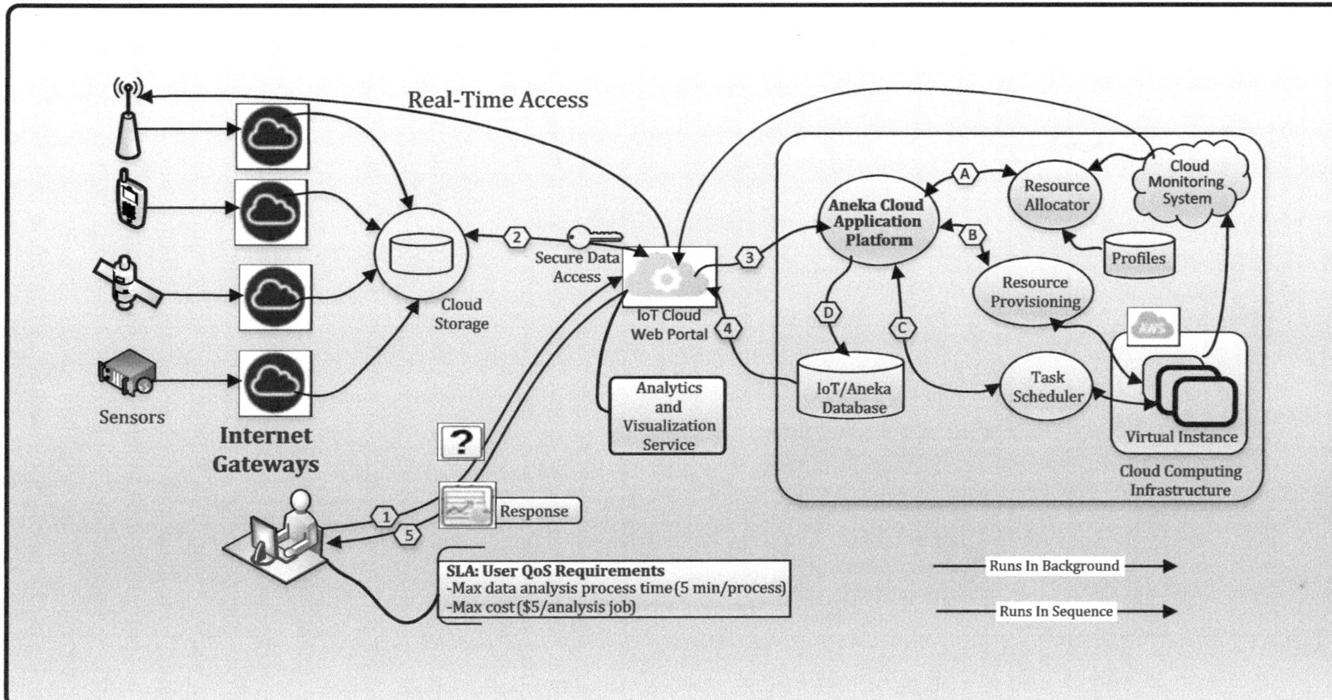

FIGURE 7.34 Model of end-to-end interaction between stakeholders in Cloud-centric Internet of Things framework. *Adapted from Gubbia, J., Buyyab, R., Marusic, S., Palaniswami, M., 2013. Internet of things (Iot): a vision, architectural elements, and future directions. Future Generation Computer Systems 29, 1645–1660.*

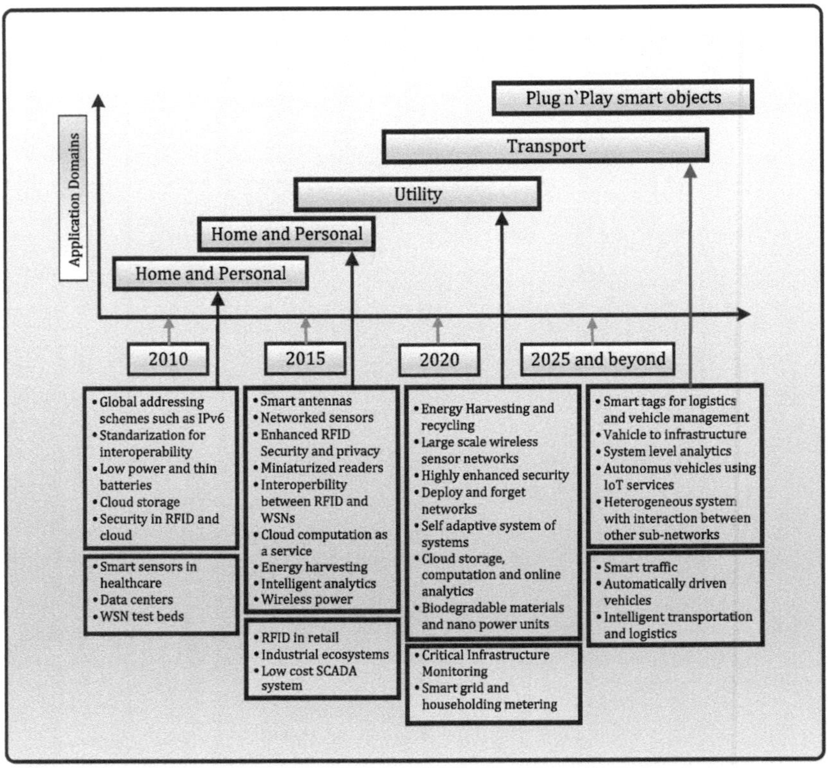

FIGURE 7.35 Road map of key technological developments envisioned in the context of Internet of Things application. *Adapted from Gubbia, J., Buyyab, R., Marusic, S., Palaniswami, M., 2013. Internet of things (Iot): a vision, architectural elements, and future directions. Future Generation Computer Systems 29, 1645–1660.*

human intelligence is relied on. Although Cloud-centric architecture may be the best where cost-based services are required, other architectures should be investigated for different application domains (Gubbia et al., 2013).

7.3.4.5.2 Energy-Efficient Sensing

Efficient heterogeneous sensing of the urban environment needs to simultaneously meet competing demands of multiple sensing modalities, with implications on network traffic, data storage, and energy utilization. Importantly, this encompasses both fixed and mobile sensing infrastructure (Rana et al., 2010) as well as continuous and random sampling. A generalized framework is required for data collection and modeling to effectively exploit spatial and temporal characteristics of the data, both in the sensing domain and in the associated transform domains. For example, urban noise mapping needs an uninterrupted collection of noise levels using battery powered nodes with fixed infrastructure and participatory sensing (Rana et al., 2010) as a key component for health and quality of life services for its inhabitants.

Compressive sensing enables reduced signal measurements without impacting accurate reconstruction of the signal. A signal sparse in one basis may be recovered from a small number of projections onto a second basis that is incoherent with the first (Donoho, 2006). The problem reduces to finding sparse solutions through the smallest l1-norm coefficient vector that agrees with the measurements. In the ubiquitous sensing context, this has implications for data compression, network traffic, and the distribution of sensors. Compressive wireless sensing uses synchronous communication to reduce the transmission power of each sensor (Bajwa et al., 2006), transmitting noisy projections of data samples to a central location for aggregation (Gubbia et al., 2013).

7.3.4.5.3 Secure Reprogrammable Networks and Privacy

Security will be a major concern wherever networks are deployed on a large scale. There can be many ways the system could be attacked: disabling the network availability; pushing erroneous data into the network; accessing personal information; etc. The three physical components of IoT, RFID, WSN, and Cloud are vulnerable to such attacks. But security is critical to any network (Navajo et al., 2010; Tang, 2009), and the first line of defense against data corruption is cryptography aggregation (Gubbia et al., 2013).

Of the three, RFID (particularly passive) seems to be the most vulnerable as it allows person tracking and object tracking, and no high-level intelligence can be enabled on the devices (Juels, 2006). These complex problems can be solved using cryptographic methods but deserve more research before they are widely accepted (Gubbia et al., 2013).

Encryption ensures data confidentiality against outsider attacks, whereas message authentication codes ensure data integrity and authenticity (Neill, 2011). Encryption does not protect against insider malicious attacks to address which noncryptographic means are needed, particularly in WSNs. New sensor applications need to be periodically installed or existing ones need to be updated. This is done by remote wireless reprogramming of all nodes in the network. Traditional network reprogramming consists solely of a data dissemination protocol that distributes code to all the nodes in the network without authentication, but this is a security threat. A secure reprogramming protocol allows the nodes to authenticate every code update and prevent malicious installation. Most such protocols (e.g., Navajo et al., 2010) are based on the benchmark protocol Deluge (Tang, 2009). Cryptographic add-ons to Deluge lay the foundation for the development of more sophisticated algorithms (Gubbia et al., 2013).

Security in the Cloud is another important area of research, which requires attention. Along with data and tools, the Cloud handles the economics of IoT, putting it under greater threat of attack. Security and identity protection becomes critical in hybrid Clouds, where private and public Clouds are used by businesses (Kaufman, 2009). Long-term memory in the context of IoT raises many privacy issues, as the data collected can be used in positive (for advertisement services) and negative ways (for defamation). Digital forgetting could emerge as one of the key areas of research to address the concerns and the development of an appropriate framework to protect personal data (Mayer-Schönberger, 2011).

7.3.4.5.4 Quality of Service

Heterogeneous networks are (by default) multiservice, providing more than one distinct application or service. This implies not only multiple traffic types within the network, but also the ability of a single network to support all applications without QoS compromise (Vera et al., 2009). There are two application classes: throughput and delay-tolerant elastic traffic (e.g., monitoring weather parameters at low sampling rates) and bandwidth and delay-sensitive inelastic (real-time) traffic (e.g., noise or traffic monitoring), which can be further discriminated by data-related applications (e.g., high- vs. low-resolution videos) with different QoS requirements. Therefore, a controlled, optimal approach to serve different network traffics, each with its own QoS needs is required (El-Sayed et al., 2008). It is not easy to provide QoS guarantees in wireless networks, as segments often constitute "gaps" in resource guarantee because of resource allocation and management ability constraints in shared wireless media.

QoS in Cloud computing is another major research area, which will require more attention as the data and tools become available on Clouds. Dynamic scheduling and resource allocation algorithms based on particle swarm optimization are now being developed, but for high capacity applications and as IoT grows, this could become a bottleneck (Gubbia et al., 2013).

7.3.4.5.5 New Protocols

The protocols at the sensing end of IoT will play a key role in complete realization. They form the backbone for the data tunnel between sensors and the outer world. For the system to work efficiently, an energy-efficient medium access control (MAC) protocol and appropriate routing protocol are critical. Several MAC protocols have been proposed for various domains with time division multiple access (TDMA) (collision free), carrier sense multiple access (CSMA) (low traffic efficiency), and frequency division multiple access (FDMA) (collision free but requiring additional circuitry in nodes) schemes available to the user (Demirkol et al., 2006). None is accepted as a standard and with more "things" available, this scenario is going to get more cluttered and require further research.

An individual sensor can drop out for a number of reasons, so the network must be self-adapting and allow for multipath routing. Multihop routing protocols are used in mobile ad hoc networks and terrestrial WSNs (Al-Karaki and Kamal, 2004). They are mainly divided into three categories, data centric, location based, and hierarchical, again based on different application domains. Energy is the main consideration for the existing routing protocols. In the case of IoT, it should be noted that a backbone will be available and the number of hops in the multihop scenario will be limited. In such a scenario, the existing routing protocols should suffice in practical implementation with minor modifications (Gubbia et al., 2013).

7.3.4.5.6 Participatory Sensing

A number of projects have begun to address the development of people-centric (or participatory) sensing platforms (Rana et al., 2010; Campbell et al., 2006; Kanjo, 2009; Santini et al., 2008). As noted earlier, people-centric sensing offers the possibility of low-cost sensing of the environment localized to the user. Therefore, it can give the closest indication of environmental parameters experienced by the user. It has been noted that environmental data collected by the user form a social currency (Kuznetsov and Paulos, 2010). This results in the generation of more timely data than in a fixed infrastructure sensor network. Most importantly, the opportunity for the user to provide feedback on his/her experience of a given environmental parameter offers valuable information in the form of context associated with a given event (Gubbia et al., 2013).

The limitations of people-centric sensing place new significance on the reference data role provided by using a fixed infrastructure IoT as a backbone. The problem of missing samples is a fundamental limitation of people-centric sensing. Relying on users to volunteer data and on the inconsistent gathering of samples across varying times and varying locations (based on a user's desired participation and given location or travel path) limits the ability to produce meaningful data for applications and policy decisions. Only in addressing issues and implications of data ownership, privacy, and appropriate participation incentives can such a platform achieve genuine end-user engagement.

Further sensing modalities can be obtained through the addition of sensor modules attached to the phone for application-specific sensing, such as air quality sensors (Honicky et al., 2008) or biometric sensors. In such scenarios, smart phones become critical IoT nodes, which are connected to the Cloud on one end and several sensors at the other (Gubbia et al., 2013).

7.3.4.5.7 Data Mining

Extracting useful information from a complex sensing environment at different spatial and temporal resolutions is a challenging research problem in AI. Current state-of-the-art methods use shallow learning methods where predefined events and data anomalies are extracted using supervised and unsupervised learning (Kulkarni et al., 2011). The next level of learning involves inferring local activities by using temporal information of events extracted from shallow learning. The ultimate vision is to detect complex events based on larger spatial and longer temporal scales based on the two preceding levels. The fundamental research problem in complex sensing environments of this nature is how to simultaneously learn representations of events and activities at multiple levels of complexity (i.e., events, local activities, and complex activities). An emerging focus in ML research has been the field of deep learning (Bengio, 2009), which aims to learn multiple layers of abstraction that can be used to interpret given data. However, the resource constraints in sensor networks create novel challenges for deep learning in terms of the need for adaptive, distributed, and incremental learning techniques (Gubbia et al., 2013).

7.3.4.5.8 Geographic Information System–Based Visualization

As new display technologies emerge, creative visualization will be enabled. The evolution from CRT to Plasma, LCD, LED, and AMOLED displays has given rise to highly efficient data representation (using touch interface), with the user able to navigate the data better than ever before. With emerging 3D displays, this area is certain to attract more research and development opportunities. However, the data that come out of ubicomp are not always ready for direct consumption using visualization platforms and may require further processing. The scenario becomes very complex for heterogeneous spatio-temporal data (Bonneau et al., 2003). New visualization schemes for the representation of heterogeneous sensors in a 3D landscape that varies temporally must be developed (Ren et al., 2010). Another challenge of visualizing data collected within IoT is that they are geo related and sparsely distributed. A framework based on Internet GIS is required to cope with the challenge (Gubbia et al., 2013).

7.3.4.5.9 Cloud Computing

Integrated IoT and Cloud computing applications enabling the creation of smart environments such as smart cities need to be able to combine services offered by multiple stakeholders and scale to support a large number of users in a reliable and decentralized manner. They must be able to operate in both wired and wireless network environments and deal with constraints such as access devices or data sources with limited power and unreliable connectivity. The Cloud application platforms need to be enhanced to support the rapid creation of applications by providing domain-specific programming tools and environments, as well as the seamless execution of applications harnessing capabilities of multiple dynamic and heterogeneous resources to meet QoS requirements of diverse users (Gubbia et al., 2013).

The Cloud resource management and scheduling system should be able to dynamically prioritize requests and provision resources such that critical requests are served in real time. To deliver results in a reliable manner, the scheduler needs to have task duplication algorithms for failure management. Specifically, the Cloud application scheduling algorithms must exhibit the following capability:

1. Multiobjective optimization: The scheduling algorithms should be able to deal with QoS parameters, such as response time, cost of service usage, maximum number of resources available per unit price, and penalties for service degradation.

2. Task duplication–based fault tolerance: Critical tasks of an application should be transparently replicated and executed on different resources so that if one resource fails to complete the task, the replicated version can be used. This logic is crucial in real-time tasks that must be processed to deliver services in a timely manner (Gubbia et al., 2013).

7.3.4.5.10 International Activities

IoT activities are gathering momentum around the world, with numerous initiatives emerging across industry, academia, and various levels of government, as key stakeholders seek to map a way forward for the coordinated realization of this technological evolution. In Europe, substantial effort is underway to consolidate the cross-domain activities of research groups and organizations, shaping M2M, WSN, and RFID into a unified IoT framework. Supported by the European Commission 7th Framework program (EU FP7), this includes the Internet of Things European Research Cluster (IERC). Encompassing a number of EU FP7 projects, its objectives are to establish a cooperation platform and research vision for IoT activities in Europe and to become a contact point for IoT research around the world. It includes projects such as CASAGRAS2, a consortium of international partners from Europe, the United States, China, Japan, and Korea exploring issues surrounding RFID and its role in realizing the IoT. IERC also includes the IoT-A project established to determine an architectural reference model for the interoperability of IoT systems and key building blocks to achieve this. At the same time, the IoT initiative (IoT-i) is a coordinated action to support the development of the European IoT community. The IoT-i project brings together a consortium of partners to create a joint strategic and technical vision for the IoT in Europe that holistically encompasses the currently fragmented sectors of the IoT domain. Simultaneously, the Smart Santander project is developing a city scale IoT testbed for research and service provision deployed across the city of Santander, Spain, as well as sites in the United Kingdom, Germany, Serbia, and Australia.

At the same time, large-scale initiatives are underway in Japan, Korea, the United States, and Australia, where industry, associated organizations, and government departments are collaborating on various programs, advancing capabilities required to achieve IoT. This includes smart city initiatives, smart grid programs incorporating smart metering technologies, and high-speed broadband infrastructure. The continuing development of RFID-related technologies by industry and consortiums includes the Auto-ID lab (founded at Massachusetts Institute of Technology and now with satellite labs at leading universities in South Korea, China, Japan, United Kingdom, Australia, and Switzerland) dedicated to creating IoT using RFID and WSNs.

Significantly, the need for consensus on IoT technical issues has seen the establishment of the Internet Protocol for Smart Objects Alliance, with more than 60 member companies from leading technology, communications and energy companies, working with standards bodies, such as IETF, IEEE, and

ITU, to specify new IP-based technologies and promote industry consensus for assembling the parts for the IoT.

Substantial IoT development activity is also ongoing in China, with its 12th Five-Year Plan (2011–15), specifying IoT investment and development to be focused on the following: smart grids; intelligent transportation; smart logistics; smart homes; environment and safety testing; industrial control and automation; health care; agriculture; finance and service; and military defense. This is being aided by the establishment of an IoT center in Shanghai (with a total investment over US $100 million) to study technologies and industrial standards. An industry fund for IoT and an IoT Union "Sensing China" has been founded in Wuxi, initiated by more than 60 telecom operators, institutes and companies, the primary drivers of the industry (Gubbia et al., 2013).

7.4 THE END OF TRADITIONAL MAINTENANCE APPROACHES: REAL-TIME DECISIONS BASED ON INDUSTRIAL BIG DATA

The term "big data" refers to the huge datasets obtained from various sources, such as science experiments, sensor systems, social network activities, telecommunications data, cameras and surveillance systems, and daily business and financial transactions. Companies collect huge volumes of data on their customers, suppliers, operations, and business transactions. Sensors embedded in roads, buildings, mobile phones and devices, vehicles, and smart energy meters generate trillions of bytes of real-time information. In addition, billions of individuals around the world contribute on a daily basis to the increasing size and availability of big data through social media sites and applications. Such huge data offer a great opportunity to understand and use them in beneficial applications. However, their sheer size raises new technological challenges in terms of storage capacity and management, organization, processing, and analysis.

With big data, businesses and organizations can learn more about their businesses and operations and translate that knowledge into improved decision-making processes and better performance (McAfee and Brynjolfsson, 2012). The availability of big data can be used to enhance performance in different application domains, such as health care, finance, transportation, education, security, and governmental planning. Some special types of big data applications need to make real-time decisions to enhance the performance of their operations or services or to increase profits. Examples of these applications are intelligent transportation, financial market trading and surveillance, military operations decision-making, smart grids, and emergency response. These applications need to deal with current and historical massive data to perform fast analysis aiding in instantaneous decision-making. Slow or untimely data collection and analysis leading to delayed decisions can significantly reduce the performance of the big data applications. Such applications have a number of technical challenges to provide real-time processing due to the massive data available for optimal or near-optimal decision-making (Mohamed and Al-Jaroodi, 2014).

7.4.1 Big Data: Analytics and Decision-Making

The term "big data" refers to datasets that typical database management systems cannot store, manage, and analyze due their massive sizes (Manyika et al., 2011). Big data sizes range from a few dozen terabytes to multiple petabytes. Big data applications generally seek new knowledge and gather intelligence from the data and convert that into business advantages (McAfee and Brynjolfsson, 2012), notably by enhancing operations and profitability and reducing risks and overheads. In addition, using big data can provide a number of new types of advanced services that have recently been introduced or will be introduced in the near future. These services enhance the quality of life and help reduce risks and threats.

Big data differ from regular data in three characteristics: volume, velocity, and variety (Mohamed and Al-Jaroodi, 2014).

- *Volume*: It was estimated that around 2.5 exabytes (one billion gigabytes) of data were created from different sources each day in 2012. This number has been doubling every 40 months since then. The expanding use of the Internet, mobile devices, and smartphones contributes to the generation of these huge data. Another source is sensors and sensor systems embedded in our lives—in roads, building, cars, different environments, and factories. These sensors collect massive amounts of information. Some are fixed, such as road and environmental sensors, while others are mobile, such as sensors installed in smartphones and vehicles such as cars and unmanned aerial vehicles.
- *Velocity*: Huge amounts of new data are created every second and are added to existing big datasets. In most big data applications, these new data need to be included in analysis and real-time decision-making. Organizing, accessing, and processing the data as they are collected is a complex technical challenge.
- *Variety*: Data are usually collected from different sources and in different formats. These sources can be messages, images, and videos posted to social networks, readings from sensors, business transactions, and economic and political news. Most of the sources of big data are relatively new with the recent introduction of online social networks and smartphones. The collected data can be structured or unstructured. They can also have different formats. Some of the data are produced in a random fashion, while some data are produced frequently or periodically. In the latter case, the data can also differ in their production frequencies (Mohamed and Al-Jaroodi, 2014).

Data generated by financial transactions, different types of sensors and meters, social media networks, and numerous other sources are growing exponentially in terms of their volume, variety, and velocity. These "3 Vs" are making datasets increasingly difficult to capture, manage, and process through conventional means. Moreover, increasing data-storage capacity has created huge collections of data. In fact, 90% of all data that exist today were created within the last 2 years (Hewitt, 2011).

Big data are high-volume, high-velocity, and high-variety information assets that demand cost-effective, innovative forms of information processing for enhanced insight, and decision-making (Gartner). As a result, efforts have been made to measure variables through ever improving digital sensors, communications, and computational power.

The following statistics highlight the developments driving the big data trend (Manyika et al., 2011):

- There were five billion mobile phones in use in 2010;
- Thirty billion different pieces of content are shared on Facebook every month;
- It costs around $450 to buy a hard drive that can store all of the world's music;
- Global growth of data is projected to increase by 40% per annum, while the growth in IT spending is projected at 5% per annum.

Ever-increasing processing power has enabled the analysis of growing amounts of data and the extraction of hidden layers of information from masses of "ordinary" data. The coupling of these can result in completely new insights, applications, products, and services. This potential for increased insight has provided business opportunities for companies in the areas of big data analysis and decision-making solutions.

Historically, data were captured by businesses in a structured manner and extracted for analysis when needed. This included, for example, customer records, shopping histories, and credit scores. Yet the rapid adoption of social media and smart devices has led to unprecedented growth in unstructured data, such as audio and video files, images, text from e-mails, blogs, and traffic information. Tools and techniques based on big data are needed to analyze these forms of unstructured data (Probst et al., 2013).

There is a growing belief among companies that remaining competitive requires the integration of ever-larger amounts of data into their decision-making process (Economist Intelligence Unit, 2011). Big data analytics and decision-making refers to the process of examining big data assets through novel and powerful tools to observe trends or derive insights for meaningful decision-making (Yiu, 2012).

Hardware solutions, software solutions, and services are available to help organizations gain significant value from big data by extracting relevant data, processing, organizing, and analyzing them and finally presenting them through intuitive visualization tools for the organization's decision makers to use. Providing insights and BI can help businesses optimize their decision-making processes with the aim of improving business performance and obtaining a competitive advantage.

Analytic and decision-making tools for big data can allow companies to discover trends and characteristics about their customers that might otherwise remain hidden, in data that were never apparent or intended in the source information. There are many facets of and channels for analytics and decision-making solutions within the realm of big data; see Fig. 7.36 (Probst et al., 2013).

Data that can be exploited by businesses can broadly be classed into the following three categories (PwC Australia, 2012):

- *Internal data*: information that a business collects through its own systems and processes. They can consist of both quantitative and qualitative information and are typically the type of data that provides the greatest insights for a business. For example, Walmart handles more than one million

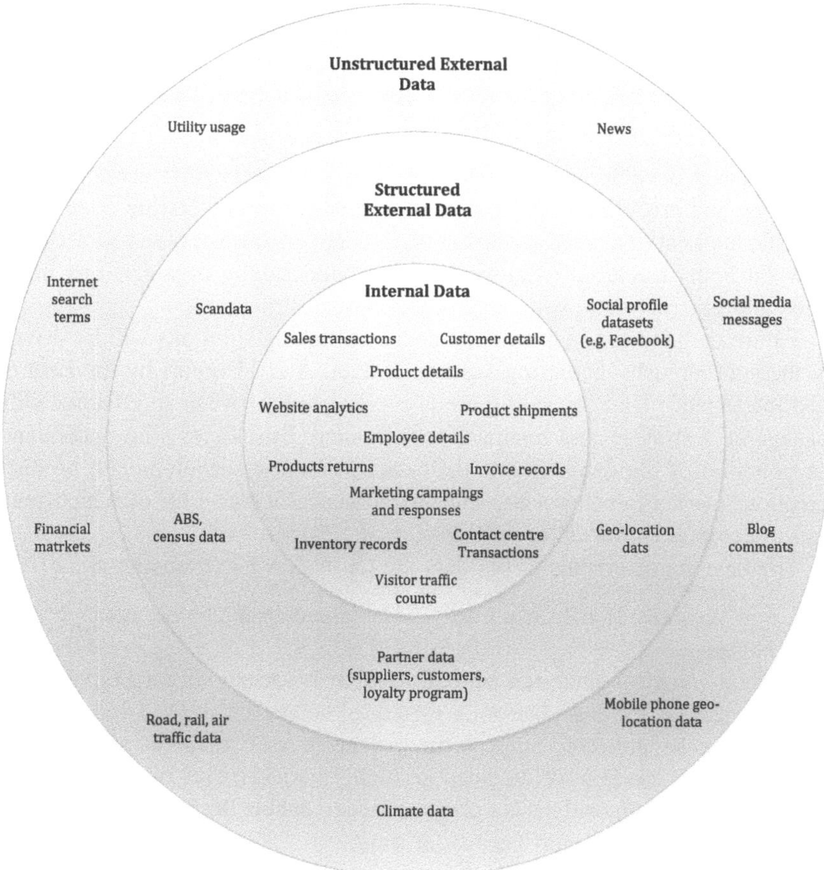

FIGURE 7.36 Examples of the different types of exploitable data by businesses. *Adapted from PwC Australia, 2012. Big Data – The Next Frontier for Innovation. http://www.pwc.com.au/consulting/assets/publications/big-data-oct12.pdf.*

customer transactions every hour; it imports these into databases estimated to contain more than 2.5 petabytes of data. This information enables the retailer to map out buyer behavior and merchandising and operational strategies.

- *Structured external data*: from third-party sources. These data often require processing to be coupled with existing internal information to yield further insights. Facebook, for example, processes 2.5 billion pieces of content and 500 + terabytes of data each day. The social network also pulls in 2.7 billion likes and 300 million photos per day and scans approximately 105 terabytes of data each half hour. These data provide a significant amount of information about user behavior.
- *Unstructured external data*: from sources outside the control of a business but having a potential impact on it. Like structured external data, these data can provide concise and targeted business insights when coupled with internal data. For example, when a search is conducted on Google, the search engine's algorithm takes into account over 200 factors, such as relevance, social media, geography, advertising, collective popularity, and so forth. This essentially makes sense of unstructured data to deliver the most relevant results to users.

The latest developments in data analytics suggest there is great value in capturing and processing information with the appropriate cutting-edge tools. Possible applications areas are business, science, government, and society and draw on both structured data, traditionally collected by organizations internally, and external data, whether structured or unstructured.

Future demand for analytics and decision-making solutions will be driven by the continuously increasing volume of data. A 2011 report by the Economist Intelligence Unit suggests there is a strong link between an effective data management strategy and financial performance. Businesses who understand the potential of big data can use their data in strategy development, product direction, market development, and operational efficiency, all of which may lead to competitive advantages (Probst et al., 2013).

The key characteristics of big data are the following (Probst et al., 2013):

- Large volumes of data can easily be amassed in terabytes, petabytes, or exabytes;
- A wide range of data can be collected in a variety of data types (e.g., numerical or text-based) from a variety of sources;
- Data can be collected in real time; datasets can arrive at a very high velocity but the flow of data can be highly variable over time;
- New powerful algorithms are programmed to enable the processing of huge amounts of data in short periods of time;
- Companies are able to build predictive models in a wide range of domains, including energy consumption patterns, inflation prediction, Internet Web-traffic optimization, traffic optimization, precommerce sales predictions, etc.

7.4.2 Real-Time Responses With Big Data

Organizations make decisions that range from one-off strategic decisions to monthly and weekly tactical decisions to high-volume, high-speed front-line operational decisions. These front-line decisions must increasingly be made in real time. Mobile devices, Cloud computing, social media, sensors, and cameras are now ubiquitous and produce massive amounts of changing data. To extract the maximum value from dynamic and perishable data, organizations need to process data much faster and take timely action. Whether marketing to potential customers, offering proactive support, detecting and preventing fraud, or managing IoT, real-time decision-making is essential.

Responding in real time requires systems to make operational decisions automatically. Agile, analytic, and adaptive decision management systems (DMSs) combine business rules and predictive analytics to render tailored recommendations. Event processing adds correlation and pattern detection on a scale of millions of events and data streams at microsecond speeds.

Event-based DMSs enrich event-based data with traditional and big data sources and determine when and why a real-time response might be required. By leveraging decision engines based on business rules and analytics, they can determine the best, most effective business response for every interaction (Taylor, 2014).

7.4.2.1 Systems for Real-Time Response

This section explores the power of DMSs, the impact of big data, the capabilities of event-based DMSs, and some common use cases for these kinds of real-time response systems (Taylor, 2014).

7.4.2.1.1 The Power of Decision Management Systems

DMSs improve profitability (better results, less fraud, lower costs), increase compliance (with policies and regulations), and deliver more precise and effective risk management. They offer organizations the ability to deliver on predictive real-time business opportunities with specific time and location content in context for 1:1 customization. These systems maximize the value of assets while allowing staff to be reallocated to higher-value decisions, and they deliver unprecedented business agility with greater automation at the heart of their operations. Building a DMS involves the following (Taylor, 2014):

- Managing decision logic for transparency and agility,
- Embedding predictive analytics for analytical decision-making,
- Optimizing results given real-world trade-offs and simulating results,
- Monitoring and improving decision-making over time.

Critical characteristics of DMSs include the following (Taylor, 2014):
Agile, analytic, adaptive: DMSs are agile, analytic, and adaptive.

- They are agile so they can be rapidly changed to cope with new regulations or business conditions.

- They are analytic, putting an organization's data to work to improve the quality and effectiveness of decisions.
- They are adaptive, learning from what works and what does not work to continuously improve over time.

Business rules, predictive analytics, optimization: Building agile, analytic, and adaptive systems requires a focus on decisions and the effective use of proven technologies—primarily business rules, predictive analytics, and optimization.

- Business rules harness best practices, expertise, policies, and regulations. They deliver agility by making it easy for business owners to make changes when they need to.
- Predictive analytics empirically explain propensity, association, segmentation, and risk to allow analytics to be embedded in systems.
- Optimization handles trade-offs and constraints, and when combined with ongoing monitoring and review of decision-making, it allows these systems to be adaptive to change.

Fit into enterprise architecture as decision services: A DMS pulls together the above technologies and deploys them as a decision service that can be consumed by business processes, business events, and applications. These decision services are at the heart of enterprise architecture, as shown in Fig. 7.37 (Taylor, 2014).

7.4.2.1.2 Impact of Big Data

The reality of big data is that more data, of more types, are arriving more quickly, making the value of DMSs increasingly clear:

- The need to handle larger data volumes can reveal the limitation of human interpretation and overwhelm decision makers. DMSs, in contrast, automate decision-making and use more advanced analytics, so they scale to larger data volumes.

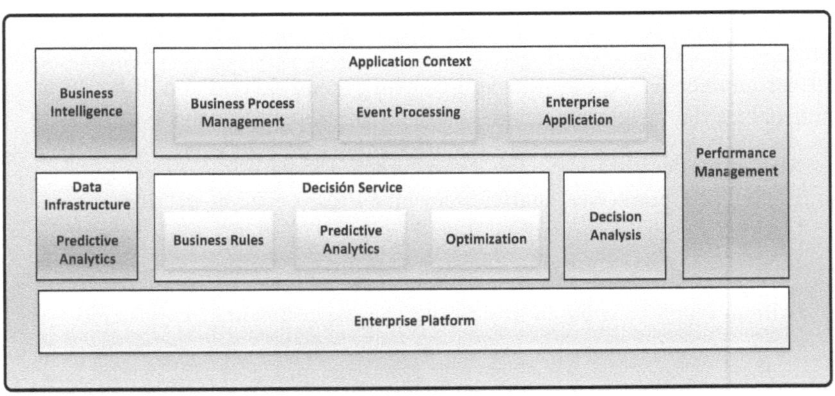

FIGURE 7.37 Decision services. *Adapted from Taylor, J., 2012. Decision Management Systems Platform Technologies Report. Decision Management Solutions, Palo Alto, CA.*

- The increasing variety of data means that organizations handle more types of data. These data come from both inside and outside an organization and can include unstructured data types, such as video, audio, or text. DMSs pull together whatever data will improve a specific decision and leverage data stored in a variety of formats.
- As these data arrive more quickly, organizations have to make decisions more quickly and deal with data "in motion"—streaming data—not just data at rest. Automating decisions in response to this fast-moving data is essential.

The increasing velocity of data adds to the general pressure on organizations to respond in real time. This is leading to an increasing focus on event-based DMSs (Taylor, 2014).

7.4.2.1.3 Event-Based Decision Management Systems

To respond in real time, organizations need to filter, correlate, and process large amounts of fast-moving data in multiple formats. Event-based DMSs deploy decision agents, not just decision services, to deliver real-time responsiveness from high-velocity data. They combine pattern detection with business decisions and recognize that business decisions must be shared between event-centric and process-centric solutions (Taylor, 2014).

7.4.2.1.3.1 From Decision Services to Decision Agents The essence of a decision service is that it makes a decision, or answers a business question, for a business process or other service. Separating out the decision-making in this way allows decision-making to evolve independently and to be reused in multiple contexts. An event-based system needs a similar capability to determine an appropriate action. Generally, however, this decision needs to be made as part of the overall response to some event and, thus, must be tightly coupled to the detection of that response. In addition, it is often more granular, with less coupling to other elements of the system. To describe this, it can be helpful to talk about decision agents rather than decision services.

7.4.2.1.3.2 From Data at Rest to Data in Motion In a decision service, all the data needed for a decision are generally either passed in as parameters or available for access from the service. This implies that the data required to make the decision are mostly "at rest" in that they are already stored somewhere and can be accessed. In an event-based DMS, some of the data may be available in this way, perhaps in an unstructured or NoSQL format, but some are not. These "in motion" data must be processed or acted on before they are saved to a data infrastructure. Data associated with event streams are critical in these systems, so decision agents need to be able to interact with data at rest and data in motion.

7.4.2.1.3.3 Pattern Detection and Business Decisions A decision service is generally invoked as part of executing a business process. As the process works its way through the various steps involved, it invokes the needed decision services. In an event-centric, real-time scenario, however, it may not be obvious what the right next step is or in what order steps should be executed. Event-based DMSs detect patterns in events, in the data in motion, to determine what the next step might need to be. In addition, these patterns become data that must be consumed in the decision-making because the specifics of that pattern are often relevant to the action selected.

7.4.2.1.3.4 Decisions Shared Across Processes and Events Many organizations need a mix of process-based and event-based systems. They have standard processes that work through a set of steps in a given order. They have situations that must be detected, so they can be responded to. And they have processes that generally work through a certain set of steps but that occasionally need to detect exceptions or abnormal conditions and respond differently in those cases. In this real-world environment, it is important that business decision-making—who is a good customer, what's the best offer for this customer, how suspicious is this fraud network—can be shared between process-based and event-based scenarios and is not embedded in either the process management or event correlation environment.

7.4.2.1.4 Use Cases for Event-Based Decision Management Systems

There are many situations where organizations must respond in real time and many use cases for event-based DMSs. There are four, however, of particular note:

- Real-time marketing,
- Proactive customer support,
- Fraud detection,
- IoT.

In all four, there is value to a real-time response, value to situational awareness, and a need for increasing sophistication in the response selected (Taylor, 2014).

7.4.2.1.4.1 Real-Time Marketing Organizations are trying to ensure that their marketing is relevant, situational, and targeted. As a result, they are dividing customers and prospects into increasingly smaller segments using analytics and then focusing in to target every prospect individually rather than relying on blanket campaigns that send the same offer to everyone. These may be up-sells, cross-sells, coupons, or retention offers. They can include geospatial precision and temporal relevance and be delivered in every channel from mobile and text to chat and call centers.

Some organizations are evolving their marketing and support systems to a next-best action approach that coordinates all possible actions and selects the one most likely to move a customer conversation along and build long-term value. The list of possible actions can include an attempt to sell a complementary product, encourage use of a service a customer already has, recommend a product fix or FAQ answer, or ask customers for clarification of their circumstances.

Many organizations are personalizing their interactions with customers and prospects using everything they know about a prospective customer, making a micro decision specific to that prospect. As customers are increasingly mobile and social, and as more of an organization becomes instrumented, the challenge is to integrate long-term trends and assessments of customers with more immediate, event-based information about them to maximize the efficacy of the offers being made.

Event-based DMSs can focus marketing efforts to make 1:1 marketing decisions in real time so that relevant and appropriate marketing responses, imagery, and content can be delivered in every channel.

7.4.2.1.4.2 Proactive Customer Support Retaining, developing, and supporting customers have always been focus areas for organizations. For large organizations and for those whose clients interact with them wholly or largely electronically, servicing and developing customers must be done by systems. The advice given to a customer and the willingness of an organization to invest in helping each customer vary, depending on the value of that customer. For these and a myriad other reasons, these systems must make decisions about how and when to help.

The same move to real time that has impacted marketing is impacting customer satisfaction. Customers working online or using a mobile application late at night need support immediately. The challenge is proactively identifying that they need this support—if an organization has to wait until customers realize they are in trouble, it may be more difficult and expensive to "fix" the problem. Detecting that a user's behavior is suggestive of someone who needs support and correlating patterns in the stream of events generated by the behavior allow proactive, automated support to be offered or a technician to be dispatched. This requires event-based DMSs.

7.4.2.1.4.3 Fraud Detection Many organizations suffer losses from fraud and abuse. These range from fraudulent claims for services that were never performed, to applications for credit from people who do not exist, to orders that include bribes and illegal payments. In every case, an organization must decide whether to accept the transaction as valid, reject it, or investigate it for fraud. These decisions are high volume and must increasingly be made in real time. Event-based DMSs can determine that a transaction or application is fraudulent, that someone is not who he or she says he or she is (identity fraud), that a supplier or provider is fraudulent, or that a transaction involves a fraud network, with multiple people collaborating so each piece looks valid, but the overall effect is fraud.

As those committing fraud have become more sophisticated, and as more fraud can be conducted electronically, it has become essential for organizations to be context-aware when assessing fraud risk. When organized gangs are conducting electronic fraud, the time between the start of an attempt and a loss can be very short. This puts enormous pressure on the fraud detection system to be both real time and able to handle rapidly arriving and changing data as part of its assessment. Add in the need to assess groupings and locations to spot abnormalities, and the value of event-based DMSs for detecting fraud becomes clear.

7.4.2.1.4.4 Internet of Things As more and more things can be identified both in the real world and in a matching virtual world, it becomes useful to connect them in an Internet-like structure. This begins by attaching RFID tags or other sensors, embedding Java on physical things, and connecting them, often wirelessly, to record where they are and what they are doing. The real world, or at least the bits with sensors on, can be tracked and a virtual representation manipulated using computers. Actuators and controllable devices increasingly allow these things to be manipulated in turn. More of these devices now embed their own computing power and are able to make their own decisions, often while collaborating with nearby objects, instead of communicating with some central network.

When things are connected using smart sensors on devices, this creates a real-time flow of information about what is going on (or not going on), and DMSs are required to determine the appropriate actions to take. Such systems make decisions about what to do (based on inputs, locations of things, etc.) and tell the things to take specific actions. This has begun with smart devices (such as smartphones in the hands of mobile customers) but will become increasingly common with physical equipment in factories and the outside world. The need to respond to this ecosystem of machines in real time with appropriate instructions will require a new generation of event-based DMSs.

7.4.3 Real-Time Big Data Analytics Applications

Real-time applications differ from regular applications in one major attribute. Real-time applications rely on instantaneous input and fast analysis to arrive at a decision or action within a short and very specific timeline. In many cases, if a decision cannot be made within that timeline, it becomes useless. As a result, it is important to make all data necessary for such decision available in a timely fashion, and the analysis must be done in a fast and reliable way. Real-time big data analytics applications include the following (Mohamed and Al-Jaroodi, 2014):

- Intelligent transportation,
- Financial market trading and surveillance,

- Crowd control,
- Military decision-making,
- Large-scale emergency response,
- Smart grid,
- Early warning for natural disasters.

7.4.4 Real-Time Big Data Analytics Challenges

There are a number of challenges facing the design, implementation, and operation of real-time big data applications. Most regular big data applications are implemented using an open-loop approach. In this approach, big data for a specific domain are analyzed to obtain new information and knowledge that can be used to enhance the operations or profitability of that domain. The time needed to analyze the big data and make decisions is generally very long and could significantly reduce the benefits and effectiveness of the applications. Unlike regular big data applications, real-time applications must initiate fast actions that are usually bounded by specific time frames dictated by the targeted domain. Real-time big data applications are commonly implemented using a closed-loop approach in which actions are based on the current and previous situations. To understand the challenges of designing, implementing, and operating real-time big data applications, we need to understand their common action steps as these can significantly impact the action completion times. Action times in real-time big data applications can be divided into five steps, as explained in the following subsections (Mohamed and Al-Jaroodi, 2014).

7.4.4.1 Real-Time Event Transfer

All current distributed application events should be transferred in real time to a place where they can be processed. These events can be transferred from their distributed sources as raw events or as filtered or aggregated events. All generated current row, filtered, and aggregated events can be transferred to a centralized processing point or to distributed intermediate processing points for preprocessing or for further filtering and aggregation before being transferred to the main decision-making unit.

The centralized approach is good if the current generated events are not huge and there are no limitations on the network resources used to transfer these events. The distributed approach is good if there are huge events such that it is impossible to transfer all the generated events to a single location. Filtering and aggregation will become important in this case. This can be done at the event sources and the intermediate points using an open-loop or a closed-loop approach. In the open-loop approach, filtering and aggregation policies are predefined, while in the closed-loop approach, filtering and aggregation policies are interactively defined based on the current events and decisions, current system and network resources, or external business policies. In both approaches, event filtering and aggregation should be done to reduce network traffic and processing time and without negatively affecting the accuracy and optimality of the decision-making in the real-time big data application (Mohamed and Al-Jaroodi, 2014).

7.4.4.2 Real-Time Situation Discovery

This step is designed to detect in real time the business or operational situations and exceptions in the current events, for instance, a significant drop in a stock price for an important company in a financial market trading application, an unexpected huge volume of traffic in a certain important area in an intelligent transportation application, or sudden high power consumption in a certain part of the city in a smart grid application. The detection process can be based on business or operational rules. These rules can be predefined, set mainly by the application developers, or interactive, defined by other modules in the real-time big data application, such as analytical or decision-making modules that can alter, drop, or add business and organizational rules to change the situation discovery process. These rules should be defined or organized to reduce the time needed to detect business or operational situations and exceptions. Real-time situation discovery processes can define and alter the policies used to filter and aggregate event transfer. In addition, they can activate analytical or decision-making processes to respond to detected business or organizational situations (Mohamed and Al-Jaroodi, 2014).

7.4.4.3 Real-Time Analytics

This involves invocations of real-time analytical services to determine the root causes of business and operational situations and exceptions. Real-time analytical processing may involve single or multiple integrated analytical services. These real-time analytical services should predict the performance and assessment of the risks involved in changing the operations and business environment. They rely on real-time information passed from the situation discovery process and other off-line stored big data, such as maps, previous transactions, situations, and decisions. Dealing with stored big data is usually challenging, but they can be abstracted into smaller datasets to significantly enhance the execution time of analytical processing. Providing abstracted information to enhance processing time can be very challenging, however. Real-time analytical services need to deploy fast algorithms that provide alternative options within a bounded time. These options can be optimal or semioptimal because of the limited time and resources available. As this part could be very processing intensive, high-performance platforms are usually needed. These platforms can be equipped with special hardware to execute special algorithms or with cluster computing systems (Mohamed and Al-Jaroodi, 2014).

7.4.4.4 Real-Time Decision-Making

Based on the results of the real-time analytics, it is possible to select the best option for improving the present business operations or profitability and determine the most appropriate actions to respond to the business or operational environment. Various configurations of decision-making can be created by defining business or operational rules developed by domain experts and derived from strategic decisions. These rules help to interactively and intelligently respond to evolving business and operational situations. A decision process checks for predefined business and operational rules to take specific business or operational actions based on the options provided by the real-time analytics modules. It can also notify responsible individuals or trigger other processes to conduct further actions. The biggest challenge is defining business and operational rules that facilitate correct and timely decision-making (Mohamed and Al-Jaroodi, 2014).

7.4.4.5 Real-Time Responses

This involves initiating, executing, and monitoring an action defined by the real-time decision-making process. For example, a company decides to buy a specific stock quantity; the real-time response process will initiate a stock buying transaction, send the transaction request, and monitor its execution. In another example, in wild fires, if the real-time big data application determines a prospective fire expansion pattern, fire fighters and all relevant resources need to be informed and actions taken to contain the expansion area before the fire spreads too far. The process usually needs to interact with various different distributed systems to perform its tasks. The interactions should be reliable and achieved in real time (Mohamed and Al-Jaroodi, 2014).

7.4.5 Big Data Techniques and Technologies

A wide variety of techniques and technologies has been developed and adapted to aggregate, manipulate, analyze, and visualize big data. These techniques and technologies draw from several fields, including statistics, computer science, applied mathematics, and economics. This means an organization that intends to derive value from big data has to adopt a flexible, multidisciplinary approach. Some techniques and technologies were developed in a world with access to far smaller volumes and variety in data but have been successfully adapted to be applicable to very large sets of more diverse data. Others have been developed more recently, specifically to capture value from big data. Some were developed by academics and others by companies, especially those with online business models predicated on analyzing big data.

In this section, we note some of the main techniques and technologies to harness big data. These are not comprehensive—the story of big data is still being written; new methods and tools continue to be developed to solve new problems (Manyika et al., 2011).

7.4.5.1 Techniques for Analyzing Big Data

In this section, we list some techniques used to analyze big data across a range of industries. Not all strictly require the use of big data—some can be applied to smaller datasets (e.g., A/B testing, regression analysis). However, all can be applied to big data, and, in general, larger and more diverse datasets can be used to generate more numerous and insightful results than smaller, less diverse ones (Manyika et al., 2011).

- *A/B testing.* A technique in which a control group is compared with a variety of test groups to determine what treatments (i.e., changes) will improve a given objective variable, e.g., marketing response rate. This technique is also known as split testing or bucket testing. An example is determining what copy text, layouts, images, or colors will improve conversion rates on an e-commerce website. Big data enable huge numbers of tests to be executed and analyzed, ensuring that groups are of sufficient size to detect meaningful (i.e., statistically significant) differences between the control and treatment groups. When more than one variable is simultaneously manipulated in the treatment, the multivariate generalization of this technique, which applies statistical modeling, is often called "A/B/N" testing.
- *Association rule learning.* A set of techniques for discovering interesting relationships, i.e., "association rules," among variables in large databases (Agrawal et al., 1993; Hajek et al., 1966). These techniques consist of a variety of algorithms to generate and test possible rules. One application is market basket analysis, in which a retailer can determine which products are frequently bought together and use this information for marketing (a commonly cited example is the discovery that many supermarket shoppers who buy diapers also tend to buy beer). Association rule learning is also used for DM (Manyika et al., 2011).
- *Classification.* A set of techniques to identify the categories in which new data points belong, based on a training set containing data points that have already been categorized. One application is the prediction of segment-specific customer behavior (e.g., buying decisions, consumption rate) where there is a clear hypothesis or objective outcome. The technique is often described as supervised learning because of the existence of a training set; it contrasts with cluster analysis, a type of unsupervised learning. It is used for DM.
- *Cluster analysis.* A statistical method for classifying objects that splits a diverse group into smaller groups of similar objects, whose characteristics of similarity are not known in advance. An example of cluster analysis is segmenting consumers into self-similar groups for targeted marketing. This is a type of unsupervised learning because training data are not used; it contrasts with classification, a type of supervised learning. It too is used for DM (Manyika et al., 2011).

- *Crowdsourcing.* A technique for collecting data submitted by a large group of people or a community (i.e., the "crowd") through an open call, usually through networked media such as the Web (Howe, 2006). This is a type of mass collaboration and an instance of using Web 2.0 (Chui et al., 2009).
- *Data fusion and data integration.* A set of techniques that integrate and analyze data from multiple sources to develop insights in ways that are more efficient and potentially more accurate than if they were developed by analyzing a single source of data. Signal processing techniques can be used to implement some types of data fusion. An example of an application is sensor data from IoT being combined to develop an integrated perspective on the performance of a complex distributed system such as an oil refinery. Data from social media, analyzed by natural language processing (NLP), can be combined with real-time sales data to determine what effect a marketing campaign is having on customer sentiment and purchasing behavior.
- *Data mining.* A set of techniques to extract patterns from large datasets by combining methods from statistics and ML with database management. These techniques include association rule learning, cluster analysis, classification, and regression. Applications include mining customer data to determine segments most likely to respond to an offer, mining human resources data to identify characteristics of most successful employees, or market basket analysis to model the purchase behavior of customers.
- *Ensemble learning.* The use of multiple predictive models (each developed using statistics and/or ML) to obtain better predictive performance could be obtained from any of the constituent models. This is a type of supervised learning.
- *Genetic algorithms.* A technique used for optimization, inspired by the process of natural evolution or "survival of the fittest." In this technique, potential solutions are encoded as "chromosomes" that can combine and mutate. These individual chromosomes are selected for survival within a modeled "environment" that determines the fitness or performance of each individual in the population. Often described as a type of "evolutionary algorithm," these algorithms are well suited for solving nonlinear problems. Examples of applications include improving job scheduling in manufacturing and optimizing the performance of an investment portfolio.
- *Machine learning.* A subspecialty of computer science (within a field historically called "artificial intelligence") concerned with the design and development of algorithms that allow computers to evolve behaviors based on empirical data. A major focus of ML research is to automatically learn to recognize complex patterns and make intelligent decisions based on data. NLP is an example of ML.
- *Natural language processing.* A set of techniques from a subspecialty of computer science (within a field historically called "artificial intelligence") and linguistics that uses computer algorithms to analyze human (natural)

language. Many NLP techniques are types of ML. One application is using sentiment analysis on social media to determine how prospective customers are reacting to a branding campaign.

- *Neural networks.* Computational models, inspired by the structure and workings of biological NNs (i.e., the cells and connections within a brain), that find patterns in data. NNs are well suited for finding nonlinear patterns. They can be used for pattern recognition and optimization. Some NN applications involve supervised learning, while others involve unsupervised learning. Examples of applications include identifying high-value customers who are at risk of leaving a particular company or identifying fraudulent insurance claims.
- *Network analysis.* A set of techniques used to characterize relationships among discrete nodes in a graph or a network. In social network analysis, connections between individuals in a community or organization are analyzed, e.g., how information travels or who has the most influence over whom. Examples of applications include identifying key opinion leaders to target for marketing and identifying bottlenecks in enterprise information flows.
- *Optimization.* A portfolio of numerical techniques used to redesign complex systems and processes to improve their performance according to one or more objective measures (e.g., cost, speed, or reliability). Examples of applications include improving operational processes, such as scheduling, routing, and floor layout, and making strategic decisions, such as product range strategy, linked investment analysis, and R&D portfolio strategy. Genetic algorithms are an example of an optimization technique.
- *Pattern recognition.* A set of ML techniques that assigns some sort of output value (or label) to a given input value (or instance) according to a specific algorithm. Classification techniques are an example.
- *Predictive modeling.* A set of techniques in which a mathematical model is created or chosen to best predict the probability of an outcome. An example of an application in customer relationship management is the use of predictive models to estimate the likelihood that a customer will "churn" (i.e., change providers) or the likelihood that a customer can be cross-sold another product. Regression is one example of the many predictive modeling techniques.
- *Regression.* A set of statistical techniques to determine how the value of the dependent variable changes when one or more independent variable is modified. Regression is often used for forecasting or prediction. Examples of applications include forecasting sales volumes based on various market and economic variables or determining what measurable manufacturing parameters most influence customer satisfaction. It is also used for DM.
- *Sentiment analysis.* Application of NLP and other analytic techniques to identify and extract subjective information from source text material. Key aspects include identifying the feature, aspect, or product about which a sentiment is being expressed and determining the type, "polarity"

(i.e., positive, negative, or neutral) and the degree and strength of the sentiment. Examples of applications include companies applying sentiment analysis to analyze social media (e.g., blogs, microblogs, and social networks) to determine how different customer segments and stakeholders are reacting to their products and actions.

- *Signal processing*. A set of techniques from electrical engineering and applied mathematics originally developed to analyze discrete and continuous signals, i.e., representations of analog physical quantities (even if represented digitally) such as radio signals, sounds, and images. The category includes techniques from signal detection theory, which quantifies the ability to discern between signal and noise. Sample applications include modeling for time-series analysis or implementing data fusion to determine a more precise reading by combining data from a set of less precise data sources (i.e., extracting the signal from the noise).
- *Spatial analysis*. A set of techniques, some applied from statistics, which analyze the topological, geometric, or geographic properties encoded in a dataset. Data often come from GISs that capture location information, e.g., addresses or latitude/longitude coordinates. Examples of applications include the incorporation of spatial data into spatial regressions (e.g., how is consumer willingness to purchase a product correlated with location?) or simulations (e.g., how would a manufacturing supply chain network perform with sites in different locations?).
- *Statistics*. The science of the collection, organization, and interpretation of data, including the design of surveys and experiments. Statistical techniques are often used to make judgments about which relationships between variables could have occurred by chance (the "null hypothesis") and which ones more likely result from some kind of underlying causal relationship (i.e., "statistically significant"). Statistical techniques are also used to reduce the likelihood of Type I errors ("false positives") and Type II errors ("false negatives"). An example of an application is A/B testing to determine what types of marketing material will most increase revenue (Manyika et al., 2011).
- *Supervised learning*. A set of ML techniques that infer a function or relationship from a set of training data. Examples include classification and support vector machines (Cortes and Vapnik, 1995). This is different from unsupervised learning (Manyika et al., 2011).
- *Simulation*. Modeling the behavior of complex systems, often used for forecasting, predicting, and scenario planning. Monte Carlo simulations, for example, are a class of algorithms that rely on repeated random sampling, i.e., running thousands of simulations, each based on different assumptions. The result is a histogram that gives a probability distribution of outcomes. One application is assessing the likelihood of meeting financial targets given uncertainties about the success of various initiatives.

- *Time-series analysis.* A set of techniques from both statistics and signal processing used to analyze sequences of data points, representing values at successive times, and extract meaningful characteristics from the data. Examples of time-series analysis include the hourly value of a stock market index or the number of patients diagnosed with a given condition every day. Time-series forecasting is the use of a model to predict future values of a time series based on known past values of the same or other series. Some of these techniques, e.g., structural modeling, decompose a series into trend, seasonal, and residual components, which can be useful for identifying cyclical patterns in the data. Examples of applications include forecasting sales figures or predicting the number of people who will be diagnosed with an infectious disease.
- *Unsupervised learning.* A set of ML techniques that finds hidden structures in unlabeled data. Cluster analysis is an example of unsupervised learning (in contrast to supervised learning).
- *Visualization.* Techniques used for creating images, diagrams, or animations to communicate, understand, and improve the results of big data analyses (Manyika et al., 2011).

7.4.5.2 Big Data Technologies

A growing number of technologies are available to aggregate, manipulate, manage, and analyze big data. In this section, we give some of them, but as noted above, the list is not exhaustive, as new technologies continue to be developed (Manyika et al., 2011).

- *Big table.* Proprietary distributed database system built on the Google File System. It is the inspiration for HBase.
- *Business intelligence.* A type of application software designed to report, analyze, and present data. BI tools are often used to read data that have been previously stored in a data warehouse or data mart. BI tools can also be used to create standard reports that are generated on a periodic basis or to display information on real-time management dashboards, i.e., integrated displays of metrics that measure the performance of a system.
- *Cassandra.* An open-source (free) database management system designed to handle huge amounts of data on a distributed system. This system was originally developed at Facebook and is now managed as a project of the Apache Software Foundation.
- *Cloud computing.* A computing paradigm in which highly scalable computing resources, often configured as a distributed system, are provided as a service through a network.
- *Data mart.* Subset of a data warehouse, used to provide data to users, usually through BI tools.

- *Data warehouse*. Specialized database optimized for reporting, often used for storing large amounts of structured data. Data are uploaded using extract, transform, and load (ETL) tools from operational data stores, and reports are often generated using BI tools.
- *Distributed system*. Multiple computers, communicating through a network, used to solve a common computational problem. The problem is divided into multiple tasks, each of which is solved by one or more computers working in parallel. Benefits of distributed systems include higher performance at a lower cost (i.e., because a cluster of lower-end computers can be less expensive than a single higher-end computer), higher reliability (i.e., because of a lack of a single point of failure), and more scalability (i.e., because increasing the power of a distributed system can be accomplished by simply adding more nodes rather than completely replacing a central computer).
- *Dynamo*. Proprietary distributed data–storage system developed by Amazon.
- *Extract, Transform, and Load*. Software tools used to extract data from outside sources, transform them to fit operational needs, and load them into a database or data warehouse.
- *Google File System*. Proprietary distributed file system developed by Google, part of the inspiration for Hadoop (Ghemawat et al., 2003).
- *Hadoop*. An open-source (free) software framework for processing huge datasets on certain kinds of problems on a distributed system. Its development was inspired by Google's MapReduce and Google File System. It was originally developed at Yahoo! and is now managed as a project of the Apache Software Foundation.
- *HBase*. An open-source (free), distributed, nonrelational database modeled on Google's Big Table. It was originally developed by Powerset and is now managed as a project of the Apache Software Foundation as part of the Hadoop.
- *MapReduce*. A software framework introduced by Google for processing huge datasets on certain kinds of problems on a distributed system (Dean and Ghemawat, 2004). It is also implemented in Hadoop (Manyika et al., 2011).
- *Mashup*. An application that uses and combines data presentation or functionality from two or more sources to create new services. These applications are often made available on the Web and frequently use data accessed through open application programming interfaces or from open data sources.
- *Metadata*. Data that describe the content and context of data files, e.g., means of creation, purpose, time and date of creation, and author.
- *Nonrelational database*. A database that does not store data in tables (rows and columns), in contrast to a relational database.
- *R*. An open-source (free) programming language and software environment for statistical computing and graphics. The R language has become a de

facto standard among statisticians for developing statistical software and is widely used for statistical software development and data analysis. R is part of the GNU Project, a collaboration that supports open-source projects.

- *Relational database.* A database made up of a collection of tables (relations); i.e., data are stored in rows and columns. Relational database management systems store a type of structured data. SQL is the most widely used language for managing relational databases.
- *Semistructured data.* Data that do not conform to fixed fields but contain tags and other markers to separate data elements. Examples of semistructured data include XML or HTML-tagged text. These data contrast with structured data and unstructured data.
- *SQL.* A computer language designed for managing data in relational databases, originally an acronym for structured query language. It includes the ability to insert, query, update, and delete data, as well as to manage data schema (database structures) and control access to data in the database.
- *Stream processing.* Technologies designed to process large real-time streams of event data. Stream processing enables applications such as algorithmic trading in financial services, RFID event processing applications, fraud detection, process monitoring, and location-based services in telecommunications. It is also known as event stream processing.
- *Structured data.* Data residing in fixed fields. Examples include relational databases or data in spreadsheets. They contrast with semistructured data and unstructured data.
- *Unstructured data.* Data that do not reside in fixed fields. Examples include free-form text (e.g., books, articles, body of e-mail messages), untagged audio, and image and video data. They contrast with structured data and semistructured data.
- *Visualization.* Technologies used for creating images, diagrams, or animations to communicate messages often used to synthesize the results of big data analyses (Manyika et al., 2011).

7.5 eMAINTENANCE AND MAINTENANCE 4.0: IMPACT OF TECHNOLOGY ON OPERATION AND MAINTENANCE KEY PERFORMANCE INDICATORS

Production and transportation systems, along with other systems such as health care or facilities, are extremely complex in both their technology and their operation and involve a wide range of human actors, organizations, and technical solutions. To operate and control such complex environments, intelligent computerized systems are often a good option, for example, computerized traffic control systems to coordinate transportation, automation systems like

SCADA to monitor and control production systems, and, obviously, advanced monitoring and diagnostic systems to perform health assessments of all systems.

Today, safety is a must for operation and maintenance but is difficult to achieve in the traditional way; computerized solutions seem the only feasible way to deal with the safety of complex systems, interacting among them to balance the growth in technical complexity with stable and acceptable dependability indexes.

eMaintenance and Maintenance 4.0 refer to the fourth generation of maintenance activity, related to Industry 4.0 (described at the beginning of the chapter) and enabled by smart systems and Internet-based solutions. Two characteristic features are computerization using CPS and intelligence based on the concept of IoT. In the scenario of smarter assets, eMaintenance and Maintenance 4.0 will provide maintenance services such as diagnosis and prognosis, predicting failures, and optimizing further decisions (Galar, 2015).

7.5.1 Maintenance 4.0

Maintenance 4.0 uses advanced technologies for predictive analytics and provides decisions based on feasibility. The list of trends for the maintenance revolution has been compiled by Process Worldwide (http://www.process-worldwide.com/); it covers the width and depth of Maintenance 4.0 and is given in the following subsections.

7.5.1.1 eMaintenance

eMaintenance is a term covering everything from simple Internet-based administration systems to advanced diagnostics and maintenance decision support. Muller et al. (2008) proposes the following definition:

> *Maintenance support … includes the resources, services and management necessary to enable proactive decision process execution. eMaintenance includes various e-technologies (e.g. ICT (Internet and Communication Technologies), Web-based, and wireless) as well as eMaintenance activities (e.g. e-Monitoring, e-Diagnosis, and e-Prognosis), in order to support production and business.*
>
> Cited in Galar et al. (2012)

According to Candell et al. (2009), eMaintenance monitors, collects, records, and distributes real-time data about a system's health. eMaintenance was introduced before Maintenance 4.0 and is applied in a broader sense. In practice, the terms overlap partly or completely (Koskinen et al., 2016). More details of eMaintenance are discussed in Section 7.5.2.

7.5.1.2 Radio-Frequency Identification

RFID is a technology for retrieving data on an object from a distance (see Section 7.3.4.2.3.1 and the following). More specifically, RFID refers to systems that read or write data to radio-frequency (RF) tags present in an RF field projected from RF reading/writing equipment. Data may be contained in one or more bits for the purpose of providing identification and other information relevant to the object to which the tag is attached. The RFID technology is useful to track such objects as vehicles or spare parts; it makes the planning of maintenance activities more efficient by providing valuable and timely information on the state of an object (Koskinen et al., 2016).

In areas ranging from inventory management to security applications, RFID is becoming a buzzword in many organizations. The technology appears to have weathered the unrealistic expectations it attracted in its earliest days and is beginning to prove its practicality in some realms—in paint shops and high-temperature manufacturing environments, for example, or for tracking individual assemblies on production lines. RFID was initially considered to have a few uses in some carefully selected areas. Now, however, it is gaining a growing reputation in many industries as a reliable and cost-effective means of identification across a wide range of applications (Electronics Enabled Products, 2006).

7.5.1.2.1 Evolution of Radio-Frequency Identification

The origins of RF-assisted tracking technology go back to the 1940s when it was pioneered in World War II to identify enemy aircraft. The technology was soon extended to include cattle and cargo tracking using radioactivity. Currently, RFID technology finds extensive applications across a broad spectrum of markets, including automated vehicle—identification systems, where it lends itself perfectly to the tracking of moving objects. Of late, the technology has become indispensable across a wide array of applications involving automated data collection, tracking, and analysis. Most recently, it has become the likely successor to bar-code technology in SCM.

The core advantage of RFID equipment is the wireless technology, which enables contactless reading without requiring a line of sight, making it suitable for use in harsh as well as ordinary environments. However, RFID tags remain expensive, and further technological refinement is required to lower their cost (Electronics Enabled Products, 2006).

7.5.1.2.2 Operational Aspects of Radio-Frequency Identification Systems

An RFID system makes use of transceivers and electronic transponders (also called tags) for wireless data collection. In this process, when tags come within range of the transceiver, the information is decoded by the transceiver. A basic RFID system generally comprises a transponder (tag) electronically programmed with product information relevant to the item to which it is affixed; a transceiver (with decoder); and a coil or antenna.

The function of the antenna is to act as a channel between the tag and the transceiver, which has ultimate control of the system's data acquisition and communication functions. The antenna activates the RF tag by emitting RF waves to read and write information from and to it. Antennas are embedded into a structure to receive tag data from persons, vehicles, or moving objects. An antenna usually comes with a transceiver and decoder to double up as reader and can be programmed for handheld operation or for operation from a fixed location. Depending on the power output and frequency, the reader emits RF waves over a distance ranging from one inch to 100 feet or more. The information stored in these tags is detected by the reader's activation signal when the RFID tag traverses an electromagnetic zone. The encoded data stored on the integrated circuit of the tag is decoded by the reader and fed into the host computer for processing.

The tag stores certain product information in its memory chip and is read by the reader emitting RF waves. The information may vary from simple data, such as serial numbers, to much more complex and critical information. RFID tags can be active or passive. Both types include an antenna and a silicon chip for processing. The difference is that an active tag features an additional component—an onboard power source—whereas a passive tag is powered by energy from the RF wave transmitted by the reader/writer.

Passive tags feature a mechanism that stores energy in a capacitor for a short duration after receiving the RF wave, just long enough to process and transmit a return signal. On entering the RF signal field, the passive tag uses a technique called "backscatter" to convert the signal into two parts, one for powering the chip and the other containing the encoded information processed by the chip. A return signal is then fed into the reader. Given the lack of active onboard power components, there is a time lag in reading passive tags. While low-frequency chips feature a read distance of about 1 m, ultrahigh-frequency chips are capable of reading signals at distances of up to 4 m or more.

Active tags feature an onboard power source and come in two variants: battery-assisted and "true" active tags. Battery-assisted tags are simply passive tags with the addition of a built-in power source. This type of tag also uses backscatter technology to reflect the signal back to the reader, while the battery assists in powering only the chip, enabling more power to be reflected back to the reader, thereby enhancing the read distance between the tag and the reader.

Active RFID tags are usually read/write and, hence, have the capability of rewriting or modifying the data. The tag is powered by an internal battery source, and memory capacity varies by application, with certain systems having up to 1 MB memory. As power is supplied by a battery, active tags generally have a longer reading range. However, active tags are larger, more expensive, and have a limited operational life.

Because passive RFID tags obtain the power required for operation from the reader, they can be extremely light and have significantly prolonged lifetimes, but they have shorter read ranges and require powerful readers. Furthermore, passive tags are mostly read-only and are programmed with unique data that cannot be modified. Conversely, because "true" active tags incorporate a power source, they are capable of transmitting signals across distances of about 40 m. Naturally, this capability makes active tags more attractive in the market (Electronics Enabled Products, 2006).

7.5.1.2.3 Radio-Frequency Identification Applications

Potential applications for RFID can be identified in many sectors of industry, commerce, and services where data are collected. The attributes of RFID are complementary to other data-capture technologies and, thus, are able to satisfy particular application requirements that cannot be adequately accommodated by alternative technologies. The most common areas of application for RFID include the following (Electronics Enabled Products, 2006):

- Manufacturing and supply chain
- Transportation and logistics
- Security and access control
- Military
- Inventory management

7.5.1.3 Virtual or Augmented Reality

Virtual or augmented reality is a term describing real-time applications that mix information from the physical world with information from the application. Augmented reality (AR) could be applied for simulation, education, and training purposes, such as in Manca et al. (2013), or for remote expert guidance of a person through complicated instructions. The cyber-physical visualization tool proposed by Penna et al. (2014) could be used for the latter purpose.

7.5.1.3.1 Virtual Reality

Virtual reality (VR) has been discussed by the most people the longest. VR involves totally immersing ourselves into a virtual world where everything we see is computer generated. VR has been around for several decades, but the quality of the experience has dramatically improved in recent years. Although earlier devices such as Virtual Boy and CyberMaxx attempted to provide pseudo-virtual worlds, today's devices, such as Oculus Rift and HTC Vive, provide a much more engaging VR experience. When we put on these VR head-mounted displays, the outside world is completely blocked, and we are looking into a fully computer-generated experience.

Many devices now available in the VR space offer a low-cost VR experience. A cell phone is required to drive the actual experience. These devices range from a Google cardboard solution for a couple of dollars to more expensive models. Many can be bought for 10 to 20 dollars, while others may be a few hundred dollars. Like the Oculus and Vive, these devices block the view of the real world and provide a virtual experience on computer screens (Jones, 2016).

7.5.1.3.2 Augmented Reality

AR has been getting more attention in the last few years. AR is as the name suggests: reality is still there, but it is augmented or enhanced in some way. Two of the most discussed AR systems today are cars and mobile phones. Another is Google Glass.

A number of cars have AR built into their systems. For example, GMC includes heads-up displays that shine onto the dashboard, providing the driver with additional information to augment the driving experience, for example, the speed of the vehicle. This virtual display of information overlays reality and supplies additional information.

Many mobile phone applications also provide AR. For example, city applications allow users to point their phones down a city street; prompts on the screen will show them what the various buildings are.

Google Glass is another AR device. Google Glass can be used to obtain additional information on the surrounding environment. This can be the type of information mentioned above for mobile phone devices, or it could be information on people users are interacting with (Jones, 2016).

7.5.1.3.3 Virtual Reality Versus Augmented Reality

VR technology facilitates the creation of real-life simulations and creates an immersive experience for any user, making him or her feel as if he or she is actually interacting with his or her digital environment. AR layers digital enhancements to enrich an existing real-life setting by appealing to the senses.

While many companies are enhancing elements of their business with these two technologies, the respective capabilities of AR and VR are helping businesses in different, but important ways (AUGMENT, 2016).

Both technologies:

- Enrich the experience of a user by offering deeper layers of interactions;
- Have the potential to transform how people engage with technology. Entertainment, engineering, or medicine is just a couple of sectors where the two technologies might have a lasting impact.

However, the two stand apart because of the following:

- VR creates a completely new environment, which is entirely computer generated. In contrast, AR enhances experiences through digital means; it offers a new layer of interaction with reality but does not seek to replace it.
- AR offers a limited field of view, while VR is totally immersive.

- With the VR goggles in place, users are essentially disconnected from the outside world. In contrast, an AR user is constantly aware of the physical surroundings, while actively engaged with simulated ones.
- VR typically requires a headed mount such as oculus rift goggles, but AR is far less demanding, requiring only a smartphone or tablet.

To this point, we have barely scratched the surface of what AR and VR can do. BCC Research recently estimated that the global market for VR and AR will reach more than $105 billion by 2020, up from $8 billion in 2015 (Puiu, 2016).

7.5.1.4 Visualization

Visualization systems such as the one intended for building maintenance described in Khosrowshahi et al. (2014) helps the user to quickly identify problem areas and assess the state of a machine by transforming big data into easy comprehensible visualizations. Different decision support and expert systems for assessing the health of assets and propose appropriate maintenance actions have existed for quite a long time, but their applicability has often been connected to management level or centralized use (Koskinen et al., 2016).

7.5.1.5 Knowledge Sharing and Networking and Assistant Systems

Knowledge sharing and networking let all people involved in maintenance share their experiences and knowledge; assistant systems allow easy interpretation of vast amounts of data for direct use, for instance, helping maintenance technicians in their daily work. Knowledge sharing using blogs, wikis, and Web portals is a development area (Koskinen et al., 2016).

7.5.2 Challenges of Maintenance 4.0

The pressure is increasing. Everything must go faster and must be optimized, and the costs must be lowered, with the smallest possible expenditures and most minimal hindrances to investment. Production is increasingly complex, with the limits continually being pushed. Production disruptions have far-reaching consequences. Downtimes as the result of a breakdown, whether human or technical in nature, cost money, time, and reputation.

Optimization approaches for economic maintenance repeatedly play an important role in strategic and operational future-oriented planning. Direct and indirect maintenance costs must be minimized, and the reliability of production, the security of the data and employees, and the output of the facilities must be maximized. The need for efficient technology is higher than ever, but the new and "promising" technologies of Industry 4.0 are increasingly being confronted by uncertainty and resistance. Nobody can say which measures will actually have which effects because there are still no long-term studies. In addition, many people feel overwhelmed by new technologies; many feel they are being controlled by them and prefer to place their trust elsewhere—in more traditional approaches, for example.

We do not need to be clairvoyant to find out if something will break down and/or could break down. In the maintenance of Western industrial companies, this has long since fallen under predictive maintenance. With Maintenance 4.0, predictive maintenance is supposed to be "given a helping hand," so to speak. Better predictions and more effective processes are the long-term goal. In Maintenance 4.0, real-time machine data, CM data, historical data from the control room, and the historical knowledge of disruption events are linked for optimal results, with the goal of being able to provide anticipative maintenance planning. Precise predictions can be made as to when a facility will suffer a disruption. This means being able to implement maintenance work at the right time and in the required scope, thereby increasing the reliability of the facilities and the efficiency of resources (Augmensys GmbH).

7.5.3 eMaintenance

The concept of eMaintenance refers to the integration of ICT with the maintenance strategy and/or plan (Li et al., 2005) to face new needs and find innovative ways of supporting production (e-Manufacturing) or business (e-business) (Yoshikawa, 1995).

A definition of eMaintenance proposed by Muller et al. (2008) takes into consideration the European Standard (EN 13306:2001) for maintenance terminology and understands eMaintenance as a component of the e-Manufacturing concept (Lee, 2003). We propose the following definition:

> *Maintenance support which includes the resources, services and management necessary to enable proactive decision process execution. This support includes e-technologies (i.e. ICT, Web-based, tetherfree, Wireless, Infotronics technologies) but also, eMaintenance activities (operations or processes) such as e-Monitoring, e-Diagnosis, e-Prognosis…*
>
> Crespo-Marquez and Iung (2008)

7.5.3.1 eMaintenance: Pervasive Computing in Maintenance

The general opinion among asset managers is that the application of IT brings dramatic results in machine reliability and O&M process efficiency, but few operations and management (O&M) managers can show or calculate the benefits of such applications. Technology providers are trying to develop more advanced tools, while maintenance departments struggle with the daily problems of implementing, integrating, and operating such systems.

Today there are two main systems deployed in maintenance departments, CMMS are the core of traditional maintenance record-keeping practices and CM systems are capable of directly monitoring asset components parameters.

However, attempts to link observed CMMS events to CM sensor measurements have been fairly limited in both approach and scalability. Moreover, information from SCADA systems, mainly used in operation for supervisory purposes, is seldom fused with either CMMS or CM (Sten-Erik et al., 2013).

During the last decade, global competition and advancements in ICT have forced production and process industries to transform and improve. The business scenario is using e-BI to perform transactions, with a focus on customers' needs for enhanced value and improvement in AM. Such business requirements compel organizations to minimize production and service downtime by reducing degradation in machine performance. Simply stated, we need proactive maintenance strategies to provide optimized and continuous process performance with minimized system breakdowns and maintenance. With the changing systems of the business world in the 21st century, a new era of data e-Services is emerging (Koskinen et al., 2016).

CMMS is primarily designed to facilitate a shift in emphasis from reactive to preventive maintenance (PM) by allowing the maintenance professionals to set up an automatic PM work order (WO) generation. CMMS can also provide historical information, which is used to adjust the PM system setup over time to minimize unnecessary repairs, while avoiding run-to-failure repairs. PM for a given piece of equipment can be set up on a calendar schedule or a usage schedule based on meter readings. A fully featured CMMS also includes inventory tracking, workforce management, and purchasing in a single package that stresses database integrity to safeguard vital information. The result is optimized equipment uptime, lower maintenance costs, and better overall plant efficiency (Galar et al., 2015).

The CM and SCADA systems are intended to accurately monitor the real-time equipment performance and condition to alert operators and maintenance professionals of any changes in performance or behavior trends. A CM and SCADA package can track a variety of measurements, and the very best CMs are expert systems able to analyze measurements such as vibration and diagnose machine faults. Such expert system analysis will put maintenance procedures on hold until absolutely necessary, thus extracting maximum equipment uptime. In addition, the best expert systems offer diagnostic fault trending whereby individual machine fault severity can be observed over time. SCADA systems provide information on the performance of the asset and give screenshots of a number of PIs at a time (commonly called now casting). This information, together with a number of alarms and incidences linked to sensor thresholds, have been historically used for process monitoring but never fused with other information.

SCADA, CMMS, and CM systems have benefits that make them indispensable to maintenance and operation improvements. CMMS is a great organizational tool but cannot directly monitor equipment conditions, whereas a CM system excels in monitoring those conditions but is not suited to organizing overall maintenance operations (Sten-Erik et al., 2013). SCADA collects huge amounts of data for supervision of the process but is not holistically integrated into AM. The logical conclusion is to combine these technologies into a seamless system to avoid catastrophic breakdowns and eliminate needless repairs to equipment that is running satisfactorily, thus optimizing the operation (Koskinen et al., 2016).

The awareness and use of eMaintenance has grown during the last few years. The hardware and software have developed rapidly, and it is increasingly possible to access the Internet wirelessly. At the same time, manufacturers are moving toward taking care of the machinery they have produced all the way through its life cycle. For this, they need cost-effective tools, and this is where eMaintenance can help. Accordingly, an increasing number of researchers are focusing on eMaintenance (Jantunen et al., 2011).

eMaintenance advantages can be slotted into the three following categories:

- Improved maintenance types and strategies.
- Improved maintenance support and tools.
- Improved maintenance activities.

7.5.3.2 Improvements in Maintenance Types and Strategies Using eMaintenance

7.5.3.2.1 Potential Improvements

Potential improvements to be achieved using eMaintenance include the following:

7.5.3.2.1.1 Remote Maintenance By leveraging wireless (e.g., Bluetooth) and Internet technologies, users may log in from anywhere and with any kind of device as soon as they get an Internet connection and a browser. Any operator, manager, or expert has the ability to remotely link to a factory's equipment through Internet, allowing him or her to take remote actions, such as setup, control, configuration, diagnosis, debugging/fixing, performance monitoring, and data collection and analysis. Consequently, the personnel retained at the customer's site can be reduced. In addition, there are facilities to diagnose problems when an error occurs.

One of the greatest advantages of eMaintenance is the ability to connect field systems with expertise centers located at distant geographical sites, allowing a remote maintenance decision-making that adds value, trims expenses, and reduces waste. This makes the development of an asset information management network a sound investment (Baldwin, 2004).

In addition, Web enablement of computerized maintenance management systems (e-CMMS) and remote CM or diagnostics avoid the expense of software maintenance, security, and hardware upgrade. Computer science experts can add new features and/or migrations without users noticing (Muller et al., 2008).

Finally, PM can be improved by machine performance monitoring (Muller et al., 2008).

7.5.3.2.1.2 Cooperative/Collaborative Maintenance eMaintenance represents the opportunity to implement an information infrastructure connecting geographically dispersed subsystems and actors (e.g., suppliers with clients and machinery with engineers) on the basis of existing Internet networks. The resulting platform facilitates cooperation between different human actors, different enterprise areas (production, maintenance, purchasing, etc.), and different companies (suppliers, customers, machine manufacturers, etc.).

An eMaintenance platform introduces an unprecedented level of transparency and efficiency (see Fig. 7.38), and it can support business process integration. As a result, there is the chance to radically reduce interfaces between personnel, departments, or even IT systems. The integration of business processes significantly contributes to the acceleration of total processes, to an easier design (lean processes), and to the synchronization of maintenance with production, maximizing process throughput and minimizing downtime costs. In general, this leads to fewer process errors, improved communication processes, shorter feedback cycles, and improved quality (Muller et al., 2008).

In short, eMaintenance facilitates the bidirectional flow of data and information into the decision-making and planning process at all levels. This automates the retrieval of the accurate information required by decision-makers

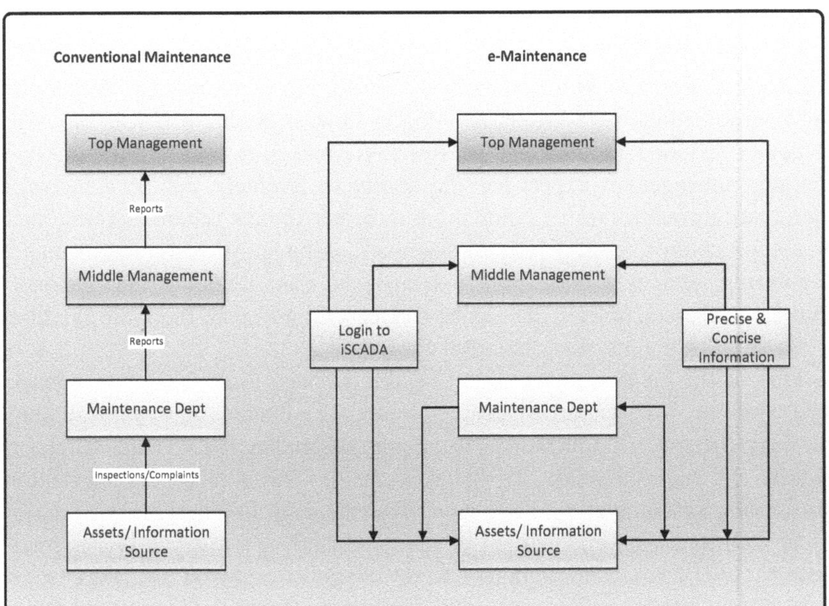

FIGURE 7.38 Implementing eMaintenance. *Adapted from Muller, A., Marquez, A.C., Iung, B., 2008. On the concept of e-maintenance: review and current research. Reliability Engineering and System Safety 93, 1165–1187.*

to determine which maintenance activities to focus resources on, so that return on investment is optimized (Moore and Starr, 2006).

7.5.3.2.1.3 Immediate/Online Maintenance The real-time remote monitoring of equipment status, coupled with programmable alerts, enables the maintenance operator to respond to any situation swiftly and optimally prepare any intervention. In addition, maintenance operators obtain expertise quickly; the feedback reaction in the local loop is accelerated, connecting product, monitoring agent, and maintenance support system. The online guidance based on the results of decision-making and the analysis of asset condition has almost unlimited potential to reduce the complexity of traditional maintenance guidance (Goncharenko and Kimura, 1999).

In this context, potential applications of eMaintenance include the formulation of policies for maintenance scheduling in real time based on up-to-date information of machinery operation history, machine status, anticipated usage, functional dependencies, production flow status, and so on (Muller et al., 2008).

7.5.3.2.1.4 Predictive Maintenance The eMaintenance platform allows any maintenance strategy; the use of a holistic approach combining the tools of predictive maintenance techniques is one of the major advantages of eMaintenance (Lee, 2003).

Potential applications in this area include equipment failure prognosis based on current condition and projected usage or remaining life prediction of components. In fact, eMaintenance provides companies with predictive intelligence tools (such as a watchdog agent) to monitor their assets (equipment, products, process, etc.) through Internet wireless communication systems to prevent unexpected breakdowns. In addition, these systems can monitor an asset's performance using globally networked monitoring systems; this allows companies to focus on degradation monitoring and prognostics rather than fault detection and diagnostics (Iung et al., 2003).

Prognostic and health management systems that can implement these capabilities will be able to reduce the overall life cycle costs of OSs and decrease the operations/maintenance logistics footprint (Roemer et al., 2005).

7.5.3.3 Potential Improvements in Maintenance Support and Tools With eMaintenance

7.5.3.3.1 Potential Improvements

Potential improvements in maintenance support and tools can be summarized as follows:

7.5.3.3.1.1 Fault/Failure Analysis The rapid development in sensor technology, signal processing, ICT, and other technologies related to CM and diagnostics increases the possibility of using data from multiple origins and sources and of different types. In addition, by networking remote plants, eMaintenance provides a multisource knowledge and data environment.

These new capabilities allow the maintenance area to improve understanding of causes of failures and system disturbances, to create better monitoring and signal analysis methods, and to improve materials, design, and production techniques. By using failure analysis, companies can move from failure detection to degradation monitoring (Holmberg et al., 2005).

7.5.3.3.1.2 Maintenance Documentation/Record The eMaintenance platform provides a transparent, seamless, and automated information exchange process to access all the documentation in a unified way, independently of its origin, whether from the equipment manufacturer, integrator, or end user. For example, a task completion form can be filled out by users and dispatched to several listeners (software or humans) registered to receive such events. Fig. 7.39 shows how this type of eMaintenance platform works, using the platform for Proteus (*plateforme reconfigurable pour l'observation, pour les télécommunications et les usages scientifiques*) (Bangemann et al., 2006).

At the device level, goods are checked out from stores against a WO or a location, and the transaction is recorded in real time. The massive data bottlenecks between the plant floor and business systems can be eliminated by converting the raw machine health data, product quality data, and process capability data into information and knowledge for dynamic decision-making (Lee, 2003).

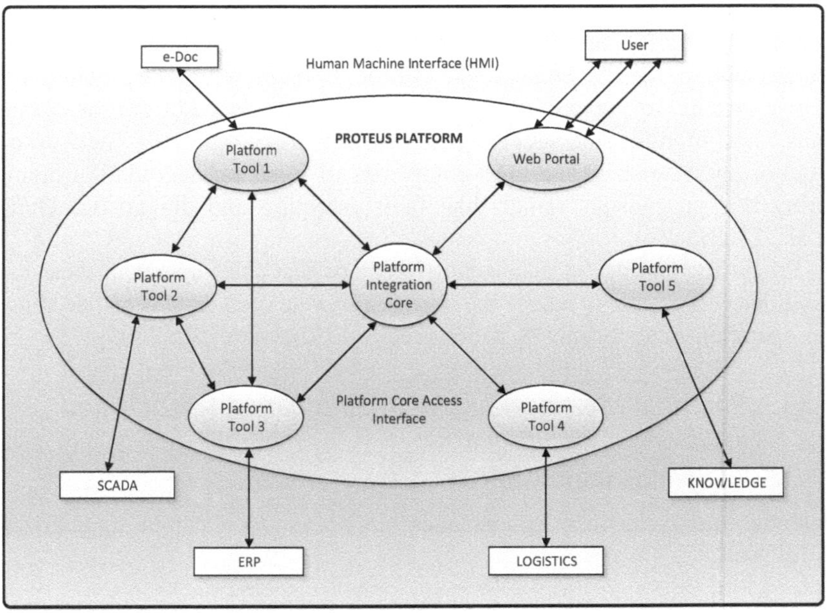

FIGURE 7.39 Proteus platform. *Adapted from Bangemann, T., Reboul, D., Scymanski, J., Thomesse, J.P., Zerhouni, N., 2006. PROTEUS - an integration platform for distributed maintenance systems. Computers in Industry (Special Issue on e-Maintenance) 57 (6), 539–551.*

7.5.3.3.1.3 After-Sales Services With the use of the Internet, Web enabled and wireless communication technology, eMaintenance can transform manufacturing companies into service businesses able to support their customers anywhere and anytime (Lee, 2001).

7.5.3.4 Potential Improvements in Maintenance Activities With eMaintenance

7.5.3.4.1 Potential Improvements

Potential improvements in maintenance activities can be summarized as follows:

7.5.3.4.1.1 Fault Diagnosis/Localization e-Diagnosis offers experts the ability to perform online fault diagnosis, share their valuable experiences with each other, and suggest remedies to operators if an anomalous condition occurs in the inspected machine. In addition, lockouts can be performed in isolation and recorded on location, thanks to wireless technology and palm computing (Wohlwend, 2005).

Consequently, the amount of time it takes to communicate a production problem to the potential expert solution provider can be reduced, the quality of the information shared can be improved, and the resolution time can be reduced. These factors help to increase the availability of equipment, reduce mean time to repair (MTTR), and significantly reduce field service resources/costs (Muller et al., 2008).

7.5.3.4.1.2 Repair/Rebuilding Using e-connections, remote operators can tap into specialized expertise rapidly without travel and scheduling delays. Downtimes could conceivably be reduced through direct interaction (trouble shooting) with source designers and specialists. In addition, diagnosis, maintenance work performed, and parts replaced are documented on the spot in structured responses to work steps displayed on the palm top (Hamel, 2000).

7.5.3.4.1.3 Modification/Improvement—Knowledge Capitalization and Management The multisource knowledge and data environment provided by eMaintenance allows efficient information sharing, leading to important knowledge capitalization and management. With the availability of tools for interacting, handling, and analyzing information about asset state, the development of maintenance engineering for PLC support, including maintenance and retirement stages (disassembly, recycling, reuse, and disposal), is becoming feasible (Goncharenko and Kimura, 1999).

7.5.4 Type of Maintenance Indicators: Leading Versus Lagging and Hard Versus Soft

PIs are used to measure the performance of any system or process. A PI is a product of several measures (i.e., metrics). When used to measure maintenance performance in an area or activity, a PI is called a maintenance performance indicator (MPI) (Wireman, 1998; Parida et al., 2003). PIs are used to find ways to reduce downtime, costs, and waste; operate more efficiently; and increase the operation's capacity. A PI compares actual conditions with a specific set of reference conditions (requirements), measuring the distance between the current situation and the desired situation (target), otherwise called the "distance to target" assessment (EEA, 1999). The list of PIs is long, and each organization's selection of PIs will reflect its corporate strategy's objectives and requirements.

PIs can be broadly classified as leading or lagging indicators. A leading indicator warns the user about the nonachievement of objectives before there is a problem. It is one of a statistical series that fairly reliably turns up or down before the general economy does. A leading indicator thus works as a performance driver and alerts the head of the specific organizational unit to ascertain the present status in comparison to the reference one. Soft or perceptual measures such as stakeholder satisfaction and employee commitments are often leading indicators in the sense that they are highly predictive of financial performance. When such measures are tracked today, it leads to less worry about tomorrow's budgets (Case, 1998).

A lagging indicator normally changes direction after the economy does. Lagging indicators are useless for prediction; the value of construction completed, for example, is outdated, as it will indicate the condition after the performance has taken place. The maintenance cost per unit or return on investment calculation could be an example of a lagging indicator. The establishment of a link between the lagging and the leading indicator makes it possible to control the process (Kumar and Ellingsen, 2000).

The complexity of some measures is an obstacle to their implementation, decreasing the likelihood of their use. In maintenance, many processes can be measured directly. Time or costs are quantities whose measurement is relatively easy. Others, such as the adequacy of repair shops and the size and type of the maintenance teams, are particularly sensitive and can only be measured with more complicated and subjective methods. This difference suggests that the indicators fall into two broad groups, "hard" and "soft." "Hard" indicators include those measurable through the extraction and exploitation of simple databases, such as CMMS or ERP databases; examples include purchase orders, energy consumption by area, etc. The operational view of the maintenance function requires certain indices for performance measurement; it does not require the tactical and strategic aspects of maintenance performance. In this case, the data collection and calculation of the indicators are fast, and measurement does not interfere in the daily work of the maintenance team. Here, a common database can be an important instrument for maintenance management decision-making (Kans and Ingwald, 2008).

While many "soft" indicators are interesting, their measurement can be rendered problematic by the absence of sources, their lack of hard objectivity, or their lack of reliability. Apart from staff and workshop size, this group includes all measures relating to elements with a strong human component, such as the impact of a training activity on the quality of repairs, or the time required for diagnosis and improvement, usually not quantified in records. Thus, the choice of measures and the indicators derived from them will be conditioned by the accessibility and reliability of the sources, with special emphasis on the soft indicators that are affected by human factors.

The people who operate the equipment are a valuable source of information. The human element is indispensable in the measurement of maintenance because of its influence on repairs. However, to assess the overall status of a maintenance system and to correct critical points, more objective tools are needed. To this end, mathematical models and some indicators can be used to assess the probability that a team is performing inspection, maintenance or repair, and determine the average time for equipment to fail after a maintenance intervention (Kumar et al., 2015).

In other words, two actors are involved in the measurement of maintenance performance: people and mathematical models. People provide information on their links to the company, morale, training, skills, and confidence, and so on; models provide information on effectiveness and efficiency related to cost or time. Combining the two leads to the attainment of the three objectives of excellence noted by Katsllometes (2004): efficiency, effectiveness, and staff involvement (see Kumar et al., 2015).

7.5.5 Maintenance Performance Indicators in the Literature

Different categories of maintenance performance measures/indicators are identified in the literature. The total productive maintenance (TPM) concept (Nakajima, 1988), launched in the 1980s, provides a quantitative metric called overall equipment effectiveness (OEE) for measuring productivity of manufacturing equipment. It identifies and measures losses in important aspects of manufacturing, namely, availability, performance/speed, and quality. This supports the improvement of equipment effectiveness and, hence, its productivity. The OEE concept has become increasingly popular and is widely used as a quantitative tool to measure equipment performance in industries (Huang and Dismukes, 2003; Muchiri and Pintelon, 2008). Arts et al. (1998) use the time horizon to classify maintenance control and PIs into three levels: strategic, tactical, and operational. Indicators proposed for operational control include the following: planned hours over hours worked, WOs executed over WOs scheduled, and PM hours over total maintenance hours (Kumar et al., 2015).

Parida and Chattopadhyay (2007) propose a multicriteria hierarchical framework for maintenance performance management that consists of multicriteria indicators for each level of management, i.e., strategic, tactical, and operational. These multicriteria indicators are categorized as equipment/process related (e.g., capacity utilization, OEE, availability, etc.), cost related (e.g., maintenance cost per unit production cost), maintenance task related (e.g., ratio of planned and total maintenance tasks), customer and employee satisfaction, and health, safety and environment. They propose indicators for each level of management in each category (Kumar et al., 2015).

Campbell (1995) classifies the commonly used measures of maintenance performance into three categories based on their focus. These are measures of equipment performance (e.g., availability, reliability, etc.), measures of cost performance (e.g., maintenance, labor, and material cost), and measures of process performance (e.g., ratio of planned and unplanned work, schedule compliance, etc.).

Coetzee (1997) outlines four categories of maintenance performance measures with detailed indicators for each category. The first is maintenance results, measured by availability, mean time to failure (MTTF), breakdown frequency, MTTR, and production rate. The second is maintenance productivity, measured by manpower utilization, manpower efficiency, and maintenance cost component over total production cost. The third is maintenance operational purposefulness, measured by scheduling intensity (scheduled tasks time over clocked time), breakdown intensity, (time spent on breakdown over clocked time), breakdown severity, WO turnover, schedule compliance, and task backlog. The fourth is maintenance cost justification, measured by maintenance cost intensity (maintenance cost per unit production), stock turnover, and maintenance cost over replacement value (Kumar et al., 2015).

Dwight (1995, 1999) classifies performance measures in a five-level hierarchy according to their implicit assumptions on the impact of the maintenance system on the business: overt (visible) bottom-line impact measurements (e.g., direct maintenance cost), profit—loss and visible cost impact measurements (e.g., total failure/downtime cost), instantaneous effectiveness measures (e.g., availability, OEE), system audits (e.g., percentage of planned work and work backlogs), and time-related performance measurements (e.g., life cycle costing and value-based performance measurement). Dwight's work looks at the variations in lag between an action and its outcome (Kumar et al., 2015).

It is clear that each author has a unique way to classify maintenance indicators. Authors also differ in their choice of indicators. However, some indicators and categories of indicators are recognized by all authors as vital for management of the maintenance function. For example, much emphasis has been placed on equipment performance in terms of number/frequency of breakdowns, MTTF, availability, and OEE. Similarly, maintenance cost—related measures are deemed important. Measures of maintenance efforts are considered important by many authors, though they use a variety of terminologies to describe them [e.g., maintenance productivity and operational purpose fullness (Coetzee, 1997), maintenance efforts (Campbell, 1995), maintenance work management (Weber and Thomas, 2006)]. Interestingly, while the literature proposes common lists of key performance indicators (KPIs), it lacks an agreed-upon methodological approach to select or derive

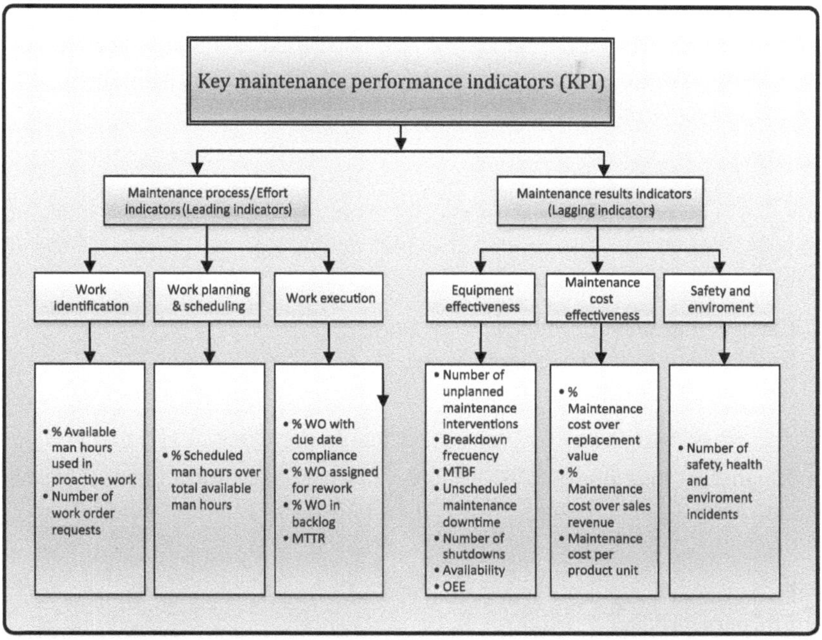

FIGURE 7.40 Key maintenance performance indicators in the literature. *Adapted from Kumar, U., Galar, D., Parida, A., Stenström, C., Berges, L., 2015. Maintenance Performance Metrics: A State of the Art Review. Aragon Institute of Engineering Research (I3A), University of Zaragoza, Zaragoza, Spain. ISBN:978-91-7439-379-8.*

them. As a result, users are left to decide the relevant KPIs for their situation. (Kumar et al., 2015).

Based on the literature, the commonly used MPIs fall into two major categories. Maintenance process or effort indicators are defined as leading indicators and maintenance results indicators as lagging indicators (as shown in Fig. 7.40). Using the definition of Weber and Thomas (2006), leading indicators monitor whether the tasks being performed will lead to the expected output, and lagging indicators monitor the outputs that have been achieved. Under maintenance process indicators and according to Muchiri et al. (2010), there are three categories of indicators: work identification, work planning and scheduling, and work execution indicators. For maintenance results, there are three categories of indicators: equipment performance, maintenance costs and safety, and environment indicators. Each category has its own PIs, as shown in Fig. 7.40 (Kumar et al., 2015).

7.5.6 Key Performance Indicators

The reason for low performance is waste in different forms and in different areas, for example, energy, raw materials, downtime, operation, maintenance, quality, etc. Identifying waste and implementing actions that reduce it will improve performance. With that in mind, in the following, we provide a number of KPIs from different areas that can be used to measure performance and identify waste. Most can be applied to individual equipment, subprocesses, and whole plants (Lindberg et al., 2015):

7.5.6.1 Energy Key Performance Indicators

Energy could be in different forms, for example, electricity, gas, coal, oil, biomass, steam, etc. The produced output could, e.g., be in units of tons/h, m^3/h, units/h, etc. The main KPIs for energy are as follows:

- Energy output/energy input.
- Energy input/produced output.

7.5.6.2 Raw Material Key Performance Indicators

Raw materials may not only be the main raw materials for a plant, they may also include water, chemicals, etc. The main KPIs for raw material are as follows:

- Raw material input/produced output,
- Emission/produced output,
- Waste deposit/produced output.

7.5.6.3 Operation Key Performance Indicators

The main operation KPI is OEE and its individual parts. The most important KPIs, including OEE are as follows:

- OEE.
- Percentage of scheduled operation time, over a time period, e.g., a day, week, month, etc.
- Percentage of actual uptime of the scheduled time, over a time period.
- Percentage of production rate of maximum rate for produced product type, over a time period.
- Percentage of full quality products of the particular production, over a time period.

7.5.6.4 Control Performance Key Performance Indicators

Control performance may influence product quality, production speed, equipment wear, etc. The main KPIs are as follows:

- Number of control loops in manual mode/total number of control loops,
- Variance of control error (set point−measured value),
- Settling time after a set point change.

7.5.6.5 Maintenance Key Performance Indicators

Too little maintenance causes an excessive number of unplanned stops, resulting in lost production and emergency maintenance. Too much maintenance causes large maintenance costs and lost production during each planned maintenance. The main maintenance KPIs are as follows:

- Maintenance costs/produced output over a time period.
- Maintenance time/produced output over a time period.
- Number of alarms over a time period.

In addition, some maintenance KPIs are the same as the operation KPIs and certain equipment KPIs presented below.

7.5.6.6 Planning Key Performance Indicators

Planning and scheduling impact how well plant capacity is used. As deriving the optimal production plan and comparing it with actual production is not within the scope of KPI calculations, we suggest a KPI based on adherence to the plan:

- Integrated sum of only positive values of planned production minus actual production over a time period. This KPI is not improved if production catches up later.

7.5.6.7 Inventory Key Performance Indicators

Large inventories are expensive, while too small inventories may cause production disturbances. The main inventory KPI is

- Throughput rate/average inventory

7.5.6.8 Equipment Key Performance Indicators

Equipment KPIs can be used to follow the condition of equipment and, in some cases, to predict when maintenance will be required. The main equipment KPIs are as follows:

- Different types of efficiencies, e.g., heat transfer rate of heat exchangers, pump/fan efficiency, drying efficiency, etc.
- Equipment wear (based on, e.g., operating hours, speed, load, startups).
- Vibration amplitude of a piece of equipment.
- Measured minus predicted performance.

REFERENCES

Agrawal, R., Imielinski, T., Swami, A., 1993. Mining association rules between sets of items in large databases. In: SIGMOD Conference 1993, pp. 207–216.

Akyildiz, I.F., Su, W., Sankarasubramaniam, Y., Cayirci, E., 2002. Wireless sensor networks: a survey. Computer Networks 38, 393–422.

Al-Karaki, J., Kamal, A., 2004. Routing techniques in wireless sensor networks: a survey. IEEE Wireless Communications 11, 6–28.

Arts, R., Knapp, G., Mann, L., 1998. Some aspects of measuring maintenance performance in the process industry. Journal of Quality in Maintenance Engineering 4 (1), 6–11.

Ashton, K., 2009. That "Internet of Things" thing. RFID Journal.

Atzori, L., Iera, A., Morabito, G., 2010. The Internet of things: a survey. Computer Networks 54, 2787–2805.

AUGMENT, August 16, 2016. The difference between virtual reality and augmented reality for enterprise. http://www.augment.com/blog/difference-virtual-reality-augmented-reality-enterprise/.

Augmensys GmbH. Maintenance 4.0 | Predictive Maintenance with Augmented Reality and OCR. Lakeside B01, 9020 Klagenfurt | Austria. http://www.augmensys.com/en/ubik-all/item/232-maintenance-4-0-predictive-maintenance-with-augmented-reality-and-ocr.

Bajwa, W., Haupt, J., Sayeed, A., Nowak, R., 2006. Compressive Wireless Sensing. ACM.

Baldwin, R.C., 2004. Enabling an e-Maintenance Infrastructure.

Bangemann, T., Reboul, D., Scymanski, J., Thomesse, J.P., Zerhouni, N., 2006. PROTEUS - an integration platform for distributed maintenance systems. Computers in Industry (Special Issue on e-Maintenance) 57 (6), 539–551.

BBC News, July 3, 2007. Warning of Data Ticking Time Bomb. http://news.bbc.co.uk/2/hi/6265976.stm.

Belissent, J., 2010. Getting clever about smart cities: new opportunities require new business models. In: Forrester Research.

Bengio, Y., 2009. Learning Deep Architectures for AI, first ed. Now Publishers Inc.

Bonneau, G.P., Nielson, G.M., Post, F., 2003. Data visualization: the state of the art. Kluwer Academic, London.

Buckley, J., 2006. The Internet of Things: From RFID to the Next-generation Pervasive Networked Systems. Auerbach Publications, New York.

Buyya, R., Yeo, C.S., Venugopal, S., Broberg, J., Brandic, I., 2009. Cloud computing and emerging it platforms: vision, hype, and reality for delivering computing as the 5th utility. Future Generation Computer Systems 25, 599–616.

Campbell, J.D., 1995. Uptime Strategies for Excellence in Maintenance Management. Productivity Press, Portland, OR, USA.

Campbell, A.T., Eisenman, S.B., Lane, N.D., Miluzzo, E., Peterson, R.A., 2006. People Centric Urban Sensing. ACM.

Candell, O., Karim, R., Söderholm, P., 2009. e-Maintenance - information logistics for maintenance support. Journal of Robotics and Computer - integrated Manufacturing Archive. Journal Robotics and Computer Integrated Manufacturing 25, 937–944.

Case, J., 1998. Using measurement to boost your unit's performance. Harvard Management Update 3, 1–4.

Castellani, A.P., Bui, N., Casari, P., Rossi, M., Shelby, Z., Zorzi, M., 2010. Architecture and Protocols for the Internet of Things: A Case Study, pp. 678–683.

Chiang, L.H., Braatz, R.D., Russell, E.L., 2001. Fault Detection and Diagnosis in Industrial Systems. Springer.

Chui, M., Miller, A., Roberts, R., February 2009. Six ways to make Web 2.0 work. McKinsey Quarterly.

Coetzee, J.L., 1997. Towards a general maintenance model. In: Martin, H.H. (Ed.), Proceedings of IFRIM'97, Hong Kong, Paper 12, pp. 1–9.

Cortes, C., Vapnik, V., September 1995. Machine learning. Support-Vector Networks 20 (3), 273–297. www.springerlink.com/content/k238jx04hm87j80g/.

Crespo-Márquez, A., Iung, B., 2008. A review of e-maintenance capabilities and challenges. Systemics, Cybernetics and Informatics. ISSN: 1690-4524 6 (1).

Dean, J., Ghemawat, S., December 2004. MapReduce: simplified data processing on large clusters. In: Sixth Symposium on Operating System Design and Implementation, San Francisco, CA.

Deloitte, 2014. Industry 4.0. Challenges and Solutions for the Digital Transformation and Use of Exponential Technologies. Audit. Tax. Consulting. Corporate Finance. March–August, 2014.

Demirkol, I., Ersoy, C., Alagoz, F., 2006. MAC protocols for wireless sensor networks: a survey. IEEE Communications Magazine 44, 115–121.

Donoho, D., 2006. Compressed sensing. IEEE Transactions on Information Theory 52, 1289–1306.

Dwight, R.A., 1995. Concepts for Measuring Maintenance Performance. New Developments in Maintenance. Moret Ernst and Young Management Consultants, Netherlands, pp. 109–125.

Dwight, R.A., 1999. Searching for real maintenance performance measures. Journal of Quality in Maintenance Engineering 5 (3), 258–275.

Economist Intelligence Unit, 2011. Big Data: Harnessing a Game-Changing Asset. A report from the economist intelligence unit sponsored by SAS. http://www.sas.com/resources/asset/SAS_BigData_final.pdf.

EEA (European Environment Agency), 1999. In: Enviromental Indicators: Typology and Overview. Technical Report No 25, Copenhagen.

El-Sayed, H., Mellouk, A., George, L., Zeadally, S., 2008. Quality of service models for heterogeneous networks: overview and challenges. Annals of Telecommunications 63, 639–668.

Electronics Enabled Products, 2006. Radio-Frequency Identification. © 2006 Per a Knowledge, ISBN 1-84402-039-8.

Galar, D., 2015. Special Session on "e-Maintenance and Maintenance 4.0". DMIN'2015.

Galar, D., Peters, R., Berges, L., Stenström, C., Kumar, U., 2012. Composite indicators in asset management. In: Department of Civil, Environmental and Natural Resources Engineering Operation, Maintenance and Acoustics. International Conference on Maintenance Performance Measurement & Management, United Kingdom/Sunderland.

Galar, D., Thaduri, A., Catelani, M., Ciani, L., 2015. Context Awareness for Maintenance Decision Making: A Diagnosis and Prognosis Approach, Measurement, vol. 67. Elsevier, pp. 137–150.

Gartner. IT Glossary–Big Data. http://www.gartner.com/it-glossary/big-data/.

Gartner Inc., 2012. Gartner's Hype Cycle Special Report for 2011. http://www.gartner.com/technology/research/hype-cycles/.

Ghemawat, S., Gobioff, H., Leung, S.-T., October 2003. The Google file system. In: 19th ACM Symposium on Operating Systems Principles, Lake George, NY.

Ghosh, A., Das, S.K., 2008. Coverage and connectivity issues in wireless sensor networks: a survey. Pervasive and Mobile Computing 4, 303–334.

Goncharenko, I., Kimura, F., 1999. Remote maintenance for IM (inverse manufacturing). In: Proceedings of the First International Symposium on Environmentally Conscious Design and Inverse Manufacturing, Tokyo, Japan, pp. 862–867.

Gubbia, J., Buyyab, R., Marusic, S., Palaniswami, M., 2013. Internet of things (Iot): a vision, architectural elements, and future directions. Future Generation Computer Systems 29, 1645–1660.

Haddad, G., Sandborn, P., Pecht, M., June 23–24, 2011. Using real options to manage condition-based maintenance enabled by PHM. In: IEEE Conference on Prognostics and Health Management (PHM), pp. 1–7.

Hajek, P., Havel, I., Chytil, M., 1966. The GUHA method of automatic hypotheses determination. Computing 1 (4), 293–308.

Hamel, W., 2000. e-Maintenance robotics in hazardous environments. In: Proceedings of the 2000 IEEE/RSJ International Conference on Intelligent Robots and Systems, Takamatsu, Japan.

Hermann, M., Pentek, T., Otto, B., 2015. Design principles for industry 4.0 scenarios: a literature review. In: Working Paper, No. 01-2015. Technical University of Dortmund.

Hewitt, B., May 27, 2011. Big data: big costs, big risks and big opportunity. Forbes Staff. http://www.forbes.com/sites/ciocentral/2011/05/27/big-data-big-costs-big-risks-and-big-opportunity/#6ce8c88c1906.

Holmberg, K., Helle, A., Halme, J., 2005. Prognostics for industrial machinery availability. In: POHTO 2005 International Seminar on Maintenance, Condition Monitoring and Diagnostics, Oulu, Finland.

Honicky, R., Brewer, E.A., Paulos, E., White, R., 2008. N-smarts: networked suite of mobile atmospheric real-time sensors. ACM 25–29.

Howe, J., June 2006. The Rise of Crowdsourcing. Wired, Issue 14.06.

Huang, S., Dismukes, J.P., 2003. Manufacturing productivity improvement using effectiveness metrics and simulation analysis. International Journal of Production Research 41 (3), 513–527.

Iung, B., Morel, G., Leger, J.B., 2003. Proactive maintenance strategy for harbor crane operation improvement. Robotics 21 (3), 313–324 (Special Issue on Cost Effective Automation. Erbe, H. (Ed.)).

Jamil, I., Jamil, R., Jinquan, Z., Ming, L., Dong, W.Y., Jamil, R., November 2013. Condition-based maintenance decision-making support system (DSS) of hydropower plant. Copyright © 2013 ISSR Journals. ISSN: 2028-9324 4 (3).

Jantunen, E., Emmanouilidis, C., Arnaiz, A., Gilabert, E., January 2011. e-Maintenance: trends, challenges and opportunities for modern industry. In: IFAC Proceedings Volumes, vol. 44 (1), pp. 453–458, 18th IFAC World Congress.

Jones, B., 2016. In: The Differences Between Virtual Reality, Augmented Reality, and Mixed Reality, March 1, 2016. http://www.codeguru.com/blog/category/technology/virtual-reality-augmented-reality-and-mixed-reality.html.

Juels, A., 2006. RFID security and privacy: a research survey. IEEE Journal on Selected Areas in Communications 24, 381–394.

Kanjo, E., 2009. Noise SPY: a real-time mobile phone platform for urban noise monitoring and mapping. Mobile Networks and Applications 15, 562–574.

Kans, M., Ingwald, A., 2008. Common database for cost-effective improvement of maintenance performance. International Journal of Production Economics 113 (2), 734–747.

Katsllometes, J., 2004. How good is my maintenance program?. In: Plant Operators Forum 2004. Colorado, Denver.

Kaufman, L.M., 2009. Data security in the world of cloud computing. IEEE Security and Privacy Magazine 7, 61–64.

Khosrowshahi, F., Ghodous, P., Sarshar, M., 2014. Visualization of the modeled degradation of building flooring systems in building maintenance. Computer-Aided Civil and Infrastructure Engineering 29, 18–30.

Koskinen, K.T., Kortelainen, H., Aaltonen, J., Uusitalo, T., Komonen, K., Mathew, J., Laitinen, J., 2016. In: Proceedings of the 10th World Congress on Engineering Asset Management (WCEAM 2015), ISBN 9783319270647. © Springer international publishing, Switzerland.

Kulkarni, R.V., Förster, A., Venayagamoorthy, G.K., 2011. Computational intelligence in wireless sensor networks: a survey. IEEE Communications Surveys & Tutorials 13, 68–96.

Kumar, U., Ellingsen, H.P., March 2000. Design and development of maintenance performance indicators for the Norwegian oil and gas industry. In: Proceedings of the 15th European Maintenance Congress: Euromaintenance 2000. The Swedish Maintenance Society (UTEK) and The European Federation of National Maintenance Societies (EFNMS), Gothenburg, Sweden, pp. 224–228.

Kumar, U., Galar, D., Parida, A., Stenström, C., Berges, L., 2015. Maintenance Performance Metrics: A State of the Art Review. Aragon Institute of Engineering Research (I3A), University of Zaragoza, Zaragoza, Spain, ISBN 978-91-7439-379-8.

Kuznetsov, S., Paulos, E., 2010. Participatory sensing in public spaces: activating urban surfaces with sensor probes. In: ACM Request Permissions.

Lee, J., 2001. A framework for next-generation e-maintenance system. In: Proceedings of the Second International Symposium on Environmentally Conscious Design and Inverse Manufacturing, Tokyo, Japan.

Lee, J., 2003. e-Manufacturing: fundamental, tools, and transformation. Robotics and Computer-Integrated Manufacturing 19 (6), 501–507.

Li, Y., Chun, L., Nee, A., Ching, Y., 2005. An agent-based platform for web-enabled equipment predictive maintenance. In: Proceedings of IAT'05 IEEE/WIC/ACM International Conference on Intelligent Agent Technology, Compiègne, France.

Lindberg, C.-F., Tanb, S.T., Yan, J.Y., Starfelt, F., 2015. Key performance indicators improve industrial performance. In: The 7th International Conference on Applied Energy–ICAE 2015. Published by Elsevier Ltd. This is an open access article under the CC BY-NC-ND license.

Lindholm, T., Yellin, F., Bracha, G., Buckley, A., 2014. The Java Virtual Machine Specification. Pearson Education.

Lopez, T.S., Ranasinghe, D.C., Harrison, M., McFarlane, D., 2012. Adding sense to the Internet of things an architecture framework for smart objective systems. Pervasive Ubiquitous Computing 16, 291–308.

Manca, D., Brambilla, S., Colombo, S., January 2013. Dipartimento di Chimica, Materiali e Ingegneria Chimica "Giulio Natta", Politecnico di Milano, Piazza Leonardo da Vinci 32, 20133 Milano, Italy. Bridging Between Virtual Reality and Accident Simulation for Training of Process-Industry Operators, vol. 55, pp. 1–9.

Manyika, J., Chui, M., Brown, B., Bughin, J., Dobbs, R., Roxburgh, C., Byers, A.H., May 2011. Big Data: The Next Frontier for Innovation, Competition, and Productivity. McKinsey Global Institute.

Mayer-Schönberger, V., 2011. Failing to forget the "Drunken Pirate". In: Delete: The Virtue of Forgetting in the Digital Age (New in Paper), first ed. Princeton University Press, pp. 3–15.

McAfee, A., Brynjolfsson, E., 2012. Big data: the management revolution. Harvard Business Review 90 (10), 60–68.

Mohamed, N., Al-Jaroodi, J., 2014. In: Real-Time Big Data Analytics: Applications and Challenges. Conference Paper, July 2014, ISBN 978-1-4799-5313-4. ©2014 IEEE.

Moore, W.J., Starr, A.G., 2006. An intelligent maintenance system for continuous cost-based prioritisation of maintenance activities. Computing Industrial [Special Issue on e-Maintenance] 57 (6), 595–606.

Muchiri, P.N., Pintelon, L., 2008. Performance measurement using overall equipment effectiveness (OEE): literature review and practical application. International Journal of Production Research 46 (13), 3517–3535.

Muchiri, P.N., Pintelon, L., Martin, H., Meyer, A.M., 2010. Empirical analysis of maintenance performance measurement. In Belgian industries. International Journal of Production Research 48 (20), 5905–5924.

Muller, A., Marquez, A.C., Iung, B., 2008. On the concept of e-maintenance: review and current research. Reliability Engineering and System Safety 93, 1165–1187.

Nakajima, S., 1988. Introduction to TPM: Total Productive Maintenance. Productivity Press, Inc., p. 129.

Navajo, M., Ballesteros, I., D'Elia, S., Sassen, A., Goyet, M., Santaella, J., 2010. Draft report of the task force on interdisciplinary research activities applicable to the future Internet. In: European Union Task Force Report.

Neill, D.B., 2011. Fast Bayesian scan statistics for multivariate event detection and visualization. Statistics in Medicine 30, 455–469.

Parida, A., Chattopadhyay, G., 2007. Development of a multi-criteria hierarchical framework for maintenance performance measurement (MPM). Journal of Quality in Maintenance Engineering 13 (3), 241–258.

Parida, A., Åhrén, T., Kumar, U., 27–29 August 2003. Integrating maintenance performance with corporate balanced scorecard. In: COMADEM 2003, Proceedings of the 16th International Congress, Växjö, Sweden, pp. 53–59.

Penna, R., Amaral, M., Espýndola, D., 2014. Visualization Tool for Cyber-Physical Maintenance Systems. IEEE, ISBN 978-1-4799-4905-2.

Price, C., Price, C.J., 1999. Computer-based Diagnostic Systems. Springer, Heidelberg, pp. 65–69.

Probst, L., Monfardini, E., Frideres, L., Clarke, S., Demetri, D., Schnabel, L., Kauffmann, A., PwC Luxembourg, September 2013. In: Big Data - Analytics & Decision Making. Case Study 8, Business Innovation Observatory. Contract No 190/PP/ENT/CIP/12/C/N03C01. European Union.

Puiu, T., September 12, 2016. The Difference Between Virtual and Augmented Reality. http://www.zmescience.com/other/did-you-know/difference-virtual-augmented-reality/.

PwC Australia, 2012. Big Data – The Next Frontier for Innovation. http://www.pwc.com.au/consulting/assets/publications/big-data-oct12.pdf.

Rana, R.K., Chou, C.T., Kanhere, S.S., Bulusu, N., Hu, W., 2010. Ear-Phone: An End-To-End Participatory Urban Noise Mapping System. ACM Request Permissions.

Ren, L., Tian, F., Zhang, X., Zhang, L., 2010. DaisyViz: a model-based user interface toolkit for interactive information visualization systems. Journal of Visual Languages and Computing 21, 209–229.

Roemer, M., Dzakowic, J., Orsagh, R., Byington, C., Vachtsevanos, G., 2005. An overview of selected prognostic technologies with reference to an integrated PHM architecture. In: Proceedings of the IEEE Aerospace Conference 2005, Big Sky, United States.

Rosenthal, D.S.H., 2015. Emulation & Virtualization as Preservation Strategies. LOCKSS Program. Stanford University Libraries. Copyright © 2015 Andrew W. Mellon Foundation.

Rothenberg, J., 1995. Ensuring the longevity of digital documents. Scientific American 272 (1).

Sang, Y., Shen, H., Inoguchi, Y., Tan, Y., Xiong, N., 2006. Secure Data Aggregation in Wireless Sensor Networks: A Survey, pp. 315–320.

Santini, S., Ostermaier, B., Vitaletti, A., 2008. First Experiences Using Wireless Sensor Networks for Noise Pollution Monitoring. ACM, Glasgow, Scotland.

Sten-Erik, B., Baglee, D., Galar, D., Sarbjeet, S., 2013. Maintenance knowledge management with fusion of CMMS and CM. In: DMIN 2013 International Conference on Data Mining, 22–25 July 2013, Las Vegas, Nevada, USA.

Sundmaeker, H., Guillemin, P., Friess, P., Woelfflé, S., 2010. Vision and challenges for realising the Internet of things. Cluster of European Research Projects on the Internet of Things—CERP IoT.

Tang, D., 2009. Event Detection in Sensor Networks. School of Engineering and Applied Sciences, The George Washington University.

Taylor, J., 2012. Decision Management Systems Platform Technologies Report. Decision Management Solutions, Palo Alto, CA.

Taylor, J., 2014. Real-Time Responses with Big Data. © 2014 Decision Management Solutions.

Tilak, S., Abu-Ghazaleh, N., Heinzelman, W., 2002. A taxonomy of wireless microsensor network models. ACM Mobile Computing and Communications Review 6, 28–36.

Toolingu, 2013. A Condition-based Maintenance. Tooling University.

Tory, M., Moller, T., 2004. Rethinking visualization: a high-level taxonomy. In: IEEE Symposium on Information Visualization, INFOVIS 2004, pp. 151–158.

Van Der Hoeven, J.R., Van Diessen, R.J., Van Der Meer, K., 2005. Development of a universal virtual computer (UVC) for long-term preservation of digital objects. Journal of Information Science 31 (3), 196–208.

Vera, E., Mancera, L., Babacan, S.D., Molina, R., Katsaggelos, A.K., 2009. Bayesian compressive sensing of wavelet coefficients using multiscale Laplacian priors. In: Statistical Signal Processing. SSP'09, IEEE/SP 15th Workshop, pp. 229–232.

Vercellis, C., 2009. Business Intelligence: Data Mining and Optimization for Decision Making. © 2009 John Wiley & Sons, Ltd., ISBN 978-0-470-51138-1.

Walther, K., Regtmeier, J., 2016. Virtualization for manufacturing and IoT. In: HARTING IT Software Development GmbH & Co. KG. Marienwerder Str. 2, 32339 Espelkamp, Germany.

Wang, K., 2016. Intelligent predictive maintenance (IPdM) system – industry 4.0 scenario. WIT Transactions on Engineering Sciences. ISSN: 1743-3533 113. http://dx.doi.org/10.2495/IWAMA150301. © 2016 WIT Presswww.witpress.com.

Weber, A., Thomas, R., 2006. Key Performance Indicators: Measuring and Managing the Maintenance Function. Ivara Corporation.

Welbourne, E., Battle, L., Cole, G., Gould, K., Rector, K., Raymer, S., 2009. Building the Internet of things using RFID the RFID ecosystem experience. IEEE Internet Computing 13, 48–55.

WIND, 2016. Virtualization and the Internet of Things. ©2016 Wind River Systems, Inc.

Wireman, T., 1998. Developing Performance Indicators for Managing Maintenance, first ed. Industrial Press, New York, USA.

Wohlwend, H., 2005. e-Diagnostics Guidebook: Revision 2.1. Technology Transfer # 01084153D-ENG, SEMATECH Manufacturing Initiative.

Woodhouse, J., 2013. Setting a Good Standard in Asset Management. Chair of Experts Panel, Institute of Asset Management and Managing Director, TWPL.

Xu, H., Guo, J., Zeng, H., Xiao, Z., October 7–10, 2007. Study on distributed cooperative maintenance decision supporting system for hydropower plant. In: IEEE International Conference on Systems, Man and Cybernetics, ISIC, pp. 2296–2301.

Yiu, C., 2012. The Big Data Opportunity. Making Government Faster, Smarter and More Personal. Policy Exchange, Clutha House, 10 Storey's Gate, London SW1P 3AY, ISBN 978-1-907689-22-2.

Yoshikawa, H., 1995. Manufacturing and the 21th century – intelligent manufacturing systems and the renaissance of the manufacturing industry. Technological Forecasting and Social Change 49 (2), 165–213.

Zorzi, M., Gluhak, A., Lange, S., Bassi, A., 2010. From today's Intranet of things to a future Internet of things: a wireless - and mobility-related view. IEEE Wireless Communications 17, 43–51.

FURTHER READING

Baglee, D., Marttonen, S., Galar, D., 2015. The need for big data collection and analyses to support the development of an advanced maintenance strategy. In: International Conference, Data Mining, DMIN'15.

Zhang, C., Ivan, J.N., El Dessouki, W.M., Anagnostou, E., 2005. Relative risk analysis for studying the impact of adverse weather conditions and congestion on traffic accidents. In: Proceedings of the 84th Annual Meeting of the Transportation Research Board, Transportation Research Board.

Chapter 8

Actuators and Self-Maintenance Approaches

8.1 INTELLIGENT (SMART) MATERIALS FOR MAINTENANCE

8.1.1 Introduction

Intelligent or smart materials are new generation materials surpassing conventional structural and functional materials in their possession of adaptive capabilities to meet external stimuli, such as loads or environment, with inherent intelligence. Rogers (1988) and Rogers et al. (1988) defines smart materials as materials possessing the ability to change their physical properties in a specific manner in response to a specific stimulus input. The stimuli could be pressure, temperature, electric and magnetic fields, chemicals, hydrostatic pressure, or nuclear radiation. The associated changeable physical properties include shape, stiffness, viscosity, and damping.

Smart material technology is relatively new and innovative. In construction, smart materials and systems could be used in "smart" buildings, for environmental control, security, and structural health monitoring, e.g., strain measurement in bridges using embedded fiber optic sensors. Magnetorheological (MR) fluids have been used to damp cable-stayed bridges and reduce the effects of earthquakes. In the aerospace industry, smart materials can be used in "smart wings," health and usage monitoring systems, and active vibration control, e.g., in helicopter blades. In marine and rail transport, possibilities include strain monitoring using embedded fiber optic sensors.

Takagi (1990) says intelligent materials respond to environmental changes at the most optimum conditions and reveal their own functions according to the environment. Smartness incorporates self-adaptability, self-sensing, memory, and multiple functionalities of materials or structures. The self-adaptation characteristics of smart structures use the embedded adaptation of smart materials, such as shape memory alloys (SMAs). By changing their properties, smart materials can detect faults and cracks, making them useful as a diagnostic tool (Takagi, 1990). This characteristic can be used to activate the smart material embedded in the host material in a proper way to compensate

e-Maintenance. http://dx.doi.org/10.1016/B978-0-12-811153-6.00008-7

for the fault. This phenomenon is called the self-repairing effect (Shaikh et al., 2015).

Smart materials have two phases depending on the temperature: the austenite phase exists at high temperatures, and the martensite phase exists at low temperatures. When the external temperature or stress condition changes, one phase will transform to the other. Smart materials exhibit many special properties during the transformations between these two phases, including shape memory effect, super elasticity effect, two-way memory effect, etc. (Shaikh et al., 2015).

With the development of material science, many new, high-quality, and cost-efficient materials have come into use in various fields of engineering. Materials have become multifunctional and now require the optimization of various characteristics and properties. The concept has recently been moving toward composite materials and, most recently, toward smart materials.

8.1.1.1 Definition of an Intelligent Material

Intelligent or smart materials have been defined as "Those materials which sense any environmental change and respond to it in an optimal manner" (Rogers et al., 1988). From this definition and using the analogy of the bionic system of humans and animals, we can define the following mechanisms as essential for any material to be made intelligent:

- A sensing device to perceive external stimuli (skin that senses thermal gradients, an eye that senses optical signals, etc.), termed a sensor function.
- A communication network by which the sensed signal is transmitted to a decision-making mechanism (e.g., the nervous system in humans and animals), termed a memory function.
- A decision-making device that has the capability of reasoning (e.g., the brain), termed a processor function.
- An actuating device, which could be inherent in the material or externally coupled with it (e.g., stiffening of muscles in humans and animals to resist deformation due to external loading), termed an actuator function.

All the above mechanisms need to be active in real time applications for the material to respond intelligently. Another important factor in this whole process is the time of response. This is the interval between the instant when the sensor senses the stimulus and that of the actuator response. An optimum time interval is crucial in the design of intelligent materials and is dependent on the type of application (Iyer and Haddad, 1993).

8.1.2 Concept of Intelligent Materials

Designing a material with sensor, processor, and actuator functions is a fundamental step in the evolution of an intelligent material to achieve a desired response adaptable to the environment. This concept is illustrated in Fig. 8.1 (Iyer and Haddad, 1993).

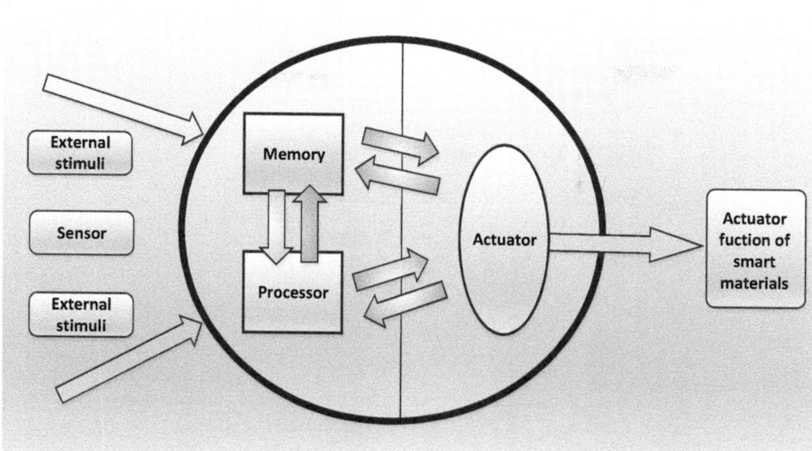

FIGURE 8.1 Concept of an intelligent material. *Adapted from Iyer, S.S., Haddad, Y.M., 1993. Intelligent Materials – an Overview. Department of Mechanical Engineering, University of Ottawa, Ottawa, Ontario, Canada, KIN 6N5.*

8.1.2.1 Sensor Function

A sensor function in a smart material is defined as the ability of the material to sense the response characteristics of itself with respect to environmental factors, such as mechanical loading, temperature, humidity, and electrical inputs. An example is a piezoelectric sensor embedded in a composite material. This sensor diagnoses a mechanical disturbance imposed on the material by generating a voltage, which can be measured and analyzed (Iyer and Haddad, 1993).

8.1.2.2 Memory and Processor Function

This mechanism stores the signals sensed and transmitted by the sensor function. The characteristics of these signals are then compared with prestored acceptable values acquired during the "training" process of the processor. The training process may be carried out using an artificial intelligence technique, e.g., pattern recognition method. Typically, this function is in the form of an executable artificial intelligence software that can produce a logical output in the form of an electrical voltage that can further be amplified and used to activate an actuator mechanism (Iyer and Haddad, 1993).

8.1.2.3 Actuator Function

This mechanism is coupled with the material. It produces an output corresponding to the signal received from the processor function. This output is usually in the form of restoring stress, strain or change in temperature, or stiffness of the actuator mechanism that is coupled with the material. This change is designed to neutralize the effect of the change in environment on the material, thereby adapting the material continuously to its environment.

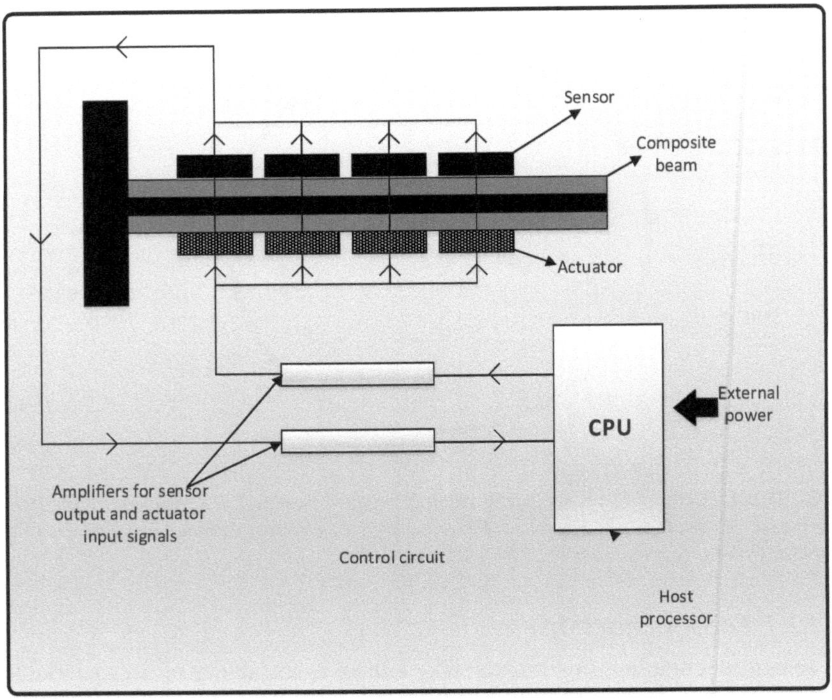

FIGURE 8.2 Incorporation of sensor, processor, and actuator functions in an intelligent composite beam. *Adapted from Iyer, S.S., Haddad, Y.M., 1993. Intelligent Materials – an Overview. Department of Mechanical Engineering, University of Ottawa, Ottawa, Ontario, Canada, KIN 6N5.*

A typical intelligent composite cantilever beam, consisting of sensor, processor, and actuator functions, is illustrated in Fig. 8.2 (Iyer and Haddad, 1993).

8.1.3 Types of Smart Materials

Smart materials with one or more property can be significantly altered in a controlled fashion by external stimuli. Thus, this material has built-in or intrinsic sensor(s), actuator(s), and control mechanism(s) by which it is capable of sensing a stimulus, responding to it in a predetermined manner and extent, in a short or appropriate time, and reverting to its original state as soon as the stimulus is removed. Several types of smart material are available and are discussed below.

8.1.3.1 Pyroelectric Material

Pyroelectrics are materials that become electrically polarized on an applied temperature change. Materials used as pyroelectrics include barium strontium titanate (Ba_1-xSrxTiO_3), lead zirconate titanate (PbZr_1-x TixO_3), barium strontium niobate (Ba1-xSrxNb_2O_6), triglycine sulfates, lithium tantalate ($LiTaO_3$), and polyvinylidene fluoride (PVDF). Ceramics are widely used because of their lower cost, including availability, ease of processing, and good stability. Their weakness is their inherent brittleness. Polymers, PVDF, and its copolymer with Trifluoroethylene provide a nonbrittle alternative but do not obtain the level of performance of ceramics. Pyroelectrics are widely used for infrared detection in surveillance and targeting applications. A pyroelectric infrared detector generally employs a thin film of pyroelectric material oriented with the electrode surfaces normal to its polarization direction. When infrared radiation is absorbed by the detector, the temperature of the pyroelectric material rises. This change in temperature alters the material's polarization, causing a change in surface charges across the material; this produces an electrical signal via the electrodes. The resulting signal is proportional to the incoming radiation.

8.1.3.2 Piezoelectric Material

Piezoelectric materials exhibit an electrical polarization with an applied mechanical stress (direct effect) or a dimensional change with an applied electric field (converse effect). They are used for both sensing and actuating devices. Lead zirconate titanate (PbZr_1-xTixO_3) is the premier piezoelectric material, as it may be doped to produce an n-type or p-type material with a range of dielectric constants to meet the requirements of numerous applications. Other piezoelectric materials include barium titanate ($BaTiO_3$), lead titanate ($PbTiO_3$), lead metaniobate ($PbNb_2O_6$), and PVDF. Polymers are generally favored for sensing applications, while ceramics are favored for actuating. Fig. 8.3 depicts the piezoelectric effect observed in lead zirconate titanate on the application of compressive forces relative to the crystal structure.

Piezoelectric is a nature technology with numerous applications throughout the military and commercial sectors. Devices using piezoelectric include adaptive optics, hydrophones, sonobuoys, fuse devices, depth sounders, thickness gauging, flaw detection, level indicators, alarm systems, strain gauges, airplane beacon locators, fetal heart detectors, and tire pressure indicators, among many others (Kaushal et al., 2016).

8.1.3.3 Electrostrictive

Piezoelectric materials exhibit a linear relationship between electric and mechanical variables. Electrostrictive materials also show a relationship between these two variables, but in this case, it is a quadratic relationship between mechanical stress and the square of electrical polarization. Electrostriction can occur in any material and is a small effect. One difference

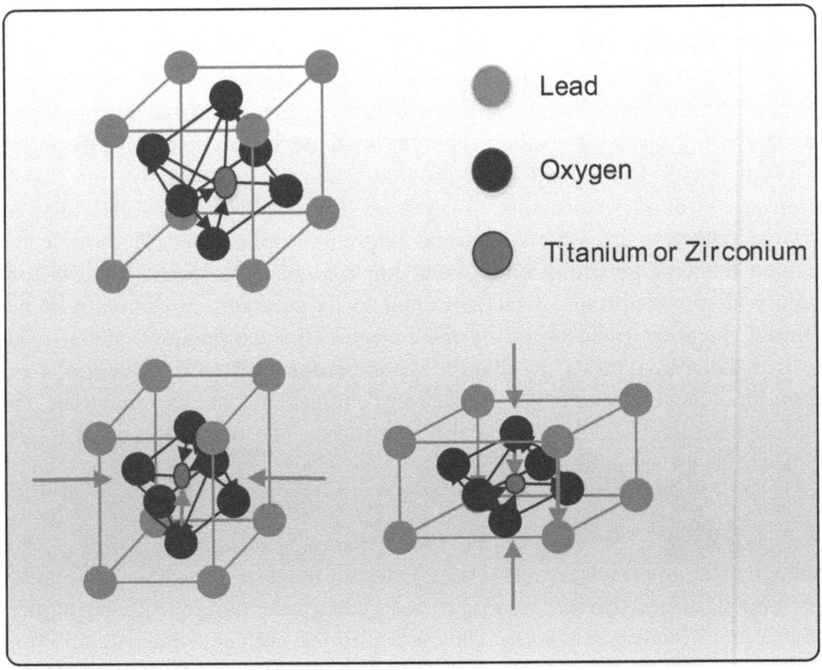

FIGURE 8.3 Piezoelectric effect. *Adapted from Kaushal, A., Vardhan, A., Rawat, R.S.S., May–June 2016. Intelligent material for modern age: a review. IOSR Journal of Mechanical and Civil Engineering (IOSR-jmce), e-ISSN: 2278–1684, ISSN: 2320-334X, 13, 3, 10–15.*

between piezoelectric and electrostrictive materials is the ability of the latter to show a larger effect in the vicinity of their Curie temperature. Electrostriction is a fourth rank tensor property observed both in centric and acentric insulators. Examples of electrostrictive materials include lead manganese niobate–lead titanate and lead lanthanium zirconate titanate (PLZT) (Kaushal et al., 2016).

8.1.3.4 Magneto-Restrictive

Magneto-restrictive materials have the material response of mechanical deformation when stimulated by a magnetic field. Shape changes are the largest in ferromagnetic and ferromagnetic materials. The repositioning of domain walls that occurs when these solids are placed in a magnetic field leads to hysteresis between magnetization and an applied magnetic field. When a ferromagnetic material is heated above its Curie temperature, these effects disappear. The magnetic dipoles of a ferromagnetic solid are aligned parallel; dipoles in a ferromagnetic solid can be aligned in parallel or in other directions (Takagi, 1990; Vardan and Vardan, 2002; Fairweather, 1998).

Magneto-restrictive materials are typically used in low-frequency, high-power sonar transducers, motors, and hydraulic actuators; along with the shape memory

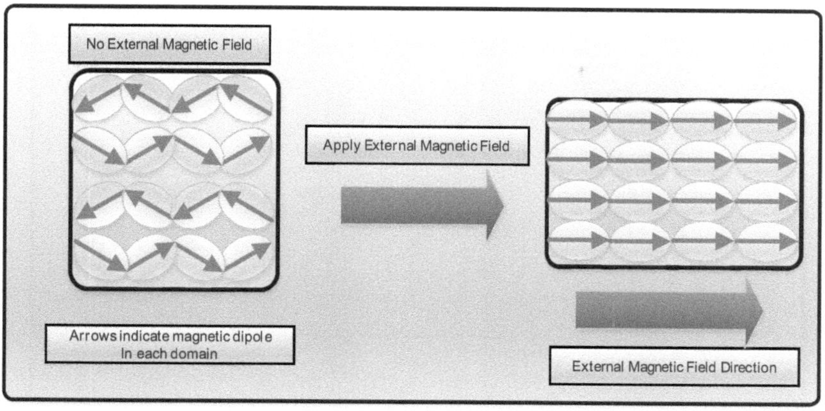

FIGURE 8.4 Internal mechanism of magneto restrictive material. *Adapted from Kaushal, A., Vardhan, A., Rawat, R.S.S., May–June 2016. Intelligent material for modern age: a review. IOSR Journal of Mechanical and Civil Engineering (IOSR-jmce), e-ISSN: 2278–1684, ISSN: 2320-334X, 13, 3, 10–15.*

alloy nitinol, magneto-restrictive materials are considered promising candidates for the active damping of vibrations. The internal mechanism of magneto-restrictive material is given in Fig. 8.4 (Kaushal et al., 2016).

8.1.3.5 Shape Memory Alloys

SMAs are a very special type of alloy. Consider a metal on which some kind of deformation force is applied: the metal becomes deformed, but when heat is applied, it returns to its original shape and size. This phenomenon is called the shape memory effect. SMAs undergo thermomechanical change as they pass from one phase to another (Kaushal et al., 2016):

- When metal is cooled below the Curie temperature, it passes into the martensitic phase.
- When the temperature increases above the critical temperature, the metal transforms from martensitic into austenitic.
- Finally, the metal regains its original shape.

Fig. 8.5 shows the shape memory effect.
In Fig. 8.5,

- "a" represents the austenitic phase of metal,
- "b" represents the deformed pattern of metal,
- "c" represents the deformed martensitic phase, and
- "a" represents the original regained position after heating.

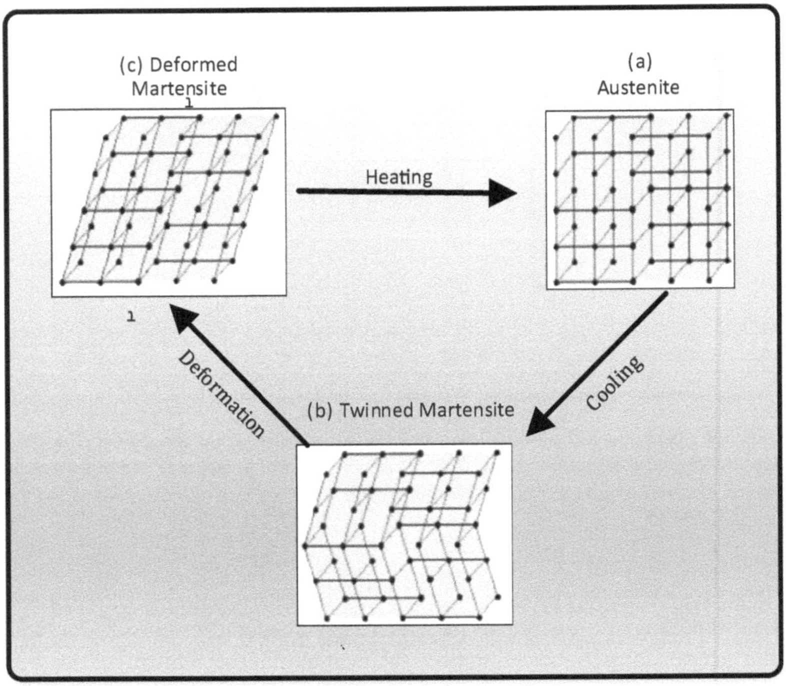

FIGURE 8.5 Mechanism of shape memory effect. *Adapted from Kaushal, A., Vardhan, A., Rawat, R.S.S., May—June 2016. Intelligent material for modern age: a review. IOSR Journal of Mechanical and Civil Engineering (IOSR-jmce), e-ISSN: 2278—1684, ISSN: 2320-334X, 13, 3, 10—15.*

The following shape memory alloys are common: nickel—titanium (nitinol family), silver—cadmium, gold—cadmium, copper—aluminum—nickel, copper—tin, copper—zinc, combinations of copper—zinc with silicon or tin or aluminum, indium—thallium, nickel—aluminum, iron—platinum, manganese—copper, and iron—manganese—silicon (Kaushal et al., 2016).

8.1.3.6 Thermoresponsive Materials

Amorphous and semicrystalline thermoplastic polymeric materials are unique because of the presence of a glass transition temperature. Changes in the specific volume of polymers and their rate of change occur at their glass transition temperatures. This transition affects a multitude of physical properties. Numerous types of indicating devices could be developed based on the stimulus-response (temperature-specific volume) behavior (Shaikh et al., 2015).

8.1.3.7 Optical Fibers

Optical fibers use intensity, phase, frequency, or polarization of modulation to measure strain, temperature, electrical/magnetic fields, pressure, and other measurable quantities. They are excellent sensors (Akhras, 2000).

8.1.3.8 pH-Sensitive Materials

pH-sensitive materials change their color as a result of changing acidity. This can be applicable for paints that can change color to indicate corrosion in the metal underneath them.

8.1.3.9 Chromogenic Systems

Chromogenic systems change color in response to electrical, optical, or thermal changes. These include electrochromic materials, which change their color or opacity on application of a voltage. A very common example is liquid crystal displays. They also include photochromic materials, which change color in response to light. This can be observed in light sensitive spectacles that darken when exposed to bright sunlight. Finally, certain paints, such as thermochromic and photochromic paints, change color on heating and on exposure to light, respectively.

8.1.3.10 Polymer Gels

Polymer gels, such as hydrogels, have a cross-linked polymer structure inflated with a solvent, such as water; the gel has the ability to swell or shrink (up to 1000 times in volume) with small changes in, e.g., temperature or pH (Susmita, 2013).

8.1.4 Classification of Smart Materials

Smart materials can be classified as active or passive. Fairweather (1998) defines active smart materials as those possessing the capacity to modify their geometric or material properties under the application of electric, thermal, or magnetic fields, thereby acquiring an inherent capacity to transduce energy. Piezoelectric materials, shape memory alloys (SMAs), electrorheological fluids, and magneto-restrictive materials are considered active smart materials and can be used as force transducers and actuators. Kumar (1991) shows that SMA has a large recovery force, of the order of 700 MPa (10^5 psi), and this can be used for actuation. Similarly piezoelectric materials, which convert electric energy into mechanical force, are "active" (Susmita, 2013). In contrast, materials that are inactive to response and functioning are called passive smart materials (Kaushal et al., 2016). Although smart, they lack the inherent capability to transduce energy. Fiber optic material is a good example of a passive smart material. Such materials can act as sensors but not as actuators or transducers (Susmita, 2013). Fig. 8.3 shows the behavior of some smart/intelligent material responses (Kaushal et al., 2016).

8.1.5 Applications of Smart Materials

Smart materials find a wide range of applications because of their varied responses to external stimuli. Their applications can be in our daily lives, or in aerospace, civil engineering, or mechatronics to name a few. They solve engineering problems with efficiency and provide opportunities to create new products that generate revenue. They encompass all fields of science and engineering.

Technical applications of smart materials include composite materials embedded with fiber optics, actuators, sensors, microelectro mechanical systems (MEMSs), vibration control, sound control, shape control, product health or lifetime monitoring, cure monitoring, intelligent processing, active and passive controls, self-repair (healing), artificial organs, novel indicating devices, designed magnets, damping aeroelastic stability, and stress distributions. Smart structures are found in automobiles, space systems, fixed- and rotary-wing aircrafts, naval vessels, civil structures, machine tools, and recreation and medical devices.

The kind of "smartness" shown by materials is generally programmed by material composition, special processing, the introduction of defects, or the modification of the microstructure, so as to adapt to the various levels of stimuli in a controlled fashion. The terms smart and intelligent are used interchangeably. Takagi (1990) defines intelligent materials as materials that respond to environmental changes at the most optimum conditions and manifest their own functions according to the environment. The feedback functions within the materials are combined with their properties and functions (Susmita, 2013).

Smart systems also sense their environment and respond but are not constructed from a single material. They may incorporate smart materials but can be constructed using traditional technology. Pacemakers are smart systems designed to respond to an irregular heart rate with an electrical impulse that regulates it. Vardan and Vardan (2002) say smart system refers to a device that can sense changes in its environment and can make an optimal response by changing its material properties, geometry, mechanical, or electromagnetic response. The sensor and the actuator functions, along with their appropriate feedback, must be properly integrated. According to Takagi (1990), a system could lose its application or become dangerous if the response is too slow or too fast.

Some common smart materials and associated stimulus responses are given in Fig. 8.6.

Smart materials and systems have a wide range of applications. Investment in their research and development is driven by factors such as legislation, reducing waste, and demand for higher quality of life. Structures such as buildings, bridges, pipelines, ships, and aircraft must be strongly designed and regularly inspected to prevent "wear and tear" damage from causing disastrous

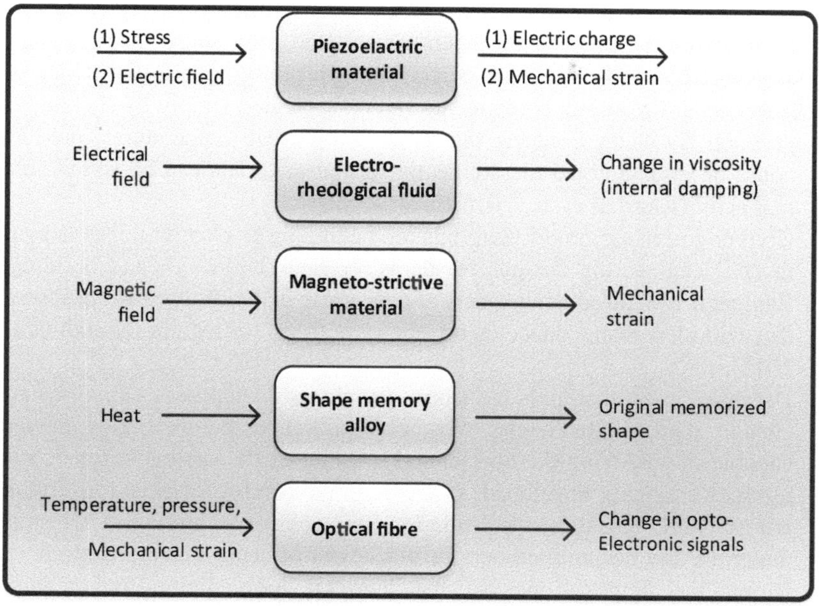

FIGURE 8.6 Behavior of smart materials and their associated stimulus-response. *Adapted from Kaushal, A., Vardhan, A., Rawat, R.S.S., May–June 2016. Intelligent material for modern age: a review. IOSR Journal of Mechanical and Civil Engineering (IOSR-JMCE), e-ISSN: 2278–1684, ISSN: 2320-334X, 13, 3, 10–15.*

failures. Inspection is expensive and time consuming, while designing to prevent damage can compromise performance. With some modern materials, damage can be internally serious but leave very little surface evidence (Susmita, 2013). This calls for more intensive use of smart materials and systems.

8.1.5.1 Some Specific Applications

- Aircraft: Sensors built into aircraft will let maintenance people know when repairs or more sophisticated inspections are required.
- Nuclear industries: Smart technology offers new opportunities in the nuclear industrial sector for safety enhancement, personal exposure reduction, life cycle cost reduction, and performance improvement. However, the radiation environments associated with nuclear operations represent a unique challenge to the testing, qualification, and use of smart materials; the use of smart materials in such facilities requires knowledge about how materials respond to irradiation (Shaikh et al., 2015).
- Waste reduction: All over the world, electronic waste is the fastest growing type of domestic waste. Hazardous and recyclable materials must be removed from the waste before disposal or processing. Manual disassembly

is expensive and time-consuming, and the use of smart materials could help to automate the process. Fasteners constructed from shape memory materials that can self-release on heating have recently been tested. Once the fasteners are released, components can be separated simply by shaking the product. By using fasteners that react to different temperatures, products could be disassembled hierarchically so that materials can be sorted automatically (Kaushal et al., 2016).

- Civil engineering: Smart materials are beginning to play an important role in civil engineering designs for dams, bridges, highways, and buildings. Engineers are introducing sheets of composite materials containing sensors that will alert maintenance engineers to the need for repairs (Shaikh et al., 2015).
- Defense: Smart materials have been developed to suppress vibrations and change shape in helicopter rotor blades. Shape memory alloy devices capable of achieving the accelerated breakup of the vortex waves of submarines are being developed, as are adaptive control surfaces for airplane wings. Some research is considering new control technologies for smart materials and design methods for placement of sensors and actuators.
- Biomedical applications: In the field of biomedicine and medical diagnostics, investigations are underway. Certain materials such as polyelectrolyte gels are being tested for artificial-muscle applications; a polymer matrix is swollen with a solvent that can expand or contract when exposed to an electric field or other stimulation. In addition, because these materials are biodegradable, they may be useful as drug-delivery systems.
- Structural engineering: Smart materials are used to monitor structures to evaluate their durability. Smart materials and structures can adapt to their surrounding environment, with the ability to move, vibrate, and demonstrate various other responses.
- Health: Biosensors made from smart materials can be used to monitor blood sugar levels in diabetics and communicate with a pump that administers insulin as required. However, the human body is a hostile environment and sensors are easily damaged. Some research on barrier materials is looking for ways to protect these sensors. Companies are developing smart orthopedic implants such as fracture plates that can sense whether bones are healing and communicate data to the surgeon (Kaushal et al., 2016).
- Automotive industry: Projects are underway to develop smart car seats that can identify primary occupants and adapt to their preferences for height, leg-room, back support, and so forth. Technology also exists to enable cars to tell owners how much air pressure there is in the tires, when the oil needs changing, and other maintenance information. Developing solid state and smart materials technologies will bring costs down.
- Magnetorheological (MR) fluid: The MR damper developed at the Intelligent Structures and Systems Laboratory uses a mixed mode configuration,

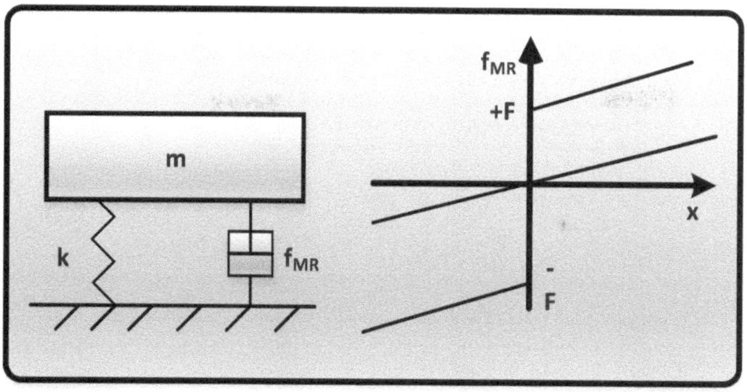

FIGURE 8.7 Simple mass-spring magnetorheological (MR) damper system. *Adapted from Shaikh, A.M., Patil, K.B., Bazar, S., 2015. Smart materials & its applications. International Journal of Emerging Technologies in Computational and Applied Sciences (IJETCAS), 15–371.*

as shown in Fig. 8.7. The MR damper has a built-in MR valve across which the MR fluid is forced. The piston of the MR damper acts as an electromagnet with coils producing the appropriate magnetic field. The MR damper has a run-through shaft to avoid an accumulator for the design of cars (Shaikh et al., 2015). Since MR fluids have a very high yield stress, they can be used to control the transmission of torque in rotating machinery. For example, they may be used in an automotive power train to transmit torque from the engine to the transmission and the vehicle. MR dampers can be used to achieve sliding mode control in vibrating systems such as seat suspensions, vehicle suspensions, washing machines, and prosthetic devices. Finally, they can be used for modeling and control of active spiral and patch antennas (Shaikh et al., 2015).

- Self-healing of polymeric and composite materials: The remarkable ability to cure wounds is one of the most important inherent properties of biological systems. A bioinspired self-healing material has a built-in capability to partially repair damage sustained during service. Material properties usually degrade over time because of the initiation and growth of damage such as microcracks. The biological world contains a range of mechanisms that continuously sense and repair damage. In the materials science field, researchers are now attempting to engineer this type of behavior in man-made materials, under the banner of "self-healing" materials. Self-healing structures are particularly important for defense platforms consisting of polymeric materials and composites. The application of self-healing materials will help in the prevention of degradation, thus improving platform safety (White et al., 2001; Kessler et al., 2003).
- Structronic systems: Active controllable smart electro/magneto material provides new dimensions and design opportunities for precision and high-performance devices, structures, mechatronic systems, and structronic systems. Integrating smart materials, sensors/actuators, computers, control

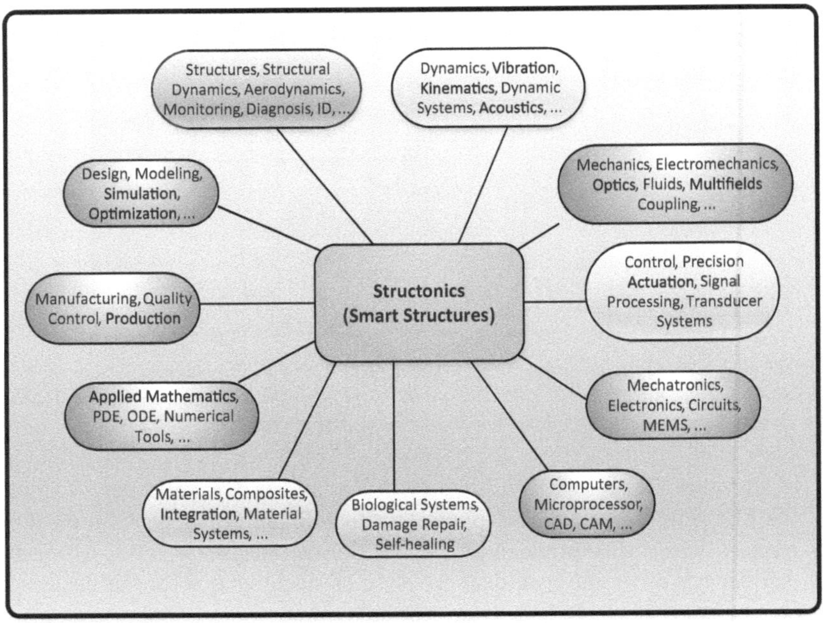

FIGURE 8.8 Structronics: Multidiscipline integration. *Adapted from Shaikh, A.M., Patil, K.B., Bazar, S., 2015. Smart materials & its applications. International Journal of Emerging Technologies in Computational and Applied Sciences (IJETCAS), 15–371.*

electronics, and artificial intelligence enhances conventional mechatronic devices or systems to a new generation of fully integrated smart structures and structronic systems. The new smart structures and structronic systems technology represents a systematic synergistic integration of conventional disciplines and applications, as shown in Fig. 8.8 (Lee and Arnold, 2004).

8.1.6 Future Trends in Smart Materials

The active materials and smart structures technology has gone past basic research and laboratory prototypes to the development of industrial applications. Until now, the main driving forces have been military and aerospace applications, but more exciting applications are being reported. These developments show the great potential of smart structures (Shaikh et al., 2015).

8.1.6.1 Issues of Smart Structures and Structronic Systems

Ongoing research is considering the following:

- Nonlinear modeling, simulation, and design tools/control structure interaction/benchmark problems;
- System integration and system design criteria, standards, tools, limits;

- Material processing, new material with enhanced engineering properties and temperature stability, restrictions, material library/database, diffusion-driven smart materials;
- New material evaluation techniques, tools;
- Product demonstrations, new mechanisms, new materials applied to MEMS;
- Applications to manufacturing;
- Material incompatibility, material/structure integration;
- Micromechanics: bonding, fracture, fatigue, nonlinear behavior, etc.;
- Health monitoring and diagnosis, nondestructive evaluation/nondestructive testing modeling tool;
- Real-time system identification and control, implementation, electronics (power efficiency);
- Biological-inspired structures, self-growth/repair;
- Education (industry, public, government, and students) network, Web information, "system" approach; and
- Distributed control of continua via structronics technology, demonstrations (Gabbert and Tzou, 2001).

8.1.7 Improvement in Intelligent/Smart Materials

Improvements may soon be possible (Kaushal et al., 2016) in the following specific areas:

- Damage recovery: Materials with the ability to restrain crack propagation caused by stress intensity around a damaged point.
- Shock behavior detection: Materials with the ability to create resistance and perform automatic detection of static or dynamic loads.
- Self-healing: Materials with the capability of self-reparability and maintainability.
- Thermal mitigation: Materials usable up to ultrahigh temperatures by suitably changing composition through transformation.

8.2 SMART DEVICES WITH ACTUATION CAPABILITIES: SMART BEARINGS

8.2.1 Smart Devices

Smart devices are electronic tools capable of operating interactively and autonomously; they are usually networked. The term may refer to many different types of pieces of electronic equipment, ranging from devices that are principally manipulated by individuals, such as smartphones, to the constitutive elements of today's "ubiquitous computing," that is, an environment with pervasive sensors and information-processing capability.

In this context, smart devices are closely related to the Internet of Things (IoT). IoT relates to a cluster of objects that are readable and/or controllable via the Internet or other technologies such as radio frequency identification; these objects sometimes communicate with each other without human interference. In some cases, smart devices can be everyday objects, simple "things" that beforehand lacked electronic components but are now embedded with sensors and microprocessors. The adjective "smart" is sometimes understood as referring to the fact that a device enjoys machine learning and adaptive capabilities, allowing it to program its activity on the basis of gathered data.

When considered through the lens of big data, the main attraction of smart devices is that they can feed data mining practices and make use of the information obtained through them; most smart devices do both, simultaneously. They are key enablers of the current data explosion (González Fuster and Scherrer, 2015).

The production of data by smart devices can occur in many different ways. Smart devices often include sensors designed for data capture. These generate continuous streams of data, such as temperature, waves, movements, or other variables. Further data might be created through information processing. In any case, smart devices generate data that inevitably include indexical data, that is, data allowing the identification of the produced data sets and the linking of data sets with other data sets. The pervasiveness of sensors and extensive routine data production might not be fully understood by individuals, who may be unaware of the presence of sensors (which are often low-cost and miniscule) and of the full spectrum of data they produce, not to mention the data processing operations treating these diverse data. Smart devices are sometimes labeled "Everyware"; that is, they embody the penetration of information technology into daily life (Greenfield, 2006).

We have already witnessed the rise of personal devices relying on the processing of vast quantities of data. The first mobile telephones with Internet connectivity and the first email-enabled mobile phone systems appeared in the 1990s. In the 2000s, smartphones opened the door to the widespread use of mobile Web applications, producing and using great quantities of data, including data about the device's location. Since 2010, tablet devices have added to the proliferation of applications.

Wearable devices, that is, accessories or clothing incorporating advanced electronic technologies, are special types of smart devices. Although their origins can be traced to the 1980s calculator watch, today's "wearables" integrate connectivity features. Modern smartwatches, for instance, typically run mobile applications and may function as mobile phones. The paradigmatic example of a smart wearable device is the Google Glass prototype (marketed from April 2013 to January 2015), an optical head-mounted display developed by Google that allows wearers to communicate with the Internet through voice commands (González Fuster and Scherrer, 2015).

8.2.2 Smart Device Paradigm

The smart device paradigm revisits the traditional client–server architecture, on which many remote laboratory implementations rely. The main differences between existing implementations and those of smart devices are, first, the complete decoupling of the server and the client and, second, the server representation as a set of well-defined services and functionalities that enable interoperability (Salzmann and Gillet, 2013; Gillet et al., 2013; Tawfik et al., 2014). Similar approaches are proposed at the sensor/actuator level to enable the plug and play mechanism for smart electronic transducers, thus providing electronic data sheets describing themselves (IEEE, 2007).

Decoupling allows client and server to live separate lives. In a traditional client-server architecture (Chen et al., 2010), the server and client share a specification that is often uniquely used by them. In the smart device paradigm, a common specification is shared by all smart devices. This reuse of a common specification and the client-server decoupling alleviates most of the problems developers are facing when a client application needs to be adapted to new operating system platforms, or if the client application must be integrated in other environments such as learning management systems (LMS), or if additional features are added to the server. Finally, interoperability with and reuse of existing applications and services become possible when laboratories share a common specification (Salzmann et al., 2015).

Smart devices provide Web services to access sensors and actuators. Traditional solutions often provide a monolithic interface without the possibility of specifically accessing a given sensor or actuator (Salzmann and Gillet, 2011). The smart device specification fully describes the smart device from a client's point of view by specifying only the interfaces, not the inner working of the laboratories, which is left to the laboratory owner's discretion. The smart device specification is indifferent to the server-side hardware but reengineers the software component by adding "intelligence" to handle complex tasks through the application programming interface (API) (Salzmann et al., 2015).

There is no assumption about communication channels for smart devices (Cascado, 2011), but the Internet is the de facto choice for online laboratories (Salzmann and Gillet, 2008; Auer et al., 2003). In addition, open Web technologies enable a broader compatibility and adoption, while proprietary technologies break the core requirement of ubiquitous access.

The smart device may not necessarily provide a user interface (UI) but often proposes a minimal client UI. Thanks to the interoperability provided by the smart device specification, client applications can be developed to operate with different smart devices, thus promoting reuse. Due to their ubiquity, Web browsers are the preferred environments to render the client UI. There is often a direct relation between the smart device sensors and actuators and the client app rendering their information. For example, an oscilloscope app renders the voltage evolution measured by a sensor of the smart device. In general, the smart device paradigm defines an ideal autonomous device that provides internal functionalities and can be accessed through well-defined services (Salzmann et al., 2015).

8.2.3 Smart Device Architecture

The smart device specification provides a set of well-defined interfaces that enable communication between the remote laboratory, external services, and applications. Fig. 8.9 illustrates a basic architecture with interaction examples that abstract the implementation of a remote laboratory, by providing a set of required and optional interfaces. The specification in the figure does not define the communication between the smart device and the remote laboratory equipment. The communication at the center left of the figure is what the smart device specifies, namely the protocols and data formats of the interfaces of the smart device (i.e., the "metadata," "client," "sensor," "actuator," and "logging" interface). Now, when we look at the far left, we see a list of clients with access to those protocols and data formats. For instance, a metadata repository can retrieve the metadata of any smart device, index it, and provide a laboratory search engine. Because the interfaces are well-defined, client apps can be reused among smart devices; for example, one data-viewer client or learning-analytics client could retrieve data from any smart device and present it to the user (Salzmann et al., 2015).

8.2.4 Smart Device Specification

This section describes the communication protocol and explains the terminology for smart devices. Then, it elaborates on smart device services and internal functionalities (Salzmann et al., 2015).

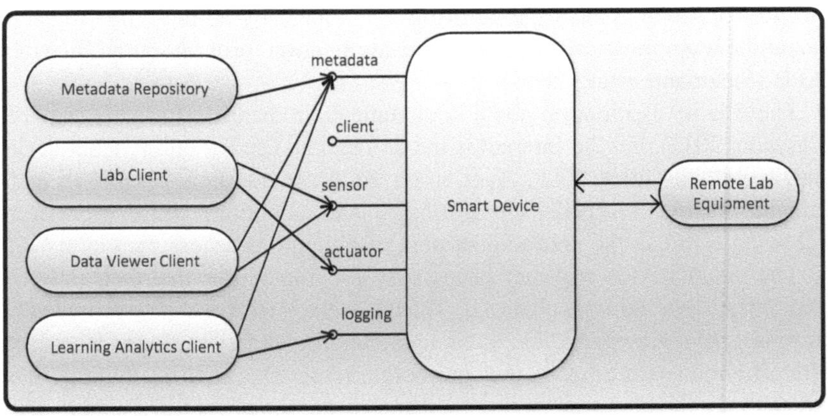

FIGURE 8.9 Unified Modeling Language Component Diagram of Clients using Smart Device Services (arrows represent calls). *Adapted from Salzmann, C., Govaerts, S., Halimi, W., Gillet, D., 2015. The smart device specification for remote labs. Ecole Polytechnique Fédérale de Lausanne (EPFL), Lausanne, Switzerland. 11, 4.*

8.2.4.1 Data Transfer Protocol

The goal of a smart device is to enable access to remote laboratories via Internet. The targeted client application is a Web-enabled client, which can run on a tablet. Open, standardized Web protocols provide the data transfer between the smart device, external services, and applications to avoid dedicated plug-ins or customer lock-in. Widely used examples are HTTP and WebSockets.

The problem with most HTTP-based Web services is that they follow a synchronous request-response schema. Hence, data can often only be "pulled" from the server, and the server cannot initiate a "push" of information to the clients. However, remote laboratory experiments often require asynchronous data transfer; for example, a lengthy experiment should be able to push its results to clients on completion. Thus, HTTP solutions are often inefficient (Loreto et al., 2011).

In contrast, WebSockets (Fette, 2011) are asynchronous by nature and allow both pushing and pulling. This provides a bidirectional, full-duplex, and efficient communication channel. Although WebSockets is a recent technology, it is supported by all modern browsers. As WebSockets support push and pull technologies efficiently, often with less programming effort than HTTP-based services, the smart device specification uses the WebSocket protocol. Only a metadata service that defines other services (discussed in a following section) is provided via HTTP GET to enable easy text retrieval (Salzmann et al., 2015).

8.2.4.2 Terminology and Concepts

- The terms sensors and actuators reflect the traveling direction of information relative to the smart device. For example, a sensor enables the reading of a thermometer. An actuator enables the setting of a value, e.g., setting a motor voltage.
- Sensors and actuators can be physical (temperature sensor), virtual (computed speed derived from a position measurement), or complex, i.e., an aggregation of sensors/actuators (the front panel buttons of an oscilloscope or a 3D accelerometer).
- Both sensors and actuators can be configured (Salzmann et al., 2015)

8.2.4.3 Metadata Service

The metadata service is a required service that is at the core of the interoperability provided by the smart device specification.

The metadata are required to do the following (Salzmann et al., 2015):

- Describe the laboratory (e.g., contact person and goals), often useful for automatic indexing by search engines.
- Describe integration with external services (e.g., authentication with a booking service).
- Describe concurrent mechanisms (e.g., are laboratory observations allowed while someone is doing an experiment?).

- Describe and define provided services (e.g., specify service request and response formats).
- Be easily extensible to enable adding extra services.

8.2.4.4 Sensor Metadata Service: getSensorMetadata

As mentioned, the sensor and actuator metadata are provided via separate services and are not in the metadata description itself. The sensor metadata service is called "getSensorMetadata." Like most smart device services, it calls up a JSON object by specifying the "method" field and an optional authentication token in case booking is required. As mentioned, this allows one WebSocket to channel multiple services. The service returns an array describing each sensor exposed to the outside world. Each sensor contains the following (Salzmann et al., 2015):

- An ID to identify the sensor, e.g., "3D-acc."
- The full name, e.g., "3D acceleration."
- The description, e.g., "the robot arm 3D acceleration."
- The WebSocket type i.e., "text" or "binary" (e.g., for video).
- The response type of the sensor service for the sensor defined as an Internet media type (Freed et al., 2014); for example, a Webcam sensor using JPEG compression uses image/JPEG.
- A measurement value array: this will contain a single value for a simple sensor such as a thermometer, but for a complex sensor such as an accelerometer, it contains more, for example, three elements for X-Y-Z acceleration. Values are described with a name and unit. As the set of possible units is almost infinite, we recommend using SI units (Taylor and Thompson, 2008) and SI-derived units (International System of Units, or SI). A last measured timestamp and a range minimum, maximum, and iteration step of the range in which the values operate safely can be added. For continuously measured values, the frequency at which the measurement is updated can be provided in Hertz (s^{-1}).
- Configuration parameters: these can be used to adjust the sensor when requesting a sensor value (see Section 8.2.4.6). Each parameter has a name and data type as a JSON schema primitive, array, or data model for complex parameters, e.g., to configure the video resolution.
- Access mode: this describes how the sensor can be accessed; for example, some sensors can be measured once (pull), while others provide a continuous data stream (push or stream). For "push" sensors, we can specify the nominal update interval and whether the user can modify the measurement frequency.

Both sensors and actuators can be configured, so information can be sent and received, even for the sensor. For example, the image resolution of a Webcam sensor can be configured. Similarly, for actuators, some aspects may be set by configuration (e.g., the gain of a power amplifier could be configured), while the actual value is set by the actuator value itself (discussed in a following section). Sensors and actuators are rarely configured, however (Salzmann et al., 2015).

Streaming video of an experiment is an essential service that a smart device can provide through a sensor. The sensor should treat the video image as an encoded image, for example JPEG encoded. Using JPEG encoding results in binary data, which should be transmitted through a binary WebSocket (recommended); alternatively, it could be BinHex'ed using a textual WebSocket prior to sending. If further processing is required at the client side, a pixmap (pixel array) could be used, but size would increase from 10% to 90% (Furht, 1995).

8.2.4.5 Actuator Metadata Service: getActuatorMetadata

Actuator metadata are provided via "getActuatorMetadata." As the service is very similar to the sensor metadata service, we will only note the difference in the service response. The input type expresses what data type can be used for a specific actuator in the actuator service. By default this is JSON, but it can be set to any Internet media type (Freed et al., 2014). This replaces the response type of the sensor metadata service (Salzmann et al., 2015).

8.2.4.6 Sensor Service: getSensorData

The sensor and actuator data services are at the core of smart device interactions and are quite similar. They handle the main data exchange between clients and the smart device. Both services, in combination with their metadata services, enable developers to create apps that can adapt to different smart devices, enabling app reuse and interoperability. Similarly, different apps could be developed for a smart device. In the case of a smart device that provides a temperature measurement every second, one app could update a text field and another could visualize the temperature evolution over time. This difference in app functionality requires absolutely no change in the smart device services. Furthermore, by using the sensor metadata service, these two proposed apps could be made interoperable and reusable with any smart device (Salzmann et al., 2015).

Sensors can be real, virtual, or complex (also see Section 8.2.4.2):

- *Real*: a physical sensor, e.g., a thermometer.
- *Virtual*: a computed sensor, e.g., a speed measurement derived from a position measurement.
- *Complex*: the aggregation of sensors/actuators, e.g., buttons on the front panel of an oscilloscope.

The data structure returned by a sensor or sent to an actuator may vary depending on the number of values and the measurement data structures. The data structure shown in Fig. 8.10 has three fields to enable flexible data representation. In the "valueNames" field, the names of the sensor or actuator

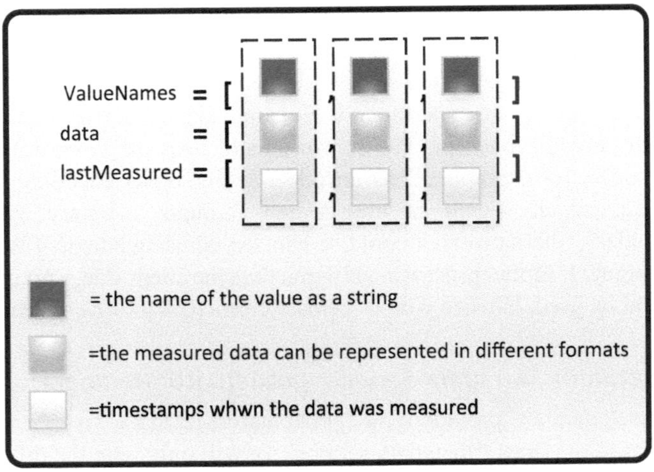

FIGURE 8.10 Sensor and Actuator data structures *Adapted from Salzmann, C., Govaerts, S., Halimi, W., Gillet, D., 2015. The smart device specification for remote labs. Ecole Polytechnique Fédérale de Lausanne (EPFL), Lausanne, Switzerland. 11, 4.*

measurement values are listed as returned by the sensor or actuator metadata services. Then, the actual data for each value are listed. Finally, the optional "lastMeasured" array contains the timestamps when a value was measured. This timestamp array should not be included when sending data to set an actuator. The data, as well as the "lastMeasured" timestamps, are listed in the same array index as the value name, as indicated by the dashed lines in Fig. 8.10.

The elements in the data array can be in different formats:

1. A single value, e.g., temperature;
2. An array of values representing a set of single values over time, e.g., temperatures over the last minute;
3. Aggregated values representing a sensor or actuator that returns multiple values, e.g., a 3D accelerometer;
4. An array of aggregated values representing a set of aggregated values over time, e.g., 3D acceleration over the last minute;
5. Complex data structures: when sensors and actuators require input and output not definable using primitive variables or arrays, e.g., for complex JSON objects or binary data. Note that flat array-based data can be more efficient to process than complex data structures interleaved with timestamps (Salzmann et al., 2015; Fig. 8.11).

As an example, a Webcam can be modeled as a single value sensor that returns a compressed image, as an array of values based on the image bitmap, or as a binary value with JPEG encoded data. The choice between the three representations is up to the laboratory owner.

A request to the getSensorData service is more complex than other services because of possible authentication, concurrency, and configuration settings.

FIGURE 8.11 Self-alignment bearing. *Adapted from Galar, D., Kumar, U., June 6–7, 2013. SMART Bearings: From Sensing to Actuation. 12th IMEKO TC10 Workshop on Technical Diagnostics. New Perspectives in Measurements, Tools and Techniques for Industrial Applications, Florence, Italy.*

An option is to use an access role from the concurrency role list. If no access role is available, the smart device can decide the role. It will decide whether these rights can be granted and react accordingly.

If the user has the controller role, the getSensorData service will return the data in the format described above (see Fig. 8.10), together with the method name, sensor ID, and access role to foster possible WebSocket reuse. But if the user is an observer and does not have access to the measured data, the service can provide extra waiting information displaying how long the user has to wait and how many people are in front of him/her (e.g., the queue size, position, and waiting time left). The sensor configuration might also be used (e.g., for a video sensor) if it is described in the sensor metadata. For example, this can be very useful to adapt to the client screen size and network speed by reducing the transmitted image resolution and compression (if configurable). Similarly, the data transmission pace can be controlled. If the user temporarily needs to throttle the video stream, the client can ask the smart device to reduce the number of images sent per second by setting the update frequency. The sending may even be interrupted by setting the update frequency to 0 Hz. It is up to the application developer to take advantage of these features (Salzmann et al., 2015).

8.2.4.7 Actuator Service: sendActuatorData

As noted above, the actuator service is very similar to the sensor service. The main difference is that the sendActuatorData service allows the user to set the desired actuator value; that is, the data model shown in Fig. 8.10 is sent in the request.

The internal functionality of the smart device should validate the sent value before applying it to the actuator. While sensors often do not have concurrency issues, the actuator access needs to be moderated. Various schemas can be implemented by the laboratory owner to internally manage the actuator access (Salzmann et al., 2015).

8.2.4.8 User Activity Logging Service: getLoggingInfo

The optional user activity logging service returns logged user actions or laboratory status information in the ActivityStream 1.0 JSON format. The ActivityStreams format is a JSON-based specification to describe a sequence of user actions with a timestamp, often used in social media platforms. To retrieve a continuous stream of real-time user activities of smart devices, the getLoggingInfo service can be called up with an optional authentication token to validate access (recommended for privacy sensitive data) (Salzmann et al., 2015).

8.2.4.9 Client Application Service: getClients

This optional service provides links to the client applications to operate the smart device. The client technology is not strongly specified. When a request is sent to the getClients service, a client app list is returned; each item has a type specifying the kind of application and a URL. The current version of the smart device specification contains the following extensible list of types: "OpenSocial Gadget," "W3C widget," "Web page," "Java WebStart," and "Desktop application" (Salzmann et al., 2015).

8.2.4.10 Model Service: getModels

This optional service can provide several models of the physical laboratory (i.e., the instrumentation) and its theoretical background. For instance, a 3D graphical model of the laboratory instrumentation can enable a client app to generate a graphical user interface with a 3D-scale object that can be manipulated. If there is a mathematical model of the experiment, a client app can be built with a local simulation. This can provide an interactive simulated version of a remote laboratory that can be used when the laboratory is already in use (i.e., to provide a better observer mode). The wide range of existing formats to express graphical and theoretical models (e.g., VRML, X3D, and MathML) means the specification is not limited and the choice of model language is left up to the laboratory owner (Salzmann et al., 2015).

8.2.4.11 Functionalities: Best Practices

Internal functionalities are implementation suggestions for smart devices. They are considered best practices as their implementation is often ad-hoc and strongly related to the connected equipment (Salzmann et al., 2015).

1. *Authentication functionality*: The smart device does not need to contain a booking system. It can make use of an external booking system. At the booked time, the user can connect to the smart device with this authentication token. The smart device then contacts the booking system to validate whether the user is currently allowed to access the smart device. Thus, integrating the booking service into the smart device requires little effort, compared to providing authentication and booking mechanisms (Salzmann et al., 2015).
2. *Self- and known state functionality*: The precise implementation of this recommended functionality is left to the laboratory owner's discretion. This functionality ensures that a remote laboratory is reset to a proper state after an experimentation session is completed or a system outage has occurred, so that the next user can properly use it. As remote experiments are supposed to be conducted at a distance, no one is expected to be around to put the laboratory/system back into a known state. Thus, the system should be as autonomous as possible; this implies an adequate and defensive software and hardware design that can adapt to "any" situation. To this end, the following procedures should be implemented in the smart device:
   - **a.** Perform automatic initialization at startup,
   - **b.** Reset to a known state after the last client disconnects, and
   - **c.** Have the potential for hardware calibration (Salzmann et al., 2015).
3. *Security and local control*: This functionality is recommended, but its implementation is left to the laboratory owner's discretion. The security of the server and its connected equipment must be ensured at all times. All commands should be validated before being forwarded to the connected equipment. This may require the addition of a local controller to track the connected equipment's state; for example, a speed increase may need to follow a ramp before being applied to a motor. Users often try to take a system to its limits; that is, not only the physical limits of a given sensor/actuator but also the signal patterns on a sensor over time may require consideration. As actuators may be connected to the Internet, it is essential to validate all applied values and to consider potential external constraints. The laboratory owner should implement the following procedures in the smart device:
   - **a.** Value validation before applying data to actuators, and
   - **b.** Actuator state validation to check that the command to be applied is safe (Salzmann et al., 2015).

4. *Logging and alarms*: This functionality logs session and laboratory information, as well as user interactions. In case of problems, alarms may be automatically triggered by this functionality. As a smart device will be typically online unattended for an extended period of time, it is essential to monitor it and have a method to perform post hoc analysis. The user action should be logged and made accessible via the user activity logging service. But extra information should also be logged, e.g., the system state and the environment (e.g., room temperature). Note that some sensors may be available internally to the smart device but not necessarily accessible via the sensor service. The following information should be tracked:
   a. User actions,
   b. The complete system state, and
   c. Its environment state.

   By definition, the smart device is connected to the Internet and has no knowledge of its clients. Therefore, action is required to prevent its abuse. A firewall or a DMZ18 may protect it from attacks. While some hostile actions may be reduced using such mechanisms, the smart device should add the following internal measures:
   a. Validate the requests sent by clients,
   b. Throttle continuous requests of a malicious client, and
   c. Log all internet connections for later analysis.

   If an unexpected event occurs, its potential danger should be assessed by the smart device, with an alarm possibly triggered (Salzmann et al., 2015).
5. *Local simulation*: When the experiment is busy or unavailable, a local simulation might be a useful alternative for waiting users. The simulation data could be read or modified by virtual sensors/actuators. A mathematical model describing the physical equipment can be made available to the client via the model service, which the client developer can use to simulate the hardware. Such simulations can require computational resources unavailable to the client. However, this computation can be done server side, with the results sent to the client using virtual sensors and actuators (Salzmann et al., 2015).

8.2.5 Smart Bearings: From Sensing to Actuation

Bearings are among the most important components in most machines, with exacting demands made on their carrying capacity and reliability. Not surprisingly, then, rolling bearings have been the subject of extensive research, with rolling bearing technology developing into a particular branch of science. The Swedish company SKF has been in the forefront from the start and has long led the field.

Among the benefits resulting from the research performed and supported by SKF is the ability to reduce the risk of premature or in-service bearing failure. By sensing the operating environment in which the bearing functions, the risk of a future failure developing can be identified and used as a basis to plan maintenance and other interventions. Moreover, by using this knowledge to trigger actions to eliminate damaging operating conditions, future failures may be avoided.

Unfortunately, a rolling bearing cannot rotate forever. Unless operating conditions are ideal and the fatigue load limit is not reached, sooner or later material fatigue will occur. The period until the first sign of fatigue appears is a function of the number of revolutions performed by the bearing and the magnitude of the load. Fatigue is the result of shear stresses cyclically appearing immediately below the load-carrying surface. After a time, these stresses cause cracks that gradually extend to the surface. As the rolling elements pass over the cracks, fragments of material break away; this is called flaking or spalling. The flaking progressively increases in extent and eventually makes the bearing unserviceable. The life of a rolling bearing is defined as the number of revolutions the bearing can perform before incipient flaking occurs. This does not mean to say that the bearing cannot be used after this point. Flaking is a relatively long, drawn-out process and makes its presence known by increasing noise and vibration levels in the bearing. Therefore, as a rule, there is plenty of time to prepare for a change of bearing.

However, a bearing occasionally does not reach its calculated lifetime, for one of many reasons: a heavier load than anticipated, inadequate or unsuitable lubrication, careless handling, ineffective sealing, or fits that are too tight, with resultant insufficient internal bearing clearance, and other unexpected physical environment factors. Each factor produces its own type of damage and leaves its own imprint on the bearing. Consequently, by examining a damaged bearing, it is possible, in the majority of cases, to form an opinion as to the cause of the damage and to take the requisite action to prevent a recurrence. These anomalies should be taken care of because of the risk of fast degradation and catastrophic consequences (Galar and Kumar, 2013).

8.2.6 Risk Assessment for Maintenance Actions

In theory, if a rolling bearing in service is properly lubricated, properly aligned, kept free of abrasives, moisture, and corrosive reagents, and properly loaded, all causes of damage are eliminated; save one, material fatigue. Historically, rolling bearing theory postulated that no rotating bearing can give unlimited service because of the probability of fatigue of the surfaces in rolling contact. The stresses repeatedly acting on these surfaces can be extremely high compared to other stresses acting on engineering structures. For example, some steels appear to have an endurance limit. This endurance limit is a level of cyclically applied, reversing stress, which, if not exceeded, the structure will accommodate without fatigue failure. The endurance limit for structural fatigue has been established by rotating beam and/or torsional testing of simple bars for various materials.

Inadequate lubrication, improper maintenance, dirtiness, etc., constitute the real risks for life reduction. The remaining useful life (RUL) with steel fatigue is still much longer than that of most of the machines where the bearings are mounted. In other words, the RUL of the bearing is much longer from a physical than a functional point of view, and risk assessment is necessary to mitigate the effects of the undesired elements that may dramatically reduce the life of the bearing (Galar and Kumar, 2013).

8.2.6.1 SMARTness in Bearings

Risk assessments are the real basis for a risk-based maintenance approach. The "S" for safety in RAMS terminology (reliability, availability, maintainability, and safety) has been a powerful driver to transfer technology from risks to people to risks to machinery. In the middle of the last century, science fiction writer Isaac Asimov devised the following three laws of robots (Asimov, 1942):

1. A robot may not injure a human being or, through inaction, allow a human being to come to harm.
2. A robot must obey the orders given to it by human beings, except where such orders would conflict with the First Law.
3. A robot must protect its own existence as long as such protection does not conflict with the First or Second Laws.

Long before today's smart machines, Asimov prioritized the behavior of machines and created the basis for risk assessment. Putting Asimov's thinking into today's context, risk assessment should primarily be performed to protect humans, whether maintainers or operators, working with productive assets. However, once the requirements to protect humans are fulfilled, the asset must protect itself. This pinpoints the concept of smart devices as protection for people and self-protection of machinery.

SKF developed the first smart bearing when the company started to commercialize the self-alignment bearing. This type of bearing has two rows of balls and a common sphered raceway in the outer ring. The bearings are self-aligning and insensitive to shaft deflections and angular misalignment of the shaft relative to the housing. They generate less friction than any other type of rolling bearing, so they can run cooler even at high speeds. They can modify their behavior according to malfunctions in the machine, correcting anomalies without human intervention, and dramatically reducing the number of shutdowns. The smartness about misalignment is the greatest advance in self-maintenance of rolling elements (Galar and Kumar, 2013).

Smart bearings must evaluate operational risks as an aid in maintenance decisions in certain conditions. These devices will always prioritize the safety of humans but will consider their own integrity when the operating condition modifies the scenario and the risk for the machine becomes dynamic.

Risk assessment will condition the real maintenance planning for a bearing according to the risks humans want to accept for their health and for the health of the machinery. Risk tolerance is up to the individual (or company) to decide in areas of economic, social, and environmental concern. For example, in vehicle health monitoring, most direct interventions (not taking shop repairs into account) are corrective actions aimed at replacing or repairing failing components. These actions are initiated based on scheduling or level of degradation, but in either case, the end user and maintainer will not be sure about the real need for action. Risk assessment provides information on the consequences to humans, machines, and economic profits of extending the scheduled replacement or permitting the component to be degraded beyond the established threshold.

Fig. 8.12 shows the dilemma of a smart maintenance system in a railway context. On the one hand, the rolling stock must reach a certain destination. On

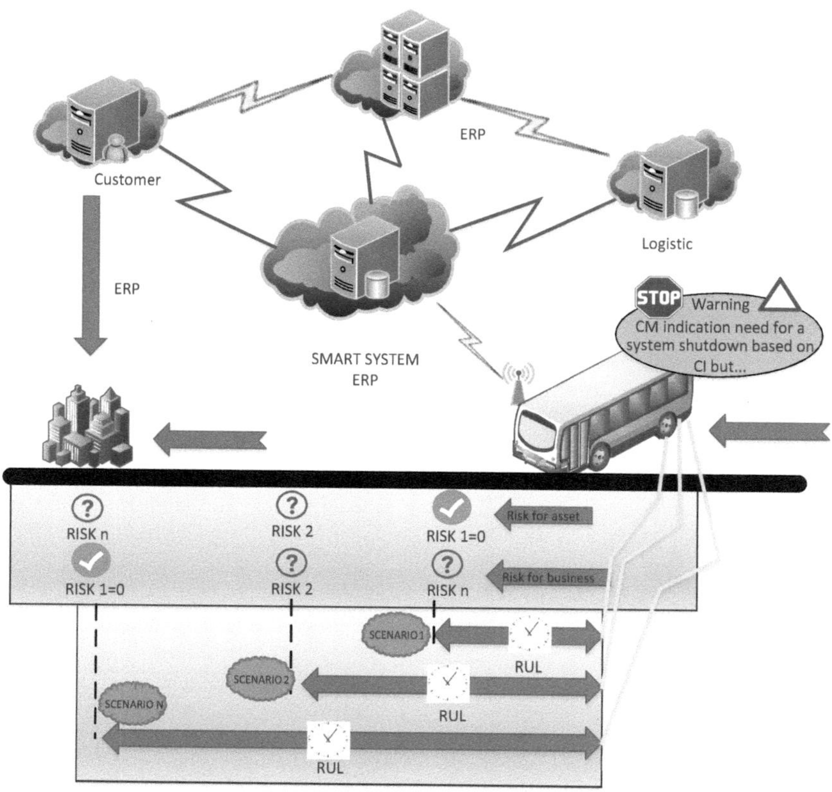

FIGURE 8.12 Smart maintenance system in railway. *Adapted from Galar, D., Kumar, U., June 6–7, 2013. SMART Bearings: From Sensing to Actuation. 12th IMEKO TC10 Workshop on Technical Diagnostics. New Perspectives in Measurements, Tools and Techniques for Industrial Applications, Florence, Italy.*

the other hand, an alarm is a potential threat to the functional integrity of the vehicle, possibly to the people traveling on it or to the business. The maintenance engineer must prioritize the maintenance actions to be taken as a function of the acceptable risk, and sacrifice either the vehicle or the business goals, as human safety cannot be neglected (Galar and Kumar, 2013).

The criteria used to evaluate risk-based decisions are moving from a purely economic perspective to consider social and environmental factors to protect humans and assets (Wolff et al., 2000). Asset maintenance practices must obviously be comprehensive if they are to encompass financial performance, environmental protection, and societal impact (Liyanage and Kumar, 2003).

8.2.6.2 Role of Risk in Smart Bearings

The most important steps in the procedure for risk assessment are summarized in Fig. 8.13. The role of models and methods for reliability evaluation and

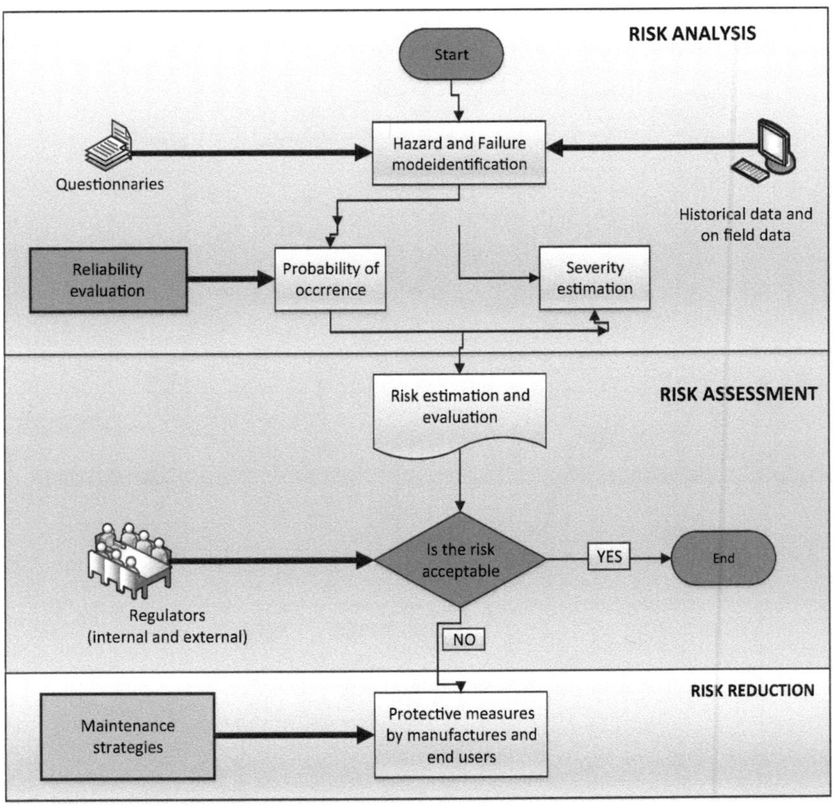

FIGURE 8.13 Flowchart of risk assessment. *Adapted from Galar, D., Kumar, U., June 6–7, 2013. SMART Bearings: From Sensing to Actuation. 12th IMEKO TC10 Workshop on Technical Diagnostics. New Perspectives in Measurements, Tools and Techniques for Industrial Applications, Florence, Italy.*

maintenance is clearly emphasized. The concept of smart maintenance proposes a plant integrity management based on risk-based maintenance as an extension of the well-known and established risk-based inspection in risky and hazardous industries such as oil and gas (API 58O and 581). This is an integrated approach to risk assessment and maintenance planning. It shows how to balance acceptable risks of failure with mission accomplishment, improving safety and health, reducing costs occasioned by repair or replacement of deteriorating equipment, and eliminating ineffective inspections or interventions (Galar and Kumar, 2013).

The smart bearing concept suggests the need for models and methods to support managers and practitioners in planning and executing maintenance actions in the framework of risk-based maintenance. Therefore, smart devices must have embedded methods for aiding in decisions on the maintenance of assets and their individual components. The assessments of smart bearings must comprise the following steps:

1. Analysis of the degradation of components;
2. Evaluation of the functional consequences for the asset of the analyzed degradation according to the scheduled use of the machine; and
3. Determination of operation of the asset based on the evaluated functional consequences.

It should be possible to optimize maintenance operations on a bearing according to its use by determining its operational risk, or in other words, the risk of disturbance of normal use. The operational risk is based on a prognosis of failure of at least one component, with this prognosis determined by analysis of the degradation of this component.

The risk assessment should include the calculation of two relevant parameters: the functional state of at least one of the components (critical), and the operational index of the asset based on the evaluated functional consequences and the functional state. The goal is to improve maintenance operations by combining diagnostic and prognostic functions for the asset's components.

The first parameter shows the degradation of at least one of the components. Preferably, it will show the degradation trend of at least one of the components and indicate the probability of breakdown.

The second parameter shows the risk. The evaluation of the functional consequences and the degradation of the asset according to its schedule are connected with the first parameter in a predefined failure model. The failure model can be determined by theoretical analysis or by using a statistical method based on observation of the behavior of similar components of the same asset or similar assets.

These techniques are based on information on potential degradation mechanisms and threats obtained by risk analysis; they are not prescriptive practices based on industrial experience (e.g., historical experience and industry guidelines for classes of equipment). These analytical models are basic, but they can support the development of ad-hoc, supportive decision-making methods for the analysts, owners, and users (e.g., safety managers and site inspectors) of many industrial and service companies (Galar and Kumar, 2013).

8.2.7 Diagnosis and Prognosis as Maintenance Decision Support System Enablers: Risk Assessment

Diagnostics are an intrinsic and valuable element of asset health assessment. Their applications and contributions to the quality of maintenance cross all aspects of asset management. Beyond the value to individual machines, diagnostics contribute to the improvement of production systems and development of new technologies in different sectors. Diagnostic tests play a role in the measurement and tracking of standards and quality of maintenance by maintainers and end users, underlying their essential and multifunctional contribution to machine health. In addition, the protection of the machine health relies on diagnostics to detect nonvisible damage that reduces the RUL of the asset.

Diagnosis and prognosis information is indispensable for decision-making by maintainers, end users, manufacturers, and even regulators. Decisions are made for asset-specific and population-wide treatments, measures, procedures, and services. Diagnostics/prognostics provide key and sometimes critical information at multiple junctures along the life cycle, from risk assessment and early diagnosis to machine follow-up and breakdown management.

The principal uses of this critical information include diagnosis, primary risk assessment (i.e., predictive and early problem identification), prognosis, selection of corrective action if needed, selected condition monitoring, and deterioration management. In short, diagnostics/prognostics provide maintainers with information essential to making appropriate decisions (Galar and Kumar, 2013).

8.2.7.1 Dilemma of Concurrent Failures

Diagnosis uses one or multiple tests, typically in combination with machine history and technician experience, to identify a particular existing problem or condition. Some tests or test combinations may identify the existence of multiple degradations and machine health problems, in addition to the primary diagnosis, providing information that can inform selection among alternative actions or the adjustment of a selected treatment. In Fig. 8.14, a centrifugal pump exhibits a vibration velocity spectrum with multiple damage of different severity, including unbalance, misalignment, looseness, etc. Fig. 8.15 shows an identical machine with unbalance in the rotor. Unfortunately, Fig. 8.14 is much more common than Fig. 8.15, so maintenance responses must be geared to existing problems, their severity, and how they are combined (Galar and Kumar, 2013).

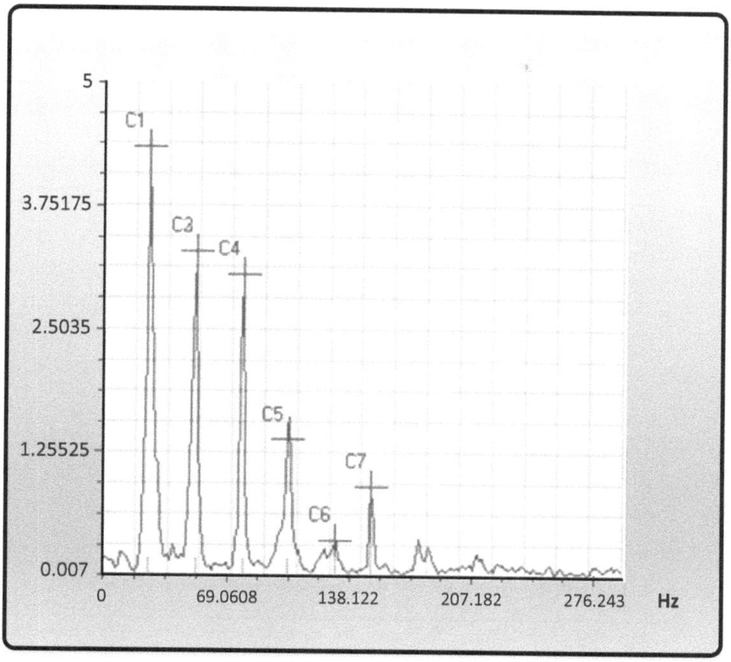

FIGURE 8.14 Asset with multiple pathologies. *Adapted from Galar, D., Kumar, U., June 6–7, 2013. SMART Bearings: From Sensing to Actuation. 12th IMEKO TC10 Workshop on Technical Diagnostics. New Perspectives in Measurements, Tools and Techniques for Industrial Applications, Florence, Italy.*

This dilemma is more dramatic in the bearing. When a rolling bearing rotates under load, the contacting surfaces of the rolling elements and the raceways normally become dull in appearance. This is no indication of wear in the usual sense of the word, and wear is of no significance to the bearing life. The dull surface in an inner or outer ring raceway forms a pattern called the path pattern. This pattern varies in appearance according to the rotational and loading conditions. By examining the path patterns in a dismantled bearing that has been in service, it is possible to get a good idea of the conditions under which the bearing has operated. By learning to distinguish between normal and abnormal path patterns, we can assess whether the bearing has run under the proper conditions.

In the majority of cases, the damage to the bearing originates within the confines of the path patterns; once their significance is understood, the appearance and location of the patterns are useful aids in diagnosing the cause of the damage.

Each of the different causes of bearing failure produces its own characteristic damage. Such damage, known as primary damage, gives rise to secondary, failure-inducing damage—flaking and cracks. Even primary damage

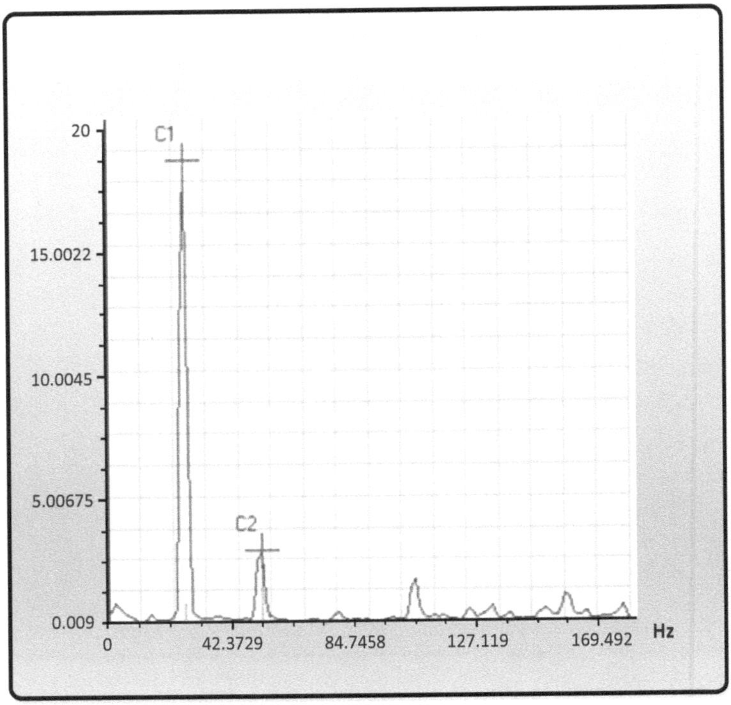

FIGURE 8.15 Asset with a single problem. *Adapted from Galar, D., Kumar, U., June 6–7, 2013. SMART Bearings: From Sensing to Actuation. 12th IMEKO TC10 Workshop on Technical Diagnostics. New Perspectives in Measurements, Tools and Techniques for Industrial Applications, Florence, Italy.*

may necessitate scrapping a bearing because of excessive internal clearance, vibration, noise, and so on. A failed bearing frequently displays a combination of primary and secondary damage. Primary damage includes wear, indentations, smearing, surface distress, corrosion, and electric current damage, as noted above, secondary damage includes flaking and cracks.

In summary, a well-known catalog of damage can be identified in a bearing. The real problem is the occurrence of several problems simultaneously overlapping and masking each other, as this makes the identification of the most severe damage extremely complicated. Proper diagnostics become essential to identify the real or most severe reason for deterioration (Galar and Kumar, 2013).

8.2.7.2 Benefits of Accurate Diagnosis

Diagnostics can detect incipient problems or determine which bearings are at increased risk of developing certain deterioration. Determination of increased risk may allow maintainers to take measures to prevent or reduce the risk of developing an undesired condition, including increased condition monitoring, operation changes, and preventive interventions.

Detection of emerging deterioration before symptoms appear or at an early symptomatic stage allows significant opportunities for early prevention and treatment. Accurate and early detection and identification of these mechanisms enable assessments of machine health status that can translate into fewer shutdowns, reduced life cycle cost, improved operation, and reduced maintenance costs. Diagnostics are evolving continually to enable more sensitive and specific detection of failure mechanisms at earlier stages via the measurement of multiple parameters. These diagnostics offer new opportunities for timely damage prevention and treatment.

Consumer expectations of diagnostics, such as rapid results, increased automation, simpler operation, and enhanced portability, continue to drive the development of smart bearings. Smart diagnostics will play a significant role in maintenance decision-making, particularly in areas where rapid and accurate response is closely tied to outcomes such as the diagnosis of an imminent failure (Galar and Kumar, 2013).

8.2.7.3 Importance of Context

The remaining useful life (RUL) of a bearing is commonly calculated following Lundberg and Palmgren (1947) by considering load and rotation speed. This equation comes from Weibull analysis and reflects a prediction of units in a large population. Today, more data are available for RUL estimations. The machine "profile" (manufacturer, environment, and operations) or other predispositions of a machine may influence the individual's response to a maintenance action.

Emerging maintenance technologies use information about asset variability to allow targeted treatment selection tailored to individual needs. For example, context diagnostics are profile-based diagnostic tests used to determine the individual benefits or losses associated with various alternatives available for various actions or decisions. Targeted treatments can allow a maintainer to avoid potentially harmful or ineffective maintenance actions for assets, resulting in improved machine health outcomes and cost savings. The move toward context-based decisions is facilitating an ideological shift within the maintenance community from a "one size fits all" treatment approach to "right maintenance actions for the right machine." Databases that compile and present such information are becoming available for scientists to study and maintainers to understand how profile variations may relate to action outcomes. With the integration of context data into maintenance practice guidelines, electronic records, and decision support systems (DSSs), managers will increasingly turn to context-based policies to make routine treatment decisions. Context technologies have the potential to yield better treatment selection and failure management strategies.

Diagnostic tests may also be used to assess damage progression or severity and the likelihood of recovery or risk of adverse outcomes. The prognostic information is frequently used to inform treatment decisions tailored to individual machine health status and needs. Prognostic assessment can include testing for certain coexisting problems (e.g., unbalance, misalignment, wear, and oil problems). The presence of coexisting failure mechanisms may inform necessary alterations in maintenance options and operation regimens. Certain chronic failures require continuous monitoring to avoid serious consequences or maintenance complications, so screening for emerging resistance to maintenance actions or cooccurring failures is essential. Commonly used for these purposes, diagnostics are instrumental in helping maintainers manage complex or later-stage machine health problems or conditions. Effective failure monitoring and management are often linked to fewer maintenance actions, reduced maintenance costs, and improved operation.

Smart bearings with contextual capabilities based on real-time condition monitoring and optimum DSS will allow maintainers to conduct rapid diagnostic in situ tests, rather than sending data to centralized laboratories. Such rapid diagnostics will provide technicians with information on machine health status and maintenance options during maintenance inspections or scheduled actions. This immediate responsiveness will reduce delays in effective decision-making and allow rapid responses to both critical situations, such as sudden breakdowns due to fast degradation mechanisms and routine or noncritical situations, thus reducing maintenance department costs (Galar and Kumar, 2013).

8.2.7.4 Network of Smart Bearings

The existence of a large population of smart bearings performing diagnosis and prognosis will have a global benefit. The information will be useful to detect problems at an individual level and at a larger population level. There are many applications for population diagnosis to determine risk, including global degradation, transmission of problems, identification of common failure mechanisms, and resistance to maintenance (preventive or corrective) actions. As novel diagnostics continue to emerge in this area, machine health threats can be characterized and contained more quickly and efficiently, affecting fewer individuals and improving health management options. Use of diagnostics for these applications will inform appropriate maintenance decisions and containment efforts to reduce the spread of the failure mechanisms along the bearing population.

Diagnostics development in this area has focused on rapid and accurate results and on portable, easy-to-use instruments. Technological advances increase flexibility and responsiveness to changing maintenance needs. Newly developed rapid detection diagnostics may help decrease the time between the introduction of a new failure mechanism and its detection, enabling a faster and more effective maintenance response. Many of these emerging diagnostics are being adapted for field use in emergency situations, ideally allowing containment efforts through self-maintenance actions (via actuators) to begin before a maintenance crew comes into the picture.

Emerging technologies, such as tribotronics, are useful for associating various products with machine health. As new markers are validated, and as the significance of various combinations of markers are better understood, these technologies are being adapted rapidly for a range of diagnostic applications. One expanding trend is multiplexing, which involves conducting tests for more than one marker in the same test sample. This testing paradigm is being developed in an array formats, with multiple multiplex tests performed on the same platform or chip. This availability of data is good, in essence, but simultaneous acquisition of large amounts of data can produce a undesired effect of prioritization of data storage and transmission rather than data analysis and information extraction. Data fusion becomes essential to combine acquired information and to weed out information that is redundant or useless.

As diagnostics become increasingly integrated and capable of generating vast amounts of data, analytical advances and ease of interpretation will better facilitate adoption and diffusion of these technologies into routine maintenance practices. For example, the tribology marker assay includes several hundred tests; its interpretation may be too complex for use in general maintenance practice without software or information processing capabilities to assist with analysis and presentation of diagnostic results. As products will soon identify many hundreds or thousands of markers, sophisticated analytical tools will be necessary to decipher the relationships between collected and processed information and the predisposition to initiate the failure mechanism.

Advances in electronic maintenance records and decision support software will assist maintainers in extracting meaning from increasingly complex diagnostic results. Computerized systems and networks are currently assisting in the processing of certain tests, and similar systems will decrease diagnostic interpretation time, allowing more rapid translation into appropriate prevention or treatment efforts.

Diagnostics have an essential role in the life cycle of an individual asset, the general population, and the environment. The significance of diagnostics is anticipated to increase as they evolve with other interventions and with machine health information technology. Cloud computing interfaces are anticipated to better link the profiles of machines, technicians, end users, manufacturers, and regulators (Galar et al., 2012). The new diagnostic interfaces have the potential to redefine, on a global scale, the relationship among these participants through timely and relevant information available at our fingertips (Wandt et al., 2012). The convergence of diagnostics and machine health information technology will be the driver of the development of more rapid, accurate, and high-throughput diagnostic information.

The smart bearing, as a self-diagnostic device, will need to incorporate advanced information technology with greater integration capabilities, ease of use, and compatibility with other instruments or information resources (e.g., electronic records or databases). Despite current advances, portable diagnostic devices, for example, instruments or field use devices, still lack features, flexibility, and capabilities to enhance the ability of maintainers and end users to capture, interpret, and use diagnostic information with speed, precision, and context. Energy consumption, transmission, and configuration are some issues the smart bearing project must overcome.

Conventional diagnostics provide minimal information, such as a numerical value to a maintainer, but even more minimal contextual information as to the potential relevance of the result at a higher plane of observation. The majority of conventional diagnostics fall short of providing rapid real-time assay results, an essential element of real-time critical diagnostic decision-making.

The recent technology developments in smart bearings will enable rapid and highly sensitive diagnostics for a wide variety of tests. The capabilities of these diagnostic technology platforms, when combined with advances in computing, telecommunications, and satellite technologies, will enhance the capacity and the potential of diagnostics, setting the stage for a paradigm shift in the generation and management of machine health information and services (Galar and Kumar, 2013).

8.2.8 Conclusions

The development of bearing technology will depend on the development of smart condition monitoring and maintenance solutions with the smart use of information technology. Risk assessment of alternative maintenance decisions based on RUL estimations seems a promising strategy to evaluate the consequences of performed or deferred maintenance actions. This methodology relies on collected information, the creation of operational scenarios, and the anticipation of potential deviations from expected behavior. Underlying technologies such as sensors and algorithms must be properly connected, preferably according to existing standards, if any.

At this point, risk assessments policies seldom appear in maintenance, and most of the existing ones are focused on protecting humans and reducing the environmental impact. It is time to redirect safety issues to include asset integrity and accomplishment of business goals in addition to human and environmental safety.

Relevant bearing technologies are already available; although they are very similar, there is a lack of standardization. Hopefully, this issue will be solved with the upcoming ISO 55000 2012a, 2012b and 2012c.

8.3 ROBOTICS IN MAINTENANCE DUTIES

Simply stated, maintenance preserves or restores a desired state of a system or facility. The maintenance process includes three major activities: inspection, planned maintenance, and disturbance handling (where disturbances are unplanned system states). Inspection is the activity in which information about state is monitored to allow prediction or early detection of disturbances.

Planned maintenance is the activity in which elements of the system are modified or replaced according to a predetermined schedule, with the aim of avoiding or reducing the frequency of disturbances. Disturbance handling is the activity in which elements of the system are modified or replaced to restore the desired state following a disturbance.

Maintenance tasks have some important differences from tasks commonly selected for industrial robots. First, maintenance often requires access to environments that are more dynamic and less predictable than is the case for many robotic applications. Second, tasks may be less predictable in maintenance, in terms of both the nature of tasks and the frequency of maintenance or latency between tasks. Finally, the cost benefits of robotic maintenance may be different from those of other robotic applications. In general, robots typically pay for themselves over their entire operational life; however, robots used in maintenance, particularly in the nuclear industry, may pay for themselves in a single application. This is particularly true if robots allow a facility to avoid shutdown and to continue operating even during maintenance (Parker and Draper, 1999).

8.3.1 Application Examples and Techniques

Robots, whether teleoperated, under supervisory control, or autonomous, have been used in a variety of applications in maintenance and repair. The following subsections describe many of these systems, focusing on applications for which working robot prototypes have been developed (Parker and Draper, 1999).

8.3.1.1 Nuclear Industry

In the nuclear industry, teleoperators have been well-used in a maintenance role for more than four decades. Several features of maintenance make it a good application for teleoperators in this arena. First, the low frequency of the maintenance operation calls for a general-purpose system capable of doing an array of maintenance tasks. Second, maintenance and repair require high levels of dexterity. Third, the complexity of these tasks may be unpredictable because of the uncertain impact of a failure. For these reasons, the choice is often between a human and a teleoperator. When the environment is hazardous, a teleoperator is usually the best selection. If humans in protective clothing can perform the same job, the benefits of having teleoperators continuously at the work site need to be weighed against the cost of suiting up and transporting humans to and from the work site. While humans are likely to be able to complete tasks more quickly than teleoperators, using teleoperators can:

1. Shorten mean time to repair by reducing the response time to failures,
2. Reduce health risks,

3. Improve safety,
4. Improve availability by allowing maintenance to take place during operations, instead of halting operations.

Maintenance is an important area for nuclear industry robotics; for example, the proceedings of the 1995 American Nuclear Society meeting on robotics and remote handling featured 124 papers, nearly a quarter of which were devoted to some aspect of maintenance. The 1997 meeting included 150 papers, with more than 40% dealing with some aspect of maintenance. And if we consider environmental recovery operations as a form of maintenance, a much larger proportion of papers at both meetings were maintenance-related (Parker and Draper, 1999).

Vertut and Coiffet (1985) list the following applications of teleoperated robots in nuclear maintenance:

1. Operation and maintenance of industrial nuclear facilities and laboratories,
2. Maintenance in nuclear reactors,
3. Decommissioning and dismantling nuclear facilities,
4. Emergency intervention.

8.3.1.2 Highways

In the developed world, highways are a critical component of the transportation network. The volume of traffic on the roadways has been steadily increasing for many years as society becomes more and more mobile. However, the funding to maintain these roadways has not been keeping pace with the traffic volume. The result is deteriorating roadways that cannot be adequately maintained. Conventional techniques of road repair lead to traffic congestion, delays, and dangers for the workers and the motorists. Robotic solutions to highway maintenance applications are attractive because of their potential to increase the safety of the highway worker, reduce delays in traffic flow, increase productivity, reduce labor costs, and increase the quality of the repairs (Parker and Draper, 1999).

Application areas for robotics include (Ravani and West, 1991) the following:

- Highway integrity management (crack sealing, pothole repair);
- Highway marking management (pavement marker replacement, paint restriping);
- Highway debris management (litter bag pickup, on-road refuse collection, hazardous spill clean-up, snow removal);
- Highway sign management (sign and guide marker washing, roadway advisories);
- Highway landscaping management (vegetation control, irrigation control);
- Highway work zone management (automatic warning systems, lightweight movable barriers, automatic cone placement and retrieval).

Although relatively few implementations in highway maintenance and repair have been attempted, some successful prototypes have been developed (Zhou and West, 1991). The California Department of Transportation and the University of California at Davis are developing a number of prototypes for highway maintenance under the Automated Highway Maintenance Technology program. Efforts are underway to develop systems for crack sealing, placement of raised highway pavement markers, paint striping, retrieving bagged garbage, pavement distress data collection, and cone dispensing. One result of this effort is a robotic system, automated *crack sealing* machine, for automatic crack sealing along roadways (Winters et al., 1994). As shown in Fig. 8.16, this machine senses, prepares, and seals cracks and joints along the highway. Cracks are sensed along the entire width of a lane using two-line scan cameras at the front of the vehicle. Sealing operations occur at the rear of the vehicle using an inverted, slide-mounted selective compliance assembly *robot* arm robot. A laser range finder at the tooling verifies the presence of the cracks and provides guidance for the sealing operation. The vehicle is able to perform this operation moving at about 1.6–3.2 km/h (1–2 miles/h). Other crack sealing prototypes have been developed at Carnegie Mellon University (Hendrickson et al., 1991) and at the University of Texas at Austin (Haas, 1996).

Earlier prototypes for highway maintenance date back to the 1980s (Skibniewski and Hendrickson, 1990). Researchers at Tyndall Air Force Base developed a rapid runway repair telerobotic system for repairing craters in runways (Nease, 1991). The objective of this work was to safely restore pavement surfaces after enemy attack to ensure subsequent future successful aircraft operations. The system was based on a John Deere multipurpose excavator enhanced for telerobotic operation. It used a four degrees of freedom joystick system with position and rate control, plus force feedback for human

FIGURE 8.16 Automated crack sealing machine developed by the University of California at Davis. *Adapted from Parker, L., Draper, J., 1999. Robotics applications in maitenance and repair. In: Nof, S. (Ed.), Handbook of Industrial Robotics, second ed. Wiley, pp. 1023–1036.*

remote control. The machine could dig, scrape, compact, break pavement, and change tools under preprogrammed, onboard control. Carnegie Mellon University also developed a robotic excavator prototype, called REX; it used topography and a computer-generated map of buried objects to generate and execute appropriate retrieval trajectories. The system used an elbow-type manipulator coupled with a master arm for manipulator setup and error recovery (Parker and Draper, 1999).

Automated pavement distress data collection vehicles have been developed as well (Zhou and West, 1991). A vehicle built by Komatsu Ltd. of Japan uses scanning lasers to examine a road surface's condition, measuring information on crack formation, wheel rutting, and longitudinal unevenness. The vehicle can travel at speeds up to 60 km/h (37 miles/h), evaluating roads up to 4 m wide with ruts and potholes up to 0.25 m deep (9.8 inches) (Parker and Draper, 1999).

The French Petroleum Studies Company has developed an automatic cone dispenser that can dispense and remove up to two rows of warning cones, for a total of 240 cones. The system operates at about 15 km/h (9.3 miles/h). The Technique Special de Securité Company has developed a mobile lane separator to place and remove concrete road marker blocks at speeds up to 30 km/h (18.6 mile/h). Systems for automatic grading have been developed by Spectra-Physics in Ohio and by Agtek Co. in California. These systems use laser guides to control the height of the grading blades, thus relieving the human operator of the need to perform manual positioning and control of the blades (Parker and Draper, 1999).

8.3.1.3 Railways

The railroad industry has recognized the economic benefits of automation, and a number of robotic maintenance and repair applications have been developed. The railway maintenance shops are the most common location of robots; they perform such activities as welding, grinding, cleaning, and painting (Martland et al., 1987).

A Toronto Transit Commission project led to the design of an automated system to clean the undersides of subway cars (Wiercienski and Leek, 1990). As shown in Fig. 8.17, this system involves the use of three industrial painting robots mounted on either side and under the subway vehicle being cleaned. An operator located remotely begins the cleaning operations after preparing the vehicle in advance. The entire system is controlled by a master computer that supervises the three individual robot controllers. A robot-mounted vision system is used to correct the robots' positions along the tracks. This system is expected to yield dramatic improvements in working conditions and work quality over the previous human worker approach. Fig. 8.17A shows the schematic layout of the robots under the railcar, and Fig. 8.17B shows the robot in operation. One robot manipulator can be seen at the right side of the photograph (Parker and Draper, 1999).

(A) (B)

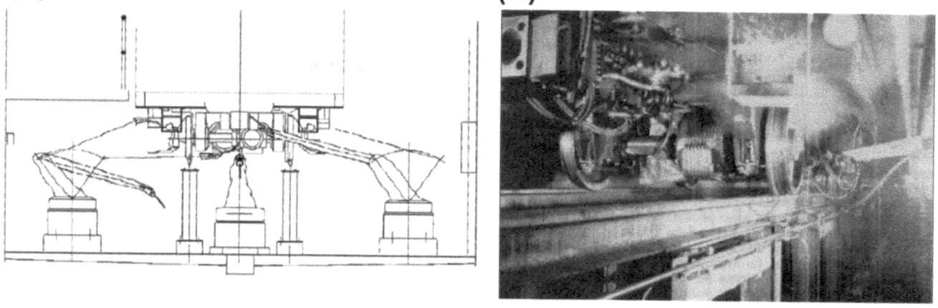

FIGURE 8.17 Robot system developed for the Toronto Transit Commission to clean the undersides of subway cars. *Adapted from Parker, L., Draper, J., 1999. Robotics applications in maitenance and repair. In: Nof, S. (Ed.), Handbook of Industrial Robotics, second ed. Wiley, pp. 1023–1036.*

Another robotic system developed by the railway industry is the RMS-2 rail grinding system (Anonymous, 1986). Developed by Speno Rail Services Co., this system has automated capabilities to sense the existing condition of the surface of the rails. Up to 99 patterns corresponding to rail contour patterns are stored. An integrated onboard computer system is used to generate onboard grinder controls to finish the rail to the appropriate, predetermined rail contour. The RMS-2 has 120 stones for grinding spread along the underside of 5 of the 12 railway cars that make up the system. The system is capable of grinding rails at speeds up to 6 miles/h (Parker and Draper, 1999).

8.3.1.4 Power Line Maintenance

Many common maintenance operations on overhead transmission lines are performed by human operators on live lines. Examples of these tasks include replacing ceramic insulators that support conductor wire and opening and reclosing the circuit between poles. These tasks are very dangerous for human workers, because of the risk of falling from a high place and the risk of electric shock. Obtaining skilled workers is difficult because of the amount of training needed and the labor requirements of the job. Performing maintenance while the lines are deenergized would alleviate some of the risks but would create other problems, as people demand interruption-free service from electric power companies.

Electric power companies have been investigating the use of robotic systems for live-line power line maintenance since the mid-1980s. Power companies in Japan, Spain, and the United States have developed teleoperated and semiautonomous approaches. One of the first systems was Teleoperator for Operations, Maintenance, and Construction using Advanced Technology (TOMCAT), developed as part of an Electric Power Research Institute (Palo Alto, California) program. The basic TOMCAT concept was first demonstrated by Philadelphia Electric in 1979 (Dunlap, 1986), with subsequent development continuing in the 1980s, leading to a system prototype. The TOMCAT system consisted of an insulated bucket truck, a Kraft 7-function manipulator bolted to the end of the truck boom, a television-viewing system for human supervisory control, and requisite control and power supplies. The operator control components were mounted on the back of the bucket truck and included a manipulator master with no force feedback (Parker and Draper, 1999).

A more recent robotic system for live-line maintenance has been developed by Kyushu Electric Power Co., Inc., in Fukuoka, Japan (Yano et al., 1995). The system configuration schematic for the dual-arm robot system is shown in Fig. 8.18. Earlier phases of this work involved the development of a two-manipulator telerobotic system; more recent phases are incorporating more autonomy, so that the system is evolving from a basic master–slave configuration to a human–robot cooperative system. Laser sensors are used on board the robot to help with position control. In the current system, a human works from a control station on the supporting truck, rather than on the elevated boom; this nearly eliminates the risk of injury due to falls (Parker and Draper, 1999).

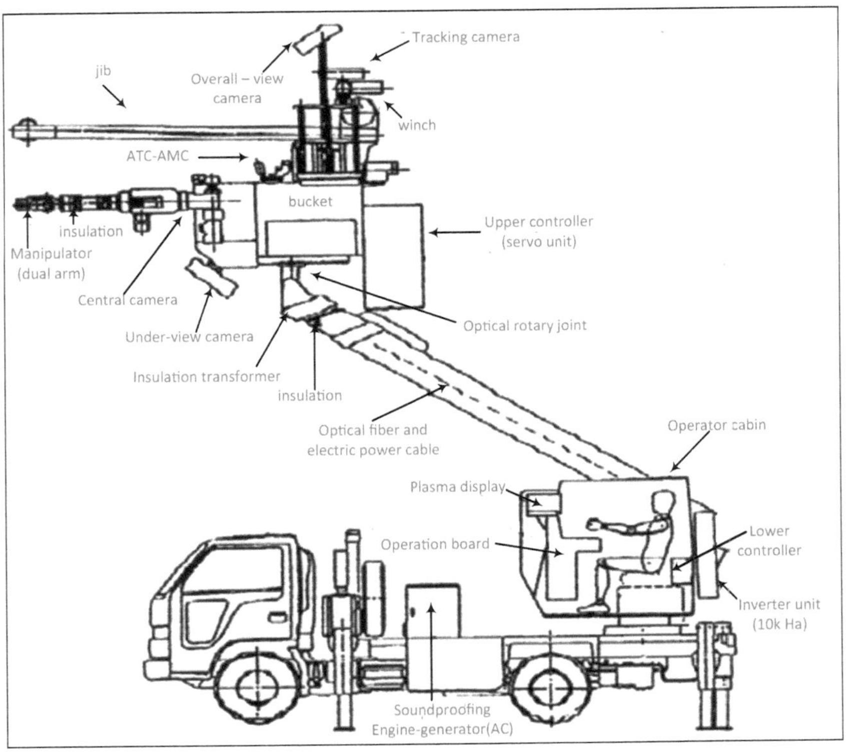

FIGURE 8.18 Dual-arm robot for live-line maintenance developed by Kyushu Electric Power Co. *Adapted from Parker, L., Draper, J., 1999. Robotics applications in maitenance and repair. In: Nof, S. (Ed.), Handbook of Industrial Robotics, second ed. Wiley, pp. 1023–1036.*

Other related work in this area includes the ROBot para Trabajos En Tension teleoperated system for live-line maintenance developed by researchers in Spain (Aracil et al., 1995) and a robot for automatic washing and brushing of polluted electric insulators (Yi and Jiansheng, 1993) developed in China.

8.3.1.5 Aircraft Servicing

Aircraft servicing applications may benefit from robotic maintenance in several areas. The size of modern multiengine jets makes inspection and coating removal and application particularly attractive in terms of improving quality and efficiency. As examples, Siegel et al. (1993) describe concepts for automating skin inspections, and Birch and Trego (1995) and Baker et al. (1996) describe stripping and painting concepts. Automated stripping and painting systems are already in place at a few US Air Force bases. A robotic assistant for rearming tactical fighter aircraft is being developed at the Oak Ridge National Laboratory for the US Air Force (Parker and Draper, 1999).

8.3.1.6 Underwater Facilities

Teleoperated robots are widely used to maintain facilities beneath the surface of the ocean, mainly in the offshore oil industry. Specific applications include repairing communications cables, pipelines, well heads, and platforms. Teleoperators have also been deployed to clean marine growth from power plant cooling systems (Edahiro, 1985), to inspect and clean steam generators (Trovato and Ruggieri, 1991), to perform underwater construction (Yemington, 1991), and to inspect and repair water conveyance tunnels (Heffron, 1990). While these efforts do not have the visibility of robotics work in industrial, space, or nuclear applications, this is an area in which robotics and remote control technology are widely used. In fact, it is perhaps the most common venue for everyday use of teleoperated robots. One publication lists 63 companies involved in building remotely operated vehicles or manipulators for subsea work and 180 commercially available, remotely controlled systems (Gallimore and Madsen, 1994). The remotely operated systems range from towed sensor arrays, to submersibles with dexterous manipulators, to large construction machinery (Parker and Draper, 1999).

8.3.1.7 Coke Ovens

Another example of robotics used in maintenance and repair operations is a robot developed by Sumitomo Metal Industries, Ltd., Japan, to repair the chamber wall of a coke oven (Sakai et al., 1988). Damage to coke ovens occurs over years of operations; repeated cycles of chamber door opening and coke pushing induce damaging changes in temperature. The result is cracks, joint separations, and chamber wall abrasion, all of which lead to gas leakage, air pollution, and structural flaws in the ovens. Effective repair will extend the life of the ovens and allow stable operation.

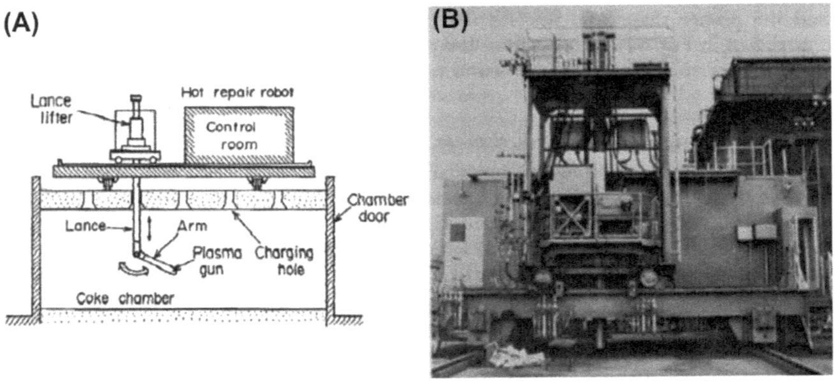

FIGURE 8.19 Hot repair robot developed by Sumitomo Metal Industries for repairing cracks and joints in the chamber wall of coke ovens during hot operation. *Adapted from Parker, L., Draper, J., 1999. Robotics applications in maitenance and repair. In: Nof, S. (Ed.), Handbook of Industrial Robotics, second ed. Wiley, pp. 1023–1036.*

Especially challenging maintenance operations involve the repair of the central portion of the oven. This type of repair is very difficult because of the inaccessibility of the area, the high temperature, and the predominance of narrow cracks. Any technology for repair in this area must involve high heat resistance components and mechanisms for external observation, resulting in repairs of high quality and durability. The solution to this repair problem was the development of a heat-resistant robot, shown in Fig. 8.19. The robot can autonomously perform individual crack repair, while being given high-level guidance through a human–machine interface. Of special benefit to the industry is the ability to perform these repairs without disturbing oven operation or incurring a large firebrick temperature drop (Parker and Draper, 1999).

8.3.1.8 Summary of Robotics Applications in Maintenance and Repair

The section has reviewed the main areas where robotics are used for maintenance and repair. Some applications have made tremendous advancements while others remain less studied. Table 8.1 summarizes the application areas, noting the importance of robotics to the maintenance tasks of inspection, planned maintenance, and disturbance handling (Parker and Draper, 1999).

8.3.2 Emerging Trends

Trends in robotic maintenance include the following:

- *Computer control/monitoring/assistance*: increasing integration of computer control over subsystems and subtasks to monitor user inputs, evaluate their appropriateness, and provide user assistance. As an example, Volpe

TABLE 8.1 Summary of Robotics Applications in Maintenance and Repair

	Maintenance Task		
Application Area	**Inspection**	**Planned Maintenance**	**Disturbance Handling**
Nuclear Industry	Growing area, especially as new facility designs incorporate remote maintenance philosophy	Well-established field, with several decades of successful robotic applications.	Much current activity related to decontamination, decommissioning, and dismantling.
Highways	Relatively new area with few current prototypes, except as packaged with crack sealing and pothole repair systems.	Relatively new area, with quickly growing interest and a huge potential impact. Several ongoing efforts should result in a number of new robot prototypes in the next 5 years.	Of significant interest, particularly for highway integrity management. A number of successful prototype systems are gradually making way into routine use. Several new efforts underway.
Railways	Few current systems and little ongoing activity.	Most common area of railway robotics but with little new activity.	Little current use.
Power Line Maintenance	Little current use.	Interest is increasing, especially for robotic techniques that work on live power lines.	Greatest area of current use, with much potential growth due to technology advances and need to remove humans from highly dangerous tasks.
Aircraft Servicing	Steadily growing area, due to recent advances in automated inspection technologies.	Steadily growing area, especially for automated stripping and painting.	Little current use.
Underwater Facilities	Steady progress over the last two decades, with continued advances.	Of increasing importance, with several new prototype systems under development.	Of increasing importance, with several new prototype systems under development.
Coke Ovens	Little current use.	Little current use.	Fair amount of activity in late 1980s. Relatively little new work in this area.

Adapted from Parker, L., Draper, J., 1999. Robotics applications in maitenance and repair. In: Nof, S. (Ed.), Handbook of Industrial Robotics, second ed. Wiley, pp. 1023–1036.

(1994) describes using model-based artificial forces to improve collision avoidance. By monitoring the position of a manipulator relative to a model of the remote work site, a repulsive force can be generated by the control system to resist movement into areas where collisions are possible or even to push the manipulator away from surfaces or features (Parker and Draper., 1999).

- *Virtual reality*: increasing integration of virtual reality interfaces into teleoperated systems. Virtual reality (VR) system and teleoperators may be considered members of a superset called Synthetic Environments Technology (Durlach and Mavor, 1995). They differ only in that a user interacts with a remote, computer-generated world in VR but a remote, real world in teleoperation. Therefore, it should be no surprise that VR techniques are being adopted for teleoperation. The application of VR to autonomous robotics is likely to increase in the future for programming and for monitoring robot operations. This may be limited to providing a virtual view of the workspace and manipulator or may be present in the form of graphical preview control. In the latter, the user completes the task in the virtual world as a means of programming the robot. This is inherently less efficient than real-time control, but proponents claim potentially greater safety and ease-of-use for graphical preview control (e.g., Small and McDonald, 1997). However, no data support that opinion. Milgram, Yin, and Grodski (1997) provide examples of an "augmented reality" interface for teleoperation in which video from a remote site is combined with partial computer models of the site (Parker and Draper, 1999).
- *Increasing sensor integration*: use of larger numbers of sensors and multisensor integration, with the goal of allowing autonomous robots to function more effectively in environments too variable for effective modeling. For example, Dudek and Freedman (1996) demonstrate how two different types of sensors, a sonar array and a laser range finder, can provide more accurate information about a work site than either type alone (Parker and Draper, 1999).
- *Responsiveness* versus *payload*: sacrificing responsiveness (the ability of a manipulator to recreate a user's trajectories and impedance in time and space) for payload in manipulators. Telerobotic manipulators seem to be increasingly less responsive but more powerful (Parker and Draper, 1999).

Table 8.2 suggests how these trends may affect each of the three maintenance tasks (inspection, planned maintenance, and disturbance handling) in future applications. Although the information is speculative, it gives some guidance in the likely development of robotic technology in maintenance in the near and intermediate future (Parker and Draper, 1999).

TABLE 8.2 Emerging Trends and Potential Impact on Robotics Activities of Inspection, Planned Maintenance, and disturbance Handling

	Maintenance Task		
Trend	**Inspection**	**Planned Maintenance**	**Disturbance Handling**
Computer control/ monitoring/assistance	Will make inspections more autonomous and at the same time more effective. Robots are more capable of accurately monitoring position and measuring sensory input than humans are (although humans better at pattern recognition).	Will assist in scheduling and reduce human involvement in tasks that can be sufficiently ordered. It is possible to remove much of the tedium from remote tasks by automating tasks or subtasks.	Most likely to improve record-keeping and to help assure quality by assisting operators in quality-critical subtasks. Also may allow oversight of user actions, which may improve safety.
Virtual reality	Graphical overlays of expected task configuration will be helpful. Multiple sensor "views" can be integrated to provide more information.	May assist in robot navigation by providing displays to enhance user situation awareness. May be used to display robot progress through planned maintenance tasks without on-site video. Maybe used to display information about maintenance campaigns (as opposed to tasks).	Forms of computer-assisted teleoperation may be beneficial; integrating graphics and video could provide information about expected state, which could help with diagnosis. Could allow online guidance for operators.
Increasing sensor integration	Will make it possible to perform inspections in multisensor fashion, and develop a picture of inspected items that integrates several sensors, including energy not sensed by humans.	Arrays of sensors may provide more complete information about a component and aid in assuring the effectiveness of planned maintenance.	Arrays of sensors may provide more complete information about a disturbance and aid in diagnosis and remediation.
Responsiveness versus payload	Not likely to have much impact.	May allow larger parts to be replaced but may make replacement more difficult under human control.	May allow application to a wider range of disturbances but may reduce efficiency.

Adapted from Parker, L., Draper, J., 1999. Robotics applications in maitenance and repair. In: Nof, S. (Ed.), Handbook of Industrial Robotics, second ed. Wiley, pp. 1023–1036.

REFERENCES

Akhras, G., 2000. Smart materials and smart systems for the future. Canadian Military Journal Autumn.

Anonymous, 1986. Toward the "smart" rail maintenance system. Railway Track & Structures 82 (11), 21–24.

Aracil, R., Penin, L., Ferre, M., Jimenez, L., Barrientos, A., Santamaria, A., Martinez, P., Tudun, A., 1995. ROBTET: a new teleoperated system for live-line maintenance. In: Proceedings of the 7th International Conference on Transmission and Distribution Construction and Live Line Maintenance, Columbus, Ohio, pp. 205–211.

Asimov, I., 1942. Runaround, Astounding Science Fiction, March. Republished in 1991, Robot Visions. Penguin.

Auer, M., Pester, A., Ursutiu, D., Samoila, C., December 2003. Distributed virtual and remote labs in engineering. In: Industrial Technology, 2003 IEEE International Conference on, vol. 2, pp. 1208–1213.

Baker, J.E., Draper, J.V., Pin, F.G., Primm, A., Shekhar, S., May 1996. Conceptual design of an aircraft automated coating removal system. In: Proceedings of ISRAM 96, Sixth International Symposium on Robotics and Manufacturing, Montpellier, France, pp. 27–30.

Birch, S., Trego, L.E., 1995. Aircraft stripping and painting. Aerospace Engineering 15, 21–23.

Cascado, D., 2011. Standards and Implementation of Pervasive Computing Applications. John Wiley & Sons, Ltd, pp. 135–158.

Chen, X., Song, G., Zhang, Y., 2010. Virtual and remote laboratory development: a Review. In: Proceedings of Earth and Space 2010, vol. 1, pp. 3843–3852.

Dunlap, J., 1986. Robotic maintenance of overhead transmission lines. IEEE Transactions on Power Delivery 1 (3), 280–284.

Durlach, N.I., Mavor, A.S., 1995. Virtual Reality: Scientific and Technological Challenges. National Research Council, National Academy Press, Washington, DC.

Edahiro, K., 1985. Development of "underwater robot" cleaner for marine live growth in power station. In: Teleoperated Robotics in Hostile Environments, H.L., Martin, D.P., Kuban (eds.), (Robotics International of the Society of Manufacturing Engineers, Dearborn, MI), pp. 108–118.

Fairweather, J.A., 1998. Designing with Active Materials: An Impedance Based Approach, first ed. UMI, New York, p. 196.

Fette, I., December 2011. The WebSocket Protocol. ISSN: 2070-1721. Isode Ltd. https://tools.ietf.org/html/rfc6455.

Freed, N., Baker, M., Hoehrmann, B., 2014. Media Types. Tech. Rep., 2014.

Furht, B., 1995. A survey of multimedia compression techniques and standards. Part I: JPEG standard. Real-Time Imaging 1 (No. 1), 49–67.

Gabbert, U., Tzou, H.S., 2001. Smart Structures and Structronic Systems. Kluwer Academic Publishers.

Galar, D., Kumar, U., June 6–7, 2013. SMART Bearings: From Sensing to Actuation. 12th IMEKO TC1O Workshop on Technical Diagnostics. New Perspectives in Measurements, Tools and Techniques for Industrial Applications. Florence, Italy.

Galar, D., Palo, M., Van Horenbeek, A., Pintelon, L., 2012. Integration of disparate data Sources to perform maintenance prognosis and optimal decision making. Insight – Non-Destructive Testing and Condition Monitoring 54 (8), 440–445.

Gallimore, D., Madsen, A., 1994. Remotely Operated Vehicles of the World. Oilfield Publications Limited, Ledbury, Herfordshire, England.

Gillet, D., De Jong, T., Sotirou, S., Salzmann, C., 2013. Personalised learning spaces and federated online labs for STEM Education at school: supporting teacher communities and inquiry learning. In: Proceedings of the 4th IEEE Global Engineering Education Conference (EDUCON). IEEE, pp. 769–773.

González Fuster, G., Scherrer, A., 2015. Big Data and Smart Devices and Their Impact on Privacy. Study for the LIBE committee. Policy Department C: Citizens' Rights and Constitutional Affairs. PE 536.455. European Parliament, Manuscript Completed in September 2015. ©European Union, Brussels.

Greenfield, A., 2006. Everyware: The Dawning Age of Ubiquitous Computing. New Riders, Berkeley, CA, p. 33.

Haas, C., 1996. Evolution of an Automated Crack Sealer: A Study in Construction Technology Development, pp. 293–305. Automation in Construction 4.

Heffron, R., 1990. The use of submersible ROVs for the inspection and repair of water conveyance tunnels. In: Water Resources Infrastructure: Needs, Economics, pp. 35–40.

Hendrickson, C., McNeil, S., Bullock, D., Haas, C., Peters, D., Grove, D., Kenneally, K., Wichman, S., 1991. Perception and control for automated pavement crack sealing. In: Proceedings of Applications of Advanced Technologies in Transportation Engineering, pp. 66–70.

IEEE, October 2007. IEEE Standard for a Smart Transducer Interface for Sensors and Actuators Wireless Communication Protocols and Transducer Electronic Data Sheet (TEDS) Formats. IEEE Std 1451.5–2007.

ISO 55000, 2012a. ISO/CD 55000.2 Asset Management: Overview, Principles and Terminology. ISO/TC 251/WG 1.

ISO 55000, 2012b. ISO/CD 55001.2 Asset Management: Management Systems - Requirements. ISO/TC 251/WG 2.

ISO 55000, 2012c. ISO/CD 55002.2 Asset Management: Management Systems - Guidelines for the Application of ISO 55001, ISO/TC 251/WG 2.

Iyer, S.S., Haddad, Y.M., 1993. Intelligent Materials – an Overview. Department of Mechanical Engineering, University of Ottawa, Ottawa, Ontario, Canada. KIN 6N5.

Kaushal, A., Vardhan, A., Rawat, R.S.S., May–June 2016. Intelligent material for modern age: a Review. IOSR Journal of Mechanical and Civil Engineering (IOSR-jmce) 13 (3), 10–15. e-ISSN: 2278–1684, p-ISSN: 2320-334X.

Kessler, M.R., Sottos, N.R., White, S.R., 2003. Self-healing structural composite materials. Composites Part A: Applied Science and Manufacturing 34, 743–753.

Kumar, S., 1991. Smart Materials for Accoustic or Vibration Control. Ph.D. Thesis. Pennsylvania State University.

Lee, H.J., Arnold, S.M., 2004. Smart materials, precision sensors/actuators, smart structures & structronic systems. Mechanics of Advanced Materials & Structures 11, 383.

Liyanage, J.P., Kumar, U., 2003. Towards a value based view on operations and maintenance performance management. Journal of Quality in Maintenance Engineering 9 (4), 333–350.

Loreto, S., Saint-Andre, P., Salsano, S., Wilkins, G., April 2011. Known Issues and Best Practices for the Use of Long Polling and Streaming in Bidirectional HTTP. RFC 6202. Internet Engineering Task Force.

Lundberg, G., Paimgren, A., 1947. Dynamic capacity of rolling bearings. In: Acta Polytechnica Mechanical Engineering Series, vol. 1 (3). Stockholm, Sweden.

Martland, C.D., 1987. Analysis of the potential impacts of automation and robotics on locomotive rebuilding. IEEE Transactions on Engineering Management, 2, 92–100.

Milgram, P., Yin, S., Grodski, J.J., 1997. An augmented reality based teleoperation interface for Unstructured environments. In: Proceedings of the ANS 7th Tropical Meeting on Robotics and Remote Systems. American Nuclear Society, LaGrange Park, IL, pp. 966–973.

Nease, A., 1991. Development of rapid runway repair (RRR) telerobotic construction equipment. In: Proceedings of IEEE National Telesystems Conference (NTC 91). Atlanta, GA, pp. 321–322.

Parker, L., Draper, J., 1999. Robotics applications in maitenance and repair. In: Nof, S. (Ed.), Handbook of Industrial Robotics, second ed. Wiley, pp. 1023–1036.

Ravani, B., West, T., 1991. Applications of robotics and automation in highway maintenance operations. In: Proceedings of 2nd International Conference on Applications of Advanced Technologies in Transportation Engineering, pp. 61–65.

Rogers, C.A., September 1988. Workshop summary. In: Proceedings U.S. Army Research Office Workshop on Smart Materials, Structures and Mathematical Issues. State University, Technomic Publishing Co., Inc., pp. 15–16, 1–12.

Rogers, C.A., Barker, D.K., Jaeger, C.A., September 1988. Introduction to smart materials and structures. In: Proceedings of U.S. Army Research Office Workshop on Smart Materials, Structures and Mathematical Issues. Virginia Polytechnic Institute and State University, Technomic Publishing Co., Inc., pp. 15–16, 17–28.

Sakai, T., Narita, Y., Yamazaki, T., Nishinaka, H., 1988. Hot repair robot for coke ovens. Transactions Iron and Steel Institute of Japan 28, 346–351.

Salzmann, C., Gillet, D., 2008. From online experiments to smart devices. International Journal of Online Engineering (IJOE) 4 (Special Issue), 50–54.

Salzmann, C., Gillet, D., 2011. Remote labs and social media: agile aggregation and exploitation in higher engineering education. IEEE EDUCON Education Engineering 307–311.

Salzmann, C., Gillet, D., 2013. Smart device paradigm standardization for online labs. IEEE EDUCON Education Engineering 1217–1221.

Salzmann, C., Govaerts, S., Halimi, W., Gillet, D., 2015. The Smart Device Specification for Remote Labs. Ecole Polytechnique Fèdèrale de Lausanne (EPFL), Lausanne,. Switzerland 11 (4).

Shaikh, A.M., Patil, K.B., Bazar, S., 2015. Smart materials & its applications. International Journal of Emerging Technologies in Computational and Applied Sciences (IJETCAS) 15, 371.

Siegel, M.W., Kaufman, W.M., Alberts, C.J., 1993. Mobile robots for difficult measurements in difficult environments: application to aging aircraft inspection. Robotics and Autonomous Systems 11, 187–194.

Skibniewski, M., Hendrickson, C., 1990. Automation and robotics for road construction and maintenance. Journal of Transportation Engineering 116 (3), 261–271.

Small, D.E., McDonald, M.J., 1997. Graphical programming of telerobotic tasks. In: Proceedings of the American Nuclear Society 7th Topical Meeting on Robotics and Remote Systems. American Nuclear Society, LaGrange Park, IL, pp. 3–7.

Susmita, K., 2013. Introduction, Classification and applications of smart materials: an Overview. Department of Chemistry, East Point College of Engineering and Technology, Bangalore-560 049, India American Journal of Applied Sciences 10 (8), 876–880. http://dx.doi.org/10.3844/ajassp.2013.876.880. ISSN: 1546–9239.

Takagi, T., 1990. A concept of intelligent materials. J. Intell. Mater. Syst. Struct. 1, 149–156. http://dx.doi.org/10.1177/1045389X9000100201.

Tawfik, M., 2014. Laboratory as a service (LaaS): a novel paradigm for developing and implementing modular remote laboratories. International Journal of Online Engineering 10 (4), 13–21.

Taylor, B.N., Thompson, A., 2008. The International system of Units (SI). National Institute of Standards and Technology. NIST Special Publication 330.

Trovato, S.A., Ruggieri, S.K., 1991. Design, development and field testing of CECIL: a steam generator secondary side maintenance robot. In: Jamshidi, M., Eicker, P.J. (Eds.), Robotics and Remote Systems: Proceedings of the Fourth ANS Topical Meeting on Robotics and Remote Systems. Sandia National Laboratory, Albuquerque, NM, pp. 121–130.

Vardan, V.K., Vardan, V.V., 2002. Microsensors, Microelectromechanical systems (MEMS) and electronics for smart structures and systems. Smart Materials and Structures 9, 953–972.

Vertut, J., Coiffet, P., 1985. Teleoperation and Robotics: Applications and Technology. Prentice-Hall, Englewood Cliffs, NJ.

Volpe, R., 1994. Techniques for Collison prevention, impact stability, and force control by space manipulators. In: Skaar, S., Ruoff, C. (Eds.), Teleoperation and Robotics in Space. AAAI Press, pp. 175–208.

Wandt, K., Karim, R., Galar, D., 2012. Context adapted prognostics and diagnostics. In: International Conference on Condition Monitoring and Machinery Failure Prevention Technologies, London, UK.

White, S.R., Sottos, N.R., Geubelle, P.H., Moore, J.S., Kessler, M.R., Sriram, S.R., Brown, E.N., Viswanathan, S., 2001. Autonomic healing of polymer composites. Nature 409, 794–797.

Wiercienski, W., Leek, A.R., 1990. Feasibility of robotic cleaning of undersides of Toronto's subway cars. Journal of Transportation Engineering, 116 (3), 272–279.

Winters, S., Hong, D., Velinsky, S., Yamazaki, K., 1994. A new robotics system concept for automating highway maintenance operations. In: Proceedings of Robotics for Challenging Environments, 374–382.

Wolff, R., Zaring, O., Furuholt, E., O'Leary, H., Gaasbeek, A.P., Thoem, T., June 26–28, 2000. Indicators for sustainable development. In: Paper Presented at the SPE International Conference on Health, Safety, and the Environment in Oil and Gas, Stavanger, Norway.

Yano, K., Maruyama, Y., Morita, K., Nakashima, M., 1995. Development of the semi-automatic Hot-line work robot system "phase II". In: Proceedings of 7th International Conference on Transmission and Distribution Construction and Live Line Maintenance. Columbus, Ohio, pp. 212–218.

Yemington, C., 1991. Telerobotics in the offshore oil industry. In: Jamshidi, M., Eicker, P.J. (Eds.), Robotics and Remote Systems: Proceedings of the Fourth ANS Topical Meeting on Robotics and Remote Systems. Sandia National Laboratory, Albuquerque, NM, pp. 441–450.

Yi, H., Jiansheng, C., 1993. The research of the automatic washing-brushing robot of 500KV DC insulator string. In: Proceedings of the 6th International Conference on Transmission and Distribution Construction and Live Line Maintenance, Las Vegas, pp. 411–424.

Zhou, T., West, T., 1991. Assessment of the state-of-the-Art of robotics applications in highway construction and maintenance. In: Proceedings of 2nd International Conference on Applications of Advanced Technologies in Transportation Engineering, pp. 56–60.

FURTHER READING

API Publication 580, 2002. Recommended Practice for Risk Based Inspection – 1st Edition.

API Publication 581, 2000. Risk Based Inspection, Base Resource Document – 1st Edition.

Wadley, H.N.G., 1996. Intelligent Processing of Materials. Characteristics and Processing of Smart Materials. Charlottesville, VA 22903.

Index

'*Note*: Page numbers followed by "f" indicate figures and "t" indicate tables.'

D

E

F

Q

R

Kolibri 360 GmbH
Ferdinand-Jühlke-Straße 7
99095 Erfurt, Deutschland
produktsicherheit@kolibri360.de